Methods in Enzymology

Volume XXXVIII
HORMONE ACTION
Part C
Cyclic Nucleotides

METHODS IN ENZYMOLOGY

EDITORS-IN-CHIEF

Sidney P. Colowick   Nathan O. Kaplan

Methods in Enzymology

Volume XXXVIII

Hormone Action

Part C

Cyclic Nucleotides

EDITED BY

Joel G. Hardman

DEPARTMENT OF PHYSIOLOGY
VANDERBILT UNIVERSITY SCHOOL OF MEDICINE
NASHVILLE, TENNESSEE

Bert W. O'Malley

DEPARTMENT OF CELL BIOLOGY
BAYLOR COLLEGE OF MEDICINE
TEXAS MEDICAL CENTER
HOUSTON, TEXAS

1974

ACADEMIC PRESS   New York   San Francisco   London
A Subsidiary of Harcourt Brace Jovanovich, Publishers

COPYRIGHT © 1974, BY ACADEMIC PRESS, INC.
ALL RIGHTS RESERVED.
NO PART OF THIS PUBLICATION MAY BE REPRODUCED OR
TRANSMITTED IN ANY FORM OR BY ANY MEANS, ELECTRONIC
OR MECHANICAL, INCLUDING PHOTOCOPY, RECORDING, OR ANY
INFORMATION STORAGE AND RETRIEVAL SYSTEM, WITHOUT
PERMISSION IN WRITING FROM THE PUBLISHER.

ACADEMIC PRESS, INC.
111 Fifth Avenue, New York, New York 10003

United Kingdom Edition published by
ACADEMIC PRESS, INC. (LONDON) LTD.
24/28 Oval Road, London NW1

Library of Congress Cataloging in Publication Data

Hardman, Joel G., joint author.
Hormones and cyclic nucleotides.
Hormone action.

(Methods in enzymology ; v. 36-   )
Includes bibliographical references and indexes.
CONTENTS: pt. A. Steroid hormones.— pt. B. Peptide hormones.—
pt. C. Cyclic nucleotides.—pt. D. Isolated cells,
tissues, and organ systems.—pt. E. Nuclear structure
and function.
    1.   Enzymes.   2.   Hormones.   3.   Cyclic nucleotides.
I.   O'Malley, Bert W.   II.   Title.
III.   Series: Methods in enzymology ; v. 36 [etc.]
[DNLM: 1.   Cell nucleus.   2.   Hormones.   3.   Nucleo-
tides, Cyclic.   W1 ME9615K v. 40 / QH595 H812]
QP601.C733 vol. 36, etc. [QP601] 574.1'925'08s
                                [574.1'92]   74-10710
ISBN 0–12–181938–8 (v. 38) (pt. C)

PRINTED IN THE UNITED STATES OF AMERICA

Table of Contents

CONTRIBUTORS TO VOLUME XXXVIII x

PREFACE xiv

VOLUMES IN SERIES xvi

EARL W. SUTHERLAND, JR. (1915–1974) xxi

Section I. Extraction and Purification of Cyclic Nucleotides

1. Rapid Tissue Fixation and Extraction Techniques — STEVEN E. MAYER, JAMES T. STULL, AND WILLIAM B. WASTILA — 3

2. Separation and Purification of Cyclic Nucleotides by Ion-Exchange Resin Column Chromatography — G. SCHULTZ, E. BÖHME, AND J. G. HARDMAN — 9

3. Separation, Purification, and Analysis of Cyclic Nucleotides by High Pressure Ion-Exchange Chromatography — GARY BROOKER — 20

4. Separation of Cyclic Nucleotides by Thin-Layer Chromatography on Polyethyleneimine Cellulose — E. BÖHME AND G. SCHULTZ — 27

5. Isolation of Cyclic AMP by Inorganic Salt Coprecipitation — PETER S. CHAN AND MICHAEL C. LIN — 38

6. Separation and Purification of Cyclic Nucleotides by Alumina Column Chromatography — ARNOLD A. WHITE — 41

Section II. Assay of Cyclic Nucleotides

7. Assay of Cyclic Nucleotides by Receptor Protein Binding Displacement — ALFRED G. GILMAN AND FERID MURAD — 49

8. Assay of Cyclic AMP by the Luciferin-Luciferase System — CHARLES A. SUTHERLAND AND R. A. JOHNSON — 62

9. Assay of Cyclic AMP by Protein Kinase Activation — STEVEN E. MAYER, JAMES T. STULL, WILLIAM B. WASTILA, AND BARBARA THOMPSON — 66

10. Quantitation of Cyclic GMP by Enzymatic Cycling — NELSON D. GOLDBERG AND MARI K. HADDOX — 73

11. The Measurement of Cyclic GMP with *Escherichia coli* Elongation Factor Tu	ROBERT F. O'DEA, JAMES W. BODLEY, LILLIAN LIN, MARI K. HADDOX, AND NELSON D. GOLDBERG	85
12. Assay of Cyclic GMP by Activation of Cyclic GMP-Dependent Protein Kinase	J. F. KUO AND PAUL GREENGARD	90
13. Assay of Cyclic Nucleotides by Radioimmunoassay Methods	ALTON L. STEINER	96
14. Determination of Cyclic GMP by Formation of [β-^{32}P]GDP	GÜNTER SCHULTZ AND JOEL G. HARDMAN	106

Section III. Biosynthesis of Cyclic Nucleotides

15. General Principles of Assays for Adenylate Cyclase and Guanylate Cyclase Activity	GÜNTER SCHULTZ	115
16. An Automated Chromatographic Assay for Adenylate Cyclase and Cyclic 3′, 5′-AMP Phosphodiesterase	MICHAEL C. LIN	125
17. Preparation of Particulate and Detergent-Dispersed Adenylate Cyclase from Brain	ROGER A. JOHNSON AND EARL W. SUTHERLAND	135
18. Preparation and Characterization of Adenylate Cyclase from Heart and Skeletal Muscle	GEORGE I. DRUMMOND AND DAVID L. SEVERSON	143
19. Adenylate Cyclase from Kidney and Bone	S. J. MARX AND G. D. AURBACH	150
20. Preparation of Vertebrate Photoreceptor Membranes for Study of Adenylate Cyclase, Guanylate Cyclase, and Cyclic Nucleotide Phosphodiesterase	J. J. KEIRNS, N. MIKI, AND M. W. BITENSKY	153
21. Preparation and Properties of Adenylate Cyclase from *Escherichia coli*	MARIANO TAO	155
22. Adenylate Cyclase from *Brevibacterium liquefaciens*	KATSUJI TAKAI, YOSHIKAZU KURASHINA, AND OSAMU HAYAISHI	160
23. Preparation of Adenylate Cyclase from Frog Erythrocytes	JACK ERLICHMAN AND ORA M. ROSEN	170
24. Method for Solubilization of Myocardial Adenylate Cyclase and Assessment of the Role of Phospholipids in Hormone Stimulation	GERALD S. LEVEY	174
25. Production of Glass Bead-Immobilized Catecholamines	J. CRAIG VENTER AND JACK E. DIXON	180

26. A Partial Purification of the β-Adrenergic Receptor Adenylate Cyclase Complex by Affinity Chromatography to Glass Bead-Immobilized Isoproterenol	J. Craig Venter and Nathan O. Kaplan	187
27. Preparation and Characterization of Guanylate Cyclase from Bovine Lung	Arnold A. White and Terry V. Zenser	192
28. Guanylate Cyclase from Sperm of the Sea Urchin, Strongylocentrotus purpuratus	David L. Garbers and J. P. Gray	196
29. Guanylate Cyclase in Human Platelets	Eycke Böhme, Regine Jung, and Ilse Mechler	199

Section IV. Degradation of Cyclic Nucleotides

30. Assay of Cyclic Nucleotide Phosphodiesterases with Radioactive Substrates	W. Joseph Thompson, Gary Brooker, and M. Michael Appleman	205
31. Assay of Cyclic Nucleotide Phosphodiesterase by a Continuous Titrimetric Technique	Wai Yiu Cheung	213
32. Cyclic 3′,5′-Nucleotide Phosphodiesterase from Bovine Heart	R. W. Butcher	218
33. Purification and Characterization of Cyclic 3′,5′-Nucleotide Phosphodiesterase from Bovine Brain	Wai Yiu Cheung and Ying Ming Lin	223
34. Purification and Characterization of Cyclic Nucleotide Phosphodiesterase from Skeletal Muscle	Robert G. Kemp and Yung-Chen Huang	240
35. Purification and Properties of Cyclic Nucleotide Phosphodiesterase from Dictyostelium discoideum (Extracellular)	Bruce M. Chassy and E. Victoria Porter	244
36. Cyclic AMP Phosphodiesterase of Escherichia coli	Larry D. Nielsen and H. V. Rickenberg	249
37. Preparation and Characterization of Multiple Forms of Cyclic Nucleotide Phosphodiesterase from Liver	Wesley L. Terasaki, Thomas R. Russell, and M. Michael Appleman	257
38. Activity Stain for the Detection of Cyclic Nucleotide Phosphodiesterase in Polyacrylamide Gels	Elihu N. Goren, Allen H. Hirsch, and Ora M. Rosen	259
39. Purification and Characterization of a Protein Activator of Cyclic Nucleotide Phosphodiesterase from Bovine Brain	Ying Ming Lin, Yung Pin Liu, and Wai Yiu Cheung	262

40. Techniques for the Formation, Partial Purification, and Assay of a Cyclic AMP Inhibitor — Ferid Murad — 273

Section V. Cyclic Nucleotide-Dependent Protein Kinases and Binding Proteins

41. Assay of Cyclic AMP-Dependent Protein Kinases — Jackie D. Corbin and Erwin M. Reimann — 287

42. Criteria for the Classification of Protein Kinases — J. A. Traugh, C. D. Ashby, and D. A. Walsh — 290

43. Preparation of Homogeneous Cyclic AMP-Dependent Protein Kinase(s) and Its Subunits from Rabbit Skeletal Muscle — J. A. Beavo, P. J. Bechtel, and E. G. Krebs — 299

44. Cyclic AMP-Dependent Protein Kinase from Bovine Heart Muscle — Charles S. Rubin, Jack Erlichman, and Ora M. Rosen — 308

45. Preparation and Properties of Cyclic AMP-Dependent Protein Kinases from Rabbit Red Blood Cells — Mariano Tao — 315

46. Cytoplasmic Hepatic Protein Kinases — Lee-Jing Chen and Donal A. Walsh — 323

47. Purification and Characterization of Cyclic GMP-Dependent Protein Kinases — J. F. Kuo and Paul Greengard — 329

48. Purification and Characterization of an Inhibitor Protein of Cyclic AMP-Dependent Protein Kinases — C. Dennis Ashby and Donal A. Walsh — 350

49. Techniques for the Study of Protein Kinase Activation in Intact Cells — Thomas R. Soderling, Jackie D. Corbin, and Charles R. Park — 358

50. The Purification and Analysis of Mechanism of Action of a Cyclic AMP-Receptor Protein from *Escherichia coli* — Ira Pastan, Maria Gallo, and Wayne B. Anderson — 367

51. The Detection and Characterization of Cyclic AMP-Receptor Proteins in Animal Cells — Gordon N. Gill and Gordon M. Walton — 376

Section VI. Synthetic Derivatives of Cyclic Nucleotides and Their Precursors

52. The Preparation and Use of Cyclic AMP Sepharose — Meir Wilchek and Zvi Selinger — 385

53. The Preparation and Use of Diazomalonyl Derivatives in Isolating and Identifying Cyclic AMP Receptor Sites — Barry S. Cooperman and David J. Brunswick — 387

54. The Preparation of Acylated Derivatives of Cyclic Nucleotides	T. Posternak and G. Weimann	399
55. The Synthesis of [^{32}P]Adenosine 3',5'-Cyclic Phosphate and Other Ribo- and Deoxyribonucleoside 3',5'-Cyclic Phosphates	R. H. Symons	410
56. Adenylylimidodiphosphate and Guanylylimidodiphosphate	Ralph G. Yount	420
57. The Synthesis of 1,N^6-Ethenoadenosine 3',5'-Monophosphate; A Fluorescent Analog of Cyclic AMP	John A. Secrist III	428
Author Index		431
Subject Index		440

Contributors to Volume XXXVIII, Part C

Article numbers are in parentheses following the names of contributors.
Affiliations listed are current.

WAYNE B. ANDERSON (50), *Laboratory of Molecular Biology, National Cancer Institute, National Institutes of Health, Bethesda, Maryland*

M. MICHAEL APPLEMAN (30, 37), *Department of Biological Sciences, University of Southern California, Los Angeles, California*

C. DENNIS ASHBY (42, 48), *Department of Biological Chemistry, Center for the Health Sciences, University of California, Los Angeles, California*

G. D. AURBACH (19), *Metabolic Diseases Branch, National Institute of Arthritis, Metabolism and Digestive Diseases, National Institutes of Health, Bethesda, Maryland*

J. A. BEAVO (43), *Department of Biological Chemistry, School of Medicine, University of California, Davis, California*

P. J. BECHTEL (43), *Department of Biological Chemistry, School of Medicine, University of California, Davis, California*

M. W. BITENSKY (20), *Department of Pathology, Yale School of Medicine, New Haven, Connecticut*

JAMES W. BODLEY (11), *Department of Biochemistry, University of Minnesota, Minneapolis, Minnesota*

EYCKE BÖHME (2, 4, 29), *Department of Pharmacology, University of Heidelberg, Heidelberg, Germany*

GARY BROOKER (3, 30), *Department of Pharmacology, University of Virginia School of Medicine, Charlottesville, Virginia*

DAVID J. BRUNSWICK (53), *Veterans Hospital, Philadelphia, Pennsylvania*

R. W. BUTCHER (32), *Department of Biochemistry, University of Massachusetts Medical School, Worcester, Massachusetts*

PETER S. CHAN (5), *Department of Cardiovascular-Renal Pharmacology, Lederle Laboratories, Pearl River, New York*

BRUCE M. CHASSY (35), *Laboratory of Microbiology and Immunology, National Institute of Dental Research, National Institutes of Health, Bethesda, Maryland*

LEE-JING CHEN (46), *Department of Internal Medicine, School of Medicine, University of California, Davis, California*

WAI YIU CHEUNG (31, 33, 39), *Laboratory of Biochemistry, St. Jude Children's Research Hospital, and Department of Biochemistry, University of Tennessee Medical Units, Memphis, Tennessee*

BARRY S. COOPERMAN (53), *Chemistry Department, University of Pennsylvania, Philadelphia, Pennsylvania*

JACKIE D. CORBIN (41, 49), *Department of Physiology, Vanderbilt University, Nashville, Tennessee*

JACK E. DIXON (25), *Department of Biochemistry, Purdue University, Lafayette, Indiana*

GEORGE I. DRUMMOND (18), *Department of Pharmacology, Faculty of Medicine, University of British Columbia, Vancouver, British Columbia, Canada*

JACK ERLICHMAN (23, 44), *Department of Molecular Biology, Albert Einstein College of Medicine, Bronx, New York*

MARIA GALLO (50), *Laboratory of Molecular Biology, National Cancer Institute, National Institutes of Health, Bethesda, Maryland*

DAVID L. GARBERS (28), *Department of Physiology, Vanderbilt University School of Medicine, Nashville, Tennessee*

GORDON N. GILL (51), *Department of Medicine, School of Medicine, University of California, San Diego, La Jolla, California*

ALFRED G. GILMAN (7), *Department of*

Pharmacology, University of Virginia, Charlottesville, Virginia

NELSON D. GOLDBERG (10, 11), *Department of Pharmacology, University of Minnesota, Minneapolis, Minnesota*

ELIHU N. GOREN (38), *Department of Medicine, Albert Einstein College of Medicine, Bronx, New York*

J. P. GRAY (28), *Department of Pharmacology, University of British Columbia, Vancouver, British Columbia, Canada*

PAUL GREENGARD (12, 47), *Department of Pharmacology, Yale University School of Medicine, New Haven, Connecticut*

MARI K. HADDOX (10, 11), *Department of Pharmacology, University of Minnesota, Minneapolis, Minnesota*

JOEL G. HARDMAN (2, 14), *Department of Physiology, Vanderbilt University School of Medicine, Nashville, Tennessee*

OSAMU HAYAISHI (22), *Department of Medical Chemistry, Faculty of Medicine, Kyoto University, Kyoto, Japan*

ALLEN H. HIRSCH (38), *Department of Molecular Biology, Albert Einstein College of Medicine, Bronx, New York*

YUNG-CHEN HUANG (34), *Department of Pharmacology Rutgers Medical School, Piscataway, New Jersey*

ROGER A. JOHNSON (8, 17), *Department of Physiology, Vanderbilt University School of Medicine, Nashville, Tennessee*

REGINE JUNG (29), *Pharmakologisches Institut der Universität Heidelberg, Heidelberg, Germany*

NATHAN O. KAPLAN (26), *Department of Chemistry, University of California, San Diego, La Jolla, California*

J. J. KEIRNS (20), *Department of Pathology, Yale University School of Medicine, New Haven, Connecticut*

ROBERT G. KEMP (34), *Department of Biochemistry, Medical College of Wisconsin, Milwaukee, Wisconsin*

E. G. KREBS (43), *Department of Biological Chemistry, School of Medicine, University of California, Davis, California*

J. F. KUO (12, 47), *Department of Pharmacology, Emory University, Atlanta, Georgia*

YOSHIKAZU KURASHINA (22), *Department of Biology, Faculty of Pharmaceutical Sciences, Kanazawa University, Kanazawa, Japan*

GERALD S. LEVEY (24), *Howard Hughes Medical Institute, Division of Endocrinology and Metabolism, Department of Medicine, University of Miami School of Medicine, Miami, Florida*

LILLIAN LIN (11), *Department of Biochemistry, University of Minnesota, Minneapolis, Minnesota*

MICHAEL C. LIN (5, 16), *Section on Membrane Regulation, Laboratory of Nutrition and Endocrinology, National Institute of Arthritis, Metabolism and Digestive Diseases, National Institutes of Health, Bethesda, Maryland*

YING MING LIN (33, 39), *Laboratory of Biochemistry, St. Jude Children's Research Hospital, and Department of Biochemistry, University of Tennessee Medical Units, Memphis, Tennessee*

YUNG PIN LIU (39), *Laboratory of Biochemistry, St. Jude Children's Research Hospital, and Department of Biochemistry, University of Tennessee Medical Units, Memphis, Tennessee*

S. J. MARX (19), *Metabolic Diseases Branch, National Institute of Arthritis, Metabolism and Digestive Diseases, National Institutes of Health, Bethesda, Maryland*

STEVEN E. MAYER (1, 9), *Department of Medicine, School of Medicine, University of California, San Diego, La Jolla, California*

ILSE MECHLER (29), *Department of Pharmacology, University of Heidelberg, Heidelberg, Germany*

N. MIKI (20), *Department of Pathology, Yale University School of Medicine, New Haven, Connecticut*

FERID MURAD (7, 40), *Division of Clinical Pharmacology, Departments of Medicine and Pharmacology, University of Virginia, Charlottesville, Virginia*

LARRY D. NIELSEN (36), *Division of Research, National Jewish Hospital and Research Center, Denver, Colorado*

ROBERT F. O'DEA (11), *Department of Pharmacology, University of Minnesota, Minneapolis, Minnesota*

CHARLES R. PARK (49), *Department of Physiology, Vanderbilt University, Nashville, Tennessee*

IRA PASTAN (50), *Laboratory of Molecular Biology, National Cancer Institute, National Institutes of Health, Bethesda, Maryland*

E. VICTORIA PORTER (35), *Laboratory of Microbiology and Immunology, National Institute of Dental Research, National Institutes of Health, Bethesda, Maryland*

T. POSTERNAK (54), *Laboratory of Biological and Special Organic Chemistry, University of Geneva, Geneva, Switzerland*

ERWIN M. REIMANN (41), *Department of Biochemistry, Medical College of Ohio, Toledo, Ohio*

H. V. RICKENBERG (36), *Division of Research, National Jewish Hospital and Research Center, and Department of Biophysics and Genetics, University of Colorado Medical Center, Denver, Colorado*

ORA M. ROSEN (23, 38, 44), *Departments of Medicine and Molecular Biology, Albert Einstein College of Medicine, Bronx, New York*

CHARLES S. RUBIN (44), *Departments of Anatomy and Molecular Biology, Albert Einstein College of Medicine, Bronx, New York*

THOMAS R. RUSSELL (37), *Department of Biochemistry, University of Miami School of Medicine, Miami, Florida*

GÜNTER SCHULTZ (2, 4, 14, 15), *Department of Pharmacology, University of Heidelberg, Heidelberg, Germany, and Department of Physiology, Vanderbilt University, Nashville, Tennessee*

JOHN A. SECRIST III (57), *Department of Chemistry, Ohio State University, Columbus, Ohio*

ZVI SELINGER (52), *Department of Biochemistry, The Hebrew University, Jerusalem, Israel*

DAVID L. SEVERSON (18), *Department of Pharmacology, Faculty of Medicine, University of British Columbia, Vancouver, British Columbia, Canada*

THOMAS R. SODERLING (49), *Department of Physiology, Vanderbilt University, Nashville, Tennessee*

ALTON L. STEINER (13), *Department of Medicine, University of North Carolina School of Medicine, Chapel Hill, North Carolina*

JAMES T. STULL (1, 9), *Department of Medicine, School of Medicine, University of California, San Diego, La Jolla, California*

CHARLES A. SUTHERLAND (8), *Department of Pharmacology, University of Virginia School of Medicine, Charlottesville, Virginia*

EARL W. SUTHERLAND* (17), *Department of Biochemistry, University of Miami School of Medicine, Miami, Florida*

R. H. SYMONS (55), *Department of Biochemistry, University of Adelaide, Adelaide, South Australia*

KATSUJI TAKAI (22), *Department of Physiological Chemistry and Nutrition, Faculty of Medicine, The University of Tokyo, Tokyo, Japan*

MARIANO TAO (21, 45), *Department of Biological Chemistry, University of Illinois at the Medical Center, Chicago, Illinois*

WESLEY L. TERASAKI (37), *Department of Pharmacology, University of Virginia School of Medicine, Charlottesville, Virginia*

BARBARA THOMPSON (9), *Department of Medicine, School of Medicine, Univer-*

* Deceased March, 1974.

sity of California, San Diego, La Jolla, California

W. JOSEPH THOMPSON (30), *Department of Pharmacology, University of Texas Medical School at Houston, Houston, Texas*

J. A. TRAUGH (42), *Department of Biochemistry, University of California, Riverside, California*

J. CRAIG VENTER (25, 26), *Division of Physiology and Pharmacology, Department of Medicine, School of Medicine, University of California, San Diego, La Jolla, California*

DONAL A. WALSH (42, 46, 48), *Department of Biological Chemistry, School of Medicine, University of California, Davis, California*

GORDON M. WALTON (51), *Department of Medicine, School of Medicine, University of California, San Diego, La Jolla, California*

WILLIAM B. WASTILA (1, 9), *Department of Pharmacology, Burroughs Wellcome & Co., Research Triangle Park, North Carolina*

G. WEIMANN (54), *Biochemical Research Center, Boehringer Mannheim G.m.b.H., Tutzing, Germany*

ARNOLD A. WHITE (6, 27), *Department of Biochemistry, University of Missouri, Columbia, Missouri*

MEIR WILCHEK (52), *Department of Biophysics, Weizmann Institute of Science, Rehovoth, Israel*

RALPH G. YOUNT (56), *Biochemistry-Biophysics Program, and Department of Chemistry, Washington State University, Pullman, Washington*

TERRY V. ZENSER (27), *The U.S. Army Medical Research Institute of Infectious Diseases, Fort Dehick, Frederick, Maryland*

Preface

The compilation of a volume on nucleoside 3',5'-cyclic monophosphates and the separation of it from other volumes dealing with nucleotide-related methodologies may seem to some like arbitrary categorization. However, the close association that these cyclic nucleotides have had with the field of basic endocrinology and hormone action and the technical problems that are peculiar to their study seemed to the editors to justify the construction of this volume and its inclusion in the series of volumes on hormones.

The study of cyclic nucleotide metabolism and action and their alteration by hormones and other factors has met with formidable technical problems from its initiation. The field grew very slowly for almost a decade after Rall and Sutherland discovered cyclic AMP in biological material in the late 1950's. During the last five years, however, cyclic nucleotide-related literature has grown at an exponential rate, in part because of the development of newer and improved methodology and in part because investigators in more and more branches of biology are finding cyclic nucleotides to be of interest.

This volume contains a broad selection of techniques that should be useful to those interested in studying the cyclic nucleotide content of intact cells and the biosynthesis, degradation, and action of cyclic nucleotides in cell-free systems. Techniques for purifying and assaying a single enzyme or nucleotide will be described in more than one article. This is largely by editorial design, not to compile a complete collection of methods, but to offer the investigator a choice of useful techniques. All of them will be reliable when properly used, but none of them will be, for all laboratories, either easier or more reliable to use than any of the others. In some cases (as, for example, with adenylate cyclase from mammalian sources) lability or lack of extensive purification of an enzyme has contributed to technical problems that differ from tissue to tissue. In other cases (as, for example, with cyclic nucleotide phosphodiesterase) certain properties of an enzyme will be found to be distinctly different from tissue to tissue or among multiple forms of the enzyme from a single tissue.

Omissions have inevitably occurred—some because potential authors were overcommitted, some because of editorial oversight, some because of the timing of new developments relative to the publication deadline. Some apparent omissions have been covered in previous volumes of "Methods in Enzymology."

We thank Drs. S. P. Colowick and N. O. Kaplan who originated the idea for and encouraged the compilation of this volume. We thank the staff of Academic Press for their help and advice. We especially thank the contributing authors for their patience and full cooperation and for carrying out the research that made this volume possible.

JOEL G. HARDMAN
BERT W. O'MALLEY

METHODS IN ENZYMOLOGY

EDITED BY

Sidney P. Colowick and Nathan O. Kaplan

VANDERBILT UNIVERSITY　　　DEPARTMENT OF CHEMISTRY
SCHOOL OF MEDICINE　　　　UNIVERSITY OF CALIFORNIA
NASHVILLE, TENNESSEE　　　　AT SAN DIEGO
　　　　　　　　　　　　　　LA JOLLA, CALIFORNIA

I. Preparation and Assay of Enzymes
II. Preparation and Assay of Enzymes
III. Preparation and Assay of Substrates
IV. Special Techniques for the Enzymologist
V. Preparation and Assay of Enzymes
VI. Preparation and Assay of Enzymes (*Continued*)
　　Preparation and Assay of Substrates
　　Special Techniques
VII. Cumulative Subject Index

METHODS IN ENZYMOLOGY

EDITORS-IN-CHIEF

Sidney P. Colowick   Nathan O. Kaplan

VOLUME VIII. Complex Carbohydrates
Edited by ELIZABETH F. NEUFELD AND VICTOR GINSBURG

VOLUME IX. Carbohydrate Metabolism
Edited by WILLIS A. WOOD

VOLUME X. Oxidation and Phosphorylation
Edited by RONALD W. ESTABROOK AND MAYNARD E. PULLMAN

VOLUME XI. Enzyme Structure
Edited by C. H. W. HIRS

VOLUME XII. Nucleic Acids (Parts A and B)
Edited by LAWRENCE GROSSMAN AND KIVIE MOLDAVE

VOLUME XIII. Citric Acid Cycle
Edited by J. M. LOWENSTEIN

VOLUME XIV. Lipids
Edited by J. M. LOWENSTEIN

VOLUME XV. Steroids and Terpenoids
Edited by RAYMOND B. CLAYTON

VOLUME XVI. Fast Reactions
Edited by KENNETH KUSTIN

VOLUME XVII. Metabolism of Amino Acids and Amines (Parts A and B)
Edited by HERBERT TABOR AND CELIA WHITE TABOR

VOLUME XVIII. Vitamins and Coenzymes (Parts A, B, and C)
Edited by DONALD B. MCCORMICK AND LEMUEL D. WRIGHT

VOLUME XIX. Proteolytic Enzymes
Edited by GERTRUDE E. PERLMANN AND LASZLO LORAND

VOLUME XX. Nucleic Acids and Protein Synthesis (Part C)
Edited by KIVIE MOLDAVE AND LAWRENCE GROSSMAN

VOLUME XXI. Nucleic Acids (Part D)
Edited by LAWRENCE GROSSMAN AND KIVIE MOLDAVE

VOLUME XXII. Enzyme Purification and Related Techniques
Edited by WILLIAM B. JAKOBY

VOLUME XXIII. Photosynthesis (Part A)
Edited by ANTHONY SAN PIETRO

VOLUME XXIV. Photosynthesis and Nitrogen Fixation (Part B)
Edited by ANTHONY SAN PIETRO

VOLUME XXV. Enzyme Structure (Part B)
Edited by C. H. W. HIRS AND SERGE N. TIMASHEFF

VOLUME XXVI. Enzyme Structure (Part C)
Edited by C. H. W. HIRS AND SERGE N. TIMASHEFF

VOLUME XXVII. Enzyme Structure (Part D)
Edited by C. H. W. HIRS AND SERGE N. TIMASHEFF

VOLUME XXVIII. Complex Carbohydrates (Part B)
Edited by VICTOR GINSBURG

VOLUME XXIX. Nucleic Acids and Protein Synthesis (Part E)
Edited by LAWRENCE GROSSMAN AND KIVIE MOLDAVE

VOLUME XXX. Nucleic Acids and Protein Synthesis (Part F)
Edited by KIVIE MOLDAVE AND LAWRENCE GROSSMAN

VOLUME XXXI. Biomembranes (Part A)
Edited by SIDNEY FLEISCHER AND LESTER PACKER

VOLUME XXXII. Biomembranes (Part B)
Edited by SIDNEY FLEISCHER AND LESTER PACKER

VOLUME XXXIII. Cumulative Subject Index Volumes I–XXX
Edited by MARTHA G. DENNIS AND EDWARD A. DENNIS

VOLUME XXXIV. Affinity Techniques (Enzyme Purification: Part B)
Edited by WILLIAM B. JAKOBY AND MEIR WILCHEK

VOLUME XXXV. Lipids (Part B)
Edited by JOHN M. LOWENSTEIN

VOLUME XXXVI. Hormone Action (Part A: Steroid Hormones)
Edited by BERT W. O'MALLEY AND JOEL G. HARDMAN

VOLUME XXXVII. Hormone Action (Part B: Peptide Hormones)
Edited by BERT W. O'MALLEY AND JOEL G. HARDMAN

VOLUME XXXVIII. Hormone Action (Part C: Cyclic Nucleotides)
Edited by JOEL G. HARDMAN AND BERT W. O'MALLEY

VOLUME XXXIX. Hormone Action (Part D: Isolated Cells, Tissues, and Organ Systems)
Edited by JOEL G. HARDMAN AND BERT W. O'MALLEY

VOLUME XL. Hormone Action (Part E: Nuclear Structure and Function)
Edited by BERT W. O'MALLEY AND JOEL G. HARDMAN

VOLUME 41. Carbohydrate Metabolism (Part B)
Edited by W. A. WOOD

VOLUME 42. Carbohydrate Metabolism (Part C)
Edited by W. A. WOOD

VOLUME 43. Antibiotics
Edited by JOHN H. HASH

EARL W. SUTHERLAND, JR.
(1915–1974)

Earl W. Sutherland, Jr.
(1915–1974)

This volume is dedicated to the memory of Earl W. Sutherland, Jr., M.D. Seldom has an individual been so closely identified with the development of a field of investigation as has Dr. Sutherland with the field of cyclic nucleotides and their relation to basic endocrinology. His pioneering work on mechanisms of hormone action was appropriately recognized when he was awarded the Nobel Prize for Physiology or Medicine in 1971. His untimely death in March of 1974 left a void in the scientific community.

Dr. Sutherland's work was driven by his intense curiosity and respect for truth. He had a remarkable instinct for asking the right question and for selecting the most straightforward of several possible approaches to its answer. Young associates were often in awe of his ability to see in data trends that were not yet apparent to others, and they frequently attached more significance to Dr. Sutherland's intuition—so often did it lead to a correct solution—than to someone else's carefully calculated arguments.

He was a creative thinker who valued an original idea, but who only critically and cautiously accepted data to support his ideas. Perhaps the lesson he tried hardest to teach his junior colleagues was, as he put it: "Never fall in love with your hypothesis." He stressed the point that an emotional investment in a concept and in its defense not only weakens objectivity but also lessens the chance to take advantage of serendipity, a factor he did not underrate in scientific discovery.

At various times, scientists from several disciplines and from no less than twelve nations worked in his laboratories. To them, he was an inspirational colleague and mentor and a generous friend. To biological scientists at large, Dr. Sutherland was a leader. His work led to insights into fundamental regulatory mechanisms that are now being explored not just by endocrinologists, physiologists, and pharmacologists, but by investigators in virtually every branch of medicine and biology.

JOEL G. HARDMAN

Methods in Enzymology

Volume XXXVIII
HORMONE ACTION
Part C
Cyclic Nucleotides

Section I

Extraction and Purification of Cyclic Nucleotides

[1] Rapid Tissue Fixation and Extraction Techniques

By STEVEN E. MAYER, JAMES T. STULL, and WILLIAM B. WASTILA

The objectives of preparing tissue for analysis of metabolite content include freezing the tissue rapidly enough to prevent changes associated with excision or other disturbances of the normal milieu of the tissue; preventing alterations during storage and during manipulation before extraction; and extracting the tissue in order to avoid alteration of metabolites by enzymes or by the extraction medium. The same objectives are sought for assay of enzymes that are transformed from nonactivated to activated states, and vice versa, by drug or hormone action on cells, e.g., phosphorylase kinase, phosphorylase, and glycogen synthase in response to epinephrine. The general aspects and many specific problems of tissue preparation for analysis are discussed in Lowry and Passonneau's "A Flexible System of Enzymatic Analysis,"[1] especially in Chapters 7 and 10 of this book.

Rapid Fixation

Freezing tissue rapidly enough to prevent artificial changes in cyclic nucleotide content may be difficult to achieve. Factors are the size and accessibility of the tissue and the activities of adenylate cyclase and cyclic nucleotide phosphodiesterase during the hazardous period between the moment the tissue is disrupted and the moment the activities of these enzymes are terminated. The sparse information available indicates that this critical period is approximately 1–3 seconds for brain, skeletal muscle, and heart, but it may be much longer for other tissues. For example, rabbit gallbladder mucosa accounts for almost all the cyclic AMP (cAMP) in the tissues of this organ. The elevation of cAMP measured in response to cholera toxin is the same whether the organ is rapidly frozen or whether the mucosa is scraped, washed, and then frozen, which takes about 2 minutes.[2] However, the stability of cAMP in this type of experiment may reflect the nature of the stimulus, not unique properties of gallbladder mucosa adenylate cyclase and phosphodiesterase. Thus, the rapidity of tissue fixation needed will be a function of the nature and intensity of the experimental stimulus that causes cAMP formation, the sensitivity required to measure a change in content and the alteration

[1] O. H. Lowry and J. V. Passonneau, "A Flexible System of Enzymatic Analysis." Academic Press, New York, 1972.
[2] R. Mertens, H. Wheeler, and S. E. Mayer, *Gastroenterology*, in press (1974).

of the rate of either cAMP formation or destruction when tissue is disrupted.

Liquid N_2 is not as satisfactory for rapid freezing of biopsy samples as is a liquid cooled below its boiling point by immersion in a liquid N_2 bath. Liquid N_2 is ordinarily at its boiling point ($-190°$). The N_2 gas formed when a tissue sample is first immersed in liquid N_2 forms an insulating layer around the tissue which significantly delays freezing. Dichlorodifluoromethane (Freon 12) which boils at $-30°$ and freezes at $-158°$, is satisfactory and readily available. The commercial can of refrigerant is held upside down, and the CCl_2F_2 is dispensed as a liquid through a valve and tube into a 50–100-ml beaker. The beaker is suspended in a Dewar flask filled with liquid N_2. The liquid CCl_2F_2 is stirred until it begins to freeze. When the tissue is immersed in the CCl_2F_2 it should be vigorously stirred to obtain the maximal rate of heat dissipation.

Tissue samples greater than 500 mg may not freeze rapidly enough in CCl_2F_2 to preserve the *in vivo* concentration of cAMP. In fact, brain in severed murine heads cooled more rapidly in liquid N_2 than in CCl_2F_2.[3] Thus for large samples liquid N_2 will be cheaper and more effective than CCl_2F_2. A method of rapidly freezing small samples in liquid N_2 when CCl_2F_2 is not available is described by Lowry and Passonneau.[1] Liquid N_2 is cooled below its boiling point by evaporation. However, no procedure for immersion of large samples, such as whole brain, canine hearts, or rabbit leg muscle, into chilled liquid is likely to prevent postmortem alterations in cAMP concentrations.

Excision of biopsy samples with scissors and forceps is less traumatic to the surrounding tissue and more flexible in terms of sample size than the clamping technique described below. Biopsy sampling is generally slower in freezing tissue, and this may be important in some tissues when relatively large samples are needed. A comparison between the biopsy and clamp techniques in which small samples (200–300 mg) were obtained from the rabbit gracilis muscle showed no difference between these sampling techniques in analysis of cAMP content, in response to isoproterenol or under control conditions.[4] A comparison was also made with cardiac tissue. Repeated biopsy samples may be taken from dog ventricle *in situ* with curved eye scissors and mosquito hemostatic forceps. Six to 15 samples can be taken from the right ventricle without significant change in ventricular function; the number depends on the size of the biopsy samples (10–100 mg) and the size of the heart. An alternative

[3] J. A. Ferrendelli, M. H. Gay, W. G. Sedgwick, and M. M. Chang, *J. Neurochem.* **19**, 979 (1972).
[4] J. T. Stull and S. E. Mayer, *J. Biol. Chem.* **246**, 5716 (1971).

biopsy technique utilizes a rapidly rotating, sharp corer mounted in a dental drill.[5] Negative pressure is applied during the cutting to retain the sample (10–30 mg) in the corer, and the sample is then rapidly ejected into the cooling liquid by positive pressure. A third method is to freeze the apex of the heart with clamps precooled in liquid nitrogen as described below. When all three techniques were applied to dog hearts under control conditions or after stimulation by norepinephrine, comparable values of cAMP concentrations were obtained (Wastila and Mayer, unpublished observations). Thus, biopsy techniques may provide adequate fixation of the tissue when compared to clamping, providing the samples are small enough to allow rapid freezing.

Wollenberger et al.[6] introduced the technique and theory of compression of organs or tissue pieces between two blocks of metal cooled in liquid N_2. Small tongs suitable for freezing rat hearts (0.7 g), preparations of isolated cardiac muscle, and flat (<3 mm) skeletal muscle *in situ*, are illustrated in Fig. 1. Larger tongs were described by Wollenberger as crushing a 1.6-g guinea pig kidney to a thickness of 0.7 mm, with the temperature reaching −20° in 0.12 second. The practical limitations of this method of fixation of larger samples are the size of the blocks needed to provide an adequate heat sink and the strength required to crush the tissue.

A method of rapidly freezing whole rats or the entire leg of an anesthetized rabbit was designed and developed by Dr. Howard E. Morgan, Pennsylvania State University. The apparatus is described in Fig. 2. Phosphorylase kinase and phosphorylase activities and cAMP concentrations in rabbit leg muscle frozen in this manner were not different from those measured in samples obtained by the freeze-clamping technique.[7]

None of the techniques already described are satisfactory where very rapid inactivation of the formation or destruction of cAMP must be combined with preservation of anatomical structure. The combination of rapid fixation and preservation of structure has been a particular problem in experiments involving brain. cAMP concentration rises rapidly in the brain after decapitation, and this is not satisfactorily arrested by freezing in liquid. A new approach to rapid tissue fixation is the use of microwave radiation to rapidly heat-inactivate brain enzymes. Schmidt et al.[8] ob-

---

[5] P. E. Pool, G. F. Norris, R. M. Lewis, and J. W. Covell, *J. Appl. Physiol.* **24**, 832 (1968).

[6] A. Wollenberger, O. Ristau, and G. Schoffa, *Pfluegers Arch. Gesamte Physiol. Menschen Tiere* **270**, 399 (1960).

[7] S. E. Mayer and E. G. Krebs, *J. Biol. Chem.* **245**, 3153 (1970).

[8] M. J. Schmidt, J. T. Hopkins, D. E. Schmidt, and G. A. Robison, *Brain Res.* **42**, 465 (1972).

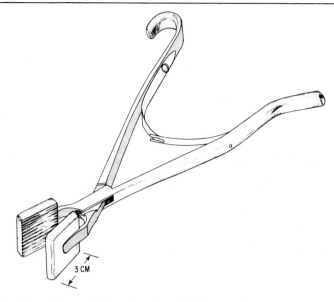

FIG. 1. Clamp for rapid freezing of small tissues and organs (up- to about 0.75 g). The tissue is tightly squeezed between the blocks, which are precooled in liquid nitrogen. The design is modified from the original described by A. Wollenberger, O. Ristau, and G. Schoffa [*Pfluegers Arch. Gesamte Physiol. Menschen Tiere* **270**, 399 (1960)]. It can be used to freeze rat organs *in situ*. The blocks are machined from silver, and the handle is a modified Kern box lock bone-holding forceps (J. Sklar Mfg. Co., Long Island City, N.Y. catalog No. 250-285).

tained nearly 100% inactivation of rat brain adenylate cyclase and phosphodiesterase after 20 seconds of irradiation of rats in a commercial oven with an output of 1250 W at 2470 MHz. Some increase in cAMP probably occurred early during irradiation. More rapid heat inactivation of acetylcholinesterase (4–6 seconds) was obtained by centering the head in the wave guide of a modified magnetron microwave tube.[9] However, changes in phosphocreatine, ATP, and lactate occur during the intense heating by microwave irradiation.[10] The freeze-blowing method, in which probes driven into the cranial vault of the rat blow brain tissue into a liquid nitrogen-cooled chamber appears to provide optimal fixation of labile metabolites and enzymes. This method of fixation therefore probably furnishes the most accurate means of measuring brain cAMP but precludes the use of brain tissue for regional studies.

[9] W. B. Stavinoha, S. T. Weintraub, and A. T. Modak, *J. Neurochem.* **20**, 361 (1973).
[10] W. D. Lust, J. V. Passonneau, and R. L. Veech, *Science* **181**, 280 (1973).

Tissue Storage

Many enzymes and metabolites in unprocessed tissues deteriorate rapidly if the samples are stored at above −35°.[1] Rabbit skeletal muscle powder stored five years at −65° has on repeated analysis yielded the same values of cAMP concentration and the activities of phosphorylase and phosphorylase kinase. Tissue samples stored in a freezer must be protected from dehydration because sublimation of H_2O from the tissue to the colder walls of the cabinet takes place. Upright deep-freezers should be avoided because of the rapid increase in temperature in the cabinet that occurs when the door is opened.

Unneutralized trichloroacetic acid extracts of tissues appear to undergo no change in cAMP content during several days at room temperature or more than 1 week at 4°. Extracts that have been neutralized should be kept at −20° or colder because bacterial or mold growth in neutralized extracts kept in a refrigerator may cause very large increases in cAMP.

Extraction Techniques

It is essential that the enzymes which may affect cAMP concentration be inactivated before the tissue sample begins to thaw (at approximately −8°). This means that dissection and weighing must be done at −15° or lower, and that the samples be exposed to this temperature for the minimum time.[1] The safest procedure is to transport samples from the freezer to the preparation area and subsequently to the homogenizer in liquid N_2. Specimen tubes or vials should not be tightly capped, since liquid N_2 trapped in the tubes will cause them to shatter when exposed to higher temperatures.

Dissection and weighing is best done in a −20° walk-in cold box with the work area well away from the door. Samples from 10 to 1000 mg may be rapidly weighed at −20° on a line-operated electric balance (Cahn Electrobalance, model RTL or DTL).

Small samples (up to 20 mg) can be homogenized directly without prior powdering. A Duall, size 20, conical ground-glass homogenizer tube (Kontes Glass Co., Vineland, New Jersey) is kept at −20° (in a 20% ethanol–H_2O slush). A volume of 10% aqueous trichloroacetic acid 10–20 times the mass of the sample is added. The sample is rapidly transferred from the liquid N_2 bath and is immediately homogenized with a motor-driven conical glass pestle rotating at about 750 rpm. Excessive friction

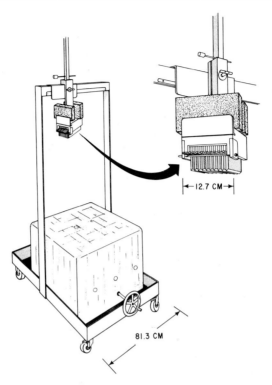

Fig. 2. An apparatus for rapid freezing of 20–200 g of tissue. It consists of a 40-cm butcher block, on which the anesthetized animal is positioned, and a knife blade assembly consisting of ten 15-cm long carpenter knives, spaced 4 mm apart, held 70 cm above the animal, and weighted with 25 kg of lead. A few minutes before use the knife blade assembly is removed, cooled in liquid nitrogen, and repositioned. After removal of the safety stop, the knife blade assembly is released and, guided by the long vertical steel rod to which it is attached, drops on the animal. When used on a skinned hindlimb of a rabbit, the impact forces muscle and crushed bone between the knives. From the rate of change of color of the cut end of muscle it was estimated that freezing takes place in 1–2 seconds. The frozen muscle wafers are driven from between the knife blades by striking the bars located between the upper ends of the knives and hinged at the distal end of the assembly. Apparatus was designed by and built for H. E. Morgan.

generating heat must be avoided. Further treatment of the homogenate is discussed in another section.[11]

Larger tissue samples should be powdered before homogenization, because whole samples, especially from tough tissues such as muscle and heart, may thaw before they are adequately homogenized. Samples up

[11] S. E. Mayer, J. T. Stull, W. B. Wastila, and B. Thompson, this volume [9].

to several grams are powdered by percussion in a stainless steel mortar. The sample, pestle, and mortar are precooled in liquid N_2 immediately before percussion. A pestle with a spherical head and mortar with a thick wall is best for small samples. The powder may be transferred to screw-cap glass vials and stored indefinitely at $-65°$. Screw-cap glass vials are particularly useful since a tight seal may be provided for storage, the sample code can be written directly on the glass with a waterproof marker, and the glass provides insulation for the sample powder during transfers. The powder is extracted with trichloroacetic acid as described above except that the powder is added to the homogenizer tube before the acid. When cAMP content is to be correlated with changes in enzyme activity, powdering of the entire specimen has the advantage that one can obtain aliquots of a homogeneous sample for the preparation of different extracts.

Unless an electrically driven or pneumatic hammer is used to strike the pestle, still larger samples (5 g or more) are not easily powdered in a percussion mortar. Pieces of muscle, 5–25 g, can be powdered in a microanalytical mill with a cryogenic attachment (National Apparatus Co., Cleveland, Ohio, catalog Nos. J3805 and J3825). Several hundred grams of skeletal muscle, frozen into wafers by the apparatus shown in Fig. 2, can be powdered by passing the wafers through a meat grinder. Both muscle and meat grinder are cooled with liquid N_2. The wafers must be added slowly to the grinder, and the grinder must not be cooled too much or it will jam. A heavy-duty motor and grinder are recommended (Hobart Manufacturing Co., Troy, Ohio, Model 4621).

[2] Separation and Purification of Cyclic Nucleotides by Ion-Exchange Resin Column Chromatography

By G. SCHULTZ, E. BÖHME, and J. G. HARDMAN

For determinations of cyclic nucleotides and related enzyme activities, ion-exchange column chromatography is the most widely used separation technique.

Cation-exchange as well as anion-exchange resins can be used for cyclic nucleotide separations. Dowex-50 columns developed with acid solvents are especially useful if the samples contain large amounts of electrolytes (e.g., tissue constituents, salts of an incubation medium or acid used for extraction) since such samples can directly be applied. The sample volume that can be applied, however, is limited.

Anion-exchange resins have also been employed for cyclic nucleotide separations. These materials have been used especially in assays of adenylate and guanylate cyclase and of cyclic nucleotide phosphodiesterase activities. Their application to the purification of cyclic nucleotides from tissue extracts can be complicated by the fact that in most cases excess electrolytes (such as those mentioned above) have to be removed before sample application to the column. For this purpose, adsorption of the nucleotides to charcoal (column or batch) under acid conditions and elution into an ammonia–alcohol mixture[1] can be used. A charcoal step can be avoided with strongly basic anion-exchange materials such as QAE-Sephadex if tissue homogenization and nucleotide extraction are not performed in acid, but in a zinc acetate–alcohol solution.[2]

The principles of cyclic nucleotide separations by thin-layer chromatography on polyethyleneimine (PEI)-cellulose[3] have been adapted to column chromatography. PEI-cellulose columns as described in this chapter are especially useful and effective for assays of cyclase and phosphodiesterase activities. However, the sensitivity of this material to electrolytes is relatively high, so that columns of it have to be preceded by an electrolyte-removing step if they are used for cyclic nucleotide purification from tissue extracts.

In addition to the ion-exchange column chromatography systems described in this chapter, the use of Dowex-1 developed with formic acid or hydrochloric acid for purification of tissue cyclic nucleotides has been reported.[4,5]

The water used for regeneration of the resins should be deionized or distilled. For the final steps of the resin regeneration and for the preparation of the elution fluids, however, water of glass-redistilled quality should be used.

Chromatography on Dowex-50

Resin Preparation

Dowex-50 resin (AG 50W-X8, 100–200 mesh) in the H$^+$ form is purchased from BioRad Laboratories. The following procedure has been used to wash new resin and to regenerate used resin, although the entire pro-

[1] K. K. Tsuboi and T. D. Price, *Arch. Biochem. Biophys.* **81**, 223 (1959).
[2] G. Schultz, J. G. Hardman, K. Schultz, J. W. Davis, and E. W. Sutherland, *Proc. Nat. Acad. Sci. U.S.* **70**, 1721 (1973).
[3] E. Böhme and G. Schultz, this volume [4].
[4] G. Brooker, L. J. Thomas, Jr., and M. M. Appleman, *Biochemistry* **7**, 4177 (1968).
[5] F. Murad, V. Manganiello, and M. Vaughan, *Proc. Nat. Acad. Sci. U.S.* **68**, 736 (1971). Also see this volume [7], [10], [12], [16].

cedure may be unnecessary for new resin that is to be used for some purposes. Batches of resin have been used and regenerated several times over a period of a year or longer without any apparent loss in resolving capacity. The stated volumes are for a 750-g batch of resin; proportionately smaller or larger volumes are used for smaller or larger batches.

Step 1. The resin is stirred for a few seconds with about 2 liters of water. After the resin has settled, the supernatant fluid and fines are decanted. This is repeated twice or until the supernatant fluid is nearly free of fines, and a slurry of the resin is then poured into a glass column with a sintered-glass bottom. A 7.5×32 cm column will hold about 750 g of resin and allow a satisfactory flow rate.

Step 2. After all the water has run through, 4 liters of 0.5 N NaOH are added in divided portions, care being taken to avoid unnecessary stirring of the resin. A dark band develops at the top and moves down the column of resin during the NaOH treatment. This band should reach the bottom of the column before all the NaOH has run through.

Step 3. The column of resin is next washed with 4 liters of water, again with care to avoid unnecessary stirring of the resin.

Step 4. The H^+ form of the resin is regenerated by washing the column of resin with 6 liters of 2 N HCl. The dark color should have disappeared before all of the HCl has run through.

Step 5. The resin is again washed with 4 liters of distilled or deionized water, and then with 4 liters of water of glass-redistilled quality.

Step 6. Finally the column of resin is washed with 2 liters of 0.1 N HCl (prepared with water of glass-redistilled quality), and a slurry of the resin in 0.1 N HCl is stored at 4°.

Separation of Cyclic AMP (cAMP) and Cyclic GMP (cGMP) from Related Nucleotides in Tissue Extracts

Dowex-50 columns offer the advantages of reproducibility and flexibility. Discernible variations in elution profiles have not been seen over a period of several years with many batches of resin. By varying the dimensions of the columns, sample volumes up to 15 ml may be processed. The sample itself may be in 0.1–0.3 N HCl or $HClO_4$ (in H_2O or 50% ethanol) with substantial variation in total ionic strength, and, as long as the elution is carried out with 0.1 N HCl, the migration of cAMP and cGMP on the resin is not appreciably affected. Disadvantages are the time required to run the columns (up to several hours, depending on the column length), the relatively large volumes required to elute the samples (usually requiring lyophilizing before analyses can be performed), and the relatively poor resolution of cyclic nucleotides from other nucleotides. In addition, Dowex-50 can catalyze the acid hydrolysis of the glycosidic bond of some nucleotides.

Inexpensive columns can be prepared from pieces of glass-tubing cut to appropriate lengths with one end constricted and tapered. The glass column should be at least twice the length of the resin bed to allow a satisfactory flow rate. A relatively uniform flow rate is important since extreme variations can alter the migration of cyclic nucleotides relative to the eluant.

A plug of Pyrex wool in the bottom of the column retains the resin, which is loaded into the column as a well-mixed thin slurry in 0.1 N HCl. A polyethylene dispensing bottle is very useful for adding the slurry. The column of settled resin is washed with 1 or 2 bed volumes

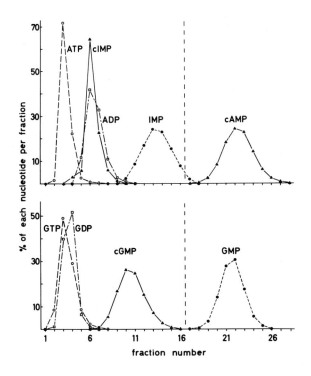

FIG. 1. Separation of cyclic nucleotides from related nucleotides by Dowex-50 column chromatography. One milliliter of a 1 mM solution of each nucleotide in 0.1 N HCl was applied to a 0.62 × 30 cm column of AG 50W-X8 (100–200 mesh) resin. Columns were eluted with 0.1 N HCl. A flow rate of approximately 0.36 ml/minute was maintained by keeping the head of HCl 25 ± 5 cm above the resin bed. Fractions 1 through 16 were 2 ml each and 17 through 28 were 4 ml each. 5'-AMP, adenosine, and guanosine were not eluted by fraction 28. Recoveries of the nucleotides varied from 70 to 95% of the amounts applied.

of 0.1 N HCl before application of the sample. It is important that the column of resin should not be allowed to dry out during the procedure.

Samples (in 0.1–0.3 N HCl or $HClO_4$) are applied to the columns of resin with Pasteur pipettes, and collection of eluate is begun after the meniscus of the sample has run into the resin bed; the sides of the column are rinsed with about 1 ml of 0.1 N HCl, and this is allowed to run into the resin bed. The column above the resin is then filled with 0.1 N HCl, care being taken not to stir the resin. The flow rate is maintained relatively constant by maintaining the level of 0.1 N HCl above the resin within narrow limits.

The dimensions of the resin bed determine the volume of sample that can be processed. To obtain resolution such as that shown in Fig. 1, 1–2.5 ml of sample may be applied to a 0.62 × 30 cm column of resin, or up to 15 ml to a 0.62 × 60 cm column. Numerous combinations of sample volumes and column dimensions can be used, depending to some extent on the degree of resolution desired. For example, for purification of cAMP for assay by methods not requiring extremely high purification[6,7] 1–2 ml of sample may be satisfactorily processed on a column only 10 cm in length. However, when cGMP is to be purified, longer columns and smaller sample volumes are necessary to separate cGMP from GDP, ADP, etc.

Krishna[8] has applied very short columns of Dowex-50 to the assay of adenylate cyclase activity. Although lacking high resolution, these columns can be very useful as a rapid preliminary purification step.

QAE-Sephadex

Preparation of Resin and Columns

QAE-Sephadex A-25 is obtained from Pharmacia in the chloride form. The resin is allowed to swell in distilled water overnight and then converted through the OH^- form to the formate form, i.e., through steps c to f of the procedure described below.

Used resin is stored at 4° and regenerated in a large glass column with sintered-glass bottom by successive treatments with the following substances:

    a. 0.1–0.2 N HCl until the pH of the effluent is 1–2
    b. Water until the pH of the effluent is 4–5

---

[6] A. G. Gilman and F. Murad, this volume [7].
[7] A. L. Steiner, this volume [13].
[8] G. Krishna, B. Weiss, and B. B. Brodie, *J. Pharmacol. Exp. Ther.* **163**, 379 (1968).

c. 0.1–0.2 N NaOH until the effluent is Cl⁻-free (as checked with AgNO$_3$ under acid conditions)
d. Water until the pH of the effluent is about 8
e. 0.2 N Formic acid until the pH of the effluent is about 2.5
f. Glass-distilled water until the pH of the effluent is about 4

The resin is stored at 4° in glass-distilled water or 50 mM ammonium formate until use.

Glass columns, 0.7 cm in inner diameter, 10 cm in length, with a plastic reservoir and plastic sieve (BioRad Laboratories, No. 731-1220) are used for purification of tissue cyclic nucleotides. Slightly conical plastic columns, 0.65–0.75 cm in inner diameter, 4 cm in length, with a reservoir (Evergreen Scientific, No. 3030-31) furnished with a fine sieve of silk, nylon, or similar material are used for separations in phosphodiesterase assays. The resin (in formate form) is loaded into the columns (without vigorous stirring to avoid the formation of fines) and washed with about 10 ml of water.

Separation of cAMP and cGMP from Related Nucleotides in Tissue Extracts

Short QAE-Sephadex columns can be used for purification of tissue cyclic nucleotides without an electrolyte-removing (charcoal) step if the following procedure[2] for tissue extraction is used.

Tissue samples of 100–200 mg wet weight are rapidly frozen and homogenized in 2 ml of a solution cooled to −20° and containing 0.1 M zinc acetate and tracer amounts of tritiated cAMP and cGMP in 50% ethanol. The homogenizer is rinsed with 2 ml of 50% ethanol which are added to the homogenate. After centrifugation (10 minutes at 50,000 g), 0.1 ml of a 2 M Na$_2$CO$_3$ solution is added to the supernatant fluid. Most of the 5′-nucleotides ($\geq$99% of ATP and GTP, $\geq$90% of AMP and GMP) are coprecipitated with the ZnCO$_3$ formed while most of the cAMP ($\geq$98%) and cGMP ($\geq$80%) are not precipitated. After centrifugation (10 minutes at 50,000 g), the supernatant fluid is passed through a very small column (0.7 × 1.5 cm) of Dowex-50 (AG 50W-X8, 100–200 mesh) in the NH$_4$⁺ form, and the tissue extract is followed by 6 ml of distilled water. Zn^{2+} is completely removed from the sample by this step whereas nucleotides are not affected.

Separation of cAMP and cGMP from each other and from related nucleotides is performed by chromatography on a 0.7 × 4 cm column of QAE-Sephadex prepared in the formate form as described above.

After application of the sample, 10 ml of distilled water are applied to the column followed by 6 ml of a 0.1 M ammonium formate solution

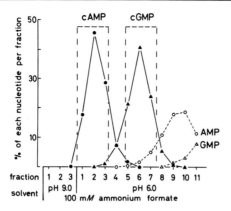

Fig. 2. Separation of cyclic nucleotides from related nucleotides by column chromatography on QAE-Sephadex A-25. After preceding $ZnCO_3$ precipitation in the presence of 45% alcohol as described in the text, nucleotides were added to the supernatant fractions and then applied to 0.7 × 4 cm columns of QAE-Sephadex in the formate form. After application of 10 ml of distilled water followed by 6 ml of 0.1 M ammonium formate, pH 9.0, cAMP and then cGMP are eluted with 0.1 M ammonium formate, pH 6.0. The fractions were 2 ml each. ADP, ATP, GDP, and GTP were not eluted by fraction 12.

adjusted to pH 9.0 with NH_4OH. The elution fluid is then changed to 0.1 M ammonium formate adjusted to pH 6.0 with formic acid. Initially, 5.5 ml of this solution are applied, and the eluate, containing cAMP, is collected. The following 3 ml of eluate are discarded. Then 5.5 ml of the same solvent are applied, and the eluate, containing cGMP, is collected (Fig. 2). The cyclic nucleotide peaks may be retarded by 1–2 ml if the column is not preceded by the $ZnCO_3$ coprecipitation step. The cyclic nucleotide-containing fractions are lyophilized for assay.[9]

Except for a small overlap of AMP with cGMP, 5′-nucleotides are effectively excluded from the cyclic nucleotide fractions. For several tissues, the purification procedure described above has given a sufficient purification of cAMP and cGMP for their assays by direct enzymatic methods.[2,10,11] The overall recoveries for cAMP and cGMP in the extraction and purification procedure described above are 60–70% and 40–50%, respectively.

---

[9] Salt residues after lyophilization can be effectively reduced by filtering before lyophilization the eluate fractions followed by 2 ml of water through short (0.7 × 1.5 cm) Dowex-50 columns in the H^+ form as suggested by Dr. N. D. Goldberg. If this step is applied, the Dowex-50 step preceding the QAE-Sephadex column becomes unnecessary.

[10] C. A. Sutherland and R. A. Johnson, this volume [8].

[11] G. Schultz and J. G Hardman, this volume [14].

Separation of Cyclic Nucleotides and Related Nucleosides

Assays of cyclic nucleotide phosphodiesterase activity are easily performed as a two-step procedure. The 5'-nucleoside monophosphate formed from labeled cyclic nucleotide by phosphodiesterase in the first step is hydrolyzed by excess 5'-nucleotidase to the corresponding nucleoside and inorganic phosphate.[12,13] A QAE-Sephadex column, 0.7 × 2 cm, can be used for a fast and effective separation of the labeled nucleoside from the residual labeled cyclic nucleotide.

After dilution of the sample to 1 ml with 20 mM ammonium formate solution, pH 7.4, containing 0.1 mM adenosine and/or 0.1 mM guanosine, the sample is applied to the column. Another 4 ml of 20 mM ammonium formate solution, pH 7.4, are used to develop the column. All 5 ml of eluate, containing more than 95% of the adenosine and guanosine, are collected and counted in a detergent-containing scintillation fluid.[14] For elution of the cyclic nucleotide fraction, the columns can be further developed with 4–5 ml of 0.2 N HCl. In this fraction, >95% of the cyclic nucleotides are eluted.

Polyethyleneimine (PEI) cellulose

Relatively small PEI-cellulose columns are suitable for fast separations of cyclic nucleotides from related compounds with high resolution. The nucleotides are eluted with electrolyte solutions. For the elution of a particular nucleotide, the same electrolyte concentration is required that is capable of eluting the compound from thin-layer chromatography plates coated with PEI-cellulose.[3] Adjustment of the solvent concentrations for optimal separations is required for each batch of PEI-cellulose. For the elution of nucleosides, bases, and uric acid, very dilute acid is used, e.g., 5 mM acetic acid.

The electrolyte content of the samples applied to the column may affect the adsorption of the nucleotides. Therefore, the electrolyte concentration of the samples should be lower than the electrolyte concentration required to elute the compound(s) of interest.

Preparation of PEI-Cellulose Columns

PEI-cellulose for column chromatography is obtained from Serva Feinbiochemica, Heidelberg, Germany, and stored at 4°. Before use, the

[12] J. A. Beavo, J. G. Hardman, and E. W. Sutherland, *J. Biol. Chem.* **245**, 5649 (1970).
[13] W. J. Thompson, G. Brooker, and M. M. Appleman this volume [30].
[14] The total nucleoside fraction can be counted, e.g., with addition of 15 ml of a "tT-21" scintillation fluid[15] (this solution contains 5 g of PPO 0.1 g of POPOP or dimethyl-POPOP in 330 ml of Triton X-100 and 670 ml of toluene).

cellulose is washed three times with 10–20 volumes of distilled water followed by three washes with 10–20 volumes of 5 mM acetic acid. Fines are removed during each wash. After final settling, one volume of cellulose is suspended in three volumes of 5 mM acetic acid. About 8 ml of this suspension are poured into a Pasteur pipette stoppered with glass wool and furnished with a plastic reservoir (10 ml) to give a column of 5.5 × 70 mm.

Fractionation of Nucleotides, Nucleosides, Bases, and Uric Acid

Samples are applied in 0.5 ml or more. The elution of the nucleotides studied generally requires 3–6 ml of eluate and is quantitative. This fraction size allows direct liquid scintillation counting of the entire fractions, e.g., with detergent-containing scintillation fluids.[14-16]

The capacity of the PEI-cellulose columns is sufficient for most studies. Pilot experiments were performed with 0.1 μmole of each compound; if 0.5 μmole of some nucleotides were applied, the peaks were still sharp.

Various nucleotides can be separated from each other and from nucleosides, bases and uric acid by stepwise elution with different eluents (see the table, columns 2–4).

Nucleosides (adenosine, inosine, guanosine, xanthosine), bases (adenine, hypoxanthine, guanine, xanthine), and uric acid are generally eluted with 10 ml of 5 mM acetic acid. More than 95% of each compound is found in the 2nd to 7th ml of the acetic acid eluate. For a complete elution of nucleosides, bases, and uric acid, a further elution step with 10 ml of water should follow (see the table, lines 1 and 2).

Cyclic nucleotides and related nucleotides are fractionated from each other by stepwise increasing concentrations of LiCl solutions. The cyclic nucleotides, cAMP, cIMP, or cGMP, are eluted by 4–5 ml of 0.02 M LiCl. If a separation of cAMP from cIMP is required, cAMP is eluted with 5 ml of 50 mM acetic acid, and then cIMP is eluted with 4 ml 0.02 M LiCl. A possible contamination of the cIMP fraction with some cAMP is prevented by an additional elution step with water after elution of cAMP. The nucleoside monophosphates, AMP, IMP, or GMP, are eluted with 5 ml of 0.2 M LiCl, nucleoside diphosphates with 4–5 ml

---

[15] R. C. Greene, *in* E. D. Bransome, "The Current Status of Liquid Scintillation Counting," p. 189. Grune & Stratton, New York, 1970.

[16] Several milliliters of aqueous solutions containing high amounts of salts can be counted in 10–15 ml of a "TTX-127" scintillation fluid.[15] This solution contains 5 g of PPO and 0.5 g of POPOP or dimethyl-POPOP in 100 ml of Triton X-100, 200 ml of Triton X-114, and 700 ml of xylene with or without addition of 100 g of naphthalene.

SOME APPLICATIONS OF POLYETHYLENEIMINE-CELLULOSE COLUMN CHROMATOGRAPHY[a]

Elution solvent	Fractionation of nucleotides			Adenylate cyclase	Guanylate cyclase	Phospho-diesterase	cAMP, cGMP from tissue
Acetic acid, 5 mM	10 NBU	10 NBU	10 NBU	10 NBU	10 NBU	10 NBU	10 NBU
Distilled water	10 —	10 —	10 —	10 —	10 —	10 —	10 —
Acetic acid, 50 mM	—	5 cAMP	—	—	—	—	1 —; 5 cAMP
Distilled water	—	10 —	—	—	—	—	10 —
LiCl, 0.02 M	1 —; 4 cAMP, cIMP	1 —; 4 cIMP	1 —; 5 cGMP	1 —; 4 cAMP	1 —; 5 cGMP	1 —; 5 cAMP, cGMP	1 —; 5 cGMP
LiCl, 0.2 M	5 AMP, IMP	5 AMP, IMP	5 cXMP, GMP	—	—	—	—
LiCl, 0.5 M	1 —; 5 ADP, IDP	1 —; 5 ADP, IDP	1 —; 4 XMP, GDP	6 AMP, ADP	6 GMP, GDP	—	—
LiCl, 1.0 M	5 ATP, ITP	5 ATP, ITP	5 GTP	5 ATP	5 GTP	—	—

[a] Numbers are volumes (ml) of elution solvent necessary for the respective separation step followed by the nucleotide(s) eluted by this step. —— indicates elutate fraction to be discarded. NBU: nucleosides (Ado, Ino, Guo, Xao), bases (Ade, Hyp, Gua, Xan), and uric acid.

of 0.5 M LiCl, and the nucleoside triphosphates with 5 ml of 1.0 M LiCl. For the fractionation of xanthosine-containing nucleotides, higher concentrations of LiCl than for the other corresponding purine nucleotides are necessary. Thus, cXMP is eluted by 0.2 M LiCl, XMP by 0.5 M LiCl.

Isolation of Cyclic Nucleotides for Assays of Adenylate and Guanylate Cyclase Activity

For assays of adenylate and guanylate cyclase activity, the respective cyclic nucleotide has to be separated from related nucleotides[17] and, if ^{3}H- or ^{14}C-labeled substrate is used, also from related nucleosides, bases, and uric acid (see the table, columns 5 and 6). After elution of nucleosides, bases, and uric acid with 10 ml of 5 mM acetic acid followed by 10 ml of water, cAMP or cGMP is eluted with 4 ml of 0.02 M LiCl. For determination of the residual substrate, nucleoside mono- and diphosphates are eluted with 6 ml of 0.5 M LiCl. Then 5 ml of 1.0 M LiCl are applied, and the eluate containing ATP or GTP is collected.

Separation of Cyclic Nucleotides from Related Nucleosides

If in assays of cyclic nucleotide phosphodiesterase activity the nucleoside monophosphate formed by phosphodiesterase is converted to the corresponding nucleoside by 5′-nucleotidase, the product of the latter enzymatic reaction, adenosine or guanosine (also inosine, xanthosine, and further possible degradation products), can be determined by collection of the 2nd to the 7th ml of the eluate obtained with 5 mM acetic acid as solvent (see Table, column 7). The remaining substrate, cAMP or cGMP, can be eluted with 5 ml of 0.02 M LiCl.

Separation of cAMP and cGMP from Related Nucleotides in Tissue Extracts

Separation of cAMP and cGMP from each other and from other nucleotides in tissue extracts follows the same principles (see the table, column 8). cAMP is eluted with 5 ml of 50 mM acetic acid, cGMP with 5 ml of 0.02 M LiCl or KCl. cGMP can also be eluted with 5 ml of 50 mM ammonium acetate. Being volatile, the latter solvent is useful when lyophilization is necessary to reduce the sample volume.[9]

It will generally be necessary to remove tissue and medium electro-

---

[17] If ^{32}P-labeled nucleotide triphosphate is used as substrate, inorganic phosphate is most effectively separated by precipitation [Y. Sugino and Y. Miyoshi, *J. Biol. Chem.* **239**, 2360 (1964)].

lytes by prepurification, e.g., another column chromatographic step, or by an electrolyte-removing step, e.g., by charcoal adsorption and desorption of the nucleotides[1] before application of tissue extracts to PEI-cellulose columns is possible.

[3] Separation, Purification and Analysis of Cyclic Nucleotides by High Pressure Ion Exchange Chromatography

By GARY BROOKER

High-pressure anion exchange chromatography has been used to measure cyclic AMP (cAMP) produced by adenylate cyclase[1] and to measure cAMP in tissues[2] and in patient urine.[3] In addition, it has been shown useful in the measurement of cyclic GMP (cGMP) and cAMP analogs.[4] The anion exchange resin used in this method is that originally described by Horvath et al.[5] Anion exchange resin is polymerized upon small glass beads and has been termed "pellicular" resin. The glass bead matrix eliminates changes in column volume when the ionic strength is altered and markedly reduces the bleed of ultraviolet absorbing materials from the resin. Our present high-pressure chromatograph system consists of a Varian LCS-1000 liquid chromatograph with a 3-meter pellicular anion exchange column. The 254 nm ultraviolet 8-μl, 1-cm pathlength flow cell has been modified as previously described[6] for increased stability and sensitivity. The analog output of the ultraviolet flow cell detector is recorded on a strip chart recorder and the chromatographic peaks digitally integrated by a Hewlett-Packard 3370B integrator with teletype output. In addition, a timer actuated solenoid has been added to automatically collect the chromatographic peak in a liquid scintillation vial.

Operation of the Equipment

In general, the ability to obtain good results at the picomole level is dependent upon optimal performance of the mechanical and electronic systems. With the modifications to the detector previously described,[6] full-scale deflection of the strip chart recorder was set to equal 2×10^{-3}

[1] G. Brooker, *Anal. Chem.* **42**, 1108 (1970).
[2] G. Brooker, *J. Biol. Chem.* **246**, 7810 (1971).
[3] M. Fichman and G. Brooker, *J. Clin. Endocrinol. Metab.* **35**, 35 (1972).
[4] G. Brooker, *in* "Methods in Molecular Biology" (M. Chasin, ed.), Vol. 3, p. 82. Dekker, New York, 1972.
[5] C. G. Horvath, B. A. Preiss, and S. R. Lipsky, *Anal. Chem.* **39**, 1422 (1967).
[6] G. Brooker, *Anal. Chem.* **43**, 1095 (1971).

absorbancy units. Under optimal conditions, after the signal from the chromatograph had passed through the maximum noise suppression filters on the Hewlett-Packard digital integrator, the noise level was 2×10^{-5} absorbancy unit.

The nongradient mode of operation has been used successfully for the analysis of cyclic nucleotides. Two HCl eluting solutions have been in repeated use in this laboratory for the last four years with no apparent damage to the chromatograph. In fact the same anion exchange column has also been used during this time. The resin in the 3-meter column was converted to the chloride form by washing the column with the 50 ml of 1 N HCl, and then to neutrality with 100 ml of distilled water. The use of a nongradient elution technique, wherein the cyclic nucleotide is just slightly retarded by the resin allows for the injection of one sample every 8–10 minutes. Column regeneration need only be performed several times a year if previously purified samples are routinely analyzed.

A column temperature of 80° has been routinely used for cyclic nucleotide analysis. Lower temperatures tend to broaden the peaks and reduce the flow rate by increasing the resistance of the column. In fact at lower temperatures the back pressure increases at the same pump setting. Since the flow characteristics of the column are temperature dependent, it is essential to maintain the column temperature constant. Large fluctuations in the column temperature can increase the instability of the detector baseline.

It is essential that for optimal operation no air bubbles or leakage from any fitting occur. We have found that operating the system for 0.5–1 hour at 1–3 times the pressure used for our analytical work tends to purge the system of any air bubbles that might have developed and, in addition, allows us to check all fittings for any obvious leakage that might not be detected at the lower pressures. Generally, we operate the system at 24–25 ml per hour, a flow rate that generates a back pressure of about 1000–1200 psi. It is easy to tell if air bubbles are present within the system, since it then does not come up to maximum pressure instantaneously when the pump is started; the baseline appears rather unstable, and a long delay occurs after the pump is turned off for the system to depressurize. Increase of the pumping rate to 2500–3000 psi for 0.5 hour usually eliminates small air bubbles; however, if this treatment does not correct the situation, more drastic purging of the system is necessary. Air bubbles seem to become easily trapped within lines in which active flow is not occurring, for example, the tube that leads to the pressure gauge. It has been found that purging the system with carbon dioxide is an effective way to eliminate air trapped within these lines. A source of CO_2 (dry ice in a filter flask with a cork and hose) is introduced into the lines in which the air bubbles are thought to reside. It is important not to purge the column itself. Once the system is reconnected and the solvent is pumped for 1–2 hours at very high pressure, the carbon dioxide then dissolves within the solvent and leaves the system free of air bubbles. All buffers should be degassed before introduction into the chromatograph, and the UV flow cell should be routinely checked by looking down into the cell to be sure that no dirt or dust particles are found within the cell itself. CAUTION: BE SURE TO PROTECT YOUR EYES FROM DANGEROUS UV RADIATION EMITTED FROM THE CELL!

Recently Varian Aerograph has redesigned the cell for more trouble-free operation. In the past, the gaskets separating the reference and sample cells eventually leaked, and liquid from the sample cell invaded the reference cell. This caused unstable baseline conditions, since this instrument operates on a dual-beam principle. The UV source is a germicidal lamp obtainable either from Varian or from General Electric, the absorption of the cell being detected by a dual-element photoresistor. The dual-element photoresistors are uniquely sensitive to changes in temperature.[6] A jacket with a circulating water supply to maintain the cell and the sample at a constant temperature has increased the stability of the unit. During operation of the system the photoresistors should have a low resistance. Optimal sensitivity occurs when resistance across pins 1 and 2 or 2 and 3 are equal and between 20,000 and 30,000 ohms. However, more noise is seen when these resistances are 80,000–200,000 ohms. In addition, it may be that the photoresistors age over a prolonged time and their sensitivity to light decreases. If the system lacks the sensitivity and stability that had been previously demonstrated, then one might suspect a leaky cell, air bubbles, or an inherent electrical problem. It is conceivable that an unstable UV source could contribute to the noise level, even though most noises should be eliminated by the dual-beam operation. It generally takes 1–2 days of constant operation for a new lamp to stabilize; in addition, it is advisable never to turn off the UV source, so that stability is maintained from day to day.

Chromatography of Cyclic Nucleotides

Tissue and urinary levels of cAMP have been determined by high pressure anion-exchange liquid chromatography after previous purification. During purification a tracer amount of ^{3}H labeled cAMP is added for the determination of purification recovery. Injection of cAMP standards gives a linear and reproducible response when based upon chromatographic peak height as shown in Fig. 1 or more recently by peak area

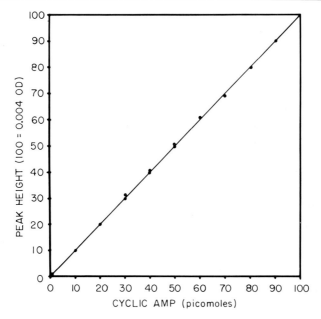

FIG. 1. Standard curve for measurement of cyclic AMP by the chromatographic system. From G. Brooker, *Anal. Chem.* **42**, 1108 (1970).

obtained with the digital integrator. The chromatographic peak for 43 pmoles of cAMP is shown in Fig. 2. The eluting solution used was HCl pH 2.20. For the analysis of cGMP, 0.2 M NaCl was added to the HCl solution. Under these conditions cAMP elutes with the breakthrough

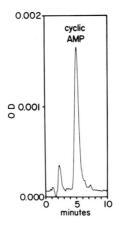

FIG. 2. High pressure chromatograph of 43 pmoles of cyclic AMP. From G. Brooker, *Anal. Chem.* **43**, 1095 (1971).

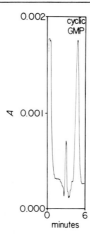

FIG. 3. High-pressure chromatograph of 50 pmoles of cyclic GMP. From G. Brooker, *J. Biol. Chem.* **246**, 7810 (1971).

point, and cGMP is eluted at about 5 minutes as shown in Fig. 3. In addition this buffer system has been used to measure cAMP analogs.[4]

The unknown samples were injected in volumes of 25 µl or less, the system was closed, and the flow started. The time-actuated automatic fraction collector was used to collect the cAMP peak in a liquid scintillation vial. Measurement of the peak height of the unknown and comparison to the peak height of the standard was used to determine the amount of nonradioactive cAMP in the sample. Measurement of tritium radioactivity in the cAMP peak was used to determine the percentage of the original sample that was measured. Division of the nonradioactive cAMP by the fractional isotopic recovery equaled the amount of cAMP present in the original tissue extract or urine specimen.

Because this method isolates cAMP free of other nucleotides and ultraviolet absorbing materials, it is an excellent method for the determination of the specific radioactivity of labeled cyclic nucleotides. Samples from tissues prelabeled with [^{14}C]adenosine can be readily determined. The [^{14}C]cAMP specific activity was also equal to the quotient of the ^{14}C radioactivity in the total cAMP peak (corrected for recovery) divided by the amount of cAMP in the original sample (corrected for recovery).

Preliminary Prepurification Procedure

Ten to 400 mg of frozen tissue obtained by freezing between clamps precooled in liquid nitrogen was placed into a 17 × 100 mm polypropylene test tube precooled in liquid nitrogen. Ten microliters of [^{3}H]cAMP tracer (0.1–1 pmoles) was added, and the tube was placed under the Polytron homogenizer; 3 ml of cold 5% trichloroacetic acid were added, and the sample was quickly homogenized at full speed for 10 seconds. The homogenate was centrifuged in a bench-top centrifuge, and the supernatant was decanted to another identical tube. The trichloroacetic acid was extracted with three 8-ml extractions of water-saturated ethyl ether. After each addition of ether, the tube was mixed on a Variwhirl vortex mixer (Van Waters & Rogers No. 5810-006) for 10 seconds. After the ether had separated, it was removed with a Teflon-tipped suction aspirator. Two-hundred microliters of 0.5 M Tris·HCl, pH 8.0 was added, and the sample was applied to an anion exchange column filled with AG-1-X2 50–100 mesh chloride form resin prepared in a disposable Pasteur pipette (0.8 × 5 cm). A small polypropylene funnel (A. H. Thomas Co. No. 5587G) was attached to the top of the column which acted as a solution reservoir above the column. The solution was allowed to pass through the column and 5 ml of water applied. After the water, 10 ml of HCl, pH 1.3, was applied and collected to elute cAMP from the column and separate it from tissue extract ATP which remained on the column.

It was necessary to eliminate ATP from the sample before treatment with Zn-Ba, to prevent the formaton of nonenzymatically produced cAMP from ATP.[2] To the cAMP fraction, 100 µl of 2 M Tris was added, followed by 2.5 ml of 5% $ZnSO_4$ and 2.5 ml of 0.3 N $Ba(OH)_2$. The precipitate was mixed and then centrifuged in a bench-top centrifuge. Ten milliliters of water were applied to the same column to ready it for reapplication of the $ZnSO_4$–$Ba(OH)_2$ supernatant. A new column was used when tissue samples had been prelabeled with [^{14}C]adenosine to determine [^{14}C]cAMP specific activity. If the samples were to be assayed by the enzyme displacement method or other enzyme method, a sheet of Whatman No. 42 filter paper was placed on the solution reservoir funnel to prevent any colloidal Zn-Ba from contaminating the Dowex resin. Once the Zn-Ba supernatant had been applied to the column, the filter paper was removed. After the Zn-Ba supernatant was applied to the column, it was followed by 30 ml of HCl, pH 2.7, followed by collection of 25 ml of HCl, pH 2.1, into a 40-ml conical centrifuge tube (Pyrex No. 8124). The flow rate for these columns was about 3 ml per minute. The cAMP fractions were evaporated to dryness under vacuum at 55° on a Buchler Evapomixer or lyophilized.

The samples were taken up and spotted on cellulose thin-layer plates to eliminate interfering substances which continuously bleed from the AG-1-X2 resin. Reference spots of cAMP (1 µmole) were spotted at each edge of the thin-layer plates. Usually, 8–12 samples were spotted on a

20 × 20 cm plate. The plates were developed for a distance of 10–15 cm with *n*-butanol:acetic acid:water, 2:1:1. The reference spots of cAMP, at each edge, were visualized with a short wavelength UV light, the area of each unknown sample cut from the plate, and the cellulose was scraped into a polypropylene test tube. Two milliliters of water were added, the sample vigorously mixed for 15 seconds, and the cellulose was centrifuged. The supernatant was transferred to a 40-ml conical centrifuge tube and extracted three times with 8 ml of ether. The sample was evaporated to dryness on the Evapomixer and the sample taken up with 30 μl of HCl, pH 2.20, and 20 μl were injected into the high-pressure chromatograph.

A slight modification of this procedure has been used to purify samples for analysis by several protein binding methods for cyclic nucleotides. Purification for the cAMP protein binding assay[7] involves elimination of the pH 1.3 and Zn-Ba steps. The pH 2.1 fraction is evaporated and assayed directly. If HCl pH 1.3 is then next added to the column, cGMP is eluted. Cyclic GMP has been measured with the radioimmunoassay[8] using this purification technique.[9]

Because authentic [^{3}H]cAMP tracer was added to each unknown sample, there is the opportunity to verify the authenticity of the cAMP peak in every sample. Each sample's tritium specific activity depends upon the amount of tritiated tracer cAMP per amount of endogenous cAMP present (ready for injection into the chromatograph). Each sample can be treated with phosphodiesterase to destroy about 50% of the cAMP present in the sample. The treated sample was then injected into the chromatograph. If the cAMP peak was authentic cAMP, then the chromatographic peak should be reduced equally. That is, the specific activity of the sample should not change if the peak absorbancy was authentic cAMP.

To accomplish this test in practice, a 10-μl fraction remaining from the purified tissue or urine sample was incubated for 10 minutes at 30° with 10 μl of a solution containing 5 m*M* $MgCl_2$, 100 mM Tris·HCl, pH 8.0, and 1 μl of phosphodiesterase. The reaction was stopped in a boiling water bath for 2 minutes, and the treated sample was then injected directly into the high-pressure liquid chromatograph.

---

[7] A. G. Gilman, *Proc. Nat. Acad. Sci. U.S.* **67**, 305 (1970).
[8] A. L. Steiner, C. W. Parker, and D. M. Kipnis, *J. Biol. Chem.* **247**, 1106 (1972).
[9] F. Murad, personal communication.

[4] Separation of Cyclic Nucleotides by Thin-Layer Chromatography on Polyethyleneimine Cellulose

By E. BÖHME and G. SCHULTZ

Thin-layer chromatography (TLC) on anion-exchange layers is a versatile method for the fractionation and purification of cyclic nucleotides and related compounds. Polyethyleneimine (PEI) cellulose was introduced and shown to be superior to other materials by K. Randerath. Although other anion-exchange celluloses and nonionic layers can be used and have been applied in the authors' laboratory, the present article is restricted to PEI-cellulose. This material is outstanding with regard to sharpness of resolution and versatile applicability. The aim of the present paper is to show the principal possibilities for the separation of cyclic nucleotides, especially adenosine 3',5'-monophosphate (cAMP)[1] and guanosine 3',5'-monophosphate (cGMP), from related nucleotides and from their possible biological degradation products, nucleosides, bases, and uric acid.

Chromatographic properties of nucleotides, nucleosides, and bases on PEI-cellulose layers with different aqueous electrolyte solutions as solvents have been described by Randerath[2,3] and Pataki.[4] The migration rates of nucleotides, nucleosides, and bases on PEI-cellulose layers are affected by the cellulose material and the amount of PEI used for the preparation of the layer. Migration rates given in this chapter have been obtained with certain batches of PEI-cellulose; with other batches from the same manufacturers, slight differences have been observed. With materials from other sources, significant differences were obtained. Therefore, adjustment of the solvent *concentrations* for optimal separations is required for each batch of PEI-cellulose.

In the authors' laboratory repeated unidirectional developments of the plates have widely been used; two-dimensional chromatography and gradient development have been avoided because of practical reasons.

---

[1] Abbreviations: Nucleoside 3',5'-monophosphates, cNMP; nucleoside 5'-monophosphates, NMP.
[2] K. Randerath, "Dünnschicht-Chromatographie," 2nd ed., Verlag Chemie, Weinheim, 1965; "Thin-Layer Chromatography," Academic Press, New York, 1966.
[3] K. Randerath and E. Randerath, see this series, Vol. 12A [40], p. 323.
[4] G. Pataki, *Advan. Chromatogr.* **7**, 47 (1968).

Materials and General Methods

PEI-cellulose for TLC is obtained from Serva[5]; 22 g of PEI-cellulose are homogenized with 0.22 g of fluorescence indicator F_{254}[6] in 105 ml of distilled water for 1 minute by use of an Ultra-Turrax[7] at 20,000 rpm or by use of another small high speed blender. The suspension is coated on 20×110 cm plastic sheets[8] essentially as described by Randerath.[2,3] The slit of the coating apparatus is set at 0.4 mm. After drying at room temperature, the sheets are cut into 20×20 cm pieces. Commercial TLC plates with similar separation properties are obtained from Schleicher & Schuell.[9] All PEI-cellulose plates are stored at 2–4°. Before use, 5–8 mm of the layer along both side margins and along the top margin of the plates are removed, and the plates are washed by ascending development with distilled water and then dried. All substances (approximately 4 nmoles) are applied spotwise along a line 3 cm from the lower margin of the sheets. The plates are developed by ascending chromatography in rectangular or sandwich[10] chambers over the entire length of the layer, i.e., approximately 16 cm from the origin. The time required per run is 60–90 minutes. Before the next development, the layers are dried in an air stream at room temperature. The compounds are detected under UV light (254 nm). Migration rates are given in hR_f, i.e., $R_f \times 100$.

Separation of Nucleotides from Nucleosides, Bases, and Uric Acid

Migration rates of nucleosides, bases and uric acid on anion-exchange cellulose layers are very similar to those observed with partition chromatography on unmodified cellulose layers. The hR_f values of these compounds are virtually unaffected by the electrolyte concentration of the solvent. Most nucleosides and bases as well as uric acid are capable of lactam-lactim tautomerism and have a weakly acidic functional group. As the dissociation of the compounds depends on the pH of the surrounding solvent, their migration rate on anion-exchange cellulose layers is affected by the pH of the ion-exchange material and of the solvent used

---

[5] Serva Feinbiochemica, Heidelberg, Germany.
[6] E. Merck, Darmstadt, Germany; also available through EM Laboratories, Inc., Elmsford, New York 10523, or Brinkmann Instruments, Westbury, New York 11590.
[7] Jahnke & Kunkel, Staufen, Germany.
[8] Sheets of Astralon, Dynamit Nobel, Troisdorf, Germany, or of Bakelite as described by Randerath.[2]
[9] TLC-ready plastic sheets, PEI-cellulose with luminescer, 0.2 mm layer, manufactured by Schleicher & Schüll, Dassel/Einbeck, Germany; also available from Schleicher & Schuell, Keene, New Hampshire 03431.
[10] CAMAG, Basel, Switzerland, or Berlin, Germany, or New Berlin, Wisconsin 53151.

for development. Chromatography, e.g., of xanthosine on PEI-cellulose layers with distilled water as solvent, results in large spots with relatively variable hR_f values between 15 and 30. If acetic acid is used as solvent, the spot size is significantly reduced and the hR_f value is increased. The migration rate of uric acid similarly increases if acetic acid is used as solvent instead of water. Considering the possible ionization of nucleosides, bases, and uric acid at neutral and higher pH values, acid solvents, e.g., acetic acid or formic acid, are used for a complete separation of these compounds from nucleotides (Table I). With acetic acid as solvent, the hR_f values of all nucleosides, bases, and uric acid are greater than 40, while the corresponding nucleotides remain at the origin or very close to it. Cyclic nucleotides, mainly cAMP, but also cIMP and cGMP, slowly migrate with acetic acid as solvent. With higher acetic acid concentrations (100 mM or more), other nucleotides, especially AMP, also begin to migrate. The greatest difference in hR_f values between cGMP and uric acid, which is the slowest compound in the group of nucleosides,

TABLE I

hR_f Values of Purine Nucleotides, Nucleosides, Bases, and Uric Acid on Polyethyleneimine-Cellulose Layers (Serva or Schleicher & Schuell*) with Water, Acetic Acid, and Formic Acid as Solvents

	Water	Acetic acid (mM)				Formic acid (mM)			
		25	50	100	500	10	50	100	500
GTP	0	0	0	0	0	0	0	0	0
GDP	0	0	0	0	0	0	0	0	0
GMP	0	0	0	4	10	0	9	16	58
XMP	0	0	0	2	5	0	4	10	50
cGMP	0	7	8	11	17	12	17	23	49
cXMP	0	0	0	3	6	0	7	14	48
Guo	55	56	58	66	62	55	55	58	65
Xao	15–30	52	57	66	63	38	54	59	64
Gua	31	36	40	45	50	33	38	44	54
Xan	42	43	46	50	47	44	41	43	47
Uric acid	8	35	42	49	45	26	37	42	47
ATP	0	0	0	0					
ADP	0	0	0	0					
AMP	0	3	9	19	35				
IMP	0	0	3	5					
cAMP	0	12	21	28	34	16	50	64	81
cIMP	0	9	10	16		26	34	39	49
Ado	54*	58	65	65					
Ino	72*	76	77	74					
Ade	44*		57	56					
Hyp	57	57	59	56	61	59	56	59	67

bases, and uric acid, occurs with 50 mM acetic acid. The largest difference in the hR_f values for cAMP and uric acid is found at a lower acetic acid concentration.

Formic acid can also be used for the separation of nucleotides from nucleosides, bases, and uric acid (see Table I). Elution of nucleotides, however, starts at lower concentrations of formic acid than of acetic acid while the migration rate of the nucleosides, bases, and urate is not significantly different for these two solvents. The hR_f difference of the cyclic nucleotides and uric acid is smaller if formic acid is used as solvent.

Relatively high hR_f differences of nucleotides and nucleosides, bases, and uric acid can be achieved by multiple development of the plates. Almost identical effects are obtained if either water or acetic acid is used as a second solvent after a first run with acetic acid.

Separation of Cyclic Nucleotides from Other Nucleotides

The migration rate of nucleotides on PEI-cellulose layers mainly depends on (a) kind of base, (b) number of phosphate groups, and (c) number of ester linkages of the phosphate group. The position of phosphate ester linkages in the nucleoside monophosphate insignificantly affects the chromatographic behavior.

Deoxyribonucleotides can be separated from ribonucleotides by the use of solvents containing borate. The deoxy derivatives of the cyclic nucleotides, however, cannot be isolated from corresponding cyclic nucleotides by the use of borate because of the diester linkage in the 3′-position.

For separation of cyclic nucleotides from the corresponding nucleotides, LiCl, NaCl, or KCl solutions can be used as solvents. The migration rates increase with decreasing numbers of phosphate groups. The hR_f values of cyclic nucleotides are higher than those of the corresponding 5′-nucleotides (Table II). cGMP is separated from guanosine- or xanthosine-containing nucleotides by 100 mM LiCl or 200 mM KCl. Similarly cAMP is separated from adenosine- or inosine-containing nucleotides by electrolyte solutions; optimal concentrations are generally somewhat lower than those required for cGMP separation.

Separation of nucleotides is also possible with volatile buffer systems as solvents. The migration rates of the nucleotides depend on the pH as well as on the concentration of these solvents. Some possibilities are indicated in Table III.

Besides ammonium acetate, sodium acetate buffer can also be used. hR_f values for adenosine- or inosine-containing substances with sodium acetate or formate as solvents are indicated in Table IV. Separation of

TABLE II
hR_f Values of Purine Nucleotides, Nucleosides, Bases, and Uric Acid on Polyethyleneimine-Cellulose Layers (Serva) with LiCl and KCl as Solvents

	LiCl (mM)					KCl (mM)			
	20	50	100	200	500	20	50	100	200
GTP	0	0	0	0	2				
GDP	0	0	0	2	12				
GMP	0	4	9	18	44	0	0	2	6
XMP	0	0	2	4	24				
AMP						0	0	6	12
cGMP	16	30	43	50	56	10	18	29	42
cXMP	0	4	12	24	52				
cAMP						12	20	32	45
Guo	58	53	56	54	53				
Xao	45	46	54	56	60				
Gua	35	30	33	32	31				
Xan	44	40	42	40	40				
Uric acid	26	28	34	34	36				

TABLE III
hR_f Values of Guanine- and Xanthine-Containing Nucleotides and Nucleosides on Polyethyleneimine-Cellulose Layers (Serva) with Ammonium Acetate and Ammonium Formate as Solvents

	Ammonium acetate				Ammonium formate			
	0.05 M		0.50 M		0.05 M		0.50 M	
	pH 5	pH 6	pH 5	pH 6	pH 3	pH 4	pH 3	pH 4
GTP	0	0	0	0	0	0	0	0
GDP	0	0	3	4	0	0	10	10
GMP	4	2	46	37	24	17	71	59
XMP	0	0	26	12	19	9		63
cGMP	32	32	46	45	30	36	60	48
cXMP	3	3	41	35	23	14	70	59
Guo	55	50	54	51	55	55	65	53
Xao	52	40	56	52	56	53	66	55
Gua	32	28	32	29	41	34	57	34
Xan	42	36	40	31	41	42	50	40
Hyp	56	52	54	52	57	57	66	54
Uric acid	31	20	35	26	41	39	51	40

TABLE IV
hR_f Values of Purine and Pyridine Nucleotides on Polyethyleneimine-Cellulose Layers (Serva) with Sodium Formate and Sodium Acetate as Solvents

	Sodium formate (mM)						Sodium acetate (mM)					Sodium acetate 60 mM, pH 5.4	
	pH 3.2			pH 4.2			pH 3.8		pH 5.4				
	10	20	50	25	50	125	5	25	30	60	130		
ATP	0	0	0	0	0	0	0	0	0	0	0	GTP	0
ADP	0	0	3	0	0	2	0	0	0	0	0	XTP	0
AMP	38	58	83	12	33	40	6	37	7	11	25	ITP	0
dAMP	40	60	83	12	35	41	8	31	7	11	26	GDP	0
IMP	20	34	67	13	38	44	5	25	8	13	31	XDP	0
cAMP	45	59	74	39	57	54	24	51	41	46	51	IDP	0
dcAMP	51	65	76	46	62	60	29	56	46	52	55	GMP	4
cIMP	46	56	73	57	74	74	24	51	56	66	72	XMP	0
NAD+	62	74	85	63	79	81	39	70	62	72	80	cGMP	31
NADH	0	6	27	3	9	25	0	2	3	8	23	cXMP	3
NADP+	0	4	18	0	3	11	0	0	0	0	0		
NADPH	0	0	18	0	0	2	0	0	0	0	0		
Acetyl-CoA	0	0	11	0	0	2	0	0	0	0	0		
ADP-Rib	0	5	26	2	7	21	0	0	2	4	14		

cAMP and cIMP from each other and from other nucleotides is possible with 60 mM sodium acetate, pH 5.4. cGMP is separated from related nucleotides by the same solvent.

With ammonium sulfate as solvent, not only anion-exchange and partition chromatography, but also "salting-out chromatography" is involved.[11] With low ammonium sulfate concentrations, the chromatographic behavior of all compounds is comparable to that observed with other electrolyte solutions as solvents. Higher concentrations of ammonium sulfate, however, cause a decrease of the migration rates of nucleosides, bases, uric acid, and nucleotides. hR_f values of the nucleotides decrease according to the number of phosphate groups (Fig. 1). The hR_f value of GMP, e.g., is lower than that of GDP if the plates are developed with a nearly saturated ammonium sulfate solution. The hR_f values are also affected by the number of ester linkages of the phosphate group, e.g., the hR_f value of cGMP is lower than that of GMP.

hR_f values for various purine and pyridine nucleotides with ammo-

[11] H. F. Walton, in "Chromatography" (E. Heftmann, ed.), 2nd ed., p. 337, Van Nostrand-Reinhold, Princeton, New Jersey, 1967; K. Randerath, *Experientia* **20**, 406, 1964; J. Neuhard, E. Randerath, and K. Randerath, *Anal. Biochem.* **13**, 211 (1965).

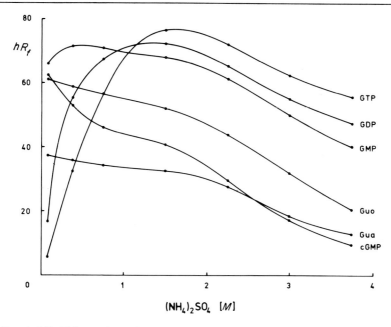

Fig. 1. hR_f Values of guanine-containing compounds on polyethyleneimine-cellulose layers (Serva) with ammonium sulfate as solvent.

TABLE V
hR_f Values of Purine and Pyridine Nucleotides on Polyethyleneimine-Cellulose Layers (Serva) with Ammonium Sulfate as Solvent

	Ammonium sulfate (M)						
	0.75	1.125	1.5	2.25	3.0		1.5
ATP	51	56	60	51	36	GTP	68
ADP	58	55	58	45	32	XTP	57
AMP	58	51	53	40	27	ITP	88
dAMP	55	46	47	32	20	GDP	70
IMP	82	77	81	71	59	XDP	59
cAMP	40	30	29	16	7	IDP	86
dcAMP	41	30	28	14	6	GMP	71
cIMP	65	57	57	42	26	XMP	59
NAD+	73	66	65	49	31	cGMP	35
NADH	59	49	45	26	12	cXMP	31
NADP+	89	83	83	68	51		
NADPH	74	65	63	43	24		
Acetyl-CoA	82	72	68	40	16		
ADP-Rib	75	67	67	51	34		

nium sulfate as solvent are given in Table V. A 1.5 M ammonium sulfate solution allows separation of cGMP from related nucleotides as well as of cAMP from adenosine- or inosine-containing nucleotides including the pyridine nucleotides.

Separation of Cyclic Nucleotides from Nucleotides, Nucleosides, Bases, and Uric Acid

Separation of cyclic nucleotides from related nucleotides and from the corresponding nucleosides, bases, and uric acid is most effectively achieved by multiple unidirectional development. First, the nucleosides, bases, and uric acid are eluted by development with acetic acid followed by a second run with acetic acid or distilled water. For a third run,

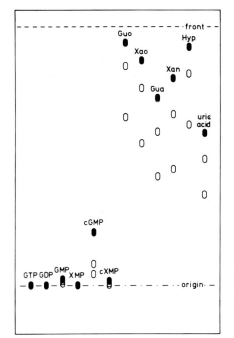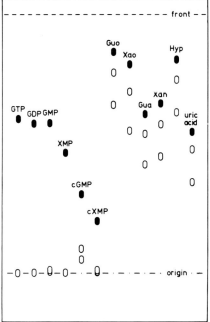

Fig. 2. Separation of cGMP from related nucleotides, nucleosides, bases, and uric acid on polyethyleneimine-cellulose layers (Schleicher & Schuell) by three successive developments with (left panel) 50 mM acetic acid, distilled water and 45 mM LiCl as solvents and (right panel) 50 mM acetic acid for the first two runs and 2.6 M ammonium sulfate (schematically). Open spots refer to positions of the compounds after the first and second run; filled spots, to final positions after the third run.

LiCl or ammonium sulfate are applied as solvent. Optimal separations of cGMP are obtained by successive developments with 50 mM acetic acid, distilled water, and 45 mM LiCl (Fig. 2) or 50 mM acetic acid, distilled water (or 50 mM acetic acid) and 2.6 M ammonium sulfate (Fig. 2). In principle, the same combinations of solvents can be used for the isolation of cAMP from related compounds.

Separation of Cyclic Nucleotides Containing Pyrimidine Bases

For separation of cyclic nucleotides containing pyrimidine bases from related nucleotides, nucleosides, and bases, development with one solvent, acetic acid or LiCl, is sufficient (Table VI). cUMP and cCMP can be isolated with 25 mM LiCl. Separation of these nucleotides from each other can be achieved by acetic acid.

Separation of Nucleoside Triphosphates

Separation of ATP or GTP from other, related nucleotides is achieved by chromatography with $LiCl^{2-4}$ (e.g., 1 M). If the nucleoside triphosphate also has to be separated from nucleosides and bases, a preceding additional development with acetic acid (0.1 M) is required. Figure 3 shows the separation of GTP from related compounds. Using the same

TABLE VI
hR_f VALUES OF PYRIMIDINE NUCLEOTIDES, NUCLEOSIDES, AND BASES ON
POLYETHYLENEIMINE–CELLULOSE LAYERS (SERVA) WITH
LiCl OR ACETIC ACID AS SOLVENTS

	LiCl (mM)				Acetic acid (mM)	
	10	25	50	75	50	100
CTP	0	0	0	0	0	0
CDP	0	0	0	0	0	3
CMP	3	7	16	22	30	47
cCMP	26	43	61	70	38	48
Cyd	90	89	90	91	89	91
Cyt	78	79	81	81	84	85
UTP	0	0	0	0	0	0
UDP	0	0	0	0	0	0
UMP	3	10	20	27	8	9
cUMP	30	53	68	76	23	23
Urd	93	95	93	92	95	93
Ura	84	85	84	83	83	83

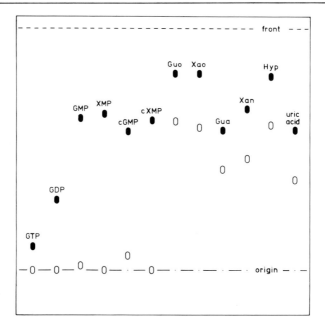

Fig. 3. Separation of GTP from related compounds on polyethyleneimine-cellulose layers (Schleicher & Schuell) by two successive runs with 0.1 M acetic acid and 1.0 M LiCl as solvents (schematically). Open spots refer to positions of compounds after the first run; filled spots, to final positions after the second run.

solvents, ATP is also separated from its derivatives. Purification of nucleoside triphosphates is also possible with other solvents that have been described above. Especially useful may be a volatile ammonium hydrogen carbonate buffer system.

Separation of Related Purine Nucleotides from Each Other

If a separation of various related purine nucleotides from each other is required, combinations of different solvent systems are used. By two successive developments with 50 mM acetic acid and distilled water, nucleosides, bases, and uric acid are separated from nucleotides. In a third run with 0.1 M LiCl, cGMP or cAMP are separated from the 5'-nucleotides. In a fourth run with 0.5 M LiCl up to 10 cm above origin, the nucleotides are separated from each other (Fig. 4). It is useful to interrupt the continuity of the layer by a line 10 cm above the origin before the fourth run. This separation system can be used for the fractionation of products enzymatically formed from a nucleotide, nucleoside, or base.

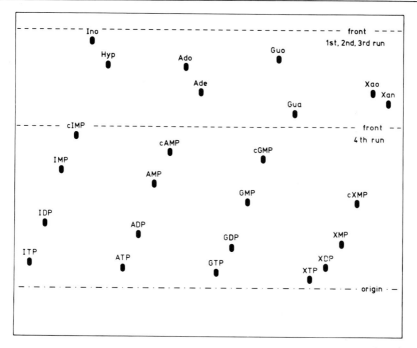

FIG. 4. Separation of related purine nucleotides from each other and from nucleosides and bases on polyethyleneimine-cellulose layers (Schleicher & Schuell). The plates were developed by four successive runs with 0.05 M acetic acid, distilled water, and 0.1 M LiCl as solvents over the entire length of the layer and with 0.5 M LiCl as a fourth solvent up to 10 cm above the origin (schematically). The spots indicate the final positions of the compounds after the fourth run.

Chromatography of Biological Material

The separation principles described above are useful tools for the isolation and purification of nucleotides. They can be applied to the isolation of cyclic nucleotides from tissue extracts and to the separation of product and residual substrate in assays of adenylate or guanylate cyclase or of phosphodiesterase activity. If adenylate or guanylate cyclase determinations are carried out with ^{3}H- or ^{14}C-labeled nucleoside triphosphates as substrates, a separation of nucleotides, nucleosides, bases, and uric acid from cyclic nucleotides is necessary; with ^{32}P-labeled substrate, a separation of the cyclic nucleotides from the respective related nucleotides and from ^{32}P$_i$ is sufficient. Inorganic phosphate runs variably fast in the separation systems described above and is most effectively separated by precipitation.[12]

[12] Y. Sugino and Y. Miyoshi, *J. Biol. Chem.* **239**, 2360 (1964).

Unidirectional development of the layers as described in this chapter for all separations, allows bandwise and, thereby, preparative separations. hR_f values obtained after bandwise application of the compounds were generally not significantly different from those found after spotwise application. The volume applicable per centimeter depends on the electrolyte content of the sample and on the thickness of the layer. Direct application of samples from enzyme assays performed in volumes between 50 and 100 µl is generally possible: 10–20 µl can be applied per centimeter, i.e., 50–100 µl upon layers 4 to 5 cm wide. For cyclic nucleotide isolation from tissue extracts, a removal of tissue electrolytes and eventually of acid is generally necessary (e.g., by a charcoal step) before the samples are applied to the thin-layer plates.

Elution from Layers

Elution of compounds from spots or bands can be performed as described by Randerath.[2,3] It is also possible to elute the compounds in batch. After detection of the compound by UV light (eventually in a reference band), the layer is scraped off while still wet and collected. An effective elution can be performed with electrolyte solutions in concentrations that are capable of inducing a migration of the compound on the plate. Volatile buffers can be useful if the eluates must be concentrated for further purification steps or assays.

Acknowledgment

The authors' studies were supported by the Deutsche Forschungsgemeinschaft.

[5] Isolation of Cyclic AMP by Inorganic Salt Coprecipitation

By Peter S. Chan and Michael C. Lin

Multivalent nucleotides can be precipitated with inorganic salts by formation of insoluble nucleotide–salt complexes or by adsorption of the nucleotides to the inorganic salt precipitates through charge interaction. This principle provides an approach to the fractionation of various nucleotides. At neutrality, cyclic AMP (cAMP) has only one negative charge, while all other adenine nucleotides, namely ATP, ADP, and 5′-AMP, are multivalent. Therefore a selective separation of cAMP from other nucleotides can be achieved.

Krishna et al.[1] have developed a sensitive assay procedure for adenylate cyclase which employs a combination of Dowex-50 column chromatography and $ZnSO_4$–$Ba(OH)_2$ precipitation. The popularity of this procedure has prompted us to study further nucleotide fractionation with other inorganic salts. The salt combinations found to be effective are $ZnSO_4$–$BaCl_2$, Na_2CO_3–$BaCl_2$, Na_2SO_4–$BaCl_2$, $ZnSO_4$–Na_2CO_3, and Na_2CO_3–$CdCl_2$,[2] in addition to $ZnSO_4$–$Ba(OH)_2$.

Reagents and Procedure

All inorganic compounds are of analytical reagent grade. [8-^{3}H]ATP, [8-^{3}H]ADP, [8-^{14}C]5'-AMP, and [8-^{3}H]adenosine were from Schwarz-Mann; [^{3}H(G)]cAMP was obtained from New England Nuclear.

Equal volumes of the two reactants in each combination are added to a test tube containing 1 ml of radioactive ATP, ADP, 5'-AMP, cAMP, or adenosine (specific activity 3–5 μCi/μmole) in 40 mM Tris·HCl buffer with 3.4 mM $MgCl_2$ at pH 7.4. After 30 seconds of thorough mixing with a Vortex mixer at 25°, the tubes are centrifuged at 2200 g for 15 minutes. Radioactivity in the supernatant is measured. If the removal of ATP, ADP, and 5'-AMP is not complete, the precipitation procedure can be repeated or the supernatant can be further subjected to paper, thin-layer, or column chromatography.

Results and Discussion

The recoveries of each adenine compound in the supernatant after precipitation are shown in the table. A few notes should be added here.

1. Zero percent indicates either less than 0.1% of radioactive nucleotide remaining in the supernatant or undetectable by paper chromatography.[3] However, often a recovery of less than 0.005% of the total radioactivity used is required; therefore, whether a particular combination of salts would provide an acceptable background after precipitation should be determined with the amount of radioactivity to be used. In order to obtain a low background, normally the precipitation procedure needs to be repeated or the supernatant is further purified with paper, thin-layer, or column chromatography.

2. The recovery of cAMP decreases with increasing amount of $ZnSO_4$–$Ba(OH)_2$. The formation of cAMP in the presence of $Ba(OH)_2$

[1] G. Krishna, B. Weiss, and B. B. Brodie, *J. Pharm. Exp. Ther.* **163**, 379 (1968).
[2] P. S. Chan, C. T. Black, and B. J. Williams, *Fed. Proc., Fed. Amer. Soc. Exp. Biol.* **29**, 616 (1970).
[3] M. Hirata and O. Hayaishi, *Biochim. Biophys. Acta* **149**, 1 (1967).

COMBINATIONS OF VARIOUS INORGANIC COMPOUNDS FOR THE PRECIPITATION OF ATP, ADP, 5'-AMP, CYCLIC 3',5'-AMP AND ADENOSINE

	Combination of inorganic compounds				% of Radioactivity remaining in the supernatant[a] ± SEM (n)[b]				
	Inorganic compounds	Molarity (M)	Volume (ml)	Resulting precipitates	[^{3}H]ATP, 1 μmole	[^{3}H]ADP, 1 μmole	[^{14}C]5'-AMP, 1 μmole	[^{3}H]Cyclic 3',5'-AMP, 0.5 μmole	[^{3}H]Adenosine, 1 μmole
1	$ZnSO_4$ $Ba(OH)_2$	0.2 0.2	0.8 0.8	$BaSO_4$ and $Zn(OH)_2$	0 (18)	0 (10)	0 (10)	81.1 ± 2.6 (8)	96.4 ± 0.6 (12)
2	$ZnSO_4$ $BaCl_2$	0.4 0.4	0.7 0.7	$BaSO_4$	0 (8)	0 (8)	0 (10)	97.3 ± 1.6 (6)	99.3 ± 0.5 (8)
3	Na_2CO_3 $BaCl_2$	0.4 0.4	0.8 0.8	$BaCO_3$	0 (8)	0 (8)	0 (8)	106.5 ± 1.2 (6)	99.9 ± 0.1 (8)
4	Na_2SO_4 $BaCl_2$	0.4 0.4	0.7 0.7	$BaSO_4$	0 (6)	0 (6)	0 (6)	103.6 ± 1.3 (6)	99.3 ± 0.7 (8)
5	$ZnSO_4$ Na_2CO_3	0.4 0.4	0.8 0.8	$ZnCO_3$	0 (8)	0 (6)	0 (6)	99.9 ± 0.1 (6)	98.1 ± 0.4 (8)
6	Na_2CO_3 $CdCl_2$	0.4 0.4	0.8 0.8	$CdCO_3$	0 (8)	0 (6)	0 (4)	99.7 ± 0.3 (5)	99.7 ± 0.3 (8)
7	$BaCO_3$ Suspension	1.0	3.0	$BaCO_3$	0 (4)	0 (4)	0 (4)	118.0 ± 1.2 (4)	103.1 ± 2.1 (4)
8	$BaSO_4$ Suspension	1.0	3.0	$BaSO_4$	0 (4)	0 (4)	0 (4)	115.7 ± 1.6 (4)	104.6 ± 2.8 (4)
9	$ZnCO_3$ Suspension	1.0	3.0	$ZnCO_3$	0 (4)	0 (4)	0 (4)	92.1 ± 1.9 (4)	95.9 ± 1.4 (4)
10	Na_2CO_3 $CaCl_2$	0.4 0.4	0.8 0.8	$CaCO_3$	0 (8)	0 (6)	67.9 ± 1.3 (4)	99.7 ± 0.4 (6)	102.4 ± 0.6 (8)
11	$AgNO_3$ NaI	0.4 0.4	0.8 0.8	AgI	0 (14)	0 (6)	0 (10)	0 (8)	0 (14)
12	$ZnSO_4$ $CaCl_2$	0.4 0.4	0.8 0.8	$CaSO_4$[c]	88.9 ± 2.4 (6)	90.7 ± 2.9 (6)	0 (10)	105.3 ± 0.7 (4)	103.6 ± 0.6 (6)

[a] The volume including 1 ml of sample and the volumes of salt solution is used as the total volume of supernatant. No corrections were made for the volume occupied by the precipitate. Specific activities of the nucleotides or nucleoside was 3–5 μCi/μmole.
[b] Figure in parentheses denotes number of determinations.
[c] Precipitate was formed only after prolonged mixing and standing for 1 hour.

is known to occur, especially at elevated temperature.[4] The combination of $ZnSO_4$–$Ba(OH)_2$ also gives a viscous solution in the presence of theophylline at concentrations of 10 mM or higher.

3. The use of $ZnSO_4$–$NaCO_3$ combination is more desirable in that it does not catalyze the formation of cAMP, nor does it give a viscous supernatant with theophylline. It also gives a complete recovery of cAMP and has been used successfully in the isolation of cyclic GMP.[5]

4. Na_2CO_3–$CaCl_2$ precipitates both ATP and ADP, while $ZnSO_4$–$CaCl_2$ precipitates only 5'-AMP after prolonged mixing and standing. All five adenine compounds tested are quantitatively precipitated by $AgNO_3$–NaI combination.

5. The concentration and volume of each solution used can be varied to suit the particular situation. For example, the same result is obtained when the concentrations of both $ZnSO_4$ and Na_2CO_3 are increased to 1 M and the volume is reduced to 0.2 ml.

6. The suspensions of $BaCO_3$, $BaSO_4$, and $ZnCO_3$ also precipitate multivalent nucleotides. Moreover, a column packed with $ZnCO_3$ can replace $ZnSO_4$–$Ba(OH)_2$ precipitation step in the assay procedure of Krishna et al.[1] with comparable results.

7. Since none of the salt combinations separates adenosine from cAMP, the use of ^{32}P-labeled ATP as substrate is preferred for the assay of adenylate cyclase to avoid the possible contamination of adenosine.

[4] W. H. Cook, D. Lipkin, and R. Markham, *J. Amer. Chem. Soc.* **79**, 3607 (1957).
[5] Personal communication from J. Hardman.

[6] Separation and Purification of Cyclic Nucleotides by Alumina Column Chromatography

By ARNOLD A. WHITE

Alumina is a useful adsorbent for the anion exchange chromatography of 3',5'-cyclic nucleotides because it effects a sharp discrimination on the basis of charge. The univalent cyclic nucleotides are eluted from alumina in high yield by a neutral aqueous medium, while the other nucleotides of higher valence are adsorbed. This behavior, coupled with the general availability and cheapness of alumina, recommends its use in the purification of cyclic nucleotides from enzyme reaction mixtures or from tissue extracts.

We have described the use of alumina columns in the assay of adenylate and guanylate cyclase.[1] Since our experience has been confined to the purification of radioactive cyclic nucleotides from cyclase reaction mixtures, we will not discuss the purification of cyclic nucleotides from tissue extracts.

Materials and Methods

Reagents

Aluminum oxide, neutral (activity I, for column chromatography, E. Merck, E. M. Reagents)

Imidazole–HCl column buffer: 0.1 M imidazole, 50 mM HCl. This is made up from 1.0 M stock solutions and has a final pH near 7.07. The imidazole is purchased from Pierce Chemical Company, and the 1.0 M HCl solution from Fisher Scientific Company

Apparatus

Columns. Our standard columns are made from 9 mm o.d. standard wall glass tubing 15 cm long with a tip opening of 2–3 mm. The top of the column opens to a reservoir made of 25 mm o.d. tubing, 65 mm long. The column tip is packed either with glass wool or a 1 inch square of household paper toweling (Teri, Kimberly-Clark Corp.). Racks were constructed of Plexiglas; each holds 40 columns and under it can be placed either an open box or a rack holding 40 scintillation vials.

Cyclase Assays

The adenylate cyclase and guanylate cyclase assays are carried out in 10 $\times$ 75 mm tubes and incubated at 30°. Final volume of the reaction mixtures for both assays is 75 μl, and both contain 50 mM Tris·HCl buffer, pH 7.6, 0.1 mg of bovine plasma albumin, 10 mM mercaptoethanol, and 20 mM caffeine. Both also utilize the same system for regenerating ATP or GTP, 15 mM creatine phosphate, and 10 units of creatine phosphokinase.

The adenylate cyclase reaction mixture contains 1.2 mM [α-^{32}P]ATP, approximately 5 $\times$ 10^5 cpm, 5 mM MgCl$_2$ and 0.2 μmole of cyclic AMP (cAMP). The guanylate cyclase reaction mixture contains 1.2 mM [α-^{32}P]GTP, approximately 5 $\times$ 10^5 cpm, 6 mM MnCl$_2$, and 0.2

---

[1] A. A. White and T. V. Zenser, *Anal. Biochem.* **41**, 372 (1971).

μmole of cyclic GMP (cGMP). Either reaction is initiated by addition of the enzyme, usually 25 μl. The reaction tubes and the enzyme preparation are incubated in the 30° water bath for 5 minutes before initiation. At the end of the reaction time, generally 10 minutes, the reaction is stopped by adding 20 μl of 0.5 M HCl, containing approximately 30,000 cpm of tritiated cyclic nucleotide, heated for 2 minutes at 100° and immediately cooled in an ice bath. The tritiated cyclic nucleotide is incorporated into the stop solution shortly before use. This addition is made in order to be able to correct for cyclic nucleotide losses on the alumina columns.

The pH of the reaction mixture after addition of 0.5 M HCl is about 1.4. The sample has to be neutralized and diluted before it is applied to the alumina. It is neutralized by adding 20 μl of 1.0 M imidazole, after which 1 ml of the imidazole–HCl column buffer is added. If extra imidazole is incorporated into the column buffer, both steps may be accomplished with a single addition.

Each of the pipettings in the assay have to be performed with precision and reproducibility. We have found that this is best accomplished with automatic micropipettes. For the enzyme preparations we use the 25-μl size SMI Micro-pettor (Scientific Manufacturing Industries, Emeryville, California). This micropipette is particularly useful for particulate preparations. For pipetting the other solutions in microliter quantities, we use either the Unimetrics Micropipetter (Unimetrics Corporation, Anaheim, California) or Eppendorf Microliter Pipets (Brinkmann Instruments Inc., Westbury, New York).

Column Preparation

The requisite number of columns are each packed with a square of paper toweling. Each packing is wet with imidazole·HCl buffer, in order to expand and fill the column tip. Approximately 1 g of neutral alumina is washed into each column with the buffer. We measure out the alumina with a spoon made from the cap of a disposable hypodermic needle container. The buffer is dispensed from a plastic wash bottle or with an automatic syringe. The measured alumina may be placed into a polystyrene weighing boat, wet with buffer, and washed into a column with about 15 ml of buffer. Another technique is to empty the measuring spoon directly into a column that is partially filled with buffer. The alumina should fall through the buffer so as to remove entrapped air.

In order to achieve reproducible columns, we pay particular attention to pH control. When an alumina column is poured, the pH of the buffer initially eluting from the column is higher than that of the buffer itself.

As the column is washed with buffer, the pH of the elutant approaches that of the buffer, and after 15–20 ml of effluent, the pH difference disappears. We recommend that the pH of the effluent be monitored each time a new lot of alumina is put into use in order to determine the volume of wash buffer required. After the columns are poured and washed, they are allowed to drain.

Chromatography

The supernatant solution from each diluted and centrifuged sample is applied to a separate column, either with a disposable Pasteur pipette or by pouring it onto the column. This application volume is allowed to completely enter the column and the displaced volume is considered the first milliliter of effluent. Subsequent treatment of the column depends upon the cyclic nucleotide to be recovered. With the alumina we are currently using about 84% of applied cyclic AMP appears in the second and third milliliter of effluent, while about 70% of applied cyclic GMP appears in the third and fourth milliliter. With some lots of alumina the cyclic GMP appears in the fourth and fifth milliliter. The 2 ml desired are collected directly into a scintillation vial. To this is added 10 ml of scintillation mixture. We have successfully used both the mixture of Bray[2] and that of Anderson and McClure.[3] The radioactivity is measured in a liquid scintillation spectrometer.

Interfering Compounds

Ammonium sulfate, sodium phosphate or pyrophosphate, and heparin all increased the ^{32}P blank. However, sodium chloride, magnesium chloride, mercaptoethanol, and dithiothreitol had no effect. It appears that divalent and polyvalent anions displace ^{32}P from the alumina. These anions have an effect that appears to be potentiated when in combination since orthophosphate and pyrophosphate combined increased the ^{32}P blank greater than the sum of their individual effects.

The polyvalent anions also increase the recovery of cyclic nucleotide. Thus incorporation of 2.0 mM pyrophosphate into the imidazole·HCl column buffer can increase the recovery of cGMP in the peak 2 ml by 8%. However, this concentration of pyrophosphate also produces a disproportionate increase in the ^{32}P blank, so that the calculated blank is higher. We have not been able to increase recoveries without also increasing blanks.

[2] G. A. Bray, *Anal. Biochem.* **1**, 279 (1960).
[3] L. E. Anderson and W. O. McClure, *Anal. Biochem.* **51**, 173 (1973).

Discussion

The purification of radioactive cyclic nucleotide from a cyclase reaction mixture should ideally result in a high yield of product with a minimal radiochemical contamination (blank). Alumina chromatography compares very well with other methods of purification, particularly with respect to simplicity and rapidity of the methodology, and high yield. However, in order to ensure low blanks, one must be aware of several factors that can affect the results.

The first of these factors is the quality of the labeled substrate. We have found that completely pure [α-^{32}P]ATP or [α-^{32}P]GTP is completely adsorbed by alumina, and that blank radioactivity is due to radiochemical impurities. As we noted earlier, these impurities are heterogeneous in nature, and contain cyclic nucleotide only in minor part.[1] This presented us with a continuously changing problem, since each lot of substrate had a blank that was not only quantitatively different, but probably also qualitatively different. Our methodology has evolved in response to this problem to the point where this variation in blank is no longer significant. In addition there has been an enormous improvement in the commercially available substrate during the last six years. Currently we are using a batch of [α-^{32}P]ATP which has a corrected blank of 0.009% of the applied radioactivity. The current batch of [α-^{32}P]GTP has a zero blank.

We have found that the blank increases during storage in the freezer of ^{32}P-labeled carrier-free nucleotides. This increase can be prevented if the substrate is diluted with cold ATP or GTP. A procedure for this dilution has been described.[4]

The second factor to be considered is the method of stopping a cyclase reaction. We had originally used EDTA or pyrophosphate to chelate magnesium or manganese ions and then heated the reaction mixture to precipitate protein. As mentioned earlier, in order to correct for losses in the subsequent chromatography, tritiated cyclic nucleotide is incorporated into the stop solution. However, we have found that when [^{3}H]cGMP was heated at 100° for 2 minutes in a simulated reaction mixture, about 4% of the total radioactivity was converted into material that did not behave chromatographically like cGMP (it could be guanine and/or guanosine). The presence of EDTA or pyrophosphate did not appreciably affect this destruction. Acidification, however, did inhibit the process; HCl being more effective than acetic acid. The conversion in the presence of HCl was about 1.3%.

[4] A. A. White, S. J. Northup, and T. V. Zenser, in "Methods in Cyclic Nucleotide Research" (M. Chasin, ed.), p. 125. Dekker, New York, 1972.

An unexpected benefit of the new stop procedure was a 50% decrease in ^{32}P blanks. It appears that heating at neutral pH will convert both [α-^{32}P]ATP and [α-^{32}P]GTP into radioactive compounds which are not adsorbed by alumina. These products could be, in part, the respective cyclic nucleotides. Heating under acid conditions inhibits this conversion.

The third factor is the chromatographic buffer. Originally, we used 50 mM Tris·HCl buffer, pH 7.6.[1] We recognized that a slightly lower pH would be desirable, since the ^{32}P blanks were lower at pH 7.0, with little decrease in ^{3}H recovery. However, because we were restricted to monovalent anions, we had difficulty in finding a buffer with the proper pK_a that was available in high purity at a reasonable cost and that did not quench a scintillation solution. Tris·HCl buffer was satisfactory in all respects except for a pK_a of 8.1. It was decided to run the columns at pH 7.6 because this was the pH of the cyclase reaction mixtures and thus there would be no pH change necessary before application to the alumina. Also Tris buffered well at pH 7.6.

These conditions were satisfactory as long as cyclase activities were not too low and the substrates had comparatively low blanks. Difficulties encountered with certain shipments of substrate led us to reexamine the procedure. Using Tris·HCl column buffer at pH 7.0 instead of 7.6 definitely lowered the blanks. Substituting imidazole·HCl buffer at the same pH decreased the blank even further. It was found that imidazole was a nearly ideal buffer for this chromatography, having all the desirable qualities of Tris plus a pK_a very close to 7. This enables a rapid equilibration of the alumina to the desired pH. At this writing we have about 8 months of experience with imidazole buffer, and conclude that its use has resulted in extremely reproducible columns, with low blanks.

The last factor to be considered is the quality of the alumina. Alumina samples from different suppliers differ markedly in behavior, particularly with respect to cyclic nucleotide yield. Some samples will almost completely retain cGMP unless they are "hydrated" by exposure to water vapor. The use of imidazole instead of Tris buffer has accentuated the variations in recovery due to the alumina. However, imidazole, because of its good buffering at pH 7.0, enables the satisfactory use of basic instead of neutral alumina. Basic alumina gives higher cyclic nucleotide recoveries, particularly of cGMP. If difficulty is encountered in obtaining a satisfactory neutral alumina, we recommend the use of basic alumina. Chromatographic alumina is basic unless labeled to the contrary.

Section II

Assay of Cyclic Nucleotides

[7] Assay of Cyclic Nucleotides by Receptor Protein Binding Displacement

By ALFRED G. GILMAN and FERID MURAD

Receptor protein binding displacement assays for cyclic nucleotides are based on competition for protein binding sites between radioisotopically labeled nucleotide and the unlabeled material to be quantified. The binding proteins utilized are those that occur naturally and that interact with the appropriate cyclic nucleotide with high affinity. These procedures offer intrinsic advantages of simplicity of operation, high sensitivity, and high specificity. These features are particularly marked in the case of the assay for adenosine 3′,5′-cyclic monophosphate (cyclic AMP, cAMP) and have led to considerable utilization of this method.

Assay for cAMP[1]

The binding protein utilized is presumably a cAMP-dependent protein kinase[2] that is purified partially from bovine skeletal muscle. Optional use is also made of a heat-stable protein inhibitor of the protein kinase that has been shown to increase the affinity of the kinase for the cyclic nucleotide.[3] Nucleotide that is bound to the receptor protein is separated from that remaining free in solution by passage through cellulose ester (Millipore) filters. These filters adsorb quantitatively large quantities of the binding protein and associated nucleotide.

Independent and variant methods utilizing different binding proteins and different techniques for isolating the protein–nucleotide complex have also been described.[4,5]

Materials

cAMP Binding Protein

This preparation was modeled after that of Miyamoto et al.[6]

Step 1. All operations are performed at 0–4°. Fresh bovine skeletal

---

[1] A. G. Gilman, Proc. Nat. Acad. Sci. U.S. **67**, 305 (1970).
[2] D. A. Walsh, J. P. Perkins, and E. G. Krebs, J. Biol. Chem. **243**, 3763 (1968).
[3] D. A. Walsh, C. D. Ashby, C. Gonzalez, D. Calkins, E. H. Fischer, and E. G. Krebs, J. Biol. Chem. **246**, 1977 (1971).
[4] G. M. Walton and L. D. Garren, Biochemistry **9**, 4223 (1970).
[5] B. L. Brown, J. D. M. Albano, R. P. Ekins, A. M. Sqherzi, and W. Tampion, Biochem. J. **121**, 561 (1971).
[6] E. Miyamoto, J. F. Kuo, and P. Greengard, J. Biol. Chem. **244**, 6395 (1969).

muscle is homogenized for 2 minutes in a Waring blender in 4 mM EDTA (Na$^+$), pH 7.0, and the resulting homogenate is centrifuged at 15,000 g for 30 minutes.

Step 2. The supernatant is adjusted to pH 4.8 by the slow addition of 1 N acetic acid. After 15 minutes the precipitate is collected at 15,000 g for 30 minutes and is discarded.

Step 3. The supernatant is adjusted to pH 6.8 by addition of 1 M potassium phosphate, pH 7.2. Solid ammonium sulfate (0.33 g/ml) is then added slowly. After 20–30 minutes the precipitate is collected at 15,000 g for 30 minutes, dissolved in approximately 5% of the original homogenate volume in 5 mM potassium phosphate, pH 7, and dialyzed overnight against two changes of the same buffer.[7]

Step 4. The dialyzed ammonium sulfate precipitate (e.g., from 250 g of muscle) is applied to a column of DEAE-cellulose (Whatman DE 11 or Sigma, 1 mEq/g; 30 × 2.5 cm), previously equilibrated with 5 mM potassium phosphate, pH 7, and the column is washed with 2 bed volumes of this buffer. The column can then be eluted with a linear gradient of potassium phosphate, pH 7, from 5 to 400 mM as shown in Fig. 1. Two prominent peaks of protein kinase and cAMP binding activity are seen. The second peak of activity is utilized after active fractions are pooled and dialyzed against 5 mM potassium phosphate, pH 7.

Step 4a. Alternatively, after the preparation is applied to the above column, the two peaks of activity can be separated by washing with two bed volumes each of 100 mM potassium phosphate, pH 7, to elute the first peak and 300 mM potassium phosphate to obtain the second fraction. This second fraction is then treated as above. Separation of the two fractions by one of these methods is recommended, since optimal conditions for their binding of cAMP are different.

Binding activity may be simply followed through the purification procedure by incubation of small aliquots (5–25 μl) of appropriate fractions with an excess of [^{3}H]cAMP in 20 mM potassium phosphate, pH 6–7, for approximately 5 minutes at 30–37° (sufficient time to establish equilibrium at this temperature). Filtration and counting of samples is done as described below. A summary of protein kinase and cAMP binding activities at various stages of the purification scheme is shown in Table I. The determinations shown were performed on dialyzed fractions, and assay of undialyzed fractions prior to the dialysis step may yield specific activities significantly lower than those shown.

Preparations that we have obtained using bovine muscles of various types obtained from slaughterhouses (and on two occasions from butcher

---

[7] All solutions used from this point in the procedure may include 2 mM EDTA (Na$^+$), pH 7.

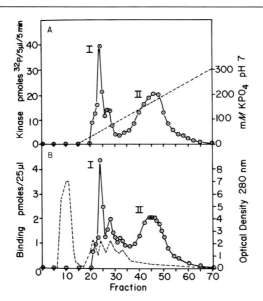

FIG. 1. DEAE-cellulose chromatography of protein kinase and cyclic AMP (cAMP) binding activities. Gradient elution was performed with potassium phosphate, pH 7, from 5 to 400 mM. Fractions collected were 20 ml. Binding activity was assayed at pH 4 with 40 nM cAMP. Kinase activity was assayed as described by E. Miyamoto, J. F. Kuo, and P. Greengard [*J. Biol. Chem.* **244**, 6395 (1969)] with histone as substrate. (A) ⊙, protein kinase activity; ---, potassium phosphate gradient. (B) ⊙, cAMP binding activity; ---, optical density at 280 nm. From A. G. Gilman, *Proc. Nat. Acad. Sci. U.S.* **67**, 305 (1970).

shops) have usually had specific activities above that shown. A good preparation from 500–1000 g of muscle yields sufficient binding protein for 10^5–10^6 assay tubes and can be completed in 2 days.

TABLE I
PROTEIN KINASE AND CYCLIC AMP (cAMP) BINDING ACTIVITY PURIFICATION[a]

Fraction	Protein kinase (pmoles/μg protein/10 min)	cAMP binding[b] (pmole/μg protein)
15,000 g supernatant	0.58	0.012
pH 4.8 supernatant	0.62	0.011
$(NH_4)_2SO_4$ precipitate	1.94	0.031
DEAE peak I	2.75	0.036
DEAE peak II	17.7	0.191

[a] From A. G. Gilman, *Proc. Nat. Acad. Sci. U.S.* **67**, 305 (1970).
[b] Assayed at pH 4.0 with 40 nM cAMP.

While we usually store binding protein at −80°, no loss of activity was observed for over 2.5 years at −20°. Repeated freezing and thawing of the protein was avoided. It has been noted, however, that highly diluted (protein less than 100 μg/ml) samples of binding protein have lost activity when stored at −20°.

Protein Kinase Inhibitor

Step 1. Fresh bovine muscle is homogenized in 2–3 volumes of 10 mM Tris chloride, pH 7.5. The homogenate is then boiled for 10 minutes and, when cool, a clear extract is obtained by filtration.

Step 2. The inhibitory activity is precipitated from the filtrate by the addition of 1/9th volume of 50% trichloroacetic acid. This precipitate is collected at 15,000 g for 30 minutes and is redissolved in water (approximately 10% of the original homogenate volume) with frequent adjustment of the pH to 7 with 1 N NaOH. This fraction is dialyzed against distilled water. A gelatinous precipitate (often copious) which forms during dialysis can be removed by centrifugation and discarded. The inhibitor fraction, while extremely crude, is readily utilized at this stage and is stored at −20°. The use of this fraction is optional, but it is recommended.[1]

The inhibitory activity can be assayed by its inclusion in a standard assay for cAMP-dependent protein kinase (see this volume [48]). The ability of the inhibitor to enhance binding of [^{3}H]cAMP to the receptor protein should be studied with subsaturating concentrations of cAMP. Other conditions are as for cAMP assay (described below). Maximally effective quantities of the inhibitor fraction are utilized.

[^{3}H]cAMP

Both generally labeled and specifically labeled [^{3}H]cAMP have been utilized with specific activities ranging from 16 to 28 Ci/mmole. If the assay is performed in the manner described, any of these preparations will yield sufficient radioactivity bound to the protein (3000–6000 cpm) to be convenient. Prolonged storage of [^{3}H]cAMP results in the appearance of radioactive impurities (particularly with the preparations of higher specific activity), and these materials can be removed by chromatography as described below or elsewhere in this volume.

Assay Method

We have chosen to perform the assay for cAMP at saturating concentrations of the nucleotide, and we have done this in order to maximize reproducibility, minimize interference, and facilitate either graphical or mathematical solution of assay data. When done in this way, the standard curve for the assay is entirely linear when the log of total counts per minute bound is plotted versus the log of total quantity of cAMP (labeled plus unlabeled) in the assay tube. The standard curve is highly reproducible from day to day, and such factors as total reaction volume become less critical. In spite of the reproducible standard curves, we routinely run a set of standards in each experiment. Some sensitivity is, however, lost when the assay is performed in this way (rather than with lower subsaturating concentrations of nucleotide). Brown et al.[8] have discussed this point thoroughly. High assay sensitivity can be obtained and the protein can be saturated if the reaction volume is kept small, and the standard assay will thus be described for a 50-μl reaction volume. If lower sensitivity is tolerable, all volumes and quantities can be increased accordingly, while keeping the concentrations the same. If increased sensitivity is necessary, the reaction volume and quantity of [^{3}H]cAMP can be decreased further and/or use can be made of less than saturating concentrations of [^{3}H]cAMP.

At room temperature, to 10×75 mm disposable tubes, are added 5 μl of sodium acetate, pH 4.0, 500 mM; 5 μl [^{3}H]cAMP (usually 1 pmole, 10,000–20,000 cpm); and 5 μl of inhibitor protein (a maximally effective quantity—usually 10–15 μg of protein). These components can be added together as a mixture of the three. Standards (typically 0.2–20 pmoles), varying aliquots of unknowns, or appropriate diluents to volume are then added. A total volume of 20 μl is typically reserved for these components. The tubes are then placed in an ice bath, binding protein is added to a total final volume of 50 μl, and the contents are mixed *gently*. The binding protein preparation is diluted prior to use with bovine serum albumin such that 10–15 μg of albumin will be added to each assay tube. This quantity of albumin has been found to enhance the maximal binding capacity of the protein under these conditions. The quantity of binding protein utilized is set arbitrarily such that not more than 30% of the total [^{3}H]cAMP is bound (to retain a sufficient concentration of unbound cAMP to ensure saturation of the protein).

The reaction tubes are allowed to incubate at 0° for at least 1 hour, by which time equilibrium is established. The observed equilibrium plateau is stable for many hours if care is taken to maintain the 0–3° temperature. The reaction mixtures are then diluted with approximately 1 ml of 20 mM potassium phosphate, pH 6, at 0–3° and are passed under gentle suction through 25-mm cellulose ester Millipore filters (catalog

[8] B. L. Brown, R. P. Ekins, and J. D. M. Albano, *Advan. Cyclic Nucleotide Res.* **2**, 25 (1972).

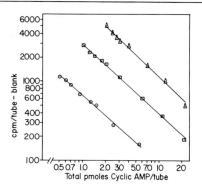

FIG. 2. Standard curves for cyclic AMP (cAMP) assay. All reactions were carried out at pH 4 and 0° in a volume of 50 μl. [³H]cAMP added per tube was ○, 0.5 pmole; □, 1.0 pmole; △, 2.0 pmoles. Inhibitor protein fraction, 14 μg, was present at the two lower cAMP concentrations, and binding protein was added at 0.5, 1.0, and 2.0 μg for the three conditions, respectively. Known quantities of unlabeled cAMP were added to achieve the total (labeled plus unlabeled) indicated content of cAMP per tube. From A. G. Gilman, *Proc. Nat. Acad. Sci.* **67**, 305 (1970).

No. HAWP025), previously rinsed with the same buffer.[9] The filter is immediately rinsed with 5–10 ml of the same buffer to remove unbound [³H]cAMP. Finally, the filters are placed in counting vials containing 1 ml of methyl Cellosolve, in which they readily dissolve, and a scintillation cocktail of toluene:methyl Cellosolve (3:1) plus fluors is added.

The counts bound to the filter are quite independent of the speed of filtration and of the volume of rinse from 3 to 20 ml. In the absence of binding protein, approximately 15–20 cpm are counted on the filter.

Three different standard curves are shown in Fig. 2. Total picomoles of cAMP plotted on the abscissa represent the sum of [³H]cAMP and unlabeled standard added to each tube. These data were all obtained at saturating concentrations of nucleotide, with 0.5 and 1.0 pmole of [³H]cAMP in 50 μl (10 and 20 nM) in the presence of the inhibitor protein and with 2.0 pmoles of [³H]cAMP (40 nM) in the absence of the inhibitor. With the most sensitive curve shown, a 20% dilution of total cpm bound is obtained with the addition of 0.10 pmole, and 0.05 pmole yields a highly significant depression.

Sample Preparation

The assay procedure is sufficiently specific that most tissues and urine can be assayed without separation of cAMP from other materials. We typically homogenize tissue samples in 5% trichloroacetic acid. Homog-

[9] The most convenient simple filter holders that we have seen are sold by Hoefer Scientific Company, 520 Bryant Street, San Francisco, California; Catalog No. FH 100 ($48 each). Samples are easily filtered in pairs with two such holders.

enization is not necessary with cell monolayers grown in tissue culture. In this case, medium is aspirated and 5% trichloroacetic acid is added. Cell fixation is extremely rapid,[10] and cAMP is extracted at least by 10–15 minutes after fixation. After separation of the trichloroacetic acid extract from the protein precipitate, HCl is added to the tissue extract to a final concentration of 0.1 N. Trichloroacetic acid is removed by extraction with diethyl ether (6 times with 2–3 volumes). Residual ether is removed by heating at 80–90° for 2–5 minutes, and the samples are lyophilized. Dried samples are redissolved in 50 mM sodium acetate, pH 4.5, prior to assay. While these extraction steps can be done with essentially no loss of cAMP, it seems prudent to monitor the recovery of cAMP in samples by the addition of [^{3}H]cAMP (typically 0.3–0.5 pmole) and subsequent counting of the sample.

Under certain circumstances, purification of samples is necessary. These include experiments where exogenous interfering compounds may have been added, situations where high salt concentrations are encountered (e.g., assay of concentrates of incubation media or of plasma and serum), or any situation in which appropriate controls demonstrate the need. We wish to point out that it is mandatory that those employing a new assay technique establish whether purification of samples is necessary under their conditions of incubation and analysis. Either of two chromatographic procedures have been utilized with equal success. Several other column and paper chromatographic methods could also be used.

Trichloroacetic acid-free extracts (1–2 ml, ideally at pH 1.5–2.0) are applied to 0.4 × 5 cm columns of Dowex 50, H$^+$. Columns are then washed with 2 ml of H$_2$O, and the cAMP fraction is collected with an additional 3 ml of H$_2$O. Samples are lyophilized and assayed.

Trichloroacetic acid-free extracts (pH > 3) can also be applied to 0.4 × 4 cm columns of Dowex 1-formate. The columns are washed with 5 ml of H$_2$O and the cAMP is collected with 3 ml of 2 N formic acid (nearly quantitative recovery can be obtained with larger volumes of 2 N formic acid). Samples are again lyophilized prior to assay.

Any such fractionation procedure obviously requires that recoveries of cAMP be monitored as described above. While more elaborate purification procedures could be required under extreme conditions, either of these procedures has been sufficient for the assay of cAMP in tissue culture media and plasma.

Discussion

The assay conditions described above are the best that were found for this protein preparation. Thus, at pH 4 the affinity of the binding

[10] A. G. Gilman and M. Nirenberg, *Nature (London)* **234**, 356 (1971).

protein for cAMP was maximal. The binding constant under these conditions was 2–3 nM in the absence of the inhibitor protein and approached 1 nM in its presence. Concentrations of [³H]cAMP in the assay of 10–20 nM are thus sufficient to saturate the binding sites and achieve the condition described above. The concentration of [³H]cAMP in the assay may be changed to achieve desired alterations in sensitivity, but, if done, it should be with an awareness of the implications of the change.

The kinetics of binding have been discussed briefly above. Equilibrium is attained within 1 hour and the plateau is stable for an extended period if careful attention is paid to control of temperature. The rate of dissociation of the protein–nucleotide complex is very slow and is first order. At 0°, a $t^{1/2}$ of approximately 7 hours was found for complex dissociation.

The specificity of the procedure is indicated by the data of Table II, which demonstrates the effect of certain nucleotides and various other compounds on the binding of [³H]cAMP. While other 3′,5′-cyclic nucleotides are most effective as competitors, none are naturally present in sufficient concentration to interfere with this procedure. The greatest problem is probably posed by ATP, which inhibited binding by 50% when present at 1 mM final concentration in the assay. Little or no interference is noted when the ATP concentration is less than 0.1 mM under the assay conditions described. There is enough cAMP in most tissues to dilute tissue extracts sufficiently to eliminate any competitive effect of ATP.

Any possible interference from ATP or any other material is best determined in a novel set of circumstances by the use of appropriate controls. These include assay of samples before and after chromatographic purification of cAMP by a method such as those described above and by treatment of tissue extracts with cyclic nucleotide phosphodiesterase.

TABLE II
Effect of Nucleotides and Related Compounds on Cyclic AMP (cAMP) Binding[a]

Compound	μM Concentration at 50% inhibition[b]	Compound	% Inhibition at 1 mM[b]
3′,5′-cIMP	0.3	UTP	30
3′,5′-cGMP	5.0	CTP	28
3′,5′-cUMP	10	5′-AMP	21
3′,5′-cCMP	30	ADP	18
GTP	700	Adenosine	0
ATP	1000	Theophylline	0

[a] From A. G. Gilman, *Proc. Nat. Acad. Sci. U.S.* **67**, 305 (1970).
[b] cAMP concentration = 40 nM; binding protein = 2 µg/200 µl. Similar data were obtained at a cAMP concentration of 20 nM in the presence of the inhibitor.

When examined, such criteria have been well satisfied. In addition, Macmanus et al.[11] have recently compared analytical results on liver with the protein binding assay and the radioimmunoassay before and after two types of chromatographic purification and phosphodiesterase treatment. There was excellent consistency throughout.

Assay for cGMP[12]

The protein binding assay for cGMP is identical in principle and nearly identical in operation to that described above for cAMP. The receptor protein utilized is a cGMP-dependent protein kinase purified partially from lobster muscle by the method of Kuo and Greengard.[13]

Materials

cGMP Binding Protein

This protein preparation is identical to that described above for the cAMP binding protein through step 3. The dialyzed ammonium sulfate precipitate is used and typically has a specific activity under binding assay conditions of 0.5–1 pmole [^{3}H]cGMP bound per milligram of protein. The specific activity of this preparation is thus considerably lower than that obtained with the cAMP binding protein, and correspondingly greater quantities of binding protein must be utilized.

Several attempts to purify further the cGMP binding protein (DEAE-cellulose columns) have resulted in large losses in activity.

In a typical preparation, approximately 300 g of lobster tail muscle (from about six 1-pound lobsters) are utilized, and sufficient protein is obtained for 10^4 assay tubes.[14] The protein can be stored at $-80°$ for extended periods without loss of activity. Some binding activity is lost (5–10%) when the protein is subjected to freezing and thawing.

[^{3}H]cGMP

[^{3}H]cGMP with a specific activity of 4.5 Ci/mmole was utilized in the original procedure and will suffice. However, vastly improved count-

---

[11] J. P. Macmanus, D. J. Franks, T. Youdale, and B. M. Braceland, *Biochem. Biophys. Res. Commun.* **49**, 1201 (1972).

[12] F. Murad, V. Manganiello, and M. Vaughan, *Proc. Nat. Acad. Sci. U.S.* **68**, 736 (1971).

[13] J. F. Kuo and P. Greengard, *J. Biol. Chem.* **245**, 2493 (1970).

[14] The rostral half of the lobster, including the claws, is not utilized in this procedure. It should, by all means, be placed in an appropriate sized vessel with a small quantity of H_2O, NaCl, and CH_3COOH (vinegar) and steamed for approximately 15 minutes. The product can be enjoyed while the centrifuge is running.

ing rates can be obtained with newer preparations with specific activities in excess of 20 Ci/mmole (ICN). These preparations are, however, rather unstable, and periodic purification and redetermination of specific activity is desirable.

Assay

To 10 × 75 mm disposable glass tubes are added 5 μmoles of sodium acetate (pH 4.0), 6–10 pmoles of [^{3}H]cGMP, and cGMP standards or unknowns. The tubes are then placed in an ice bath, and sufficient cGMP binding protein is added to achieve a final reaction volume of 100 μl. Approximately 200 μg of the protein preparation is usually required to achieve a reasonable quantity of bound ^{3}H-cGMP. Reactions are incubated at 0–3° for at least 75 minutes, by which time equilibrium is complete. Samples are then treated exactly as described above for the cAMP assay.

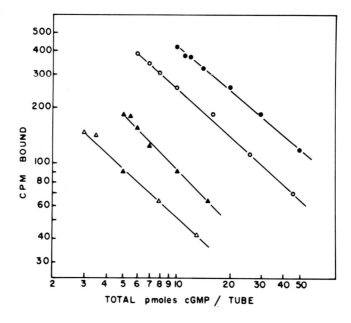

Fig. 3. Standard curves for the cyclic GMP (cGMP) assay. Tubes containing either 10 (●) or 6 (○) pmoles of [^{3}H]cGMP in 100 μl or 5 (▲) or 3 (△) pmoles in 50 μl were incubated as described. Known amounts of unlabeled cGMP were added to some tubes as indicated. A 260-μg amount of binding protein was used in 100-μl incubations, and 130 μg in 50-μl reactions. Total cGMP per tube represents labeled plus unlabeled nucleotide. From F. Murad, V. Manganiello, and M. Vaughan, *Proc. Nat. Acad. Sci. U.S.* **68**, 736 (1971).

A variety of standard curves generated by this method are shown in Fig. 3. Again, when plotted logarithmically, lines are straight, parallel, and close to theoretical. Such curves allow the estimation of 0.5–1.0 pmole of cGMP.

Sample Preparation

While the specificity of this assay procedure is sufficiently high to allow estimation of cGMP concentrations in unpurified urine and extracts of a few tissues, the concentrations of cGMP in most tissues are too low to be determined without separation of this cyclic nucleotide from interfering materials. cAMP, 5'-AMP, and high concentrations of salt appear to be the major sources of interference that were identified. ATP, ADP, adenosine, GTP, GDP, 5'-GMP, and guanosine at 1 mM concentrations had little or no effect on the binding of 0.1 μM cGMP.[12] The following procedure has been useful.

Tissues are homogenized in 5% trichloroacetic acid, protein is removed, and the trichloroacetic acid is extracted as described above. Tissue extracts (pH > 3) are then applied to 0.5 × 3 cm columns of Dowex 1-X8 formate (200–400 mesh).[15] The columns are washed with 10 ml of H_2O. cAMP and 5'-AMP are eluted with 10 ml of 2 N formic acid, and this fraction can be collected for assay of cAMP if desired. cGMP is eluted with 14 ml of 4 N formic acid. ATP is not eluted under these conditions. Column eluates are lyophilized and dissolved in 50 mM sodium acetate, pH 4.5, for assay. Recoveries of both cAMP and cGMP from this column are nearly quantitative, and cGMP can be monitored by addition of tracer quantities of [^{3}H]cGMP at the time of homogenization.

Discussion

Under the conditions described for binding of cGMP to the protein kinase, some variability in the binding constant has been observed with different protein preparations. Values ranging from 2 to 10 nM have been observed. Saturating concentrations of [^{3}H]cGMP have been utilized in this assay procedure for the reasons discussed above for the cAMP method, and, in addition, to maximize the quantity of isotope that can be bound.

Two major problems impose certain limitations on the general utility of this procedure. First, the sensitivity of this method is somewhat lower than that of cAMP, while tissue concentrations of cGMP are usually

---

[15] This particular size mesh is crucial for this separation.

lower. While this problem is readily circumvented if sufficient tissue is available (approximately 100 mg), this limitation must be realized. Second, the number of counts bound to the cGMP-dependent protein kinase in an assay tube necessitates extended counting times. While the situation has been improved by the availability of [^{3}H]cGMP of higher specific activity, significant further improvement must come from better sources and higher specific activities of cGMP receptor proteins.

Simultaneous Assay of cAMP and cGMP[16]

Since the reaction conditions and the procedures for the isolation of the protein–nucleotide complexes for both the cAMP assay and the cGMP assay are essentially identical, and since each cyclic nucleotide produced little or no interference in the assay of the other, we developed a procedure for the simultaneous determination of both cyclic nucleotides in the same incubation mixture. All necessary materials are prepared as described for the individual methods.

Assay Method

The combined assay may be conducted in 100 μl of 50 mM sodium acetate, pH 4.0, containing [^{3}H]cGMP (10 pmoles) and [^{32}P]cAMP (5–10 pmoles). Increased amounts of [^{32}P]cAMP can be used as radioactive decay occurs, since the sensitivity of the assay is more limited by that for cGMP and since most samples contain greater quantities of cAMP. Reactions are initiated by addition of a mixture of the two binding proteins, and incubation and filtration proceed as described above. Filters are counted for their ^{3}H and ^{32}P content, with resultant standard curves as shown in Fig. 4. Unlabeled cAMP produces a nearly theoretical decrease in the [^{32}P]cAMP bound, and unlabeled cGMP produces similar dilution of [^{3}H]cGMP bound in the combined assay, just as shown previously when the assays were performed individually.

Discussion

As shown in Fig. 4, cGMP does not interfere with the binding of cAMP in the simultaneous (or individual) assay unless very high concentrations of cGMP are present. cAMP causes a slight inhibition of binding of cGMP at equimolar concentrations (not shown), but further inhibition is not observed to a total cAMP:cGMP ratio of 4:1. Up to 30 pmoles of unknown cAMP can therefore be added under assay conditions.

[16] F. Murad and A. G. Gilman, *Biochim. Biophys. Acta* **252**, 397 (1971).

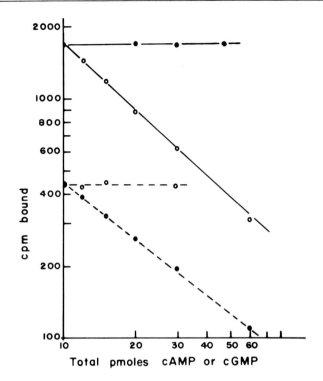

FIG. 4. Standard curves for simultaneous assay of cyclic AMP (cAMP) and cyclic GMP (cGMP). Incubations (100 μl) were performed with 10 pmoles [^{3}H]-cGMP, 10 pmoles [^{32}P]cAMP, 260 μg of cGMP binding protein, and 1.3 μg of cAMP binding protein. Solid lines represent amount of [^{32}P]cAMP bound and broken lines represent amount of [^{3}H]cGMP bound. To some tubes (—○— or ---●---) standard solutions containing either 2, 5, 10, 20, or 50 pmoles each of cAMP and cGMP were added. To other tubes, standard solutions of only cAMP (2, 5, or 20 pmoles, --○--) or cGMP (10, 20, 40 pmoles, ——●——) were added. From F. Murad and A. G. Gilman, *Biochim. Biophys. Acta* **252**, 397 (1971).

Simultaneous assay of cAMP and cGMP in the same incubation can save considerable technical time when large numbers of urine samples are analyzed for cyclic nucleotides. The simultaneous assay has not been applied to biological materials other than urine because of materials in tissues that interfere with the cGMP assay.

Acknowledgments

These studies were supported in part by United States Public Health Service grants NS10193 and AM15316 and the Virginia Heart Association. Ferid Murad is the recipient of USPHS Research Career Development Award AM70456.

[8] Assay of Cyclic AMP by the Luciferin–Luciferase System

By Charles A. Sutherland and R. A. Johnson

The luminescence assay for adenosine 3',5'-monophosphate (cAMP)[1] is based on its enzymatic conversion to adenosine 5'-triphosphate (ATP)[2,3] and the subsequent determination of ATP by its luminescent reaction with firefly luciferin (LH_2) and luciferase (E).[4]

Principle. The reaction upon which the luminescence assay for cAMP is based is the following.[4]

$$E + LH_2 + MgATP \rightleftharpoons E \cdot LH_2AMP + MgPP_i \quad (1)$$
$$E \cdot LH_2AMP + O_2 \rightarrow E + CO_2 + AMP + products + light \quad (2)$$

cAMP is converted to ATP by simultaneous incubation with cyclic 3',5'-nucleotide phosphodiesterase, adenylate kinase, pyruvate kinase, phosphoenolpyruvate, and a catalytic amount of ATP. The ATP content of each sample is then determined according to reactions (1) and (2). The reaction of LH_2 and ATP with E produces light which reaches maximum intensity within 1 second and thereafter decays by a process that is dependent upon several factors. The peak light intensity is directly proportional to the initial ATP concentration.

Instrumentation. Because of the nature of the reaction of ATP and LH_2 with E, it is desirable, though not necessary, to measure the peak light intensity. Any system will suffice which permits the rapid mixing of the reagents and monitoring of the light emission by a sensitive photomultiplier coupled to an appropriate recording device. For example, a fluorometer may be used in conjunction with a short-rise-time recorder. Liquid scintillation counters may also be used to measure light emission in this assay. Alternatively, there are photometers designed by Aminco (Chem-Glow photometer) or DuPont (Luminescent Biometer) principally for use with luminescent reactions. The following procedure has been used with the DuPont instrument and may require slight modification with other instruments.

---

[1] R. A. Johnson, J. G. Hardman, A. E. Broadus, and E. W. Sutherland, *Anal. Biochem.* **35**, 91 (1970).
[2] B. M. Breckenridge, *Proc. Nat. Acad. Sci. U.S.* **52**, 1580 (1964).
[3] N. D. Goldberg, J. Larner, H. Sasko, and A. G. O'Toole, *Anal. Biochem.* **28**, 523 (1969).
[4] J. L. Denburg, R. T. Lee, and W. D. McElroy, *Arch. Biochem. Biophys.* **134**, 381 (1969).

Reagents

(A) Tris·Cl, pH 7.5, 50 mM
(B) Mg(C$_2$H$_3$O$_2$)$_2$, 5 mM; KCl, 10 mM; Tris·Cl, 50 mM, pH 7.5; dithiothreitol (DTT), 5 mM; disodium ethylenediamine tetraacetate (EDTA), 0.1 mM; bovine serum albumin (BSA, Sigma No. A 4378) 1 mg/ml
(C) Mg(C$_2$H$_3$O$_2$)$_2$, 5 mM; Tris·Cl, 50 mM pH 7.5; DTT, 5 mM; EDTA, 0.2 mM; BSA, 1 mg/ml
(D) Mg(C$_2$H$_3$O$_2$)$_2$, 5 mM; KCl, 10 mM; Tris·Cl, pH 7.5, 50 mM
(E) Luciferase (E) from *Photinus pyralis*. Available as a crystallized powder complex with synthetically prepared crystalline luciferin from DuPont de Nemours & Co., Inc.
(F) Adenylate kinase (AK), EC 2.7.4.3; from Boehringer/Mannheim No. 15355, EMAB, 360 U/mg
(G) Pyruvate kinase (PK), EC 2.7.1.40; from Boehringer/Mannheim No. 15744, EPAU, 150 U/mg
(H) Cyclic 3',5'-nucleotide phosphodiesterae (PDE), 25 U/ml[5] Some commercial sources of PDE may be contaminated with excessive nucleotidase and therefore require further purification prior to use.
(I) cAMP standards and unknowns taken up in (A) above.

Procedure. Prior to assay, cAMP samples must be purified free of contaminating AMP, ADP, and ATP. Column and thin-layer chromatography systems and coprecipitation have been used for this purpose and are described elsewhere.[6-9]

The assay is most sensitive when used with enzymes from which the contaminating adenine nucleotides have been partially removed. This can be done by several methods. In the method described in the next section AK and PK are purified simultaneously in the ratio to one another in which they are to be used and are frozen until needed. When an assay is to be run, an appropriate volume of that decontaminated PK-AK stock solution is diluted with 6 volumes of B. This solution is divided into 2 parts and PDE (1 U/ml) is added to one of them. Alternatively, one can use untreated enzymes at concentrations of 60 μg/ml (AK) and 80 μg/ml (PK) dissolved in B.

All assay tubes are kept in an ice bath while the appropriate reagent

[5] G. A. Robison, R. W. Butcher, and E. W. Sutherland, "Cyclic AMP." Academic Press, New York, 1971.
[6] R. A. Johnson, *Methods Mol. Biol.* 3, 1 (1972).
[7] G. Schultz, E. Böhme, and J. G. Hardman, this volume [2].
[8] E. Böhme and G. Schultz, this volume [4].
[9] P. S. Chan and M. C. Lin, this volume [5].

additions are made. These same tubes (6 × 50 mm disposable glass tubes) are later used as cuvettes. Six assay tubes are run for each unknown sample. Two of these tubes contain 50 μl of unknown, 50 μl of PK-AK solution, and 50 μl of A. These tubes permit estimation of the amount of contaminating adenine nucleotides present in the unknown and in the PK-AK reagent. Two more tubes contain 50 μl of PK-AK-PDE solution, 50 μl of unknown and 50 μl of A. The readings obtained from these tubes are due to the cAMP present in the unknown plus the nucleotide contamination in the PDE. The last two tubes for each unknown contain 50 μl of unknown, 50 μl of PK-AK-PDE and 50 μl of A containing a known amount of cAMP (5–50 pmoles). The added cAMP serves as an internal standard which allows for correction of assay interference due to impurities in the unknowns and for correction of small errors due to inactivation of E-LH_2.

The standard curve points are run in duplicate. In addition to the tubes containing cAMP and PK-AK-PDE, there must be included tubes containing only PK-AK (50 μl) and A (100 μl). When compared with the PK-AK-PDE blank tubes, these PK-AK tubes provide for estimation of the level of nucleotide contamination in the PDE. The final volume in all tubes is 150 μl. The loaded tubes are then capped and incubated for 3–4 hours at 35°. Overnight incubation at room temperature has also been used. The reaction is terminated by freezing. The incubated tubes may be stored at −20° for several days before continuing with the final step.

One hour prior to determination of ATP levels, a solution of C plus E-LH_2 (25–30 mg/ml) is prepared allowing 20 μl per assay tube plus some extra for rinsing the injection syringe. E-LH_2 must be injected at a uniform velocity in order to get good reproducibility. This may be accomplished by using a 50 μl, flat-tipped Hamilton syringe mounted in a spring-loaded holder (Shandon Repro-Jector). The E-LH_2 solution must be kept at 0–2° at all times. Aliquots of E-LH_2 solution are withdrawn as needed. The enzyme solution must not be injected back into the stock solution as E is rapidly inactivated by the agitation. In the absence of such factors, E activity is lost at the rate of 1–2% per hour.

All tubes are placed in a freezing bath (dry ice–ethanol). They are then serially transferred to a room temperature bath and then to the Biometer. Precise timing in this operation is less important than careful temperature equilibration. A cold tube may yield 2–3 times the light intensity of a tube held too long in a worker's hand. Following E-LH_2 injection, the peak intensity of the light generated is expressed in arbitrary units by direct digital readout and is directly proportional to the initial ATP concentration.

Enzyme Decontamination. Several methods have been used to remove nucleotide contaminations from PK and AK. Treatment of these enzymes individually with either charcoal or anion-exchange resin has been previously reported. The method described below involves the combined chromatography of PK and AK on QAE-Sephadex (A-25 from Pharmacia).

Sixty milliliters of QAE-Sephadex are equilibrated with D and then added to 200 ml of D supplemented with 5 mM PEP, 5 mM DTT, 0.1 mM EDTA and BSA (25 µg/ml). This solution is mixed with a stirring bar for 5 minutes. The resin is poured into a 50×1 cm water-jacketed column. Water warmed to 35° is circulated through the jacket when the enzymes are loaded onto the column.

AK (8 mg) and PK (11 mg) are suspended in 3 ml of B and incubated at 35° for 30 minutes. The enzyme mixture is then applied to the QAE-Sephadex column, followed with 2 ml of B and eluted from the column with D at a flow rate of about 10 ml per hour. AK and PK are not retarded by the column. The first 14 ml of eluate are discarded and the next 8 ml are collected into 12 ml of B modified to contain 10 mM DTT, 0.2 mM EDTA and 0.2 mM PEP. This collecting solution is kept at 0–2° as the enzyme fraction is collected from the column. The resulting AK-PK reagent is divided into 2-ml aliquots, frozen in liquid N_2, and stored at $-70°$. These aliquots may be thawed and refrozen several times without significant loss of enzyme activity. Each new batch of enzyme should be characterized to the extent of determining the enzyme stock dilution and incubation time required for subsequent assay. A 6-fold dilution of enzyme used in a 3–4 hour incubation is usually adequate for this preparation.

Assay Characteristics. Assay specificity for cAMP is assured by three factors. The primary factor is the purification of the cAMP sample prior to assay. Further specificity is provided by the selectivity of PDE for 3'5'-cyclic nucleotides and by the high specificity of luciferase for ATP.

With the above purification method, the assay blank can be lowered by 50–70% to the equivalent of about 0.2 pmole of cAMP (1.3×10^{-9} M). The adenine nucleotides associated with PDE may constitute 5–50% of this remaining contamination. At least 90% of both the AK and PK enzyme activity that is added to the QAE sephadex column is recovered in the eluate.

The usual variation between duplicate assay samples is 5%. In data pooled from several experiments, the coefficient of variability was 4%. The standard curve is linear with a slope of 1 over at least 5 orders of magnitude. The procedure can be readily adapted to the measurement of other substrates that are convertible to ATP or to the assay of enzymes that metabolize substrates convertible to ATP.[10]

[10] B. Weiss, R. Lehne, and S. Strada, *Anal. Biochem.* **45**, 222 (1972).

[9] Assay of Cyclic AMP by Protein Kinase Activation

By STEVEN E. MAYER, JAMES T. STULL, WILLIAM B. WASTILA, and BARBARA THOMPSON

This assay of cyclic AMP (cAMP) is based on the rate of casein phosphorylation catalyzed by skeletal muscle protein kinase which is dependent upon the cyclic nucleotide concentration.[1] The reaction [Eq. (1)] is carried out in the presence of trichloroacetic acid-treated tissue extract, [γ-^{32}P]ATP, purified casein, and partly purified rabbit skeletal muscle protein kinase.

$$[\gamma\text{-}^{32}\text{P}]\text{ATP} + \text{casein} \xrightarrow[\text{protein kinase}]{\text{cAMP}} [^{32}\text{P}]\text{casein} + \text{ADP} \qquad (1)$$

The phosphorylated casein is isolated on filter paper disks,[2] and the protein-bound ^{32}P is measured. The advantages of this assay are (1) no purification of cAMP beyond extraction of the tissue with trichloroacetic acid is required; (2) the amount of ^{32}P incorporated into casein is a linear function of the cAMP concentration over a 25-fold range with a minimum of 10–25 nM cAMP in the tissue extract; (3) the procedure is rapid (about 120 assays can be processed a day by one person) and sensitive (about 0.05 pmole can be determined); and (4) it is readily applied to the assay of adenylate cyclase.

Method

Materials

cAMP-dependent protein kinase is partially purified by a minor modification[1] of the method reported by Walsh *et al.*[3] This procedure is rapid and yields an enzyme preparation with a reproducible, highly sensitive activation by cAMP. Fresh, unfrozen muscle (2.4 kg) is ground twice in a chilled meat grinder, weighed, and homogenized with 2.5 volumes of ice-mold 4 mM EDTA, pH 7.0, in a Waring blender. All subsequent operations are carried out at 0–4°. The homogenate is centrifuged at 13,000 g for 20 minutes. The supernatant fraction is adjusted to pH 5.5 with 1 N acetic acid, and the precipitate is removed by centrifugation.

---

[1] W. B. Wastila, J. T. Stull, S. E. Mayer, and D. A. Walsh, *J. Biol. Chem.* **246**, 1996 (1971).
[2] R. J. Mans and G. D. Novelli, *Arch. Biochem. Biophys.* **94**, 48 (1961).
[3] D. A. Walsh, J. P. Perkins, and E. G. Krebs, *J. Biol. Chem.* **243**, 3763 (1968).

The clear supernatant solution, readjusted to pH 6.8, is fractionated by the slow addition of 32.5 g of ammonium sulfate per 100 ml of protein solution. The precipitate, collected by centrifugation, is dissolved in 5 mM potassium phosphate buffer, pH 7.0, containing 2 mM EDTA. The dissolved precipitate is dialyzed 18 hours against three changes (12 liters each) of the same buffer. The dialyzed protein solution is centrifuged at 78,000 g for 1 hour, and the precipitate is discarded. The enzyme is adsorbed on a column (24 × 4.5 cm) of DEAE-cellulose (Sigma, 0.88 meq/g, coarse) equilibrated with 5 mM potassium phosphate buffer, pH 7.0, containing 2 mM EDTA. The column is washed with 800 ml of the same buffer, and the enzyme is then eluted with 30 mM potassium phosphate buffer, pH 7.0, containing 2 mM EDTA. Purification results in an approximately 100-fold increase in specific activity, based on the enzymatic activity in the pH 5.5 supernatant fraction. The enzyme, in 0.5-ml aliquots, may be stored at −65° for at least 6 months and at 4° for 1 month after thawing.

Casein is purified and dephosphorylated by a modification of the method of Reimann et al.[4] Casein (Matheson, Coleman and Bell purified grade), 5 g, is mixed with 40 ml H_2O. The pH is adjusted slowly to 9.5 with 1 N NaOH and the solution is heated in boiling H_2O for 10 minutes with addition of NaOH to maintain pH 9.5. The solution is cooled by allowing it to stand at room temperature and diluted to 60 ml. The pH is adjusted to pH 5.7 by the dropwise addition of 1 N HCl over a period of 60–90 minutes. After centrifugation for 20 minutes at 45,000 g the infranatant solution is carefully separated from floating particulate matter. The pH is adjusted to 5.9, the centrifugation is repeated, and the concentration is adjusted to 60 mg/ml. The solution is stable frozen. Care must be taken in making the pH adjustments and avoiding turbidity in the final product.

[γ-^{32}P]ATP is prepared by the method of Glynn and Chappell.[5] Details are described elsewhere in this volume.[6] The incubation mixture contains 20–40 mCi [^{32}P]P$_i$. Separation of the [γ-^{32}P]ATP formed is based on its adsorption on DEAE-Sephadex and elution with triethylammonium bicarbonate (TEAB).[7] All operations must be conducted at 0–5° to prevent ATP hydrolysis. A 5-ml bed volume (0.9 × 8 cm) of DEAE-Sephadex A-25 is converted to the bicarbonate form by washing with 50 ml of 1 M NH_4HCO_3 and then 100 ml of H_2O. TEAB, 1 M is prepared

---

[4] E. M. Reimann, D. A. Walsh, and E. G. Krebs, *J. Biol. Chem.* **246**, 1986 (1971).
[5] I. M. Glynn and J. B. Chappell, *Biochem. J.* **90**, 147 (1964).
[6] G. Schultz and J. G. Hardman, this volume [14].
[7] This procedure was suggested by Dr. Mehran Goulian, Department of Medicine, University of California, San Diego.

fresh before use. Triethylamine (20 ml) is added to about 80 ml of H_2O at 0°. CO_2 is bubbled through the solution until the pH is less than 8 (about 1 hour) and the solution is diluted to 144 ml yielding 1 M TEAB. The reaction mixture used in the preparation of $[\gamma\text{-}^{32}P]ATP$ is applied to the column and allowed to percolate by gravity. The column is then eluted with 25 ml of 0.1 M TEAB containing 20 mM potassium phosphate, pH 7.0 at a rate of about 1 ml per minute. This fraction contains $[^{32}P]P_i$. TEAB, 40 ml of a 1 M solution, is then added. The first 3 ml are discarded and the remainder collected in a 500-ml lyophilizing flask. This fraction contains the $[\gamma\text{-}^{32}P]ATP$. The elution can be followed more precisely with a Geiger tube monitor shielded so that it is sensitive only to the effluent from the column passing through plastic tubing (bore, 1–2 mm). The eluate is frozen and lyophilized. The powder in the flask is dissolved in 8 ml of methanol, frozen, and again lyophilized. This is repeated once more to remove all the TEAB. The $[\gamma\text{-}^{32}P]ATP$ is dissolved in 5 ml of H_2O and stored at $-20°$. The yield is 85–95% of the ^{32}P added to the incubation with >99% of the label in the γ-position of ATP.

Filter paper disks. Whatman No. 3 MM, 2.3 cm in diameter (Arthur H. Thomas Co. No. 5271-510) are washed in distilled H_2O to remove loose fibers and dried. Each disk is prepared for use by being pierced centrally with a pin on a Styrofoam plate (made from packing material) and then numbered in pencil. The pierced disks are then transferred to a methacrylate plate (28 × 10 × 0.64 cm) with three rows of 10 holes, 2.5 cm apart. The holes are to hold the pins.

Tissue extracts. Details of tissue fixation and extraction are described elsewhere in this treatise.[8] For the standard procedure described below, 5–20 mg of frozen tissue are transferred to a size 20 Duall ground-glass homogenizer tube (Kontes Glass Co., Vineland, New Jersey) at $-20°$. Ten volumes ice-cold 10% trichloroacetic acid are added, and the mixture is immediately homogenized. After it has stood for 10 minutes at 0°, it is centrifuged for 20 minutes at 3000 g at 0–3°. The supernatant solution is transferred to a 6 × 50-mm tube and extracted six times with 2 volumes of water-saturated diethyl ether. Following the last ether extraction the solution is heated in a bath at 100° for 1–2 minutes until the odor of ether has disappeared. The cAMP in the neutralized extract is stable indefinitely at $-20°$. Perchloric acid cannot be substituted for trichloroacetic acid as a protein precipitant. Neutralized extracts after perchlorate treatment strongly inhibit protein kinase.

Reagents. The incubation reagent contains the following ingredients:
KF, 50 mM

---

[8] S. E. Mayer, J. T. Stull, and W. B. Wastila, this volume [1].

Ethylene glycol bis(β-aminoethyl ether)-N,N'-tetraacetic acid (EGTA), 2.5 mM

Theophylline, USP, free base, 5 mM

Na β-glycerophosphate, 250 mM (Sigma Grade II; 1.5 M solution is filtered through fluted paper and then through an 0.22 μm Millipore filter. It is then recrystallized three times by the addition of ⅗ volume of 95% ethanol and dried under vacuum; the product has 4 H_2O, formula weight = 288)

KH_2PO_4, 76 mM. The pH is now adjusted to 6.0 with HCl.

Casein (purified as described above), 18 mg/ml

The pH of the incubation solution is readjusted to 6.0 with KOH. It may be stored at $-20°$.

The other reagents are as follows:

Mg-ATP: 100 mM magnesium acetate and 12 mM [γ-^{32}P]ATP, specific activity 2–5 $\times$ 10^7 cpm/μmole

cAMP standards: 10, 15, 25, 35, 50, 75, 100, 150, 200, 250, 300, and 500 nM

cAMP-dependent protein kinase: This is diluted so that the rate of ^{32}P incorporation is constant for at least 60 minutes with the highest concentration of cAMP standard needed. The dilution generally requires 100–200 μg of enzyme per milliliter during the incubation.

Stopping reagent: 50 mg/ml bovine plasma albumin (Sigma, Cohn Fraction V) and 100 mM ethylene diaminetetraacetic acid (EDTA), pH 7

Perchloric acid (PCA), 5%; ethanol, 95%. To conserve these solutions only the first PCA wash is discarded each time. The second is then used as the first in the next assay, etc. The alcohol can be reused 5 or 6 times.

Procedure

Step 1. In 6 $\times$ 50 mm borosilicate tubes (Kimble), pipette rapidly and in the order indicated: 25 μl of H_2O, cAMP standards or tissue extract, 75 μl of incubation reagent, and 5 μl of protein kinase; mix gently.

Step 2. Transfer to a 30° bath at timed intervals (10 or 15 seconds) with addition to each tube of 10 μl of Mg-ATP. Incubate precisely for a period of 30–120 minutes. The time depends on the activity of the protein kinase, the specific activity of the [γ-^{32}P]ATP, and the concentrations of cAMP.

Step 3. Add 75 μl stopping reagent at timed intervals and place the tubes in an ice bath.

Step 4. Aliquots, 100 μl, of the stopped reaction mixtures are spotted on filter paper disks.[9] Preparation of the disks is described above. The methacrylate plate containing 30 disks is immedately transferred to a plastic refrigerator dish (31 × 13 × 10 cm) containing the following: two pieces, 15 cm long, of vinyl tubing, 2 cm in diameter, to support the plate; a magnetic stirring bar; and 1.3 liters of 5% PCA. The acid is stirred for 30 minutes at room temperature on a magnetic stirrer.

Step 5. The tray of disks is further washed and then dried: 20 minutes in 5% PCA at room temperature, 30 minutes in 5% PCA at room temperature, 2 minutes in 95% ethanol, 2–5 minutes under an infrared heat lamp.

Step 6. The dried disks are removed from the pins with forceps and placed into liquid scintillation counting vials containing 7 ml of a solution of Omnifluor (New England Nuclear Corporation, Boston), 4 g/liter, in toluene. The samples are counted in a liquid scintillation spectrometer. The counted disks are removed from the vials with forceps, the background count of the vials is checked, and the vials are reused. Radioactivity on the disks does not dissolve in the counting solution.

Sensitivity, Reproductibility, and Validation of the cAMP Assay

The sensitivity of the assay is limited by the blank counts, the precision of duplicate standards and unknown determinations, and the slope of the standard curve. The latter is determined by the quality of the protein kinase preparation and inhibitors of the reaction in the samples assayed. This is discussed further below. The blank is composed of nonprotein bound ^{32}P that cannot be washed off the paper disks and of cAMP-independent protein kinase activity. A standard curve from a single experiment that was designed to obtain maximum sensitivity is shown in Fig. 1.

The linearity of the curve can be extended to higher concentrations of cAMP at a sacrifice of sensitivity by using less protein kinase and a shorter incubation time. The reproducibility of cAMP concentrations measured in liver, heart, skeletal muscle, adipocytes, mast cells, cultured fibroblasts, gall bladder mucosa, urine, toad bladder, brain, *Escherichia coli*, etc., samples is with a standard error of 5% or less of the mean value. Recovery of known amounts of cAMP added to extracts from such tissues averages 93 ± 2%.

[9] The partial drying of the disks at this point causes variability. If the time lapse from the application of the sample to the first disk until that disk is immersed in 5% PCA is 8 minutes or less, variability is low. If time is greater than 8 minutes, then the disks should be allowed to dry completely by standing in a moving stream of air (as in a hood) for 30–60 minutes.

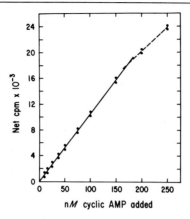

FIG. 1. A cyclic AMP standard curve from an experiment designed for maximum sensitivity and precision. The specific activity of the ATP was 7.66×10^7 cpm/μmole and protein kinase concentration 216 μg/ml. Incubation was for 1 hour. Duplicate determinations of blanks and standards varied by 2% or less. The mean blank was 2146 cpm. The lower limit of determination of cyclic AMP in 25 μl of a mast cell extract was 0.25 (± 0.013) pmole.

For each new tissue to which the method is applied, the validity of the procedure should be tested. The most important condition that must be met is the elimination of factors in the trichloroacetic acid extract that interfere with the protein kinase reaction. These factors may be unknown substances in the tissue or high ionic strength, both of which depress the reaction. Extraction of tissue with 10 volumes of trichloroacetic acid is effective against the former (except brain, which requires about 15 volumes). If the ionic strength is still sufficiently high to affect recovery of added cAMP, the cAMP solutions used to construct the standard curves must be prepared in a medium with a composition similar to that of the unknowns. This requirement is most likely to be encountered in assaying cAMP in bacteria lyophilized in their nutritive media and in animal cells suspended in an isotonic medium such as Krebs-Ringer solution.

The validiy of the procedure is tested by measuring recovery of endogenous and added cAMP in serial dilutions of tissue extracts. The concentration–dilution relationship should be linear and extrapolate to zero on both axes. All measurable protein kinase-stimulating activity of extracts with or without added cAMP should disappear after incubation with cyclic nucleotide phosphodiesterase.[1] Interference by cGMP is negligible in vertebrate and bacterial extracts. It is a potential source of error in sea urchin sperm and probably other invertebrate tissues that contain more than 1 μM cGMP.

Dealing with Problems Encountered in the Assay. *High blank and poor sensitivity* are the results of a poor preparation of muscle protein kinase. Only those fractions of the enzyme eluted from the DEAE column that contain minimal cAMP-independent activity and maximum dependence upon cAMP should be used in the assay. *Variable blanks* are most likely to be due to contaminated pipettes or nonuniform drying of all the disks on a plate. In, for example, a 120 disk assay, there should be at least 4 blanks per plate of 30 disks. Samples that give *counts lower than blanks* indicate one or both of two problems. One is that the ionic strength of the unknown samples is significantly higher than that of the standards. The latter should then be prepared in a medium similar to that of the unknowns. The second is pH. If the buffering capacity of the unknown samples is sufficient to alter the pH of the incubation mixture 0.1 pH unit, or more then casein precipitates (pH < 5.9), or protein kinase activity is reduced (pH < 6.1). The samples must then either be further diluted or their pH adjusted to about 6.

Modifications of the Assay

Adenylate cyclase activity may be determined by measuring the cAMP formed without separation of the reaction products or use of labeled ATP during the initial incubation.[1] The cyclase reaction is terminated by heating for 3 minutes at 100° or with trichloroacetic acid. Adenylate cyclase incubation media contain substances that interfere with the cAMP determination; consequently cAMP standards are prepared in the same medium as that used for the cyclase reaction. Thus the effect of dilution of the [γ-^{32}P]ATP specific activity by ATP required for adenylate cyclase activity is canceled. Recovery of cAMP added to the cyclase reaction step is 95–100% provided that phosphodiesterase activity is virtually completely inhibited. This is generally accomplished by adding a potent inhibitor such as papaverine or 1-methyl-3-isobutyl-xanthine (500 μM). Assay of adenylate cyclase in whole homogenates of brain, which contains much phosphodiesterase, may require greater concentration or more potent phosphodiesterase inhibitors.

Dilute Samples. Determination of cAMP in medium after incubation of cells that extrude cAMP may be difficult if the medium volume is large relative to the number of cells and if ionic strength is appreciable. Such samples may be assayed after adsorption of cAMP onto anion exchange resin, elution with HCl[10] and concentration by lyophilization.

Small Samples. The procedure described above requires about 5 mg

---

[10] G. Schultz, E. Böhme, and J. G. Hardman, this volume [2].

of tissue if duplicate determinations (25 µl each) on 10% trichloroacetic acid extracts are made. The assay procedure has been scaled down by a factor of 4 for measurement of cAMP in 1 mg of cat ventricle papillary muscles. The isolation of ^{32}P-labeled casein on filter paper disks is the same. The specific activity of the [γ-^{32}P]ATP is increased 2–5-fold. This permits the determination of about 0.05 pmoles of cAMP.

[10] Quantitation of Cyclic GMP by Enzymatic Cycling

By NELSON D. GOLDBERG and MARK K. HADDOX

The analytical procedure to be described is similar to the one first utilized to provide evidence that cyclic 3′,5′-guanosine monophosphate (cGMP) is a naturally occurring constituent of mammalian tissues[1] and subsequently that the biological role of cGMP is quite different and probably antagonistic to that of cAMP in certain biological systems.[2-4]

The procedure consists of the following steps:

1. Extraction of cGMP from biological material.
2. Isolation of cGMP from other tissue constituents by column and thin-layer chromatography.
3. Conversion of cGMP to 5′-GMP with phosphodiesterase (PD).

$$\text{cyclic GMP} \xrightarrow{\text{PD}} 5'\text{-GMP}$$

4. Conversion of 5′-GMP to GDP with ATP-GMP phosphotransferase (GMP kinase, GK). Total conversion is ensured by converting the diphosphate products to triphosphates with aid of creatine phosphokinase (CK).

$$\text{GMP} \xrightarrow{\text{GK}} \text{GDP} \xrightarrow{\text{CK}} \text{GTP}$$
$$+$$
$$\text{ATP} \xleftarrow{\text{CK}} \text{ADP}$$

5. Magnification of the GTP (or GDP) generated by an enzymatic cycling system composed of succinate thiokinase (STK) and pyruvate

[1] N. D. Goldberg, S. B. Dietz, and A. G. O'Toole, *J. Biol. Chem.* **244**, 4458 (1969).
[2] N. D. Goldberg, M. K. Haddox, D. K. Hartle, and J. W. Hadden, *Proc. Int. Congr. Pharmacol. 5th* Vol. 5, p. 146 (1972).
[3] J. W. Hadden, E. M. Hadden, M. K. Haddox, and N. D. Goldberg, *Proc. Nat. Acad. Sci. U.S.* **69**, 3024 (1972).
[4] N. D. Goldberg, M. K. Haddox, R. Estensen, J. G. White, C. Lopez, and J. W. Hadden, *in* "Cyclic AMP in Immune Response and Tumor Growth" (L. Lichtenstein and C. Parker, eds.). Springer-Verlag, Berlin and New York, 1973.

kinase (PK), which generates a proportional amount of pyruvate greater than 10^3 times the GTP (or GDP) present.

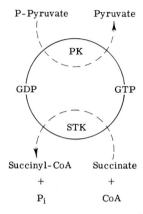

6. Conversion of pyruvate to lactate with lactate dehydrogenase (LDH) and formation of a stoichiometric amount of DPN^+.

$$\text{Pyruvate} + \text{DPNH} + \text{H}^+ \xrightarrow{\text{LDH}} \text{lactate} + \text{DPN}^+$$

7. Conversion of DPN^+ to a highly fluorescent product by treatment with strong alkali and quantitation of fluorescence.

The specificity of the method for cGMP derives from the chromatographic purification procedure and the selectivity of GMP kinase for GMP as the nucleoside monophosphate substrate and of succinate thiokinase for GTP (and ITP) as the nucleoside triphosphate substrate. The level of sensitivity achievable with this procedure is limited by the blank inherent in the system. Under the assay conditions to be described, 10^{-13} mole of cGMP can be detected reliably with a variation in reproducibility of about 5%. This provides for the quantitation of this cyclic nucleotide in 10–100 mg (wet weight) samples of most tissues and microliter amounts of urine.

Sampling and Extraction Procedures

The tissue levels of cGMP have been found to undergo rapid changes in some systems seconds after exposure to hormones or as a result of other treatments. Experimental procedures that involve removal of tissue samples from the intact animal could, by themselves, lead to rapid changes in cGMP levels. No procedure yet devised has been shown to completely eliminate the possibility of such artifacts, but at this time there is general agreement that quick-freeze procedures are probably the most acceptable. The use of Wollenberger-type clamps,[5] stainless steel or aluminum blocks cooled in liquid nitrogen, which compress and freeze the sample uniformly and efficiently, are, therefore, recommended.

If the anatomical integrity of the specimen is to be retained, or when working with an isolated organ *in vitro*, the sample can be rapidly frozen by immersing it in Freon-12 or isopentane cooled to $-150°$ with liquid nitrogen.

Dissection, weighing, or other preparation of the sample before extraction of the cyclic nucleotide should be carried out at temperature below $-20°$. The extraction procedure used should minimize losses due to enzymatic hydrolysis or continued enzymatic or chemical generation of the cyclic nucleotide. For example, phosphodiesterase has been shown to promote cGMP hydrolysis at $0°$[6] at approximately one-third the rate seen at $30°$.

The following extraction procedures have proved to be efficient and reliable:

a. Four volumes of ice-cold 10% TCA are added to a known aliquot of frozen powdered tissue contained in a vessel that can also be used to carry out the subsequent homogenization and centrifugation. The TCA-tissue mixture is brought to $-20°$ in an ethanol–dry ice bath and the sample homogenized for about 2 minutes until the "slush" that results is melted. After 10 minutes on ice the homogenate is centrifuged (10,000 g for 20 minutes), a known aliquot of the supernatant is removed, and the denatured pellet is used to quantitate the protein content of the sample, if that is desired.

b. A known weight of frozen tissue sample is powdered at liquid nitrogen temperature and layered over approximately 3 volumes of frozen 3 M PCA. The tubes are then transferred to an ethanol–dry ice bath maintained at $-12°$, and the PCA solution is thawed while the tissue, which remains frozen, settles into the liquefied acid solution. After intermittent agitation for 5 minutes the suspension is brought to $0°$ and 2 ml of ice cold distilled water added. The sample is centrifuged (10,000 g for 20 minutes) to separate the denatured protein from the clear acid extract.

c. When working with cultured or suspended cell systems, 10% TCA can be added directly to the sample at appropriate times after a given treatment.

---

[5] A. Wollenberger, O. Ristau, and G. Schoffa, *Pflueger's Arch. Gesamte Physiol. Menschen Tiere* **270**, 399 (1960).

[6] R. F. O'Dea, M. K. Haddox, and N. D. Goldberg, *Pharmacologist* **12**, 291 (1970).

TCA is removed from the acid supernatant fraction by three successive extractions with 10 volumes of water-saturated ether. Traces of ether in the acid extract are removed by aspirating the vapors above the liquid phase after heating of the extract to 60° in a water bath.

PCA is removed from the acid supernatant fraction by neutralization with an appropriate volume of 2 M $KHCO_3$. The resulting precipitate (potassium perchlorate) is removed by centrifugation.

Neutralized acid "blanks" should also be prepared during the extraction procedures. These are carried through the purification procedure in the same manner as tissue-containing samples and are used for detection of possible contamination of reagents used in the extraction or analytical procedures.

Once the activities of the tissue enzymes are arrested, the cGMP is relatively stable. Neutralized samples can be stored at −80° for an indefinite period of time without any substantial degradation of the cyclic nucleotide.

To monitor the recovery of cGMP during the extraction and subsequent procedures [^{3}H]cGMP (ca. 2000 cpm) should be added to the extraction media (i.e., TCA or PCA). The tritiated cyclic nucleotide should be purified before use; radioactive impurities amounting to as much as 15% in many commercial preparations are often present. The purification may be carried out by chromatography on silica gel thin-layer plates according to the procedures outlined below, by the method outlined in this volume [11] by O'Dea et al. using alumina and BioRad AG1-X8 (formate) or by a previously described procedure employing DEAE-cellulose (bicarbonate).

Urine samples, which normally do not contain 5'-nucleotides, may be prepared for analysis by simply heating the sample (90° for 2 minutes) immediately after collection. The sample is then chilled and clarified by centrifugation (10,000 g for 15 minutes). cGMP in the heat-denatured urine sample may be quantitated without further purification of the urine.

Separation of cGMP from Interfering Substances

Column Chromatography

The neutralized acid extract is applied to a 0.5 × 2.5 cm column of BioRad AG 1-X8, (formate form) (200–400 mesh), which is then washed with 5–10 ml of H_2O and then 8 ml of 1 N formic acid. cGMP is eluted with 7–9 ml of 4 N formic acid and the eluate is evaporated to dryness. This procedure is similar to one which has been described by Murad *et* al.[7] "Blank" columns to which samples of neutralized acid ("blanks") are applied should also be included. If the PCA extraction procedure is employed, formic acid should be added to the extracts (50 mM final concentration) before they are applied to the columns. This will bring about the release of CO_2 from unreacted $KHCO_3$ that may be present and prevent the formation of bubbles in the column.

Thin-Layer Chromatography

Silica gel thin-layer plates with fluorescent indicator (20 × 20 cm, 250 µm thick from Analtech Inc.) are first washed by development with water, then air dried and activated (90° for 30 minutes). Plates are inscribed with vertical channels 15 mm wide and with a horizontal line approximately 2.5 cm from the top to halt migration of the solvent.

The evaporated column eluates are reconstituted with 30 µl of H_2O and applied with either a microsyringe or disposable micropipette 3 cm from the bottom of the plate. In addition to tissue extracts, samples (H_2O) representing column purification "blanks" are also applied.

The mobility of cGMP is determined by chromatographing an amount of cGMP (approximately 0.1 µmole), that is detectable by ultraviolet absorption, in tissue extract, and in water in channels at one end of the plate.

The plates are developed in ascending fashion at room temperature in a covered glass tank with freshly prepared solvent composed of redistilled isopropanol, NH_4OH, H_2O in a ratio of 7:1.5:1.5. The solvent is allowed to migrate to the horizontal line 2.5 cm from the top of the plate (4–5 hours). The plates are then air dried and stored in a desiccated chamber until the operation of recovering the cGMP is carried out. The R_f values for cGMP chromatographed in tissue extract (ca. 0.50) and in water (ca. 0.55) are usually slightly different so that the ultraviolet-detectable cGMP chromatographed in tissue extract represents the more useful "marker." The segment of each channel corresponding to the area to which authentic cyclic GMP migrates in the "marker" channel is vacuumed into a Swinnex-13 filter holder (Millipore Corp. catalog No. SX001300) containing a Teflon Millipore filter (Mitex No. LSWP 01300), which serves to trap the adsorbent. The silica gel is eluted with ethanol by connecting the open end of a Luer stub adapter (15 gauge, No. A1030, Caly-Adams, Inc.), attached to the Swinnex holder, to a 5-ml hypodermic syringe (with the aid of a segment of polyethylene tubing) which has been filled with 2 ml of 95% ethanol. The ethanol solution is allowed

[7] F. Murad, V. Manganiello, and M. Vaughan, *Proc. Nat. Acad. Sci. U.S.* **68**, 736 (1971).

to flow slowly (0.5 ml/min) through the adsorbent trapped by the filter and into a 5-ml conical centrifuge tube. The eluate is then evaporated to dryness under reduced pressure at room temperature, reconstituted with 200 µl of H_2O which is used to thoroughly rinse the walls of the tube and evaporated to dryness again. About 80% of the cyclic nucleotide is recoverable by this elution procedure.

Enzymatic Detection of cGMP

Composition of the reagents for each of the analytical steps described below is shown in the table.

Conversion to 5′-GMP

The evaporated eluates obtained from the silica gel thin-layer chromatographic procedure are reconstituted with 50 µl of the reagent for

COMPOSITION OF THE REAGENTS FOR EACH ANALYTICAL STEP

Enzyme step	Reagent[a]	Reagent volume (µl)	Time (min)	Temperature (°C)
1. cGMP → 5′-GMP	1. Tris · HCl, pH 7.7, 50; $MgCl_2$ 2; EDTA, 0.5; reaction initiated with 0.8 µg of phosphodiesterase	10	30	30
2. 5′-GMP → GDP	2. Tris · HCl, pH 7.7, 222; $MgCl_2$, 12; KCl, 185; ATP, 0.035; creatine phosphate, 1.75; GMP kinase, 200 µg/ml; creatine phosphokinase, 0.25 µg/ml	4	60	37
3. GDP ⇌ GTP	3. Tris · HCl, pH 7.7, 50; $MgCl_2$, 5; P-pyruvate, 8; sodium succinate, 8; CoA, 4.9; pyruvate phosphokinase, 1.25 µg/ml; succinate thiokinase, 360 µg/ml	2	60	37
4. Pyruvate → lactate	4. EDTA, 125; DPNH, 5; lactate dehydrogenase, 0.44 µg/ml	4	15	25
5. DPNH destruction	5. HCl, 5 N	2	1	25
6. Strong alkali treatment	6. NaOH, 8 N	100	30	38
7. DPN^+ detection	7. H_2O	1000		

[a] Concentrations in the reagent are given in terms of millimolarity.

step 1 shown in the table. One aliquot of 10 µl is prepared for counting in the scintillation spectrometer to determine recovery. Three aliquots (10 µl each) are transferred to 7×75 mm (o.d.) glass tubes incubating in an ice bath and 1 µl of purified phosphodiesterase (Boehringer and Sons) previously diluted 1:1 with 50 mM Tris·HCl, pH 7.7 added to only two of the three samples. After incubation at 30° for 30 minutes, the reaction is terminated by placing the tubes in a 90° water bath for 2 minutes. If urine is to be analyzed, 1 µl is diluted to 10 µl with the step 1 reagent. The subsequent steps are identical for the analysis of purified tissue extracts or urine samples.

The phosphodiesterase preparation used should be analyzed for contaminating cGMP or other guanine nucleotides. If the preparation does prove to be a source of contamination, treatment with charcoal should be employed. The process is as follows: activated coconut charcoal [50–200 mesh (Fisher Chemical Co.)] is acid washed by five successive sedimentations in 0.1 N HCl, followed by repeated washes with water until a pH of 3 to 4 is achieved. The charcoal is then neutralized by resuspending it in about 10 volumes of 25 mM Tris·HCl, pH 7.5, at least three times followed by a final wash with water before it is dried at 90°. To each 100 µl of enzyme solution (resuspended in Tris buffer as described above) 10 mg of the treated charcoal is added and the mixture is swirled in a 7×75 mm o.d. tube intermittently about 15 minutes at 0°. The charcoal is then sedimented by centrifugation for 5 minutes at 5000 g. This treatment of the enzyme effectively removes approximately 90% of the contaminating nucleotides, but has little or no effect on the enzyme activity.

The possibility of error due to 5'-nucleotidase in the phosphodiesterase preparation should also be evaluated routinely by carrying 5'-GMP standards through the entire procedure both with and without exposure to the enzyme. Including these 5'-GMP standards also provides a means of determining whether the conversion of 5'-GMP to GDP is complete (step 2). Step 1 reagent alone ("buffer blank") and three concentrations of standardized cGMP added to the step 1 reagent ("buffer standards") are always included.

Eluates obtained from channels on the silica gel representing blanks are also analyzed. Extra tubes containing the step 1 reagent alone should be carried through the procedure so that GTP standards can be added at step 3 to aid in determining the efficiency of the cycling and conversion systems. Several internal standards, which are added to tissue extracts after thin-layer chromatography or to urine after boiling, should also be included.

Conversion to GDP

To each tube 4 μl of the reagent for the second step, described in the table, is added and the reaction allowed to proceed for 1 hour at 37°. At the end of the incubation, the rack of tubes is returned to the ice bath for the addition of the next reagent.

The ATP-GMP phosphotransferase purified by the procedure of Meich and Parks[8] or obtainable from Boehringer and Sons has never been found to be a source of guanine nucleotide contamination. However, when used in excess, it can be inhibitory to the enzymatic cycling system. Therefore, it is recommended that only the amount of enzyme required to provide a complete conversion of GDP in a reasonable period of time be used. This can be determined by carrying out a time course with various concentrations of the enzyme under the same conditions as those described for the assay.

Although the GMP kinase is highly specific for GMP, preparations purified by the procedure cited may have some detectable activity with 5'-AMP, which is less than 1% of the rate with 5'-GMP. Using creatine phosphokinase to recycle ADP to ATP (and to convert any other nucleoside diphosphates to triphosphates) eliminates the possibility of pyruvate formation that may arise from ADP (generated from AMP contaminating the ATP) or other nucleoside diphosphates other than GDP.

The major source of contamination in this step is guanine nucleotide reactive material present in commercial preparations of ATP to the extent of about 0.05 to 0.5%. Conventional ion-exchange chromatographic purification of ATP does not remove the interfering substance(s). Cha and Cha[9] have described an effective procedure for purifying ATP (and CoA, see step 3) free from this contaminant. Purification involves treating the ATP with succinate thiokinase and nucleoside diphosphatase, which leads to the degradation of both GTP and GDP to 5'-GMP. The 5'-GMP is then separated from ATP by DEAE-cellulose chromatography.[9] However, it is also possible to obtain ATP suitable for the assay procedure by analyzing various commercially available preparations for their guanine nucleotide content (using the analytical procedure described here) and selecting a preparation that is contaminated by no more than 0.05%. By carefully selecting the ATP in this manner and reducing the ATP concentration in the reagent to as low as 5 μM, the blank contribution by this component is tolerable (i.e., ca. 1.5×10^{-8} M with respect to cGMP standards).

[8] R. P. Meich and R. E. Parks, Jr., *J. Biol. Chem.* **240**, 351 (1965).
[9] S. Cha and C.-J. M. Cha, *Anal. Biochem.* **33**, 174 (1970).

Enzymatic Cycling

After the tubes have been thoroughly chilled in an ice bath, 2 µl of ice-cold cycling reagent (step 3 of table) are added to each tube and the cycling reaction carried out for 60 minutes at 37°. At the end of the incubation, the tubes are transferred to an ice-water bath and chilled for 10 minutes.

If the succinate thiokinase or pyruvate kinase is stored in saturated $(NH_4)_2SO_4$, it must be resuspended in Tris·HCl buffer. This can easily be accomplished by centrifuging an aliquot of the enzyme at 12,000 g for 15 minutes, removing the $(NH_4)_2SO_4$ supernatant solution with a micropipette, and resuspending the protein pellet with a volume of 50 mM Tris·HCl, pH 7.7, equivalent to the volume of $(NH_4)_2SO_4$ solution removed.

Coenzyme A (CoA), like ATP, is contaminated by "guanine nucleotide-like" compounds but the contaminants can be removed by ion-exchange chromatography. The procedure which was originally described by Cha et al.[10] is as follows: reduced CoA (approximately 100 µmoles) is oxidized with Lugol's solution (i.e., by titration to a faint orange color) and applied to a DEAE-cellulose column in the bicarbonate form (2.5 × 30 cm). After washing with several bed volumes of 50 mM triethylammonium bicarbonate, the oxidized CoA is eluted with a linear gradient (0.05 to 0.5 M) of this salt. The fractions containing the oxidized CoA are pooled and lyophilized to remove the triethylammonium bicarbonate, which is volatile. Dithiothreitol in 10-fold excess of the oxidized CoA (determined spectrophotometrically at 260 nm, and assuming an A_m of 15.4 × 10³) is used to reduce the CoA which can then be stored at −80° for further use.

The process of balancing the concentrations of enzymes in the cycling system can be carried out by continuous monitoring of the cycling rate in a system containing all the components described in step 3 of the table in the presence of lactate dehydrogenase (0.5 µg/ml) and DPNH (5 µM in a fluorometer or 0.1 mM in a spectrophotometer).

The equations derived by the Chas[9,11] for the kinetics of enzymatic cycling systems are very helpful for establishing guidelines to determine the optimal ratio of activities of the enzymes. The basic equation used[9] to approximate the cycling rate is as follows:

$$\text{cycles/min} = \frac{V}{S_0} = \left(\frac{K_m + S_0}{V_{max}} + \frac{K'_m + S_0}{V'_{max}}\right)^{-1}$$

[10] S. Cha, C.-J. M. Cha, and R. E. Parks, Jr., *J. Biol. Chem.* **242**, 2577 (1967).
[11] S. Cha and C.-J. M. Cha, *Mol. Pharmacol.* **1**, 178 (1965).

where V is the reaction velocity in terms of the rate of formation of pyruvate (μmoles/ml min); S_0 is the sum of the concentrations of the cycling substrate (i.e., GDP and GTP); K_m and K'_m are the apparent values of K_m (μmoles/ml) determined for GDP (ca. 1200 μM) with pyruvate kinase and GTP (ca. 10 μm) with succinate thiokinase, respectively; V_{max} and V'_{max} are the apparent maximal velocities (μmoles/ml per minute) for the pyruvate kinase (with saturating GDP) and succinate thiokinase (with saturating GTP), respectively.

This cycling system can easily be adjusted to provide a 2000- to 4000-fold amplification at 37° in 1 hour. The rate of pyruvate generation is linear with time, until the concentration of CoA becomes rate limiting (i.e., at about 0.10 mM), and proportional to the concentration of GTP (GDP) over a range from below 10^{-9} to 10^{-6} M.

Conversion of Pyruvate to Lactate

The tubes are thoroughly chilled after the cycling reaction, then 4 μl of a solution containing EDTA, lactate dehydrogenase, and DPNH (step 4 of the table) are added to each tube to stop the cycling reactions (i.e., chelation of Mg^{2+}) and to reduce the pyruvate present to lactate. The lactate dehydrogenase reaction is allowed to proceed for 15 minutes at room temperature, then 2 μl of 5 N HCl are added to destroy the unreacted DPNH. This process is complete within 1 minute at room temperature.

Detection of DPN^+

The DPN^+ present is converted to a highly fluorescent product by the addition of 100 μl of 8 N NaOH, followed by an incubation period of 30 minutes at 38°. An aliquot of each reaction (predetermined by testing representative samples) is transferred to a fluorometer tube (10 × 75 mm) and thoroughly mixed with 1 ml of distilled water. The fluorescence is then determined in a fluorometer (Farrand, Model A-2 or A-3 with primary filter No. 5860 and secondary filters Nos. 3387, 5563, and 4303).

Because most commercial preparations of DPNH are heavily contaminated with DPN^+ (10–15%) it is essential that the stock DPNH solution be made up in dilute alkali (50 mM bicarbonate buffer, pH 10) and heated before each use at 60° for about 20 minutes to destroy the oxidized pyridine nucleotide present. The concentration of DPNH in the stock solution should be no greater than 10 mM. For further details of the stability of pyridine nucleotides the work of Lowry and his co-workers should be consulted.[12,13] The strong alkali treatment of DPN⁺ provides almost a 10-fold increase in the intensity of fluorescence relative to the native fluorescence of a comparable concentration of the reduced form. For this reason only 20–50 μl of the strong alkali-treated reaction is usually required for the final reading in the fluorometer. In most commercial fluorometers the proportionality between DPNH concentration and fluorescence is linear up to, but not beyond, 2×10^{-5} M DPNH (or its fluorescent equivalent).

Interpretation of Results

The concentration of DPN⁺ in the aliquot of each sample carried through the entire procedure without phosphodiesterase treatment [(—) PD blank] is substracted from the average concentration present in the two aliquots exposed to this enzyme. Ordinarily there is no significant variation among the (—) PD values from the different samples. If this blank is especially high for a given sample it is advisable to repeat the analysis on that sample. Ordinarily the (—) PD blank for the system (i.e., reagent alone) should be equivalent to a concentration of 1 to 4×10^{-8} M cGMP (all cGMP concentrations are based on the level existing at the step 1 reaction volume).

The (—) PD blanks from channels on thin-layer chromatography plates representing blank neutralized acid solution or blank column eluate do not vary appreciably from plate to plate, nor are they usually significantly different from the (—) PD blank values obtained from the tissue samples. If significant and consistent variations in this value from different types of samples occur (i.e., tissue samples, buffer blank, thin-layer or column chromatography, etc.) it is an indication of contamination from some component associated with that operation.

A net increment in the DPN⁺ generated as a result of phosphodiesterase addition to buffer blanks [i.e., (+) PD blank] represents possible cGMP contamination of the components of the reagent for step 1 or contamination of the phosphodiesterase itself with cGMP. Analyzing heat-denatured aliquots of the phosphodiesterase with and without the active enzyme should help determine whether the phosphodiesterase contains such a contaminant. On rare occasions, thin-layer silica gel plates may be contaminated with cGMP which would be apparent from high (+) PD values obtained from samples representing the plate blank.

The values (fluorescence units) obtained from cGMP standards, after

[12] O. H. Lowry, J. V. Passonneau, D. W. Schulz, and M. W. Rock, *J. Biol. Chem.* **236**, 2746 (1961).
[13] O. H. Lowry, N. R. Roberts, and J. I. Kapphahn, *J. Biol. Chem.* **244**, 1047 (1957).

subtracting the appropriate blanks, are proportional to the cyclic nucleotide concentration and are conveniently used to construct a linear standard curve to determine the concentrations in the tissue extracts. However, because of the relatively large amount of tissue represented by the reconstituted eluates in the analytical procedure, the curves constructed from (tissue) internal standards sometimes differ (in slope) from those based on buffer standards. It is essential, therefore, that a number of internal standards be included so that correct, absolute values may be calculated.

The tissue concentration can be calculated from the internal standard curve by correcting for recovery and for concentrating cGMP during the purification process. A typical calculation could be represented as follows:

$$\frac{\Delta(10^{-7}\,M)(4.8)(0.104)(1.9)}{50} = \frac{\text{moles} \times 10^{-7}\,\text{cGMP}}{\text{kg tissue (wet weight)}}$$

where 50 is the number of fluorescent units equivalent to a $10^{-7}\,M$ solution of cGMP (at the step 1 reaction volume); Δ is the units of fluorescence obtained for the unknown tissue sample; 4.8 represents the tissue dilution of 100 mg of frozen powdered tissue extracted with 400 μl of 10% TCA (i.e., assuming that only 80% of the tissue weight is water and only this portion contributes to the total final volume); 0.104 is the correction for the concentration resulting from reconstituting the extract after purification in a total volume of 50 μl; and 1.9 is the correction for the loss of [^{3}H]cGMP (52.6% recovery).

The variation encountered when a single sample is subjected to repeated analysis is less than 15%, and the variation between duplicate samples in a single experiment is no greater than 5%. The specificity of this system for cGMP has been tested repeatedly. Concentrations of cAMP or 5'-AMP as great as 0.1 mM are not detectable. Cytidine and uridine nucleotide polyphosphates are also not reactive. Because of the specificity of the GMP-kinase, cIMP, and 5'-IMP would not interfere and inosine di- or triphosphate, which could serve as a substrate for the cycling system, would be removed from tissue extracts by the chromatographic purification procedures should these nucleotides occur in tissue.

[11] The Measurement of Cyclic GMP with *Escherichia coli* Elongation Factor Tu

By ROBERT F. O'DEA, JAMES W. BODLEY, LILLIAN LIN, MARI K. HADDOX, and NELSON D. GOLDBERG

Elongation factor Tu (EF-Tu) is a soluble factor from *E. coli* which promotes the binding of aminoacyl-tRNA to ribosomes in the process of peptide chain elongation.[1] EF-Tu is easily obtainable in a reasonably good state of purity, is relatively stable, exhibits a high degree of specificity for GDP (and GTP) and a dissociation constant for GDP in the range of 10^{-9} M (10^{-7} M for GTP). These properties of EF-Tu have been used to advantage to devise a procedure for the quantitation of endogenous tissue cGMP. The procedure can also easily be modified to measure 5'-GMP, GDP, or GTP in tissue extracts. The principle of the assay for the detection of cGMP is as follows: cGMP is converted to GDP with the aid of cyclic nucleotide phosphodiesterase and ATP-GMP phosphotransferase. The GDP generated and [α-^{32}P]GDP compete for complex formation with EF-Tu. The radioactive nucleotide–protein complexes retained by cellulose acetate filters are quantitated by liquid scintillation spectrometry.

The specificity of the assay derives from three steps in the procedure: (a) the chromatographic purification of tissue extracts, (b) the selective conversion of GMP catalyzed by ATP-GMP phosphotransferase, and (c) the selective binding by EF-Tu of GDP (and GTP).

Preparation of Extracts

Preparation of EF-Tu

Highly purified EF-Tu can be prepared from *Escherichia coli* (B cells obtainable from Grain Processing Corporation, Muscatine, Iowa) by the method of Miller and Weissbach[1] or Arai et al.[2] The procedure involves preparation of a high speed supernatant fraction from broken cells followed by partial purification of an EF-Tu–GDP complex by ammonium sulfate fractionation, chromatography on DEAE Sephadex A-50, and Sephadex G-100. The pooled fractions containing EF-Tu–GDP can be stored at −20°C in a solution containing 10 mM Tris·HCl, pH 7.4, 10

---

[1] D. L. Miller and H. Weissbach, *Arch. Biochem. Biophys.* **141,** 26 (1970).
[2] K.-I. Arai, M. Kawakita, and Y. Kaziro, *J. Biol. Chem.* **247,** 7029 (1972).

mM magnesium acetate and 50% glycerol. According to the procedures cited EF-Tu–GDP can be prepared routinely with an activity of over 8000 units/mg where 1 unit is defined as the exchange of 1 pmole of GDP in 10 minutes at 37°.

The GDP in complex with the purified EF-Tu is removed according to the method of Weissbach et al.[3] by prolonged dialysis against a Mg^{2+}-free buffer containing 10 mM Tris·HCl, pH 7.4, 10 mM NH_4Cl, 5 mM DTT, 1 mM EDTA, and 50% glycerol. The rate of GDP removal by dialysis can be monitored by first incubating the purified EF-Tu–GDP complex with 40 pmoles of radioactive GDP (per 1.5 nmoles of EF-Tu–GDP complex) in the reagent just described for 10 minutes at 37° before commencing the dialysis and removing small aliquots from the solution undergoing dialysis to determine the amount of radioactive GDP remaining as the process of GDP removal proceeds. To facilitate sampling of the dialyzing solution the open end of the dialysis pack can be secured to the end of a piece of glass tubing through which a micropipette may be inserted. The glass tubing is suspended from a rubber stopper inserted in the top of a flask.

The dialysis is conducted at 0–2° against 250 ml of the reagent described. The latter is changed daily. The $t_{1/2}$ for the removal of GDP at 0–2° is 29–30 hours. The procedure can be accelerated by conducting the dialysis at 37° ($t_{1/2} = 1.5$ hours); however, at the higher temperature a larger percentage of EF-Tu activity is lost (i.e., 70–90% at 37° vs <40% when performed at 0–2°). The EF-Tu freed of over 95% of GDP can be stored at 0–2° in the dialysis buffer fortified with 10 mM magnesium acetate. Under these conditions there is negligible loss of activity after storage for as long as a month.

Preparation of [α-^{32}P]*GDP*

[α-^{32}P]GDP can be obtained from various commercial sources but because [α-^{32}P]GTP is available at much higher specific activities and at a lower cost it is preferable to use it as the starting material from which to generate [α-^{32}P]GDP. Any enzyme capable of converting GDP from GTP can be used for this purpose (i.e., phosphoenolpyruvate carboxykinase, hexokinase, etc.). The procedure we have employed is as follows: 10 nmoles of [α-^{32}P]GTP is incubated for 30 minutes at 37° in a reaction mixture containing 10 mM NH_4Cl, 1 mM DTT, 4 mM β-mercaptoethanol, 150 μg of high salt-washed ribosomes, and 40 pmoles of soluble trans-

---

[3] H. Weissbach, D. L. Miller, and J. Hachmann, *Arch. Biochem. Biophys.* **137**, 262 (1970).

fer factor G (G factor) in a total volume of 125 μl. The G factor and ribosomes can be prepared according to methods described previously by Bodley et al.[4] The reaction is terminated by the addition of 20 μl of 30% formic acid, and centrifuged for 2 minutes at 8000 g to remove the denatured protein precipitate. The supernatant is neutralized by the addition of an appropriate volume of concentrated NH_4OH. The rate of conversion of GTP to GDP can be determined by monitoring the [α-^{32}P]GDP generated. The latter can be isolated and quantitated by chromatographing the neutralized supernatant on polyethyleneimine cellulose thin-layer plates (Beckman Instruments) employing 0.75 M phosphate buffer as the eluent ($R_{f_{GDP}}$ = 0.33). The conversion of tri- to diphosphate under these conditions varies between 85 and 95%.

cGMP, used in the preparation of standard solutions (Sigma Chemical Company or Boehringer-Mannheim Corp), should be purified free of a 5'-GMP reactive material. One procedure that can be used is described below for the purification of cGMP from tissue extracts. The purified cGMP can be standardized fluorometrically according to the procedure outlined by Goldberg et al.[5] Standard GDP (obtained from Sigma Chemical Company) solutions can be calibrated fluorometrically by a similar procedure. All nucleotides should be stored (preferably at −80°) as the neutral aqueous solutions.

Extraction of Tissues

cGMP is efficiently extracted from tissue samples with either trichloroacetic (TCA) or perchloric acid (PCA) by procedures which are discussed in Goldberg and Haddox (this volume [10]).

Determination of Recovery

In order to monitor the recovery of cGMP during both the extraction and purification procedures approximately 0.15 pmole of purified [^{3}H]cGMP (specific activity 24 Ci/mmole) (ca. 2000 cpm) should be added to the acid extracting media prior to homogenization or following homogenization of the sample.

Purification of Tissue Extracts

To remove potentially interfering guanine nucleotide intermediates such as 5'-GMP, GDP or GTP, TCA (after ether extraction) or PCA

[4] J. W. Bodley, F. J. Zieve, L. Lin, and S. T. Zieve, J. Biol. Chem. **245**, 5656 (1970).
[5] N. D. Goldberg, S. B. Dietz, and A. G. O'Toole, J. Biol. Chem. **244**, 4458 (1969).

(after neutralization with $KHCO_3$) extracted tissue samples of volumes up to 5ml are applied to dry neutral alumina (Baker Chemical) columns (4 × 5 mm) and eluted with 10 ml of 25 mM Tris formate, pH 7.5. The entire breakthrough volume from the extract and the Tris formate eluate are allowed to flow directly through a Dowex 1-X8 (100–200 mesh) formate column (5 × 25 mm). The Dowex 1-formate is then washed with 10 ml of H_2O, then with 10 ml of 1 N formic acid to remove cAMP, and finally with 10 ml of 4 N formic acid to remove cGMP. The latter is a modification of procedures previously described by White and Zenser[6] and Murad et al.[7] Before purification of PCA extracts, the samples are acidified by making them 50 mM with respect to formate. After partial purification on the alumina and Dowex 1-formate columns, the formic acid eluate containing the cGMP is reduced to dryness on a Buchler Evap-o-Mix. Each sample is then reconstituted with 200 μl of 50 mM Tris·HCl buffer, pH 7.4 containing 2 mM $MgCl_2$ and 0.25 mM EDTA.

Treatment with 5'-Nucleotidase

To guard against any residual traces of 5'-GMP that may be present, the samples are incubated for 10 minutes at 37° with 7–10 μg of snake venom 5'-nucleotidase (*Vipera russelli*, Sigma Chemical Company). The enzyme activity is destroyed by immersing the tubes in a 90° water bath for 3 minutes.

Conversion of cGMP to GDP

cGMP is first converted to 5'-GMP in 20 μl of either the 5'-nucleotidase-treated samples or standard solutions of cGMP prepared in an identical reagent by the addition of 2 μl (2 μg) of purified beef heart cyclic nucleotide phosphodiesterase (Boehringer Mannheim Corporation) and incubating the reaction mixture at 30° for 30 minutes. Phosphodiesterase activity is destroyed by immersion of the samples in a 90° water bath for 3 minutes. Blank reactions consisting of the complete reaction mixture (tissue extract or cGMP standards) minus phosphodiesterase are carried through the incubation, heat denaturation step, and all subsequent procedures. It is also desirable to develop a set of blank reactions by adding cGMP to reagent containing phosphodiesterase previously subjected to the heat denaturation procedure. The latter is recom-

---

[6] A. A. White and T. W. Zenser, *Anal. Biochem.* **41**, 372 (1971).
[7] F. Murad, V. Manganiello, and M. Vaughan, *Proc. Nat. Acad. Sci. U.S.* **68**, 736 (1971).

mended because some preparations of phosphodiesterase can enhance the binding of GDP to EF-Tu as much as 10%. After the heat inactivation of phosphodiesterase, the tubes are cooled in an ice-water bath, and 5 µl of a reagent containing 125 mM Tris·HCl, pH 7.4, 12 mM MgCl$_2$, 0.025 mM ATP, and 1 µg of ATP-GMP phosphotransferase (Boehringer Mannheim Corporation) is added, and an incubation is conducted for 30 minutes at 37° to effect conversion of 5'-GMP to GDP. The two enzymatic conversions can be carried out in a single analytical step at 30° or 37° followed by heat denaturation. Whether the enzyme conversions are performed sequentially or simultaneously, the tubes are chilled in an ice-water bath prior to the initiation of the competitive binding reaction.

GDP Binding Reaction with EF-Tu

The unlabeled GDP generated by the previous step is determined by competition with [α-^{32}P]GDP for binding to EF-Tu in the following manner. Twenty-five (25) µl of a reagent containing 20 mM Tris·HCl, pH 7.4, 20 mM magnesium acetate, 20 mM NH$_4$Cl, 2 mM dithiothreitol, 0.1% gelatin, 0.4–0.6 pmole [α-^{32}P]GDP and an amount of dialyzed EF-Tu that will bind 0.2–0.4 pmole of the diphosphate are added to each tube and allowed to incubate 10 minutes at 37°. Under these conditions the binding and/or exchange of GDP and [α-^{32}P]GDP to EF-Tu is complete within 2 minutes. The reaction mixture is then diluted with 2.5 ml of 10 mM Tris·HCl, pH 7.4, containing 10 mM magnesium acetate and immediately poured over a Millipore filter (type HAWP 25 mm, 0.45 µm) previously soaked in the same buffer. The reaction tube is rinsed twice with 2.5 ml of the buffer, and the washings are passed through the filter. The filter is then washed once with a 5-ml aliquot of the buffer. The entire procedure from dilution of the sample to removal of the filter requires no more than 30 seconds. The filters are dried under an infrared lamp then placed in 2 ml of a solution containing 8 g of 2-(4'-butylphenyl)-5-(4'-biphenyl)-1,3,4-oxadiazole, 0.5 g of 2-(4'-biphenyl)-6-phenylbenzoxazole, and toluene (1 liter) (Beckman TLA fluoralloy mixture). The radioactivity retained on the filters can be determined in any liquid scintillation spectrometer.

Sensitivity

A linear standard curve from which tissue cGMP (or GDP) concentrations can be determined can be constructed on log paper by plotting the detectable radioactivity retained on the filters (counts per minute) vs the amount of standard cGMP (or GDP) added. Such a curve is linear from 10^{-14} to 10^{-11} mole of the cyclic nucleotide. A major determinant of the maximum sensitivity of the procedure to detect cGMP (i.e., GDP) is the specific activity of the [α-^{32}P]GDP used in the competitive binding reaction. The level of sensitivity when using [α-^{32}P]GDP with specific activity of 15 Ci/mmole is 0.1 pmole of cGMP (or GDP). However, with [α-^{32}P]GDP with specific activity greater than 100 Ci/mmole, which is now commercially available as [α-^{32}P]GTP, a level of 0.02 pmole of cGMP is detectable. The variation among samples is less than 5%.

[12] Assay of Cyclic GMP by Activation of Cyclic GMP-Dependent Protein Kinase

By J. F. Kuo *and* Paul Greengard

Principle

The assay method is based upon the ability of low concentrations of cyclic GMP (cGMP) to activate cyclic GMP-dependent protein kinases, which catalyze the phosphorylation of substrate proteins (e.g., histone, protamine) by ATP. The reaction is depicted as follows:

$$\text{Histone} + [\gamma\text{-}^{32}\text{P}]\text{ATP} \xrightarrow[\text{cGMP, Mg}^{2+}]{\text{cGMP-dependent protein kinase}} [^{32}\text{P}]\text{histone} + \text{ADP}$$

Under the standard assay conditions, the extent of histone phosphorylation is directly proportional to the amount of cGMP present in the incubation tube, thus providing a direct, sensitive and specific method for the measurement of the cyclic nucleotide in tissue and body fluid samples.

Assay of cGMP

Preparation of Reagents

Histone Mixture. Dissolve 10 mg of histone mixture (calf thymus, Schwarz/Mann) in 20 ml of 0.01 N HCl. Adjust the pH of the solution to 6.0 with 5 N NaOH, add water to make a final volume of 50 ml. The resultant solution should be clear.

[γ-^{32}P]ATP. In our laboratory, this is prepared from ortho-^{32}P according to the method of Post and Sen.[1] It can be directly obtained, however, from commercial sources (e.g., ICN, New England Nuclear, etc.).

Trichloroacetic Acid-Sodium Tungstate-H_2SO_4 Precipitating Solution. Slowly dissolve 60 g of NaOH pellets, with stirring, in 5 liters of 5% trichloroacetic acid (the pH of the resultant solution should be about 2) followed by dissolving 12.5 g of $NaWO_4 \cdot 2H_2O$. Readjust the pH to 2 with 100% trichloroacetic acid, and finally add 10 ml of concentrated (36 N) H_2SO_4. The resultant solution remains clear for at least several months.

Cyclic GMP-Dependent Protein Kinase. The enzyme is prepared from either the lobster tail muscle through the step of DEAE-cellulose column chromatography[2] or the silkmoth larval fat body through the step of ammonium sulfate precipitation,[3] as described in this volume.[4]

Preparation of Biological Materials for Cyclic GMP Assay

Tissues or cells (about 5 mg to 2 g wet weight) are homogenized in glass homogenizers (or alternatively with a sonicator) in 2–3 volumes of ice-cold 5% trichloroacetic acid. If the cyclic nucleotide levels are to be measured in body fluids, one-ninth volume of 50% trichloroacetic acid is added with mixing. A 5-μl aliquot of [^{3}H]cGMP (about 0.5 pmole; 10^3 cpm) is added to the homogenate for the purpose of determining the recovery of tissue cGMP. The precipitate is removed by centrifugation, and the supernatant solution (usually 0.5–1.5 ml) is charged onto an 0.5 × 2.0 cm column of AG1-X8 (200–400 mesh, formate form, BioRad), the resin having been previously washed with water. The column is then washed with 6 ml of 0.5 N formic acid, and the eluate, which contains cAMP, is discarded. (If cAMP is also measured in the same tissue or cell samples, this eluate may be saved for that purpose.) cGMP is then eluted from the column with 3 ml of 4 N formic acid, and this eluate is lyophilized. The dried material is taken up in 0.2 ml of water and quantitatively spotted on a thin-layer chromatographic plate (SilicAT TLC-7GF, Mallinckrodt). The plate is developed using a solvent system consisting of isopropyl alcohol: H_2O:2.9% NH_4OH (7:2:1, v/v). cGMP is then eluted with 1-ml aliquots of absolute ethanol from the area on the plate corresponding to the spot of authentic compound. Overall recovery of tissue cGMP is between 70 and 80%. Aliquots (usually 0.3–1.0

---

[1] R. L. Post and A. K. Sen, this series, Vol. 10, p. 773.
[2] J. F. Kuo and P. Greengard, *J. Biol. Chem.* **245**, 2493 (1970).
[3] J. F. Kuo, G. R. Wyatt, and P. Greengard, *J. Biol. Chem.* **246**, 7159 (1971).
[4] J. F. Kuo and P. Greengard, this volume [47].

ml) of the alcoholic eluates of cGMP are dried at 65°, *in vacuo*, in the test tubes (1.3 × 10.0 cm) in which the cyclic nucleotide is to be assayed.

Standard Assay Procedure

For the assay of cGMP, the following components are added to the incubation tubes (some of which contain lyophilized sample), in the order given: sodium acetate buffer (0.1 M), pH 6.0, 0.100 ml; magnesium acetate (2 μmoles), 0.020 ml; cGMP standard or sample (up to 20 pmoles), 0.020 ml; histone mixture (40 μg), 0.020 ml; water or any other additive,

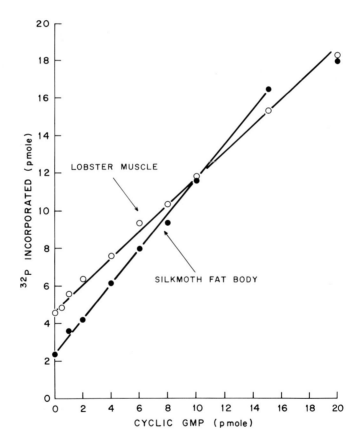

FIG. 1. Standard curves for the measurement of cyclic GMP with cyclic GMP-dependent protein kinase prepared from either lobster tail muscle or cecropia silkmoth larval fat body. One picomole of ^{32}P incorporated represents about 2000 cpm. Modified from J. F. Kuo, T. P. Lee, P. L. Reyes, K. G. Walton, T. E. Donnelly, Jr., and P. Greengard, *J. Biol. Chem.* **247,** 16 (1972).

0.014 ml; cGMP-dependent protein kinase (10–100 units), 0.020 ml; [γ-^{32}P]ATP (1 nmole, containing about 1–2 × 10^6 cpm), 0.006 ml.

The tubes are kept in ice during the addition of the enzyme. One unit of enzyme is defined as that amount of activity which transfers 1 pmole of ^{32}P from [γ-^{32}P]ATP to histone in 5 minutes at 30° under the assay conditions. The reaction is commenced by the addition of radioactive ATP, which is conveniently delivered from a 0.1-ml Hamilton microsyringe (with fused needle) attached to a repeating dispenser. The tubes are incubated for 5 minutes at 30°, with shaking, and the reaction is terminated by the addition of 2 ml of ice-cold trichloroacetic acid–tungstate–H$_2$SO$_4$ precipitating solution. Two-tenths milliliter of 0.6% bovine serum albumin is added as a carrier protein, and the contents of the tubes are mixed by vigorous addition of another 2 ml of the precipitating solution. The mixture is centrifuged, and the supernatant solution is removed by aspiration. The precipitate is dissolved in 0.1 ml of 1 N NaOH, and 2 ml of the precipitating solution is added. The procedure of centrifuging,

TABLE I
Levels of Cyclic GMP (cGMP) and Cyclic AMP (cAMP) in Various Biological Materials[a,b]

Sample	cGMP level (pmoles/mg protein)	cAMP level (pmoles/mg protein)	Ratio cGMP to cAMP
Rat brain (whole)	0.23	11.37	0.020
Rat cerebrum	0.20	12.66	0.016
Rat cerebellum	4.29	6.25	0.686
Rat heart (whole)	0.29	2.43	0.119
Rat lung	5.81	8.39	0.692
Rat liver	0.14	2.70	0.052
Rat skeletal muscle	0.18	3.48	0.052
Rat adipose tissue	0.29	4.35	0.066
Calf atrium	0.38	7.76	0.049
Calf ventricle	0.18	13.13	0.014
Silkmoth fat body	1.11	2.66	0.417
Lobster tail muscle	0.26	1.83	0.142
Human urine	935.5[c]	19100.0[c]	0.049

[a] Taken from J. F. Kuo, T. P. Lee, P. L. Reyes, K. G. Walton, T. E. Donnelly Jr., and P. Greengard, *J. Biol. Chem.* **247**, 16 (1972).

[b] All data represent means of triplicate assays and were obtained from tissue samples of individual rats except for adipose tissue, cerebrum, and cerebellum, for which tissues were pooled from four rats. A single sample of human urine was collected from an adult male, weighing 100 kg, in the morning.

[c] Values are expressed as picomoles per milliliter.

TABLE II
Regulation by Various Agents of Cyclic GMP (cGMP) and Cyclic AMP (cAMP) Levels in Tissues[a]

Tissue and incubation condition	cGMP (pmoles/mg protein)	cAMP (pmoles/mg protein)
Rabbit cerebral cortex slices		
Experiment 1 (15 min incubation)[b]		
Control	0.68	20.0
Histamine, 1 mM	1.12	250.5
Experiment 2 (3 min incubation)[c]		
Control	0.89 ± 0.18	
Acetylcholine, 1 μM	1.77 ± 0.54	
Bethanechol, 1 μM	1.67 ± 0.65	
Tetramethylammonium, 100 μM	1.06 ± 0.17	
Experiment 3 (3 min incubation)[c]		
Control	0.78 ± 0.16	
Bethanechol, 1 μM	2.26 ± 0.20	
Atropine, 1 μM	0.99 ± 0.14	
Bethanechol, 1 μM, + atropine, 1 μM	0.87 ± 0.16	
Hexamethonium, 100 μM	0.85 ± 0.08	
Bethanechol, 1 μM, + hexamethonium, 100 μM	1.77 ± 0.31	
Rat heart ventricular slices		
Experiment 1 (1 min incubation)[b]		
Control	0.32 ± 0.08	4.5 ± 0.4
Acetylcholine, 0.3 μM	3.15 ± 0.12	4.3 ± 0.2
Isoproterenol, 0.3 μM	0.29 ± 0.01	12.8 ± 0.9
Glucagon, 1 μM	0.32 ± 0.02	8.6 ± 0.5
Acetylcholine, 0.3 μM, + isoproterenol, 0.3 μM	1.72 ± 0.53	7.5 ± 0.8
Acetylcholine, 0.3 μM, + glucagon, 1 μM	1.62 ± 0.45	5.9 ± 0.2
Experiment 2 (0.5 min incubation)[c]		
Control	0.60 ± 0.09	
Acetylcholine, 1 μM	1.86 ± 0.54	
Atropine, 1 μM	0.78 ± 0.03	
Acetylcholine, 1 μM, + atropine, 1 μM	0.72 ± 0.15	
Tetramethylammonium, 100 μM	0.67 ± 0.15	
Guinea pig ileum slices (0.5 min incubation)[c]		
Control	1.03 ± 0.03	
Bethanechol, 1 μM	5.37 ± 0.51	
Atropine, 1 μM	1.10 ± 0.06	
Bethanechol, 1 μM, + atropine, 1 μM	1.08 ± 0.18	
Hexamethonium, 10 μM	1.05 ± 0.11	
Bethanechol, 1 μM, + hexamethonium, 10 μM	6.05 ± 0.50	
Tetramethylammonium, 1 μM	1.34 ± 0.12	
100 μM	1.40 ± 0.14	

[a] Data presented are either the means of assays on duplicate incubations or means ± SE of assays on triplicate incubations.

[b] Taken from J. F. Kuo, T. P. Lee, P. L. Reyes, K. G. Walton, T. E. Donnelly, Jr., and P. Greengard, *J. Biol. Chem.* **247**, 16 (1972).

[c] Taken from T. P. Lee, J. F. Kuo, and P. Greengard, *Proc. Nat. Acad. Sci. U.S.* **69**, 3287 (1972).

removing the supernatant, dissolving the precipitate in alkali, the reprecipitating the protein is repeated once more. The protein is finally collected by centrifugation and dissolved in 0.1 ml of 1 N NaOH, and the radioactivity is counted in 6 ml of scintillation fluid (made by dissolving 8 g of Omnifluor, New England Nuclear, in 2 liters of toluene and 2 liters of ethylene glycol monoethyl ether). The amount of cGMP present in the sample is determined from a standard curve obtained by assaying known quantities of cGMP.

Figure 1 shows a typical standard curve for the cGMP assay. Some results of cGMP assays, together with cAMP values obtained from the same tissue samples, with and without treatment with various agents, are presented in Tables I and II.

Concluding Remarks

The limit of sensitivity of the procedure for assaying cGMP with cGMP-dependent protein kinase is about 0.5 pmole. The sensitivity can be improved to about 0.3 pmole of cGMP if the incubation volume is reduced to about 0.1 ml. The preliminary purification of tissue cGMP is found to effectively separate this cyclic nucleotide from cAMP (which is usually present in amounts 5–100 times that of cGMP in most tissues; the ratios are even higher if the tissues have been exposed to certain agents) and remove any substances, such as ATP, ADP, originally present in the crude extracts, which might interfere with the assay for cGMP based upon its ability to activate the enzyme. The assay method described here has been shown to be useful in determining the levels of cGMP in a wide variety of tissues under various experimental conditions.[5,6]

Acknowledgment

This work was supported by Grants HL-13305, NS-08440, and MH-17387 from the United States Public Health Service. One of us (J. F. K.) is Recipient of Research Career Development Award (GM-50165) from the United States Public Health Service. This paper is also Publication No. 1158 of the Division of Basic Health Sciences, Emory University.

---

[5] J. F. Kuo, T. P. Lee, P. L. Reyes, K. G. Walton, T. E. Donnelly, Jr., and P. Greengard, *J. Biol. Chem.* **247**, 16 (1972).
[6] T. P. Lee, J. F. Kuo, and P. Greengard, *Proc. Nat. Acad. Sci. U.S.* **69**, 3287 (1972).

[13] Assay of Cyclic Nucleotides by Radioimmunoassay Methods

By ALTON L. STEINER

Radioimmunoassay is a relatively simple, sensitive, and specific method for measuring cyclic nucleotides in tissues and body fluids. Radioimmunoassays have been developed for four cyclic nucleotides: cyclic 3′,5′-adenosine monophosphate (cAMP); cyclic 3′,5′-guanosine monophosphate (cGMP); cyclic 3′,5-inosine monophosphate (cIMP), and cyclic 3′,5′-uridine monophosphate (cUMP).[1-3] cAMP and cGMP can be measured simultaneously by radioimmunoassay,[4] and at least two commercial sources (Schwarz/Mann BioResearch, Orangeburg, New York and Collaborative Research, Waltham, Massachusetts) have developed radioimmunoassay kits for the measurement of cAMP and/or cGMP. This review will focus on the radioimmunoassays for cAMP and cGMP.

Preparation of Materials for Cyclic Nucleotide Radioimmunoassay

Synthesis of 2′-O-Succinyl Cyclic Nucleotide

Because of the relatively low molecular weight of the cyclic nucleotides, it was unlikely that antibodies could be made against them directly. Consequently, 2′-O-succinyl derivatives of cAMP (ScAMP) and of cGMP (ScGMP) were synthesized so that the cyclic nucleotide could be conjugated to protein. The cyclic nucleotides were succinylated at the 2′-O position with succinic anhydride by a modification of the method of Falbriard et al.[5] for synthesizing monocarboxylic acid derivatives of cAMP, and the free carboxyl group of this derivative was then conjugated to protein. The method of synthesis of 2′-O-succinyl cyclic nucleotide is essentially the same as reported previously.[2] Cailla and Delaage[6] have

---

[1] A. L. Steiner, D. M. Kipnis, R. Utiger, and C. W. Parker, *Proc. Nat. Acad. Sci. U.S.* **64**, 367 (1969).

[2] A. L. Steiner, C. W. Parker, and D. M. Kipnis, in "Role of Cyclic AMP in Cell Function," Advances in Biochemical Psychopharmacology (P. Greengard and E. Costa, eds.), Vol. 3, p. 89. Raven, New York.

[3] A. L. Steiner, C. W. Parker, and D. M. Kipnis, *J. Biol. Chem.* **247**, 1106 (1972).

[4] R. E. Wehmann, L. Blonde, and A. L. Steiner, *Endocrinology* **90**, 330 (1972).

[5] J.-G. Falbriard, T. Posternak, and E. W. Sutherland, *Biochim. Biophys. Acta* **148**, 99 (1967).

[6] H. Cailla and M. Delaage, *Anal. Biochem.* **48**, 62 (1972).

recently described the synthesis of 2'-O- and N^6-succinyl cyclic nucleotide derivatives of cAMP.

2'-O-Succinyl-cAMP (ScAMP)

Morpholino-N,N'-dicyclohexylcarboxyamidine (0.76 mmole) was dissolved in 7.5 ml of hot anhydrous pyridine (dried over solid KOH), and 0.7 mmole of cAMP (free acid) was added slowly over the following 30–60 minutes. After cooling, 10 mmoles of succinic anhydride were added, and the suspension was stirred at room temperature for 18 hours. Unreacted succinic anhydride was hydrolyzed by the addition of 3.75 ml of water, and the reaction mixture was allowed to stand for an additional 2 hours at 4°. Thin-layer chromatography of the reaction mixture on cellulose with butanol–glacial acetic acid–H_2O (12:3:5, v/v) demonstrated 2'-O-succinyl-cAMP which ran ahead (R_f 0.42) of cAMP (R_f 0.30). Pyridine was removed by repeated rotary evaporation at 40° under reduced pressure, and the residue was dissolved in 10 ml of water after the pH was adjusted to 4.5. ScAMP was purified by chromatography on a column (1.5 cm × 44 cm) of Dowex AG 50-W X8 100–200 mesh H^+ form using distilled H_2O as the eluent at a flow rate of 30 ml per hour. Fractions (8 ml) were collected, then monitored by ultraviolet absorption at 258 nm; the peaks were checked by thin-layer chromatography. Succinic acid appeared in the first 50 ml, ScAMP eluted in tubes 30 to 45 and cAMP in tubes 50 to 65. The fractions containing ScAMP were pooled and lyophilized. Thin-layer chromatography of this product indicated >98% ScAMP with maximal absorption at 258 nm. After brief treatment with 0.1 N NaOH, the product reverted quantitatively to cAMP, confirming that the succinyl substitution was exclusively at the 2'-O position.[5]

2'-O-Succinyl-cGMP (ScGMP)

Guanosine nucleotides are very insoluble in anhydrous pyridine, but partial solubilization is achieved by forming the trioctyl ammonium salt. ScGMP was prepared by adding 0.07 mmole of cGMP (free acid) and 3 eq of trioctylamine to 1.5 ml of hot anhydrous pyridine. After cooling. 1 mmole of succinic anhydride was added to the suspension, and the reaction mixture was stirred for 18 hours at room temperature. Then 1.5 ml of water were added, and the reaction was left for 2 hours at 4°. ScGMP was purified by thin-layer chromatography on cellulose with butanol–glacial acetic acid–water (12:3:5,v/v). ScGMP (R_f 0.42) migrated ahead of cGMP (R_f 0.31), absorbed maximally at 255 nm, and on treatment with 0.1 N NaOH completely reverted to cGMP.

Preparation of Antigen

Succinylated cyclic nucleotides were coupled to protein [human serum albumin (HSA)], keyhole limpet hemocyanin, or poly-L-lysine polymers. ScAMP (10 mg) was allowed to react with 20 mg of HSA and 10 mg of 1-ethyl-3-(3-dimethylaminopropyl) carbodiimide-HCl (EDC) in aqueous solution at pH 5.5. After incubation of this mixture at 24° for 16 hours in the dark, the conjugate was dialyzed against phosphosaline buffer (0.01 M sodium phosphate, 0.15 M sodium chloride, pH 7.4) for 48 hours. The dialyzed conjugate ScAMP-albumin exhibited an absorption maximum at 260 nm. On the basis of the spectrum of ScAMP-albumin and unconjugated HSA, and assuming a molar extinction coefficient of 15,000 for ScAMP, the conjugate was estimated to contain an average of 5 to 6 cyclic AMP residues per albumin molecule. ScAMP was also coupled to poly-L-lysine and keyhole limpet hemocyanin using the same method with similar results. Using essentially identical conditions, Weinryb[7] found 4–5 residues of nucleotide per albumin molecule.

Immunization and Bleeding Schedule

Randomly bred rabbits were immunized with 0.25 mg of protein conjugate emulsified in Freund's complete adjuvant (FCA) and injected into each foot pad. Booster injections totaling 0.25–0.30 mg were injected subcutaneously either into the foot pads or the back at intervals of 4–6 weeks, and the animals were bled 10 days later. Serum was separated by centrifugation, diluted 1:100 with 0.05 M acetate buffer (pH 6.2), and stored in small aliquots at −20°.

Weinryb[7] has immunized goats with an initial injection of 0.1 mg of nucleotide-HSA conjugate in FCA into each of four surgically exposed lymph nodes in the posterior cervical and high inguinal regions. They were boosted 6 weeks later with a total of 1 mg of conjugate in FCA subcutaneously in the regions of the initial incisions. Subsequent booster injections were given subcutaneously every 6 weeks, each with a total of 1 mg of conjugate in saline, and the animals were bled 10 days to 2 weeks after the booster injections. High antibody titers were found as early as after the second boost.

[7] I. Weinryb, *in* "Methods in Cyclic Nucleotide Research" (M. Chasin, ed.), p. 29; Vol. 3 of Methods in Molecular Biology. Dekker, New York.

Synthesis of Tyrosine Methyl Ester Derivatives of 2'-O-Succinyl Cyclic Nucleotides: ScAMP-TME and ScGMP-TME

Radioactive derivatives of the cyclic nucleotide of high specific activity were synthesized by tyrosination of the succinylated compounds and subsequent iodination of the tyrosine moiety. The synthesis of these compounds by the mixed carboxylic–carbonic acid reaction using ethyl chloroformate is recommended.[8] This method of synthesis achieves significantly higher yields of ScAMP-TME than synthesis with N,N^1-dicyclohexylcarbodiimide as described in our initial publication.[1] Cailla and Delaage[6] also prefer the mixed carboxylic-carbonic acid reaction and have described the reaction conditions and products in detail.

The synthesis of succinyl cyclic nucleotide tyrosine methyl ester was performed in two steps: (1) One equivalent (5 μmoles) of the succinylated cyclic nucleotide was dissolved in 0.1 ml of dimethylformamide at 0° with three equivalents of trioctylamine. ScGMP remains as a fine suspension, while ScAMP readily dissolves. One equivalent of ethyl chloroformate in dimethylformamide was added, and the reaction was carried out at 0° for 15 minutes. (2) Two equivalents of both tyrosine methyl ester hydrochloride and trioctylamine were added in 0.1 ml of dimethylformamide, and the reaction was continued at room temperature for an additional 2–3 hours with continuous stirring. The tyrosinated product was isolated by thin-layer chromatography on cellulose with butanol–glacial acetic acid–H_2O (12:3:5, v/v). The new nitrosonaphthol-positive derivative (R_f 0.57) migrated ahead of the succinylated cyclic methyl ester hydrochloride (R_f 0.65) and behind unreacted tyrosine methyl ester hydrochloride (R_f 0.65). The tyrosinated derivatives exhibited an absorption maximum in water identical with that of the parent cyclic nucleotide The yield of the tyrosine methyl ester derivatives of cAMP and cGMP ranged from 40 to 55% on repeated synthesis.

Preparation of Radioactive [^{125}I]- or [^{131}I]Succinyl Cyclic Nucleotide Tyrosine Methyl Ester

The procedure employed has not been modified from the conditions described previously.[2,3] Succinyl cyclic nucleotide tyrosine methyl ester was iodinated with ^{125}I or ^{131}I by the method of Hunter and Greenwood.[9] Approximately 1–3 μg of the derivative (in 50 μl of water) was added to 40 μl of 0.5 M phosphate buffer, pH 7.5. After the addition of 0.5–1.0

---
[8] J. P. Greenstein and M. A. Winitz, *in* "Chemistry of the Amino Acids," Vol. 2, p. 978. Wiley, New York.
[9] W. M. Hunter and F. C. Greenwood, *Nature (London)* **194**, 495 (1962).

mCi of ^{125}I or ^{131}I, 50 μl of a solution of chloramine-T (35 mg/10 ml of 50 mM phosphate buffer) was added and the reaction run for 45 seconds. The iodine was then reduced by the addition of 100 μl of a solution of sodium metabisulfite (24 mg/10 ml 0.05 M phosphate buffer).

The iodinated cyclic nucleotide derivatives were purified either by column chromatography on Sephadex G-10 or by thin-layer chromatography on cellulose. The reaction mixture was applied to a 0.9 cm × 9 cm Sephadex G-10 column previously washed with 1 ml of 3% human serum albumin in phosphosaline buffer, pH 7.5, and eluted with phosphosaline buffer (flow rate 40 ml per hour). Three distinct peaks of radioactivity were found: peak 1 (void volume) has not been identified, peak 2 (9–12 ml) was free iodine, and peak (22–32 ml) was [^{125}I]succinyl cyclic nucleotide. The iodinated tyrosine methyl ester derivatives isolated in peak 3 cochromatographed with their respective uniodinated compounds on thin-layer chromatography using the previously described solvent system. All iodinated compounds had a specific activity of >150 Ci/mmole. The iodinated ligands were diluted to 50 mM acetate buffer, pH 6.2 (3 to 4 × 10^6 cpm/ml) and stored as small aliquots at −20°. The ^{125}I material retained full immunoreactivity for periods up to 2 months or longer, provided it was stored at −20° in small aliquots and not subjected to repeated freezing and thawing. The ^{131}I derivatives were stable for 3–4 weeks.

Preparation of Tissues, Blood, and Urine

Frozen tissue samples are homogenized at 4°C in 0.5 ml of cold 6% trichloroacetic acid (TCA). TCA supernatants are extracted four times with 5 ml of ethyl ether saturated with water. The extracted aqueous phase is evaporated under a stream of air, and the residue is dissolved in 50 mM sodium acetate buffer, pH 6.2, and used directly in the immunoassay.

Blood is collected in heparinized tubes and centrifuged immediately at 2500 g for 5 minutes at 4°. Because of an unidentified substance(s) in plasma which occasionally interferes in the cAMP, but not in the cGMP, radioimmunoassay, we routinely separate plasma cAMP from interfering substances by Dowex column chromatography. Extracts of plasma are divided into two fractions: To 0.5 ml of plasma (for cAMP assay) is added an equal volume of 10% TCA. The supernatant fraction is applied to a column of Dowex-50 W-X8 (100–200 mesh) 4 cm × 8 mm and the column eluted with water. cAMP elutes in the 4th to 8th ml. This fraction is dried at 60° under a stream of air and resuspended in 50 mM sodium acetate buffer, pH 6.2. For cGMP assay,

0.5 ml of 10% TCA is added to an equal volume of plasma. The TCA supernatant is treated in a manner identical to that of tissue extracts. Urine samples of 2–10 µl are added directed into the immunoassay.

Immunoassays

Procedure

cAMP and cGMP immunoassays are performed in 50 mM sodium acetate buffer, pH 6.2. Each tube contains (in order of addition) 50–100 µl of cyclic nucleotide standard or unknown solution, 100 µl of antibody in appropriate buffer at a dilution sufficient to bind 40–55% of the labeled marker, 100 µl of the ^{125}I-labeled marker (approximately 15,000 cpm and representing 0.01 pmole of ligand), and 100 µl containing 500 µg of rabbit γ-globulin as carrier, in a final volume of 600 µl. The most commonly employed method for separating bound and free ^{125}I-labeled ligand used in our laboratory is ammonium sulfate precipitation. After 2–18 hours of incubation, 2.5 ml of 60% $(NH_4)_2SO_4$ solution is added. The tubes are centrifuged at 4° for 15 minutes, and the precipitate is counted in a gamma spectrometer. All analyses are carried out in triplicate.

In the simultaneous assays of cAMP and cGMP, the immunoassay procedure is identical, except that specific cAMP and cGMP antibodies are added in the 100-µl antibody aliquot. [^{131}I]ScAMP-TME and [^{125}I]ScGMP-TME (approximately 15,000 cpm each) are added in the 100-µl aliquot. The precipitate is then counted in a dual-channel spectrometer equipped with a punched paper tape printout. A computer program for use on a NCR Century 200 system using intermediate FORTRAN has been written for analysis for both the single and simultaneous radioimmunoassays.

Cyclic [^{3}H]nucleotide can be used instead of the iodinated labeled compounds in either the cAMP or cGMP assay. Significantly more antibody is required, since cyclic [^{3}H]nucleotide binds 200 less avidly than succinyl cyclic nucleotide tyrosine methyl ester.[2]

The separation of bound from free labeled compounds can also be performed by the second antibody method,[1] precipitation with polyethylene glycol[10] or by filtration on cellulose ester (Millipore) filters.[7]

Cyclic Nucleotide Immunoassays

The selection of appropriate antisera is most important for successful cyclic nucleotide immunoassay. We screen antibodies after each booster

[10] B. Desbuquois and G. D. Aurbach, *J. Clin. Endocrinol. Metab.* **33,** 732 (1971).

injection, both for sensitivity and specificity, checking in particular cross reactivity with ATP. Since tissue concentrations are approximately 10,000- and 100,000-fold greater than cAMP and cGMP, respectively, it is necessary to select antisera which cross react <0.02% with ATP in the cAMP immunoassay and <0.002% in the cGMP immunoassay to avoid chromatographic preparation of tissues. We initially immunized a group of 10 rabbits for each cyclic nucleotide immunoassay and found several antisera in each group which met these cross-reactivity requirements. Since tissue concentrations of cGMP are usually an order of magnitude less than cAMP, the cross-reactivity of cGMP in the cAMP immunoassay can be 1% or higher and yet chromatographic steps will be unnecessary. For the cGMP immunoassay, the cross-reactivity of the antisera and cAMP should be minimal. Several cGMP antibodies require at least 10,000-fold greater concentrations of cAMP to produce displacement of marker equal to that of cGMP.

A standard immunoassay curve for cAMP is shown in Fig. 1. The antisera currently used were made at Schwarz-Mann Bioresearch, Orangeburg, New York, and shows a linear displacement of [^{125}I]ScAMP-TME by unlabeled cAMP, plotted as a semilogarithmic function from 0.025 to 5 pmoles. The cross-reactivity of this antibody with ATP, 5'-AMP and cGMP is shown in Fig. 2. ATP in a several millionfold higher concentration failed to cause significant cross reactivity with the cAMP antibody. The cross reactivity of cGMP is 0.01%. The sensitivity and specificity of several cAMP antisera are shown in Table I.

The immunoassay for cGMP is sensitive to 0.03 pmole cGMP. This degree of sensitivity allows measurement of cGMP in triplicate on 10–20 mg of most tissues. Cross-reactivity of the cGMP antibody with all

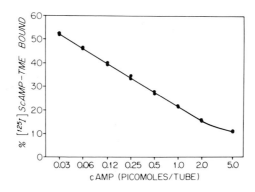

FIG. 1. Standard immunoassay curve for cAMP. Reaction conditions are described in text. Reaction volume was 600 μl. Antibody was kindly supplied by Schwarz-Mann BioResearch, Orangeburg, New York.

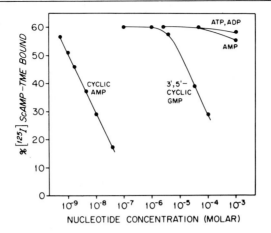

FIG. 2. The inhibition of [^{125}I]ScAMP-TME binding to cAMP antibody by various nucleotides.

purine and pyrimidine nucleotides is minimal (<0.002%) except for cIMP which reacts at the 1% level.

Tissue Measurement of Cyclic Nucleotides by Single and Simultaneous cAMP and cGMP Radioimmunoassay

The concentrations of cAMP and cGMP in various rat tissues are shown in Table II. The values are in the same range as those reported by others using enzymatic techniques or competitive protein binding radioassay.[11-13]

The intrassay coefficient of variation for a number of tissues in both the cAMP and cGMP immunoassays is in the 5–8% range, while the interassay coefficient of variation is 14% for cAMP and 17% for cGMP.

Tissue values for cAMP and cGMP when measured by simultaneous immunoassay are identical to those found by individual radioassay. Since the tissue concentration of cAMP relative to cGMP is in general an order of magnitude greater, it is important to select a cGMP antibody that shows minimal cross reactivity with cAMP. With certain antibodies, even a 1000-fold increase in cAMP relative to cGMP causes no significant change in the amount of cGMP measured.

[11] N. D. Goldberg, J. Larner, H. Sasko, and A. G. O'Toole, *Anal. Biochem.* **28**, 523 (1969).
[12] E. Ishikawa, S. Ishikawa, J. W. Davis, and E. W. Sutherland, *J. Biol. Chem.* **244**, 6371 (1969).
[13] A. G. Gilman, *Proc. Nat. Acad. Sci. U.S.* **67**, 305 (1970).

TABLE I
SENSITIVITY AND SPECIFICITY OF VARIOUS CYCLIC AMP (cAMP) ANTISERA[a]

Antisera	Maximal sensitivity		Relative binding affinity	
	Serum dilution	cAMP (pmoles tube)	ATP (%)	cGMP (%)
RC A-1	1:5,000	1	0.002	0.01
RC A-3	1:5,000	1	0.0001	0.005
RC A-7	1:5,000	0.25	0.002	0.01
SM-381[b]	1:5,000	0.025	0.00001	0.01
SM-291[b]	1:4,000	0.05	0.01	0.10
LCA-1[c]	1:40,000	0.25	0.0001	1.0
LCA-2[c]	1:40,000	0.25	0.0001	1.0

[a] Expressed as the minimal concentration of cAMP which causes linear displacement in the immunoassay.

[b] cAMP antisera obtained from rabbits after 4 boosts of ScAMP-albumin at monthly intervals kindly supplied by Schwarz-Mann BioResearch, Orangeburg, New York.

[c] cAMP antisera obtained from rabbits after 3 boosts of ScAMP-albumin at monthly intervals kindly supplied by Drs. G. Liddle and N. Kaminsky of Vanderbilt University.

A convenient verification of the reliability of a cyclic nucleotide determination is measurement of the amount of immunologically reactive cyclic nucleotide before and after hydrolysis by cyclic nucleotide phosphodiesterase. After such treatment, greater than 90% of the immunologically reactive cAMP is hydrolyzed in various rat tissues and human urine and plasma. In certain rat tissues (liver, kidney, cortex, skeletal muscle) a blank of 15–40% is found. Tissues values for cGMP are corrected by

TABLE II
CONCENTRATION OF cAMP AND cGMP IN VARIOUS RAT TISSUES[a]

	cAMP	cGMP
Liver	960 ± 98	15 ± 2
Kidney cortex	980 ± 92	38 ± 4
Skeletal muscle	360 ± 52	18 ± 1.6
Lung	1250 ± 110	56 ± 6
Jejunum	1010 ± 97	120 ± 11
Pituitary[b]	880 ± 105	9.0 ± 1.1

[a] Values represent mean ± SEM expressed as picomoles per gram wet weight of tissue.

[b] Hemipituitaries were incubated for 2 hours at 37% in TC 199.

subtracting the "blank" from the total cGMP determined. The cause of the "blank" in these tissues has not been determined.

Checking against Interfering Substances

Certain substances can affect binding in both the cAMP and cGMP assays. EDTA 1–0.1 mM will interfere with certain antisera in the cAMP, but not the cGMP immunoassay. As noted above, occasional human plasma samples will give a falsely high value for cAMP; consequently, chromatography of plasma extracts is recommended before measurement of cAMP by radioimmunoassay. Weinryb[7] has noticed that extracts of rat cerebral cortex and lipocytes can enhance binding in the cAMP radioimmunoassay by 10–20%. This "antibody binding enhancing factor" was removed by Dowex-50 H$^+$ column chromatography, but not by 12% TCA or heating to 90°. This phenomenon apparently occurs with only certain cAMP antibodies, since this author has not observed enhancement of binding with extracts of rat cerebral cortex or lipocytes.

Cailla et al.[14] have recently shown that ScAMP can be synthesized in 100% yield in water. ScGMP can also be made under the same conditions with the same results (unpublished observations of the author). This important modification of the synthesis of 2'-O-succinyl cyclic nucleotides not only simplifies the synthesis of these compounds for the purpose of antibody production, but also succinylation of tissue samples can be performed with 100% yield prior to assay, thus increasing the sensitivity of the cAMP and cGMP radioimmunoassays by 100-fold to the 10^{-15} to 10^{-16} moles range.

Acknowledgments

This work was supported by Research Grant AM-05675 from the U.S. Public Health Service. This work was done in part during the tenure of the Otto C. Storm Investigatorship of the American Heart Association.

---

[14] H L. Cailla, M. S. Racine-Weisbuch, and M. A. Delaage, *Anal. Biochem.* **56**, 394, 1973.

[14] Determination of Cyclic GMP by Formation of [β-^{32}P]GDP

By GÜNTER SCHULTZ and JOEL G. HARDMAN

Principle[1]

$$\text{Cyclic GMP} \xrightarrow{\text{phosphodiesterase}} \text{GMP} \quad (1)$$

$$\text{GMP} + [\gamma\text{-}^{32}\text{P}]\text{ATP} \xrightarrow{\text{GMP kinase}} [\beta\text{-}^{32}\text{P}]\text{GDP} + \text{ADP} \quad (2)$$

$$[\gamma\text{-}^{32}\text{P}]\text{ATP} \xrightarrow{\text{myosin}} \text{ADP} + {}^{32}\text{P}_i \quad (3)$$

$$\text{Precipitation of } {}^{32}\text{P}_i \text{ and determination of } [\beta\text{-}^{32}\text{P}]\text{GDP} \quad (4)$$

Materials

Cyclic nucleotide phosphodiesterase is purified[2] from bovine hearts with a specific activity of about 0.5 μmole min^{-1} mg protein^{-1} if measured with 1 μM cGMP as substrate.[3]

ATP:GMP phosphotransferase (*GMP kinase*, EC 2.7.4.8) prepared from pig brain with a specific activity of about 10 U/mg is commercially available from Boehringer Mannheim Corp. The enzyme is dialyzed overnight against 500–1000 volumes of 10 mM N-Tris(hydroxymethyl)methyl-2-aminoethane sulfonate (TES) buffer, pH 7.4, and stored frozen at $-20°$ until use. If the enzyme solution becomes hazy during dialysis, it is centrifuged, and the supernatant fluid is used. GMP kinase with similar properties and about the same specific activity can be prepared from calf thymus.[4]

Myosin is prepared from rabbit skeletal muscle[5] with two additional precipitations. The specific activity is at least 0.4 U/mg protein. The enzyme is stored in a 250 mM KCl solution containing 50% (v/v) glycerol at $-20°$ and diluted to an appropriate protein concentration with 500 mM KCl before use.

Glyceraldehyde-3-phosphate dehydrogenase (EC 1.2.1.12) from rabbit muscle (80 U/mg) and *3-phosphoglycerate kinase* (EC 2.7.2.3) from yeast (180 U/mg) are obtained from Boehringer Mannheim Corp. Ali-

---

[1] G. Schultz, J. G. Hardman, K. Schultz, J. W. Davis, and E. W. Sutherland, *Proc. Nat. Acad. Sci. U.S.* **70**, 1721 (1973).
[2] R. W. Butcher and E. W. Sutherland, *J. Biol. Chem.* **237**, 1244 (1962).
[3] J. A. Beavo, J. G. Hardman, and E. W. Sutherland, *J. Biol. Chem.* **245**, 5649 (1970).
[4] H. Shimoni and Y. Sugino, *Eur. J. Biochem.* **19**, 256 (1971).
[5] S. V. Perry, this series, Vol. 2, p. 582 (1955).

quots of the crystalline enzyme suspensions are centrifuged (about 15 minutes at 40,000 g), and the pellets are dissolved in 10 mM Tris(hydroxymethyl)aminomethane (Tris)·HCl, pH 7.5, and stored frozen until use.

[γ-^{32}P]ATP of the required high specific activity and purity is enzymatically prepared by a modification of published procedures.[6,7] Into a conical 1.5-ml plastic tube (Eppendorf) are pipetted in the given order: (a) 10 (or 15) μl of a 300 mM Tris solution; (b) 100 (or 150) μl of a solution containing 3–4 mCi of carrier-free inorganic ^{32}P in 0.02 N HCl (obtained from International Chemical and Nuclear Corp.). The pH of the mixture of (a) and (b) is about 8; (c) 50 μl of a solution containing 200 nmoles of 3-phosphoglycerate, 20 nmoles of NAD$^+$, 1000 nmoles of MgCl$_2$, 200 nmoles of EDTA, 600 nmoles of cysteine, 10–20 nmoles of ATP, about 1 μg of 3-phosphoglycerate kinase, and about 1 μg of glyceraldehyde-3-phosphate dehydrogenase in 25 mM Tris·HCl, pH 8.0. This solution is prepared by mixing stock solutions (stored at −20°) of appropriate concentrations of the substrates and cofactors, of ATP and of the enzymes with a freshly prepared and neutralized solution of cysteine.

After 15–20 minutes of incubation at room temperature (20–25°), the incorporation of ^{32}P into ATP is checked. For this purpose, a small aliquot (<1 μl) of the incubation medium is transferred with the tip of a piece of fine plastic tubing into a vial containing about 1 ml of 0.5 M KH$_2$PO$_4$ in 0.1 N HCl. After mixing, 100 μl of this solution are taken out for counting. A few milligrams of Norit A are then added to the remaining mixture. After stirring for 30–60 seconds on a Vortex mixer and (low speed) centrifugation, the amount of ^{32}P is determined in another 100-μl aliquot of the supernatant fluid. The fraction of the radioactivity that is not absorbed by the charcoal (i.e., cpm in second aliquot/cpm in first aliquot) mainly represents the ^{32}P that is not incorporated into ATP. After 15–20 minutes of incubation, more than 50% of the ^{32}P are usually incorporated under the above conditions. After 30–45 minutes of incubation, when 70–85% of the P$_i$ are incorporated, the enzymatic reaction is stopped. A part of the incubation mixture (20–50 μl) is then applied in a 4 cm wide band to a thin-layer plate coated with polyethyleneimine cellulose.[8] The rest of the medium is frozen at −20°, and another aliquot is usually chromatographed a month later. The plate is developed at 2–4° with 0.8–1.0 M LiCl as solvent. The R_f of ATP should be about 0.1. ATP is detected in a reference band by UV light. The layer

---

[6] I. M. Glynn and J. B. Chappell, *Biochem. J.* **90,** 147 (1964).
[7] R. L. Post and A. K. Sen, this series, Vol. 10, p. 773 (1967).
[8] E. Böhme and G. Schultz, this volume [4].

with the ATP band is scraped off the plate and transferred to a 1.5-ml conical plastic vial. ATP is eluted by shaking or mixing the vial for a few minutes with 1 ml of a 2 M KCl solution. The vial is centrifuged, the supernatant fluid is collected, and the elution is twice repeated. After addition of TES buffer, pH 7.5, and of ethanol to final concentrations of 5 mM and 45–50%, respectively, the eluate is stored at $-20°$. Here about 90% of the KCl is precipitated. The supernatant fluid is used for the assay after appropriate dilution with 10 mM TES buffer, pH 7.5, and addition of unlabeled ATP.

If [γ-^{32}P]ATP is prepared as described above, specific activities between 100 and 250 Ci/mmole are obtained, and more than 99.7% of the ^{32}P can be split off by incubation with myosin and subsequently be precipitated as inorganic phosphate. All commercial [γ-^{32}P]ATP preparations tested so far have not been satisfactory for this assay of cGMP because of high myosin-resistant blanks.

General Procedure of the Assay

Into glass or plastic vials are pipetted according to the table: (1) 50 μl of purified tissue extract or (for the standard curve) water; (2) 50 μl of buffer solution A or B containing 260 mM KCl, 7.8 mM MgCl$_2$, 0.52 mM EDTA and 130 mM TES buffer, pH 7.5; solution B additionally contains 0.7–1 μg of phosphodiesterase per 50 μl; (3) 10 μl of H$_2$O (for tissue samples) or cGMP standard solution (as internal standard for tissue samples or for standard curve).

TYPICAL cGMP ASSAY[a]

	No. of tubes	Buffer A (without PDE) 50 μl	Buffer B (with PDE) 50 μl	cGMP standard (pmole) 10 μl	Distilled water 10 μl
Tissue samples	2	+	−	−	+
	2	−	+	−	+
	2	−	+	0.2	−
Standard curve	3	+	−	−	+
	3	−	+	−	+
	2	−	+	0.1–1.0	−

[a] Tissue samples are incubated (in duplicate) without addition of phosphodiesterase, with phosphodiesterase, and with phosphodiesterase and internal standard (0.2 pmole of cGMP). The standard curve consists of standards from 0.1 to 1.0 pmole of cGMP (incubated in duplicate) and of blanks with and without phosphodiesterase (incubated in triplicate).

The tubes are incubated for 30 minutes at 30°, stoppered and then heated for 3–5 minutes in a boiling water bath. After short centrifugation in order to collect condensed water from the upper part of the tubes, to each tube are added 10 µl of a solution containing GMP kinase (about 2 µg of the thymus enzyme or about 4 µg of the pig brain enzyme) and 10 µl of a solution containing 15–20 nCi [^{32}P]ATP and up to 5 pmoles of carrier ATP.

After 2 more hours of incubation, 20 µl of a solution containing about 10 µg of myosin and 80 mM CaCl$_2$[9] are added to each tube. After 30 minutes at 30°, the incubation is stopped by the addition of 500 µl of a solution containing 2 mM KH$_2$PO$_4$ and 1 µM GDP, and the tubes are transferred to an ice bath.

P$_i$ is then precipitated[10] by addition of 500 µl of a fresh mixture of 1 volume 1.2 N perchloric acid containing 40 mM ammonium molybdate and 1 volume of a 60 mM triethylamine solution adjusted to pH 5 with HCl. After low speed centrifugation for a few minutes, 1 ml of the supernatant fluid is transferred to another tube. The transfer of any particles must be avoided. In the second vial, the precipitation step is repeated by the addition of 200 µl of 2 mM KH$_2$PO$_4$ and 1 µM GDP. After short centrifugation, 1 ml of the supernatant fluid is transferred to a liquid scintillation vial. After addition of 15 ml of a 0.01–0.1% aqueous solution of 7-amino-1,3-naphthalene disulfonate, the Čerenkov radiation of the [^{32}P]GDP is measured in a liquid scintillation spectrometer.

Concentrations of cGMP in unknowns are calculated from the standard curve. Corrections are applied for variations in values obtained with the internal standard included with each sample unless the variations exceed 20% of the real value.

Assay Sensitivity

The slope and shape of standard curves obtained by plotting the ^{32}P counted in the supernatant fluid versus cGMP depend on the amount of unlabeled ATP used in the assay.[1] Standard curves linear up to 1 pmole of cGMP per tube are usually obtained with about 5 pmoles of ATP per tube. The blank counting rate observed in the absence of cGMP is then doubled by about 0.1 pmole of cGMP. Although the fractional phosphorylation of GMP is reduced if the amount of unlabeled ATP is decreased, the amount of ^{32}P incorporated into GDP and the slope of

[9] The reconversion of [^{32}P]GDP + ADP → GMP + [^{32}P]ATP during the incubation with myosin is inhibited by Ca^{2+}.
[10] Y. Sugino and Y. Miyoshi, *J. Biol. Chem.* **239**, 2360 (1964).

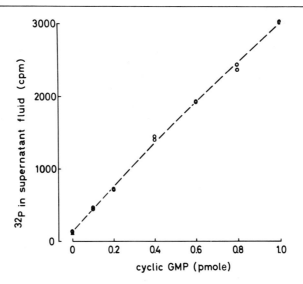

Fig. 1. Standard curve for cGMP. Various amounts of cGMP were incubated with phosphodiesterase, pig brain GMP kinase and [^{32}P]ATP as described in the text. ▲, blank incubated without addition of phosphodiesterase.

the standard curve are increased. Using 1–2 pmoles of ATP per tube, slightly parabolic curves for the range between 0.05 and 1 pmole per tube are obtained. The blank counting rate is doubled by <0.05 pmole of cGMP with most preparations of GMP kinase (Fig. 1).

The sensitivity of the assay depends not only on the specific activity and purity of the [^{32}P]ATP, but also on the blanks caused by the GMP kinase and PDE. The blank caused by the nucleotide content of the GMP kinase varies from batch to batch. While the counting rate observed in the absence of cGMP has generally been higher with the commercial enzyme from pig brain, more recent batches of this enzyme have yielded blanks that were as low as those obtained with many preparations of the calf thymus enzyme. The nucleotide content of the GMP kinase has not been effectively reduced by treatment of the enzyme with anion-exchange resins or charcoal.

If the samples are heated in boiling water before [^{32}P]ATP and GMP kinase are added, the blank caused by phosphodiesterase is usually small. However, with some phosphodiesterase preparations high blanks were observed if this enzyme was added simultaneously with GMP kinase and [^{32}P]ATP or if the boiling step after incubation with phosphodiesterase was omitted.

If [^{32}P]ATP is prepared as described above, labeled contaminants in the preparation that cannot be degraded by myosin and precipitated by the P_i-precipitation step are usually very small (<0.3%). Therefore, the contribution of the ATP-induced blank to the overall blank is low.

The sensitivity of the assay has recently been improved by almost an order of magnitude by the following modifications. After incubation with phosphodiesterase and boiling, the samples are incubated with GMP kinase and [^{32}P]ATP for about 16 hours in the presence of 0.3 mM dithiothreitol and 0.1 mg/ml of bovine serum albumin. GMP kinase should be prepared from calf thymus at least through the DEAE cellulose step.[4] Not more than 0.1–0.2 pmole of [^{32}P]ATP should be used per assay tube.

Adjusting the Enzyme Amounts in the Assay

Since the sensitivity of the assay is a result of the blanks contributed by each step of the procedure, the enzyme amount used for the three successive enzymatic steps must carefully be adjusted.

Cyclic nucleotide phosphodiesterase should be used in about a 2- to 3-fold excess over the amount that causes a maximal, i.e., complete, hydrolysis of cGMP. For each batch of *GMP kinase*, a pilot experiment has to be performed using various amounts of GMP kinase in the absence and in the presence of cGMP. With a given incubation time and temperature, the blank caused by the GMP kinase in the absence of cGMP increases linearly with the amount of enzyme. The amount of [^{32}P]GDP formed from cGMP increases hyperbolically with increasing amounts of enzyme. Maximal sensitivity of the assay is obtained with the amount of GMP kinase that gives a maximal ratio of counts observed in the presence of cGMP over counts observed in the absence of cGMP. *Myosin* is used in a 3- to 10-fold excess over the enzyme amount that is required to split [^{32}P]ATP maximally.

Assay Specificity and Possibly Interfering Factors

The specificity of the assay for cGMP is assured by (1) the chromatographic purification of cGMP from the tissue extracts, (2) the substrate specificity of the phosphodiesterase for 3′,5′-cyclic nucleotides, and (3) the substrate specificity of the ATP:GMP phosphotransferase for GMP. When various nucleotides were tested in large amounts,[1] a small amount of ^{32}P was transferred from [^{32}P]ATP by GMP kinase to AMP and coenzyme A (or contaminants of the preparations used) yielding myosin-stable products, but not to other nucleoside monophosphates or AMP-containing substances. The formation of labeled products of AMP and CoA did not require the presence of phosphodiesterase. The addition of high amounts of various nucleoside monophosphates did not affect the formation of [^{32}P]GDP from GMP.

With unlabeled GDP added, amounts of [^{32}P]GDP are formed which can be as high as with equimolar amounts of GMP. This effect is probably caused by an exchange reaction of the terminal phosphates of ATP and GDP catalyzed by the GMP kinase (and contaminating amounts of GMP and ADP). The labeling of GDP by this reaction may be used for the determination of small amounts of GDP and (after enzymatic transformation to GDP) of GTP.

The formation of [^{32}P]GDP from [^{32}P]ATP and GMP is strongly inhibited by small amounts of *ADP*, the other product of the enzymatic reaction.

Thus, in the purification of cyclic GMP from tissue extracts, cGMP should be well separated from GMP, GDP, cAMP, AMP, ADP, and ATP.

Besides ADP, the only other agents that are known to inhibit the formation of [^{32}P]GDP are heavy metals. If Zn^{2+} is used in the extraction and purification procedure, e.g., for a $ZnCO_3$ coprecipitation step, it should be considered that not all Zn^{2+} is removed through a purification procedure using anion-exchange resins, and the cGMP-containing fraction should be passed through a short column of Dowex-50.[11]

---

[11] G. Schultz, E. Böhme, and J. G. Hardman, this volume [2].

Section III

Biosynthesis of Cyclic Nucleotides

[15] General Principles of Assays for Adenylate Cyclase and Guanylate Cyclase Activity

By Günter Schultz

Adenylate cyclase and guanylate cyclase catalyze apparently analogous reactions. The substrates appear to be complexes of the respective nucleoside triphosphate and a divalent cation, i.e., Me·ATP or Me·GTP. Additional free Me^{2+} appears to be necessary for the maximal activity of both enzymes. Inorganic pyrophosphate is a product of the adenylate cyclase reaction, but this has not yet been established for the guanylate cyclase reaction.

The activity of the cyclases can be determined using either unlabeled or labeled ATP or GTP. Determinations carried out using unlabeled nucleoside triphosphate as substrate are more laborious since the product formed must then be assayed by one of the methods for cAMP or cGMP described elsewhere in this volume. Cyclase assays employing radioactively labeled substrate require cyclic nucleotide purification and liquid scintillation counting. They are more widely used though are not without significant pitfalls.

In the crude preparations with which most cyclase determinations are performed, adenylate and guanylate cyclase activities are usually small compared to the activities of enzymes that are capable of interfering with the assay. Such interfering enzymes are various nucleoside triphosphatases and cyclic 3':5'-nucleotide phosphodiesterases, but nucleotidases and deaminases may also cause problems. The products of the reactions, which these contaminating enzymes catalyze, may cochromatograph with the cyclic nucleotides. Therefore, an effective separation of the cyclic nucleotide from the substrate and all its possible degradation products is imperative.

In the following description of the principles of adenylate and guanylate cyclase assays, the procedures utilizing labeled substrate will be emphasized. Factors relating to assay sensitivity and precision will be discussed. However, the possible interaction of hormone receptors or regulatory factors, e.g., trace metals, are not considered.

Nucleoside Triphosphate Concentration and Volume

With a given enzyme amount and incubation time, the substrate concentration and the assay volume (and, accordingly, the substrate specific activity) are important determinants of the sensitivity of an assay using labeled nucleoside triphosphate. The maximal sensitivity of an assay with labeled substrate is generally achieved under conditions that allow a maximal fractional conversion of the labeled substrate to the product.

In experiments in which amounts of adenylate or guanylate cyclase, e.g., enzyme distributions, are to be determined, the assays should be performed preferably[1] with a saturating substrate concentration at 30° in a buffered medium of optimal or physiological pH and otherwise optimal conditions. For studies on the regulation of enzyme activity, however, it may become desirable to vary one or more of these conditions.

For practical purposes, we can assume zero-order kinetics of cyclic nucleotide formation with respect to substrate ($v \cong V_{\max}$) if $[S] \geq 10\,K_m$, i.e., with $[NTP] \geq 1$ mM for many adenylate and guanylate cyclase preparations. Under this condition, the accumulation of product is theoretically proportional to the incubation time and to the amount of enzyme over a relatively broad range.

A further increase of the substrate concentration over 10 times the K_m value will not detectably increase v further, but will decrease the fractional conversion of labeled substrate to the product and thus decrease the sensitivity of the assay. Impurities of the labeled substrate that cochromatography with the isolated product are not affected by the addition of unlabeled substrate. Thus, the contribution of a substrate impurity to the counting rate in the product fraction would be increased under these conditions.

The sensitivity of the assay may be increased by decreasing $[S]$. In the region of mixed-order kinetics there will be a significant increase in the fractional amount of substrate converted to product since v does not decrease proportionally with decreasing $[S]$.

The fractional conversion of substrate will increase until $[S]$ becomes $<0.1\,K_m$. Then the product formation follows first-order kinetics with respect to substrate. The fractional conversion of substrate and thereby the sensitivity of the assay become virtually constant since v is lowered proportionally with $[S]$.

When, for example in kinetic studies, low substrate concentrations are used for adenylate and guanylate cyclase determinations, the increased probability of interference by enzymatic and nonenzymatic tissue constituents (e.g., endogenous substrate, possible inhibitors, and nucleoside triphosphatases) should be considered. Under low substrate condi-

---

[1] Recommendations of the Enzyme Commission of the International Union of Biochemistry 1964, *in* "Enzyme Nomenclature," Comprehensive Biochemistry, (M. Florkin and E. H. Stotz, eds) 2nd ed., Vol. 13. Elsevier, Amsterdam, 1965.

tions, it is especially recommended to monitor the substrate concentration remaining at the end of the incubation period.

The *assay volume* is a factor that effectively determines the assay sensitivity at any given enzyme amount and substrate concentration. By keeping [S] and the total amount of radioactivity constant, while decreasing the assay volume, one increases the specific activity of the substrate. In the decreased volume, the number of substrate molecules converted by a certain amount of enzyme is unchanged; the fraction of the labeled substrate converted to the product, however, is increased in proportion to the volume reduction, i.e., in proportion to the increase in specific activity of the substrate. Therefore, adenylate and guanylate cyclase assays should be performed in a volume that is large enough to ensure accurate and easy pipetting, but that is small enough to yield the sensitivity needed with a given enzyme preparation. Keeping the assay volume relatively small also has the advantage that the total amounts of nucleotides and electrolytes may be so small that anion-exchange procedures using small columns or thin-layer chromatography plates can be used effectively for isolation of the cyclic nucleotide. The total cyclic nucleotide fraction eluted from the column or thin-layer chromatography plate can then directly be counted in a detergent-containing scintillation fluid.

In assays of adenylate and guanylate cyclase activity using unlabeled nucleoside triphosphates as substrate at V_{max} conditions, [S] and assay volume are not restricted. The substrate amount can be increased (by increased concentration or volume) without changing the assay sensitivity. Therefore, it is easier to avoid reduction of [S] by nucleoside triphosphatases. The use of a nucleoside triphosphate-regenerating system (see below) is usually not required. The sensitivity of this type of cyclase assay only depends on the sensitivity of the final assay used for determining the amount of cyclic nucleotide formed by the cyclase preparation.

Enzyme Concentration

Working with a given substrate concentration, assay volume and incubation time, the fractional conversion of the substrate can be increased by using a higher enzyme concentration. However, the strong possibility that the amount of interfering enzymes, e.g., nucleoside triphosphatases, will simultaneously be increased, should not be overlooked. When only small amounts of adenylate or guanylate cyclase are available, a small assay volume should be considered a very effective means to obtain the highest possible sensitivity.

Labeled Nucleoside Triphosphates

^{32}P-, ^{14}C- or ^{3}H-labeled nucleoside triphosphates can be used as substrates. If ^{14}C- or ^{3}H-labeled ATP or GTP are used, the separation of the cyclic nucleotide from other labeled compounds has to be done somewhat more carefully than with ^{32}P-labeled nucleoside triphosphates (see below and elsewhere in this volume[2,3]).

The use of [^{14}C]ATP and [^{14}C]GTP is probably the most expensive alternative, but blank problems are usually very small. Repurification of commercial [^{14}C]ATP or [^{14}C]GTP preparations has not been necessary in our hands. α-^{32}P- and ^{3}H-labeled ATP and GTP obtained from various commercial sources may contain impurities which cochromatograph with the cyclic nucleotides. Significant impurities have been found in some preparations of high specific activity [8-^{3}H]GTP. Some preparations of tritiated nucleoside triphosphates also contain significant amounts of ^{3}H-nucleoside diphosphate and tritiated water. To reduce these impurities, which may result either in reduced assay sensitivity or inaccurate results because of erroneously assumed specific activity, purification of α-^{32}P- and ^{3}H-labeled nucleoside triphosphates before use may be necessary.

The lability of the tritium label under certain conditions, especially at elevated pH,[4] should not be overlooked when tritiated nucleoside triphosphate is used as substrate. [^{3}H]ATP labeled in the 2-position is more stable than, and therefore preferable to, [8-^{3}H]ATP.

The amount of labeled nucleoside triphosphate required per assay tube mainly depends on the specific activity of the adenylate or guanylate cyclase studied. If about 200–250 dpm in the purified cyclic nucleotide fraction are desired for counting and if the recovery of the cyclic nucleotide is almost complete after purification and the fractional conversion of substrate to cyclic nucleotide is 0.1–1%, then 0.01–0.1 μCi of labeled substrate would be required per assay tube. Since in most adenylate and guanylate cyclase preparations linear product formation is found over only a short period of time, the conversion of substrate to product is often <0.1% and the use of a higher amount of labeled nucleoside triphosphate (up to 1 μCi per tube) is then indicated. It should be kept in mind, however, that increasing the amount of ^{3}H- or ^{32}P-labeled nucleoside triphosphate usually increases the blank proportionally that is caused by impurities cochromatographing with the cyclic nucleotide.[5] For this rea-

[2] G. Schultz, E. Böhme, and J. G. Hardman, this volume [2].
[3] E. Böhme and G. Schultz, this volume [4].
[4] R. Monks, K. G. Oldham, and K. C. Tovey, "Labelled Nucleotides in Biochemistry," Radiochemical Centre (Amersham) Revue 12 (1972).
[5] K. G. Oldham, *Int. J. Appl. Radiation and Isotopes* 21, 421 (1970).

son and because of the cost involved, the use of high amounts of labeled substrate is often not a satisfactory way to increase the assay sensitivity.

Reducing Substrate Degradation

Most preparations used for adenylate cyclase and guanylate cyclase determinations contain high activities of various ATP- and GTP-degrading enzymes. The enzymatic degradation of ATP or GTP (except at very high concentrations) is often so rapid with mammalian cyclase preparation that an accurate determination of adenylate and guanylate cyclase activity is not possible. So far two techniques have been used to reduce the nucleoside triphosphate degradation through noncyclase pathways.

The ATP and GTP concentration can be kept relatively constant over longer periods of time by the addition of a nucleoside triphosphate-regenerating system. Phosphoenol pyruvate with pyruvate kinase (with addition of potassium ions) or creatine phosphate with creatine kinase have been used for this purpose. While these nucleoside triphosphate-regenerating systems are capable of retarding the ATP and GTP degradation effectively, the property of the phosphate-containing substances to bind divalent cations can complicate kinetic studies (see following paragraph) especially if one considers that the concentration of these phosphates, and hence the concentration of free Me^{2+}, changes with the length of the incubation time. The use of nucleoside triphosphate-regenerating systems is additionally complicated by the following facts. A strong inhibition by phosphoenol pyruvate has been shown for several cyclase preparations. Therefore, creatine phosphate and creatine kinase are generally preferable. However, creatine phosphate is capable of accelerating the nonenzymatic formation of cGMP from GTP.[6]

Nucleoside triphosphate degradation can also be reduced by addition of nucleoside triphosphatase inhibitors. To our knowledge, ouabain has not been used as ATPase inhibitor in adenylate cyclase studies. Azide (10–20 mM) has successfully been used as an inhibitor of some types of nucleoside triphosphatases in adenylate and guanylate cyclase studies.[6a] Additionally, however, azide appears to be capable of stimulating adenylate cyclase activity more directly.[6a] Fluoride should not be used to inhibit ATP degradation because of its known stimulatory effect on most adenylate cyclases and its metal-binding property.

The nucleoside triphosphate derivatives, AMP-PNP and GMP-PNP,

---

[6] H. Kimura and F. Murad, *J. Biol. Chem.* **249,** 329 (1974).

[6a] J. P. Gray, D. L. Garbers, and J. G. Hardman, R. A. Johnson and S. J. Pilkis, unpublished observations, 1973.

are virtually unaffected by most nucleoside triphosphatases[7] but do serve as substrates for adenylate and guanylate cyclase, respectively.[8] However, the apparent kinetic constants for some cyclases that have been obtained with these derivatives are different from those obtained with the natural substrates.[8]

Divalent Cations

The presence of divalent cations is required for the formation of cAMP and cGMP. While Mn·ATP is generally a better substrate for adenylate cyclase than is Mg·ATP, all mammalian guanylate cyclases studied so far appear to require Mn·GTP (or Fe·GTP) as substrate.[9] For full expression of adenylate and guanylate cyclase activities, the presence of additional free cations, Mg^{2+} or Mn^{2+}, is also required. Other metals, Ca^{2+} for example, may additionally be involved in the regulation of adenylate and guanylase cyclase activity.

The concentration of free Me^{2+} results from the total Me^{2+} concentration and from the concentrations of those compounds in the medium that bind the metal. In addition to the high affinity of nucleoside triphosphates for Me^{2+}, the Me^{2+}-binding properties of other nucleotides, phosphoenol pyruvate, creatine phosphate, buffer substances, and other compounds should be considered. The most important binding constants have been compiled.[10] If two divalent cations are used simultaneously, their competition for a common binding agent, e.g., the nucleoside triphosphate, should also be considered. For the calculation of free Me^{2+} and Me·NTP concentrations in the presence of two Me^{2+} or two binding agents, see, for example, the approach used in studies with another enzyme.[11] Since it is possible to consider only some, but not all, of the Me^{2+} interactions, one usually obtains only a rough estimation of free Me^{2+} and Me·NTP concentrations.

Reducing Product Degradation

Essentially all crude adenylate and guanylate cyclase preparations contain cyclic nucleotide phosphodiesterase activity which can signifi-

[7] R. G. Yount, D. Babcock, and D. Ojala, *Biochemistry* **10**, 2484 (1971); R. G. Yount, this volume [56].
[8] M. Rodbell, L. Birnbaumer, S. L. Pohl, and H. M. J. Krans, *J. Biol. Chem.* **246**, 1877 (1971); T. D. Chrisman and J. G. Hardman, unpublished observations, 1973.
[9] J. G. Hardman, T. D. Chrisman, J. P. Gray, J. L. Suddath, and E. W. Sutherland, *Proc. 5th Int. Congr. Pharmacol. 1972*, Vol. 5, p. 134. Karger, Basel, 1973.
[10] "Biochemist's Handbook" (C. Long, ed.), p. 97. Van Nostrand-Reinhold, Princeton, New Jersey, 1961; "Stability Constants of Metal-Ion Complexes" with Suppl. 1, Special Publ. Nos. 17 and 25, The Chemical Society, London, 1964 and 1971.
[11] O. A. Moe and L. G. Butler, *J. Biol. Chem.* **247**, 7308 and 7319 (1972).

cantly affect the net rate of cyclic nucleotide accumulation during the incubation period. To minimize an inaccurate determination of adenylate or guanylate cyclase activity, two precautions have commonly been used.

The addition of a phosphodiesterase inhibitor, e.g., a methylxanthine derivative, can reduce the cyclic nucleotide degradation by >90%. The addition of very high concentrations of methylxanthines should be avoided since theophylline and caffeine have been shown to inhibit the adenylate cyclase in some tissues.[12,13] The use of lower concentrations of more potent phosphodiesterase inhibiting methylxanthines, e.g., 1-methyl-3-isobutylxanthine (SC-2964), or of other potent phosphodiesterase inhibitors is probably preferable.

In adenylate and guanylate cyclase assays using labeled substrate, it is also possible to reduce the interference of phosphodiesterase activity by the addition of unlabeled cyclic nucleotide. Possible cyclic nucleotide degradation through other pathways would also be reduced by the addition of unlabeled cyclic nucleotide. Very high concentrations (>1 mM) of cyclic nucleotide, however, should be avoided since cAMP has been shown to inhibit some adenylate cyclase preparations.[12] The combined use of unlabeled cyclic nucleotide and of a phosphodiesterase inhibitor in lower concentrations is probably the optimal precaution to minimize cyclic nucleotide degradation in radioactive adenylate and guanylate cyclase assays.

The degree of cyclic nucleotide degradation during the incubation period should be checked with labeled cyclic nucleotide in each cyclase assay system. In assays of guanylate cyclase activity in supernatant fractions, the cGMP breakdown during the incubation period can be relatively large, and a correction for the enzymatic product loss may be necessary.[14]

Protection of Cyclase Activity

In cell-free preparations, adenylate and guanylate cyclase activities are often very unstable especially under assay conditions employing elevated temperature, e.g., 37°. Therefore, product formation appears to be proportional to the incubation time only over a short period of time, generally not longer than 10–30 minutes. With longer incubation times, a reduced rate of cyclic nucleotide formation has often been observed which cannot be explained by reduced substrate levels. In particulate preparations, e.g., purified plasma membranes, as well as with some soluble prep-

[12] K. H. Jakobs, K. Schultz, and G. Schultz, *Naunyn-Schmiedeberg's Arch. Pharmacol.* **273**, 248 (1972).
[13] H. Sheppard, *Nature (London)* **228**, 567, 1970.
[14] E. Böhme, *Eur. J. Biochem.* **14**, 422 (1970).

arations, the addition of albumin may help to stabilize the enzyme activity.

Since many adenylate and guanylate cyclase preparations have been shown to be very sensitive to low concentrations of heavy metals, the addition of a sulfhydryl group-protecting agent, e.g., dithiothreitol or dithioerythritol, is generally recommended. Their effects should be checked, however, for the particular enzyme being studied, since these agents may reduce the formation of cGMP in fractions from human platelets under certain conditions.[15]

"Linearity" of Cyclic Nucleotide Formation

For each adenylate or guanylate cyclase preparation and for each experimental condition, it should be established that the studies are performed under conditions where the accumulation of cyclic nucleotide is proportional to the incubation time and to the amount of enzyme used. The period of time, during which the rate of cyclic nucleotide formation is constant and the product accumulation is linear with the incubation time, is mostly short. It is usually easier to obtain a linear product accumulation when the assays are performed with high substrate concentrations than when performed under conditions that result in higher assay sensitivity. The two major reasons for nonlinear plots, breakdown of the initial substrate concentration and cyclase inactivation, have already been mentioned and should be considered in choosing the conditions for subsequent assays.

While most adenylate and guanylate cyclase preparations lose activity with longer incubation times, platelet guanylate cyclase has been shown to gain activity with prolonged incubation for unknown reasons.[15] Since enzyme activation and reduction of the substrate concentration may simultaneously occur and result in apparently and fortuitously linear time-product plots, determination of the substrate remaining at the end of the incubation period is desirable. The residual substrate can so easily be monitored that this check should routinely be performed, especially if experimental conditions are changed.

Termination of Cyclase Reaction

The enzymatic reactions can be stopped by various means. Denaturation of the proteins by addition of acid (e.g., perchloric, trichloroacetic or hydrochloric acid) can be helpful. However, acid interferes with many

---

[15] E. Böhme, R. Jung, and I. Mechler, this volume [29].

separation procedures applying anion-exchange materials, even if the samples are neutralized and/or diluted.

Heating of the samples for several (2–5) minutes in a boiling water bath is a widely used technique, but some precautions should be considered. An accelerated formation of labeled product during heating can be avoided by the addition of an excess of unlabeled substrate before boiling. A nonenzymatic, Me^{2+}-catalyzed cyclic nucleotide formation that has been described,[6,17] e.g., for $Ba \cdot ATP$,[16] $Mn \cdot ATP$,[17] and $Mn \cdot GTP$[6,17] can be avoided by the addition of a Me^{2+} chelator, e.g., EDTA, and by reducing the pH for the boiling step.

The enzymes can also be inactivated by the addition of heavy metals, e.g., Zn^{2+}. The Zn^{2+} can additionally be used for a first purification step. Most 5'-nucleotides are coprecipitated with $ZnCO_3$ formed by subsequent addition of a carbonate.[2,18] However, increased nonenzymatic formation of cGMP from GTP has also been described in the presence of Zn^{2+}.[6]

The possibility of a nonenzymatic cyclic nucleotide formation during the termination step should be excluded by appropriate control experiments.

Purification, Proof of Identity, and Determination of Product

In adenylate and guanylate cyclase assays using labeled nucleoside triphosphate as substrate, the necessary purification of the product depends on the isotope. If α-^{32}P-labeled ATP or GTP is used, separation of cAMP and cGMP, respectively, from all other purine nucleotides and inorganic phosphate is required. With ^{3}H- or ^{14}C-labeled ATP or GTP as substrate, additional separation from all purine nucleosides, bases, and uric acid is necessary. The purification of the cyclic nucleotide generally has to be quite rigorous because of the small amount of cyclic nucleotide formed from the nucleoside triphosphate compared to the usually much larger amounts of other labeled products.

Since the first publication of a procedure using ion-exchange chromatography and coprecipitation of 5'-nucleotides for the determination of adenylate cyclase activity,[19] several other chromatographic techniques have been described. One ion-exchange chromatography step (column or thin-layer chromatography plate) and an additional precipitation step may result in a sufficient cyclic nucleotide separation. The sufficiency of

---

[16] W. H. Cook, D. Lipkin, and R. Markham, *J. Amer. Chem. Soc.* **79**, 3607 (1957); D. Lipkin, R. Markham, and W. H. Cook, *J. Amer. Chem. Soc.* **81**, 6075 (1959).

[17] J. P. Gray, Thesis, Vanderbilt University School of Medicine, 1971.

[18] P. S. Chan and M. C. Lin, this volume [5].

[19] G. Krishna, B. Weiss, and B. B. Brodie, *J. Pharmacol. Exp. Ther.* **163**, 379 (1968).

the purification procedure, however, has to be proved. The identity of the apparent product should be shown by obtaining a constant specific activity of the product (i.e., labeled cNMP/unlabeled cNMP, or in double-labeled assays: label$_1$ in cNMP/label$_2$ in cNMP) through one or two additional purification steps. The identity of the product should furthermore be assured by incubation of the apparent product with highly purified cyclic nucleotide phosphodiesterase (commercial preparations may not be sufficiently pure) and by identifying the product of this enzymatic reaction as the respective 5'-nucleotide monophosphate by chromatographic means. Another procedure involves measuring the amount of cyclic nucleotide by a specific assay which is based on independent principles.

The determination of the amount of product formed in a radioactive cyclase assay is easily performed by measuring, i.e., counting, the fraction of substrate converted to cyclic nucleotide. The fraction of cyclic nucleotide recovered after the purification steps is usually >50% and must be determined and corrected for. If unlabeled cyclic nucleotide is added to the incubation medium (before or after incubation), the recovery can be determined by measuring the UV absorption (e.g., at 260 nm) in the recovered fraction.[20] Addition of a cyclic nucleotide labeled with an isotope other than that of the nucleoside triphosphate used as substrate can also be used for the determination of the recovery through the purification procedure.

In cyclase assays using unlabeled nucleoside triphosphate as substrate, the necessary purification of the cyclic nucleotide depends on the type of assay used for the determination of the cyclic nucleotide formed.

Determination of Residual Substrate

In assays using radioactively labeled nucleoside triphosphate as substrate, the residual substrate concentration can easily be determined by certain anion-exchange chromatographic procedures. Using PEI-cellulose or other anion-exchange resin columns,[2] the nucleoside triphosphate fraction can be eluted from the same column used for product purification. A correction for the loss during the purification can be performed, for example, by photometrically measuring the recovered amount of the excess nucleoside triphosphate added before heating the samples.[20] If

---

[20] If phosphodiesterase inhibitors are used to prevent cyclic nucleotide degradation, the possible absorption of light by these agents should be considered. In the chromatographic procedure, these agents should be separated from cyclic nucleotide and nucleoside triphosphate.

[8-³H]ATP is used as substrate, the instability of tritium in this position of the molecule at higher temperatures and neutral or alkaline pH should be considered. Since most of the ATP or GTP are coprecipitated by $ZnCO_3$, such a precipitation step has to follow the ion-exchange separation step in the cyclic nucleotide fraction so that the nucleoside triphosphate fraction is not affected.

The residual substrate concentration should not be decreased so far that the activity measured in the assay is significantly affected if one considers the kinetic properties of the particular enzyme.

[16] An Automated Chromatographic Assay for Adenylate Cyclase and Cyclic 3',5'-AMP Phosphodiesterase

By MICHAEL C. LIN

The establishment of cyclic 3',5'-AMP (cAMP) as a second messenger in numerous hormonal regulations[1] has increased the need for efficient assays for adenylate cyclase and for cyclic 3',5'-AMP phosphodiesterase. Adenylate cyclase preparations normally contain various enzymatic activities, such as ATPase, phosphodiesterase, and other phosphatases; therefore, numerous by-products other than cAMP are formed during the assay. The success of the procedure depends on the complete separation of cAMP, which often amounts to as low as 0.01 to 0.1% of the ATP substrate, from other adenine nucleotides. Adenylate cyclase was first measured by an assay which employed coupled enzyme systems.[2] Subsequently, methods employing thin-layer chromatography,[3] paper chromatography,[4,5] columns of alumina,[6,7] or a combination of ion exchange and $ZnSO_4$–$Ba(OH)_2$ precipitation[8,9] have been widely used. More recently, a method for the direct determination of picomole amounts of cAMP by high pressure anion exchange chromatography has been de-

[1] G. A. Robison, R. W. Butcher, and E. W. Sutherland, *Annu. Rev. Biochem.* **37**, 149 (1968).
[2] E. W. Sutherland, T. W. Rall, and T. Menon, *J. Biol. Chem.* **237**, 1220 (1962).
[3] H.-P. Bär and O. Hechter, *Anal. Biochem.* **29**, 476 (1969).
[4] M. Rabinowitz, L. Desalles, J. Meisler, and L. Lorand, *Biochim. Biophys. Acta* **97**, 29 (1965).
[5] M. Hirata and O. Hayaishi, *Biochim. Biophys. Acta* **149**, 1 (1967).
[6] A. A. White and T. V. Zenser, *Anal. Biochem.* **43**, 227 (1971).
[7] J. Ramachandran, *Anal. Biochem.* **41**, 372 (1971).
[8] G. Krishna, B. Weiss, and B. B. Brodie, *J. Pharmacol. Exp. Ther.* **163**, 379 (1968).
[9] M. Rodbell, *J. Biol. Chem.* **242**, 5744 (1967).

ELUTION POSITIONS OF ADENINE NUCLEOTIDES ON ANION EXCHANGE RESIN[a]

M^b Cl⁻ molarity	T Retention time[c] (min)			
	Cyclic AMP	AMP	ADP	ATP
0.20	50.0	3.7	41.2	194.9
0.22	43.0	2.2	27.7	121.0
0.24	40.0	1.0	19.2	76.3
0.26	36.7	0	12.9	48.5
0.28	34.2	0	10.2	37.2
0.30	35.9	0	7.7	27.2

[a] A column (0.9 × 6.5 cm) of Dowex 1-X8 was used. Flow rate was 130 ml per hour at 25°. The elution of nucleotides was continuously monitored with a capillary flow cell at 260 nm (22) with a Zeiss PMQ II spectrophotometer equipped to record absorbance linearly.
[b] Total Cl⁻ concentration in 20 mM Tris buffer, pH 8.5.
[c] Retention time = elution time at center of peak minus breakthrough time.

scribed[10] and used for phosphodiesterase assay with liver tissue.[11] The automated procedure described in this section was developed to provide a convenient means for assaying adenylate cyclase activity in "solubilized" preparation from bovine brain; the same technique has proved adaptable for the assay of phosphodiesterase activity.

Separation of Nucleotides on Anion-Exchange Resin

Nucleotides were among the first compounds of biochemical interest to be studied by ion-exchange chromatography.[12] For the assay of adenylate cyclase, it would save time to have AMP, ADP, and ATP all eluted at the void volume, to be followed by the compound to be measured, cAMP. This result has been accomplished by further experimentation with the ionic strength of the elution medium. The trend of the results is shown in the table. There is nearly a 10-fold increase in the rate of elution of ATP when the NaCl concentration undergoes a relatively small increase from 0.2 to 0.3 M, whereas the retention volume of cAMP only decreases by 30%. This sensitivity of multivalent solutes in the system to ionic strength can be expressed approximately in the terms that the retention volume is inversely proportional to $[Cl^-]^n$, where n is the num-

[10] G. Brooker, *Anal. Chem.* **42**, 1108 (1970).
[11] S. N. Pennington, *Anal. Chem.* **43**, 1701 (1971).
[12] W. E. Cohn, in "Ion Exchangers in Organic and Biochemistry" (C. Calmon and T. R. E. Kressman, eds.), p. 345. Wiley (Interscience), New York, 1957.

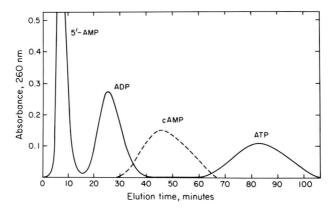

Fig. 1. Elution pattern of adenine nucleotides with low ionic strength eluent. A column (0.9 × 6.5 cm) of Dowex 1-X8 was used. The nucleotides, 0.2 μmole each, were eluted with 0.02 M Tris-chloride, 0.24 M NaCl, 0.06% Brij 35, pH 8.5, at 25° and a flow rate of 130 ml per hour. The elution position of cyclic AMP (cAMP) was determined in a separate experiment.

ber of charges on the nucleotide.[13] The exponential relationship corresponds to the one treated by Boardman and Partridge[14] in their discussion of the sensitivity to ionic strength of the multipoint attachment of proteins to ion exchange resins. According to this principle, the relative position of various nucleotides can be changed by changing the ionic strength of the eluent from 0.24 M NaCl (Fig. 1) to 0.6 M NaCl (Fig. 2). To sharpen the peak of cAMP, 1 ml of 1 N HCl has been injected at 22 minutes. Recovery of cAMP, measured either as radioactivity or absorbance at 260 mm, is about 95%. Adenosine is eluted midway between the two peaks; if ^{14}C is used rather than ^{32}P, there is some overlap of radioactivity from adenosine in the cAMP peak.

Materials

The quaternary base resin was Dowex 1-X8 (AG 1-X8, minus-400-mesh, Cl⁻ form from Bio-Rad Labs). Theophylline was obtained from Mann. ADP, AMP, adenosine, and adenine were obtained from Sigma. ATP, cyclic 3',5'-AMP, 5'-AMP, digitonin, and dithiothreitol (all A grade) were obtained from Calbiochem; the purity of ATP and cAMP was checked by column and paper chromatography. [8-^{14}C]AMP and [8-^{14}C]cAMP were obtained from New England Nuclear. [α-^{32}P]ATP

---

[13] The advice of Dr. Stanford Moore on the behavior of multivalent solutes in ion-exchange chromatography is deeply appreciated.

[14] N. K. Boardman and S. M. Partridge, *Biochem. J.* **59**, 543 (1955).

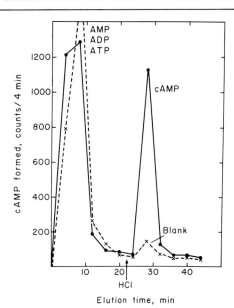

Fig. 2. Elution pattern in a typical assay for adenylate cyclase from bovine brain cortex. The column of Dowex 1-X8 was 0.9 × 12 cm. The eluent was 0.1 M Tris-chloride, 0.6 M NaCl, 0.06% Brij 35, at pH 7.2, with 1 ml of 1 N HCl injected at 22 minutes. The sample of solubilized enzymes assayed contained 100 μg of protein.

was obtained from ICN Corporation. Bovine serum albumin was from Armour. The detergent Brij 35 was from Atlas Powder Co.

Preparation of Adenylate Cyclase from Bovine Brain

Fresh bovine brain was obtained from the slaughterhouse and was kept on ice. Brain cortex was collected by removing as much white matter as possible. Fresh or frozen brain cortex was weighed and was homogenized using 5 strokes of a Teflon–glass homogenizer with 9 ml, per gram of tissue, of 50 mM Tris·HCl[15] (containing 5 mM MgSO$_4$) pH 7.6 at 0°. The homogenate was centrifuged at 30,000 g for 20 minutes at 2°. The pellet was rinsed and either resuspended in the same buffer for assays as the particulate fraction or extracted three times at 0 to 4° by homogenization with 3 ml, per gram of tissue, of 2% neutral aqueous digitonin (prepared as described by Kaplan et al.[16]) containing 5 mM dithiothreitol and 10 mM NaF. The insoluble residue was removed by centrifugation

[15] The molarity refers to the concentration of Tris.
[16] N. O. Kaplan, S. P. Colowick, and E. F. Neufeld, J. Biol. Chem. **205**, 1 (1953).

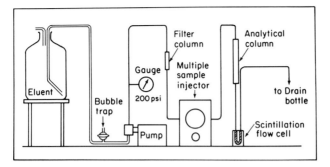

Fig. 3. Diagram of chromatographic apparatus.

at 48,000 g for 40 minutes. The clear digitonin extracts were combined, and the protein was precipitated in the cold between 30 and 80% saturation of $(NH_4)_2SO_4$, with centrifugation at 5000 g. The precipitate was redissolved in Tris·HCl buffer (50 mM) 2–5 ml per grams of tissue), pH 7.6, containing 5 mM Mg^{2+}, 5 mM dithiothreitol, and 10 mM NaF. After dialysis against the same buffer, the enzyme solution was cleared by centrifugation at 48,000 g. This preparation, representing about 20% of the total protein of the cortex, could be stored at −20° without loss of activity.

Assay Procedure for Adenylate Cyclase

The assay medium contained 0.2 mM [α-^{32}P]ATP (specific activity 5 μCi/μmole), 5 mM $MgCl_2$, 10 mM NaF (omitted, if hormone response was to be studied), 10 mM theophylline or 1 mM cAMP, 1 mM dithiothreitol, 50 mM Tris·HCl, pH 7.6 and 20–500 μg of protein in a final volume of 0.5 ml. (The protein concentration of the enzyme preparation was determined by the ninhydrin method[17,18] after 5–6 hours of alkaline hydrolysis.[19] Bovine serum albumin was used as standard). After 15 minutes of incubation at 37°, the reaction was terminated by boiling the tubes for 2 minutes. To the cooled solution were added 0.5 ml of 0.2 mM cAMP as carrier and 0.2 ml each of 1 M $ZnSO_4$ and 1 M Na_2CO_3.[20] After the mixture was well mixed, the precipitate was removed by centrifugation and a 1-ml sample of the supernatant was taken for chromatographic analysis (Fig. 3). Ten 1-ml samples were loaded (by suction) alternately with 1 N HCl into an ROSV-1.0 rotary sample injector with

[17] S. Moore and W. H. Stein, *J. Biol. Chem.* **211**, 907 (1954).
[18] S. Moore, *J. Biol. Chem.* **243**, 6281 (1968).
[19] R. G. Fruchter and A. M. Crestfield, *J. Biol. Chem.* **240**, 3868 (1965).
[20] P. S. Chan and M. C. Lin, this volume [5].

20 sample loops (Chromatronix, Inc.). Samples and HCl were injected alternately every 22 minutes with the aid of a Chromatronix valve drive unit attached to a Microflex reset timer (Eagle Signal Corp.). The eluting solution consisting of 0.1 M Tris-chloride, 0.6 M NaCl, 0.06% Brij 35 (added as 30% solution),[21] at pH 7.2 was placed in a reservoir 18 inches above bench level and was pumped with a Milton Roy MiniPump at a rate of 130 ml per hour at 50–100 psi. A nonionic detergent, Brij 35, is included in the eluent to reduce the tendency for retention of radioactivity by the anthracene in the flow cell, thus to reduce the background count.

The chromatographic tube and the connecting 22-gauge Teflon lines on the effluent side were similar to those used by Crestfield[22]; the columns can be assembled from parts available from manufacturers of amino acid analyzers (e.g., Beckman/Spinco), including an adjustable column head for use with a sample injector and a filter column for the buffer line. The filter (0.9 × 4 cm of Dowex 1-X8) was inserted between the pump and the sample injector. The analytical column was 0.9 × 12 cm of Dowex 1-X8 and was operated at room temperature (i.e., 25°). The radioactivity was continuously monitored by a Nuclear-Chicago Model 6350, 877260, 8704, 8437 Chroma/Cell detector assembly with a 2-ml flow cell.[23] The counts were printed out every 4 minutes. The efficiency of the counting system for ^{32}P was about 50%.

On overnight runs, a shut-down timer (Gra-Lab Universal Timer, Dimco-Gray Co., Dayton, Ohio) was used to turn off the pump and the scintillation counter after the last chromatogram.

Radioactivity eluted from 24 to 32 minutes represents the total quantity of cAMP. (A control chromatogram with [^{14}C]cAMP was run to establish the exact time of elution.) Included in the series was a sample containing no enzyme, which determined the background (usually 100–150 counts per 8 minutes) in the same manner. The results were expressed as total counts of cAMP, or were converted to nanomoles of cAMP from the specific activity of the substrate.

The columns can be used repeatedly for several weeks, until the pressure required for rapid flow rate is more than 100 psi. The resin in the filter column is then replaced; if the pressure is still too high, the porous plate and the top centimeter of resin in the analytical column are replaced. Repacking of the analytical column is rarely required.

---

[21] S. Moore and W. H. Stein, *J. Biol. Chem.* **211**, 893 (1954).
[22] A. M. Crestfield, *Anal. Chem.* **35**, 1762 (1963).
[23] Each flow cell and the connecting leads should be checked for leaks for 1 hour on the bench at 15 psi back pressure before being inserted in the counter.

Sensitivity and Reproducibility of the Method

The specific activity of the ATP substrate is 5 μCi/μmole or 5 cpm/pmole, therefore, this procedure allows the detection of 50 pmoles of cAMP with reasonable accuracy. The activities of the cyclase preparations used in this study vary from 2 to 15 nmoles per milligram of protein per 30 minutes, depending on the purity of the preparations. No correction has been made for the 95% recovery of cAMP in the procedure. At the level of activity, where the counts in the cAMP peak are about 1000 with the background counting being about 150, as illustrated by the analysis graphed in Fig. 2, a series of assays on the same sample has a reproducibility of 100 ± 5%. The precision of the assay is lower when the net counts are in the 200–400 range.

The amount of cAMP is reasonably proportional to the amount of sample added (Fig. 4A); the counts also increase linearly with time during the first 15 minutes (Fig. 4B), but there is a small falloff from linearity during the second 15 minutes. This detectable decrease is probably the result of the use of the relatively low substrate concentration of 0.2 mM ATP and destruction of cAMP by phosphodiesterase action. The concentration of ATP has been lowered from that normally used[4,6] in order to minimize the breakthrough peak of labeled ATP. At higher ATP concentrations a longer ion exchange column and hence a slower analysis would have to be used to provide adequate separation of cAMP.

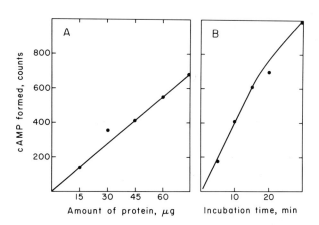

Fig. 4. Linearity of response in the adenylate cyclase assay. (A) Proportionality to the amount of adenylate cyclase. A solution of the $(NH_4)_2SO_4$ precipitate obtained after digitonin extraction was used as source of enzyme. (B) Proportionality to the incubation time. The sample of solubilized enzyme assayed contained about 100 μg of protein. Each point represents the average of triplicate determinations.

Identification of the Product as cAMP

The purity of the product being measured in the adenylate cyclase assay was checked by paper chromatography.[5] After $ZnSO_4$–Na_2CO_3 precipitation, an aliquot of the sample was applied on Whatman 3 MM paper. The solvent was 1 M ammonium acetate:95% ethanol, 3:7 (v/v). The front migrated 20–25 cm (ascending) after 10–12 hours at 25°. The R_f values of some of the constituents of the mixture are ATP 0.05, ADP 0.08, AMP 0.15, IMP 0.13, adenosine 0.62, inosine 0.62, theophylline 0.73, and cAMP 0.46. The spots of nucleotides and nucleosides were detected with ultraviolet light, cut out and placed in vials containing 15 ml of scintillation fluid.[24] The radioactivity was measured with a Nuclear-Chicago liquid scintillation counter, Model 720. The efficiency was about 73% for ^{14}C. When paper chromatography was used to identify cAMP, the ion-exchange column and the paper chromatograms gave comparable results. Additional tests were made to identify the compound eluted from 24 to 32 minutes as cAMP. The same results were obtained with partially purified adenylate cyclase whether ^{14}C- or ^{32}P-labeled ATP was used as substrate. Partially purified soluble phosphodiesterase active against cyclic 3′,5′-AMP[8] abolished completely the radioactive peak eluted at the cAMP position.

Assay for Cyclic 3′,5′-AMP Phosphodiesterase from Bovine Brain by the Same Procedure

With cyclic [^{14}C]AMP as substrate, the diesterase activity can be determined by measuring 5′-AMP eluted at the breakthrough volume. Either soluble or membrane bound diesterase was used.[25] Brain cortex was homogenized in Tris buffer centrifuged as described above for the preparation of adenylate cyclase. The supernatant was used as the source of soluble diesterase. For the preparation of membrane-bound enzyme, the pellet was resuspended and washed 2–3 times with the same buffer (Tris·HCl at pH 7.6, 5 mM $MgCl_2$, 1 mM dithiothreitol) in order to remove water-soluble diesterase. The precipitate was then extracted 3–4 times with 3 volume of neutral 2% aqueous digitonin. The clear extracts obtained after 40 minutes of centrifugation at 48,000 g were combined. The protein was precipitated between 30% and 60% saturation of $(NH_4)_2SO_4$, redissolved, and then dialyzed against the Tris·HCl buffer.

[24] K. Takahashi, W. H. Stein, and S. Moore, *J. Biol. Chem.* **242**, 4682 (1967).
[25] W. Y. Cheung, *Biochemistry* **6**, 1079 (1967).

The assay medium contained 0.2 mM [^{14}C] cAMP (specific activity 50 μCi/mmole), 5 mM MgCl$_2$, 10 mM NaF, 2 mM dithiothreitol, 50 mM Tris·HCl, pH 7.6 and 5 to 100 μg of protein in a final volume of 0.5 ml. After 10 minutes of incubation at 37°, the reaction was terminated by boiling the tubes for 2 minutes. Then 0.7 ml of 0.2 mM 5'-AMP was added as carrier. The tubes were centrifuged if the mixture was cloudy, and 1-ml samples were taken for chromatographic analysis.

Samples of 1 N HCl were loaded alternately in the sample injector. The interval between injections was 10 minutes. The eluent was the same as that used for the adenylate cyclase assay. The radioactivty of 5'-AMP was eluted from 4 to 8 minutes and the counts were printed out every 4 minutes. The efficiency of the counting system for ^{14}C was about 38%. Optimal conditions for the chromatographic assay of cyclic 3',5'-AMP phosphodiesterase requires further study; the responses to amount of enzyme and to time of incubation are given in Fig. 5. In a typical assay, 10–20% of the cAMP is converted to 5'-AMP. The inclusion of 10 mM NaF in the assay medium inhibits 5'-nucleotidase activity by 90%. If the addition of NaF is undesirable, the total amount of both 5'-AMP and adenosine formed during assay can be measured simultaneously by slight modification of the procedure.

In order to elute both 5'-AMP and adenosine at the breakthrough volume, a shorter column (0.9 × 3.2 cm) is required. The ionic strength of eluent is lowered to 0.25 M NaCl (with 50 mM Tris·HCl and 0.1% Brij 35 at pH 7.6) to allow sufficient retardation of the cAMP peak. The elution temperature is raised from 25° to 50° to sharpen the peak of adenosine. With this modification, the breakthrough peak of 5'-AMP

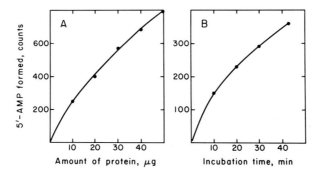

Fig. 5. Linearity of response in the cyclic 3',5'-AMP phosphodiesterase assay. The solubilized preparation of the membrane-bound enzyme was used. (A) Proportionality to the amount of phosphodiesterase. (B) Proportionality to the incubation time. Each point represents the average of duplicate determinations.

and adenosine is now eluted from 2 to 6 minutes and 1 ml of 1 N HCl can be injected at 6 minutes to wash off the cAMP.

Discussion

The uniqueness of this method is its automation, and thus the minimal effort by the investigator. The essentially complete recovery of cyclic AMP makes it unnecessary to use two radioisotopes to correct for loss. A reusable ion-exchange column and the use of a flow cell for radioactivity measurement obviates the need for handling and disposal of numerous radioactive columns and counting vials. Another important aspect of this procedure is its versatility. The same method with slight modification can be used for phosphodiesterase assay. The principles of the use of Dowex 1-X8 for the chromatography of nucleotides are sufficiently well known to make it possible to extend the ion-exchange method to the assay of several additional enzymes that act upon nucleotides. The relative position of nucleotides can be changed simply by varying the ionic strength of the eluent. The formation of adenosine, 5'-AMP, ADP, or phosphate from their respective precursors can be measured for the assay of various phosphatases. Since guanine nucleotides behave similarly on the anion-exchange resin, this procedure can be adapted for enzymes utilizing guanine nucleotides as substrate. If the UV absorbance of the product provides sufficient sensitivity for detection, then a recording spectrophotometer can be used in place of the flow cell and scintillation counter.

The main drawback of this method is the low sensitivity. The method as it stands now has a sensitivity of 50 pmoles of cAMP, which may not be low enough for cyclases from other sources. The amount of radioactivity of the substrate is limited to such extent that the tailing of the ATP peak does not increase the background significantly. Possible improvements in sensitivity of the method include a higher temperature to facilitate the elution of nucleotides, a longer column or slower flow rate to permit the use of a higher concentration, and more radioactivity of the substrate. The tailing of ATP is likely due to the tendency of anthracene in the flow cell to retain the nucleotides, therefore, any reduction of this tendency should improve the separation of radioactive peaks of ATP and cAMP. The sensitivity of the method can also be improved by slowing the rate of passage of cAMP peak through flow cell and thus increasing the counting time.

The time per analysis can be shortened to about 30 minutes by using two timers for sample injection. A 72-sample injector could be used to permit an overnight load of 36 samples without attendance of the worker. It is the hope of the author that this automated procedure, in spite of its drawbacks would add another dimension to the available assay procedures for adenylate cyclase.

Acknowledgment

This research, supported in part by Grants GM 7256 and NS 9639 from the United States Public Health Service, was done in the laboratories of Drs. Stanford Moore and William H. Stein at the Rockefeller University, New York, New York. Their advice is gratefully acknowledged.

[17] Preparation of Particulate and Detergent-Dispersed Adenylate Cyclase from Brain

By ROGER A. JOHNSON and EARL W. SUTHERLAND[1]

Adenylate cyclase catalyzes the formation of adenosine 3',5'-monophosphate (cAMP)[1a] from ATP, and has been well characterized in numerous tissues.[2-4] A fuller understanding of the properties of the enzyme have been impaired in part by the insolubility and crude particulate nature of most mammalian cyclase preparations and in part by the lability of the enzyme. Accordingly, these factors have also hampered the purification of the enzyme. The purification of adenylate cyclase may be facilitated by the use of detergents since detergents have been shown to be useful in dispersing the enzyme from several mammalian sources, and the "solubilized" enzyme has been studied in a few instances.[2-10]

Of the various tissues studied to date, neural tissues have exhibited the greatest adenylate cyclase activities,[2-4] and accordingly may provide an excellent source of high activity enzyme for its eventual purification.

[1] Deceased.
[1a] Abbreviations used: cAMP, adenosine 3',5'-monophosphate; DTT, dithiothreitol; BSA, bovine serum albumin; EDTA, ethylenediaminetetraacetic acid; CDTA, cyclohexanediaminetetraacetic acid; EGTA, ethyleneglycol bis(β-aminoethylenether)N,N,N',N'-tetraacetic acid.
[2] E. W. Sutherland, T. W. Rall, and T. Menon, *J. Biol. Chem.* **237**, 1220 (1962).
[3] G. A. Robison, R. W. Butcher, and E. W. Sutherland, "Cyclic AMP," Academic Press, New York, 1971.
[4] J. P. Perkins, *Advan. Cyclic Nucleotide Res.* **3**, 1 (1973).
[5] R. A. Johnson and E. W. Sutherland, *J. Biol. Chem.* **248**, 5114 (1973).
[6] L. R. Forte, *Biochem. Biophys. Acta* **266**, 254 (1972).
[7] M. C. Lin, *Fed. Proc., Fed. Amer. Soc. Exp. Biol.* **30**, 1206 (1971).
[8] N. I. Swislocki, *Fed. Proc., Fed. Amer. Soc. Exp. Biol.* **31**, 911 (1972).
[9] G. S. Levey, *Biochem. Biophys. Res. Commun.* **38**, 86 (1970).
[10] G. S. Levey, *Biochem. Biophys. Res. Commun.* **43**, 108 (1971).

In addition to providing a source of high activity enzyme, the brain adenylate cyclase exhibits some characteristics which appear distinct from those cyclases isolated from other tissues.[2-10]

Assay of Adenylate Cyclase Activity

Adenylate cyclase activity is determined by measuring the amount of cAMP formed from ATP by either the luminescence assay for cAMP[11] or the cAMP-binding assay of Gilman.[12] The reaction mixture contains ATP (usually 4 mM), divalent cation (Mn^{2+}, Mg^{2+}, or Co^{2+}, usually 8 mM), 50 mM glycylglycine buffer (from either Sigma or Calbiochem), pH 7.5, 10 mM theophylline, and, unless otherwise indicated, 1 mg/ml BSA (cystalline, lyophilized from Sigma), and 1 mM DTT (from Calbiochem). The enzyme fraction is then diluted between 5- and 20-fold with the appropriate homogenizing medium prior to assay. The diluted enzyme constitutes 10% of the assay reaction volume of 0.5 ml with a final protein concentration usually of 20–40 μg/ml. The reaction period is 5 minutes at 37°. Under these conditions of low protein and high substrate concentrations, an ATP regenerating system was found to be unnecessary.[5] The reaction is terminated by the addition of 0.2 ml of 0.3 N HCl (or 1.5 N $HClO_4$) containing 5000–10,000 cpm [G-^{3}H]cAMP (24 Ci/mmole, from New England Nuclear) for measuring loss of sample during its fractionation and purification prior to determination of the cAMP.

The assay of adenylate cyclase based on the determination of labeled cAMP derived from labeled ATP (usually [α-^{32}P]ATP or [2-^{3}H]ATP) can also be useful if appropriate care is taken to separate the cAMP formed from the ATP and the by-products of ATP degradation.[13]

Preparation of Particulate Adenylate Cyclase

Male Sprague-Dawley rats weighing 150–250 g are sacrificed by decapitation. The whole cerebellum or cerebrum is immediately removed, taking perhaps 30–60 seconds, and placed in ice-cold homogenizing medium; the tissue weight is then determined. No attempt is made to separate the cortical layer from the rest of the brain. The tissue is then homogenized on ice or at 4° in 9 volumes (relative to tissue weight) of homogenizing medium with three passes of a glass–Teflon motor-driven homogenizer according to the scheme shown in Fig. 1. The third homogenate is designated the "washed particulate" preparation.

[11] R. A. Johnson, *Advan. Cyclic Nucleotide Res.* **2**, 81 (1972).
[12] A. G. Gilman, *Proc. Nat. Acad. Sci. U.S.* **67**, 305 (1970).
[13] G. Schultz, this volume [15].

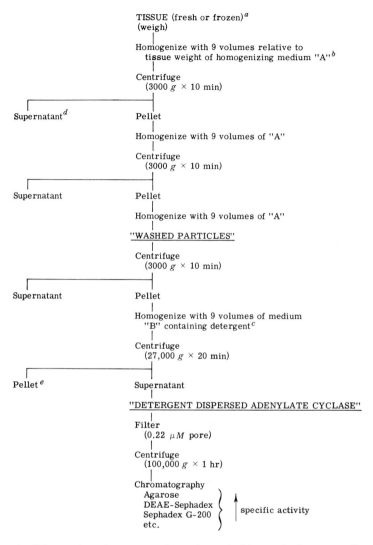

FIG. 1. Scheme for the preparation of particulate and detergent dispersed adenylate cyclase from rat brain.

a Either freshly excised cerebellum or cerebrum can be used. Frozen brain may also be used if it is quickly frozen following excision and then stored at $-70°$. Prior to homogenization the frozen brain is pulverized with a mortar and pestle on dry ice.

b Homogenizing medium "A": 0.25 M sucrose, 0.1 M glycylglycine buffer, pH 7.5, and 3 mM DTT.

c Homogenizing medium "B": 0.1 M glycylglycine buffer, pH 7.5, 3 mM DTT, and 1% (w/v) Lubrol-PX.

d The first supernatant fraction may contain as much as 35–40% of the total activity of the whole homogenate.

e Significant activity remains in the pellet derived from the dispersion procedure.

In other experiments lower specific activities resulted when the initial 3000 g for 10 minutes centrifugation step was substituted with a spin at 600 g for 10 minutes and the resulting supernatant fraction used for the preparation of washed particulate and detergent dispersed enzyme. Also, if instead of the glass–Teflon motor-driven homogenizer an Ultra-Turrax homogenizer was used, lower activity resulted.

In some instances 1 mM EDTA and 2 mM MgCl$_2$ have been included in the homogenizing medium, though these additions now seem to be unnecessary.

Dispersion of Particulate Brain Adenylate Cyclase

"Solubilized" adenylate cyclase is prepared from the washed particulate enzyme by dispersion with detergent (see Fig. 1). The particles are collected by centrifugation at 3000 g for 10 minutes. The pellet is then resuspended and rehomogenized in 9 volumes of homogenizing medium containing Lubrol-PX (from ICI America), a nonionic detergent. This homogenate is then centrifuged immediately at 27,000 g for 20 minutes and the resulting supernatant fraction is removed and contains the dispersed adenylate cyclase.

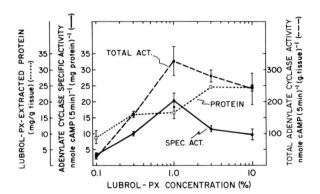

FIG. 2. Dispersion of adenylate cyclase from cerebellum by Lubrol-PX. The data represent the mean ± SEM of adenylate cyclase activity determined on from three to six separately homogenized and dispersed tissue preparations as shown in Fig. 1, involving from one to five cerebellums each, at each concentration of Lubrol-PX in the homogenizing medium tested. The homogenizing medium contained 1 mM EDTA and 2 mM MgCl$_2$ but did not contain DTT. Homogenizing medium "B" contained sucrose as well. Each preparation was assayed for cyclase activity by duplicate incubations. Protein concentrations ranged from 5.4 to 30 μg/ml. Total activity (●----●); specific activity, (●——●); and protein (□....□). Data from R. A. Johnson and E. W. Sutherland, *J. Biol. Chem.* **248**, 5114 (1973).

The effects of several concentrations of Lubrol-PX on the dispersion of protein and adenylate cyclase from washed particulate cerebellar preparations are shown in Fig. 2. The optimal concentration of Lubrol-PX for the dispersion of the enzyme is 1% (w/v). Further increasing the detergent concentration results in increased dispersion of protein, decreased cyclase specific activity, and decreased total dispersed activity.

The presence of DTT in the homogenizing medium affects this dispersion by Lubrol-PX. At a concentration of 0.3 mM DTT a measurable increase in specific activity of the dispersed enzyme can be observed; with an optimal increase at 3 mM DTT. Part of this stimulatory effect of DTT is due to a stabilizing effect on the enzyme during its assay as will be described below. In addition, DTT also increases the amount of detergent-solubilized protein. Accordingly, DTT is routinely included in the homogenizing medium.

The dispersion of adenylate cyclase from cerebellum and cerebrum is compared in the table. In both cases about half the protein and most

COMPARATIVE DISPERSION OF ADENYLATE CYCLASE FROM CEREBELLUM AND CEREBRUM[a]

Preparation	Protein (mg/g tissue)	Adenylate cyclase activity		Total[b] (units/g tissue)
		Control	Lubrol-PX,	
		[nmole (5 min)$^{-1}$(mg protein)$^{-1}$]		
Cerebellum				
Washed particles	49.8 ± 1.3	1.50 ± 0.12	11.8 ± 0.1	117
Lubrol-PX supernatant	23.3 ± 1.7	24.2 ± 1.7	—	113
Cerebrum				
Washed particles	50.9 ± 1.9	1.05 ± 0.03	7.23 ± 0.22	74
Lubrol-PX supernatant	29.6 ± 0.4	10.7 ± 0.4	—	63

[a] Washed particulate and Lubrol-PX dispersed preparations of adenylate cyclase were prepared as shown in Fig. 1. The homogenizing media contained 1 mM EDTA and 2 mM MgCl$_2$ and homogenizing medium "B" also contained 0.25 M sucrose. The adenylate cyclase activities represent the mean ± SEM of activity determined with triplicate incubations on two separately homogenized tissue preparations, each derived either from four pooled cerebellums or from one cerebrum. One unit of enzyme activity equals 1 nmole of cAMP formed per minute. Protein values represent the average ± range of the two preparations. The reaction mixture contained protein concentrations of 50–52 µg/ml for the particulate preparations and 20–27 µg/ml for the dispersed preparations. Data from R. A. Johnson and E. W. Sutherland, *J. Biol. Chem.* **248**, 5114 (1973).

[b] Total activity was calculated from the specific activity of washed particles assayed in the presence of 0.1% Lubrol-PX and from the control activity of the Lubrol-PX supernatant preparation.

(>85%) of the total enzyme activity of the washed particles was dispersed by detergent into the 27,000 g for 20 minutes supernatant fraction. The total adenylate cyclase activity in both washed particulate and in the dispersed preparations of cerebellum was consistently greater than that found in the cerebrum.

Characteristics of "Particulate" Brain Adenylate Cyclase

Particulate brain adenylate cyclase is not consistently stimulated by catecholamines. Fluoride stimulates the enzyme with maximum stimulation occurring with about 4–5 mM F$^-$; higher concentrations are less effective and can inhibit the enzyme. The enzyme activity can be enhanced by freezing and thawing,[14] presumably by rupturing incompletely homogenized cells or vesicles, and also by detergents.[5,14] The stimulatory effect of detergents is effectively limited to nonionic detergents, Triton X-100, Lubrol-WX, and Lubrol-PX being most effective. Lubrol-PX exerts its stimulatory effect at an optimal concentration of 0.1% in the reaction mixture. The stimulatory effect of Lubrol-PX, on either cerebellar or cerebral adenylate cyclase, is enhanced (about +70%) by the inclusion of DTT in the reaction mixture, although DTT is without effect on the enzyme in the absence of detergent. Furthermore, in the presence of 0.1% Lubrol-PX fluoride no longer exerts its stimulatory effect and at concentrations of 10 mM and gretaer is inhibitory; 10 mM F$^-$ inhibits 20%.[5,15]

Characteristics of "Solubilized" Brain Adenylate Cyclase

Solubility. Detergent-dispersed adenylate cyclase is visibly transparent, although often it has a very slightly yellow-red color, presumably due to small amounts of hemoglobin contamination. The dispersed enzyme loses little (4%) of its activity on passage through a 0.22-μm pore diameter filter (Millipore) and loses slightly more (13%) of its activity when the filtered enzyme is centrifuged at 100,000 g for 1 hour.

Stability. Detergent dispersed adenylate cyclase is relatively stable to rapid freezing and thawing. After 4 times frozen and thawed, no activity is lost; after 16 times, about 25% of the initial activity may be lost. At −70° the enzyme is stable to storage for at least 3 months.

When the enzyme is preincubated, in its homogenizing medium, at 0° for 4 hours, 30% of the initial activity is lost. However, this loss is

[14] J. P. Perkins and M. M. Moore, *J. Biol. Chem.* **246**, 62 (1971).
[15] R. A. Johnson and E. W. Sutherland, *Fed. Proc., Fed. Amer. Soc. Exp. Biol.* **30**, 220 (1971).

prevented by the inclusion of BSA and DTT in the homogenizing medium. With preincubation of adenylate cyclase at 37° (in its homogenizing medium containing 1% Lubrol-PX), there is a rapid loss of enzyme activity (half-life ca. 6 minutes) followed by a slower rate of inactivation (half-life ca. 4 hours). In the presence of BSA and DTT the overall rate of inactivation is decreased by a factor of about 2. However, in the assay reaction mixture (i.e., 0.1% Lubrol-PX) the enzyme is not rapidly inactivated if both BSA and DTT are present. In the absence of either agent the dispersed enzyme loses activity with time at 37°. The inclusion of both agents in the reaction mixture permits the reaction to be linear for up to 20 minutes. The effects of BSA and DTT are more than additive. The minimal concentrations of BSA and DTT for a linear 5-minute reaction period is 1 mg/ml and 1 mM, respectively.

It is evident that because of the lability of the enzyme in the presence of 1% detergent at 37°, some caution must be exercised in the freezing and thawing of the enzyme for repetitive use. Consistent enzyme activities are obtained when the stored enzyme aliquots (at −70°) are thawed by swirling in cold tap water (ca. 20–22°) until the last traces of ice remain and then the enzyme is placed in ice (0°). For storage the enzyme aliquots are rapidly frozen in liquid N_2.

Interaction with Cations and Substrate. In the presence of 4 mM ATP maximal activity is obtained with 6–8 mM Mg^{2+} and half-maximal activity with 4–5 mM Mg^{2+} or 3–4 mM Mn^{2+}. Invariably the enzyme activity in the presence of Mn^{2+} is greater than that seen with equimolar Mg^{2+}, although the maximal activity with Mn^{2+} may vary from 10% to 50% greater than that with Mg^{2+}.[5]

At lower ATP concentrations (e.g., 0.5 mM) the difference between the effects of Mn^{2+} and Mg^{2+} are more distinct.[16] The maximal activity is obtained with 0.5 mM Mn^{2+} and is about twice the maximal activity obtained with Mg^{2+}, at 4 mM. Half-maximal activity is observed with about 0.3–0.4 mM Mn^{2+} or about 1 mM Mg^{2+}.

In the presence of 8 mM $MgSO_4$ dispersed cerebellar adenylate cyclase exhibits an apparent K_m for ATP of about 20 μM. These experiments were done in the absence of an ATP-regenerating system, but under the conditions of the assay at a protein concentration of 40 μg/ml, [γ-^{32}P]ATP hydrolysis was only 14% in 5 minutes at 10 μM ATP.

Calcium inhibits the dispersed brain adenylate cyclase[5] in the presence of 8 mM Mg^{2+}, 50% inhibition being observed at concentrations between 100 and 200 μM Ca^{2+}. However, in the presence of 8 mM Mn^{2+}, Ca^{2+} is without effect at concentrations up to 1 mM and inhibits only

[16] R. A. Johnson and E. W. Sutherland, unpublished observations (1973).

40% at 10 mM Ca^{2+}. No concentration of calcium tested (between 100 nM and 10 mM) has a stimulatory effect either alone or in combination with Mg^{2+}.

Dispersed brain cyclase is also markedly inhibited (>70%) by 100 μM Cu^{2+}, Fe^{2+}, or Zn^{2+} and to a lesser degree by 100 μM Sr^{2+} (-25%) or Co^{2+} (-10%).

The effects of several metal binding agents and chelators on the activity of dispersed brain adenylate cyclase have also been studied.[5] Cysteine, 8-OH-quinoline, 1,10-phenanthroline, ethylenediamine, CDTA, and EDTA inhibit adenylate cyclase by no more than 20% at concentrations up to 1 mM. However, EGTA, at concentrations as low as 10–30 μM, markedly inhibit the enzyme.[5,17] Both washed particulate and detergent dispersed adenylate cyclase from either cerebellum or cerebrum are inhibited by these low concentrations of EGTA.[5,16–21] The concentration exhibiting half-maximal inhibition is about 30 μM, but is dependent on the protein concentration in the assay reaction mixture. At higher protein concentrations higher EGTA concentrations are required for inhibition.[5]

The inhibition of brain adenylate cyclase by EGTA appears to be a function of free chelator concentration in that certain metals (e.g., Co^{2+}, Ca^{2+}, or Mn^{2+}) can either *prevent* or *reverse* the inhibition by EGTA. From experiments using EGTA-pretreatment and subsequent gel filtration on Sephadex G-25 (to separate enzyme from EGTA), it is apparent that the inhibitory effects of EGTA are readily reversible, but suggest that the factor(s) with which the chelator interacts is(are) not readily removed from the preparation. There are significant amounts of bound, not readily dissociable, metal in the detergent dispersed brain adenylate cyclase preparations with which the chelator may interact.[5,16,17] While the data suggest the participation of some metal in addition to added Mg^{2+} in the activity of brain adenylate cyclase, it remains possible that the highly specific inhibition by EGTA of the enzyme occurs in some manner independent of its metal binding properties.

Contaminating Enzyme Activities. Dispersed adenylate cyclase prepared in the described manner contains other contaminating en-

---

[17] R. A. Johnson and E. W. Sutherland, *Fed. Proc., Fed. Amer. Soc. Exp. Biol.* **32**, 568 (1973).

[18] L. S. Bradham, D. A. Holt, and M. Sims, *Biochim. Biophys. Acta* **201**, 250 (1970).

[19] L. S. Bradham, *Biochim. Biophys. Acta* **276**, 434 (1972).

[20] J. P. Perkins, personal communication (1972).

[21] Under conditions with which the brain adenylate cyclase is inhibited, EGTA (10 μM to 1 mM) has no inhibitory effect on the adenylate cyclase of either intact or detergent dispersed partially purified plasma membranes from rat liver. R. A. Johnson, S. J. Pilkis, and E. W. Sutherland, unpublished observations (1972).

zymes.[2,15,16] Significant ATPase activity [2–3 μmoles (5 min)$^{-1}$(mg protein)$^{-1}$ with 4 mM ATP] and 3′,5′-cyclic nucleotide phosphodiesterase activity (3–6 nmoles(30 min)$^{-1}$(mg protein)$^{-1}$ with 1 μM cAMP) have been observed.[16] Both these enzymes are capable of interfering with the determination of adenylate cyclase activity and should be considered in the design and interpretation of experiments. Protein kinase activities, both cAMP-dependent and -independent, have also been observed.[16]

[18] Preparation and Characterization of Adenylate Cyclase from Heart and Skeletal Muscle

By GEORGE I. DRUMMOND and DAVID L. SEVERSON

Assay Method

The assay depends upon the conversion of labeled ATP to labeled cyclic AMP (cAMP) according to the general methods described by Schultz[1] in this volume. [U-^{14}C]ATP (18–22 μCi/μmole) or [α-^{32}P]ATP (15–20 μCi/μmole) are used as substrate. For both cardiac[2,3] and skeletal muscle[4] preparations the assay system (final volume 150 μl) is fortified with an ATP regenerating system comprised of 20 mM phosphoenolpyruvate and 130 μg/ml of pyruvate kinase together with 5.5 mM KCl. In the case of heart extracts, destruction of labeled product by cyclic nucleotide phosphodiesterase is prevented by the addition of 8 mM theophylline and 2 mM unlabeled cAMP. Theophylline is omitted from the assay of skeletal muscle extracts because it is inhibitory. Assays are conducted at pH 7.5, the temperature is 37°, and the time of incubation is 10 or 20 minutes. The reaction is terminated by immersing the tubes in a boiling water bath for 3 minutes. Controls are carried in which all components except substrate are present; after boiling, labeled ATP is added and the tubes are again placed in a boiling water bath for 3 minutes. Denatured protein is removed by centrifugation. When [U-^{14}C]ATP is employed, labeled cAMP can be isolated by paper chromatography provided substrate concentration does not exceed 0.3 mM. An aliquot (100 μl) of each clear supernatant is applied to Whatman

[1] G. Schultz, this volume [15].
[2] G. I. Drummond and L. Duncan, *J. Biol. Chem.* **245**, 976 (1970).
[3] G. I. Drummond, D. L. Severson, and L. Duncan, *J. Biol. Chem.* **246**, 4166 (1971).
[4] D. L. Severson, G. I. Drummond, and P. V. Sulakhe, *J. Biol. Chem.* **247**, 2949 (1972).

No. 3 MM paper and chromatographed (descending) for 22 hours at 27° with 1 M ammonium acetate–95% ethanol (15:35) as the developing solvent. Authentic markers (0.05 μmole) of ATP, ADP, 5'-AMP, cAMP, adenosine, and inosine are included. After drying, the areas containing cAMP are cut out, placed in 18 ml of scintillation fluid (4 g of 2,5-diphenyloxazole and 50 mg of 1,4-bis[2-(5-phenyloxazolyl)]benzene dissolved in 1 liter of toluene), and radioactivity is determined by scintillation spectrometry. The amount of [^{14}C]cAMP present is calculated from the counts present (after correcting for radioactivity in the boiled controls) and the specific activity of the substrate. Separation of the product by paper chromatography should not be used for substrate concentrations above 0.3 mM, or when [α-^{32}P]ATP is used as substrate. In these cases labeled product is determined by chromatography of the supernatant fluid on Dowex AG 50W-X4 followed by precipitation with $Ba(OH)_2$–$ZnSO_4$ according to the method of Krishna et al.[5] (see Schultz[1]). Specific activity is defined as picomoles of cAMP formed per minute per milligram of protein. Protein is determined by the method of Lowry et al.[6]

Enzyme Preparation

As in most tissues, adenylate cyclase in heart and skeletal muscle is almost entirely bound to cellular membranes which sediment at low gravitational forces.[7,8] It is generally recognized that the enzyme is located predominantly in plasma membranes of most cells and tissues. Evidence exists, however, that some adenylate cyclase activity is present in cardiac sarcoplasmic reticulum.[9] It has been studied in a microsomal fraction of rabbit muscle,[10] and Rabinowitz et al.[11] have reported that the enzyme in this tissue is distributed largely in mitochondrial and microsomal fractions. The precise quantitative cellular localization of the enzyme is not settled. Most studies on kinetic properties and hormonal stimulation have been performed on washed particulate preparations sedimenting at low gravitational forces: 1000–2000 g.

[5] G. Krishna, B. Weiss, and B. B. Brodie, *J. Pharmacol. Exp. Ther.* **163**, 379 (1968).
[6] O. H. Lowry, N. J. Rosebrough, A. L. Farr, and R. J. Randall, *J. Biol. Chem.* **193**, 265 (1951).
[7] E. W. Sutherland, T. W. Rall, and T. Menon, *J. Biol. Chem.* **237**, 1220 (1962).
[8] F. Murad, Y.-M. Chi, T. W. Rall, and E. W. Sutherland, *J. Biol. Chem.* **237**, 1233 (1962).
[9] M. L. Entman, G. S. Levey, and S. E. Epstein, *Biochem. Biophys. Res. Commun.* **35**, 728 (1969).
[10] K. Seraydarian and W. F. H. M. Mommaerts, *J. Cell. Biol.* **26**, 641 (1965).
[11] M. Rabinowitz, L. Desalles, J. Meisler, and L. Lorand, *Biochim. Biophys. Acta* **97**, 29 (1965).

[18] ADENYLATE CYCLASE (HEART AND SKELETAL MUSCLE) 145

Heart. Hearts from adult guinea pigs or rabbits are perfused with physiological saline to thoroughly remove blood and placed in 10 mM Tris·HCl pH 7.5 at 4°. Ventricular tissue is excised, blotted, weighed, and homogenized in 10 volumes of the above buffer for 1 minute in a Sorvall Omni-Mixer. Connective tissue is removed by passing the homogenate through a nylon sieve (pore size 1 mm^2); it is then centrifuged at 1000 g for 15 minutes at 4°. The precipitate, which contains at least 95% of the activity, is washed twice by resuspending in buffer (original volume) and centrifuging. The resulting pellet is suspended in 5 volumes of buffer (based on tissue weight) in a Potter-Elvehjem homogenizer. The preparation should be used immediately. The specific activities of such preparations are usually 60–65 pmoles per minute per milligram of protein (basal activity) and 450–500 when assayed in the presence of 8 mM fluoride. Dog heart is equally active. Rat heart is less active (15 pmoles per minute per milligram). Hearts of clam, crab, frog, trout, and turtle are much less active.[2] A small (2- to 4-fold) enrichment of activity can be achieved by extracting washed particle preparations with LiBr. Thus, the suspension of washed particles is brought to 0.4 M LiBr by the addition of a 4 M solution of this salt. The suspension is stirred slowly at 4° for 45 minutes and then centrifuged at 1000 g for 20 minutes. The precipitate is washed twice by suspending in 10 mM Tris·HCl, pH 7.5 followed by centrifugation. The final pellet is suspended in 5 volumes of 10 mM Tris·HCl, pH 7.5 (based on original tissue weight). Such preparations can be lyophilized and the dry powders retain their activity for several months at −18°. Specific activities are in the range 100 to 150 pmoles per minute per milligram (basal activity) and 600 to 1000 pmoles per minute per milligram when assayed in the presence of 8 mM fluoride.

Skeletal Muscle. We have found that sarcolemma prepared from rabbit muscle contains a good yield of adenylate cyclase and have used this preparation for kinetic studies.[4] Hind leg muscle of an exsanguinated rabbit is thoroughly cleared of fat and connective tissue, minced with scissors and homogenized in 5 volumes of 50 mM CaCl$_2$ in a Sorvall Omni-Mixer for 15 seconds at maximal velocity. The homogenate after filtration through a nylon seive (pore size 1 mm^2) is centrifuged at 2000 g for 10 minutes. The sediment is washed four times by suspending in 10 volumes of 10 mM Tris·HCl, pH 8.0 (based on tissue weight) and centrifuging at 2000 g for 10 minutes. The washed particles are suspended in 5 volumes of this buffer (based on tissue weight) and 4.0 M LiBr is added to produce a final concentration of 0.4 M. The suspension is stirred slowly at 4° for 4–5 hours, then diluted 4-fold with 10 mM Tris·HCl, pH 8.0 and centrifuged at 2000 g for 10 minutes. The sediment is washed once with Tris·HCl buffer as described above. The residue is then suspended in 8 volumes (based on tissue weight) of 25% KBr containing 10 mM Tris·HCl pH 7.5 (density 1.21) and the suspension is centrifuged at 25,000 g for 30 minutes. The pellet is washed twice by suspending in 8 volumes of 10 mM Tris·HCl pH 8.0 (based on tissue weight) and centrifuging at 2000 g for 10 minutes. The final sediment (sarcolemma) is suspended by use of a Potter-Elvehjem homogenizer in 10 mM Tris·HCl pH 7.5 to a protein concentration of 2–5 mg/ml. Such preparations should be used immediately. The final preparation (before suspension by homogenization) when examined by phase-contrast microscopy consists of empty, transparent saclike structures. Marker enzyme activities and lipid composition are consistent with that of sarcolemma in a high degree of purity.[4] Adenylate cyclase activities of crude homogenates of rabbit muscle (prepared in 10 mM Tris·HCl, pH 7.5, rather than in 50 mM $CaCl_2$) are in the range of 50 pmoles of cAMP per minute per milligram (F^- stimulated activity). Specific activities of the sarcolemmal membrane preparation vary between 900 and 1200 pmoles per minute per milligram. Purification is usually in the range of 20- to 25-fold with a yield of 30–40%. The enzyme present in such membranes is relatively unstable. Storage at $-4°$ for 18 hr or even at $-80°$ leads to 50% loss in activity. Membrane preparations may be lyophilized; however, the dried powders lose 50% of their activity in 1 week when stored at $-18°$.

Properties of the Cardiac and Skeletal Muscle Enzyme

Effect of ATP and Metal Ions. The K_m of the cardiac enzyme for ATP is 0.08 mM. At pH 7.5 the true substrate is considered to be Mg·ATP. Concentrations of Mg^{2+} as high as 10 mM are required to saturate the enzyme, the K_a for this cation is 2–3 mM. The effect of this added Mg^{2+} was to increase V_{max}; it did not affect the K_m for substrate.[2] These observations indicated that Mg^{2+} may bind to a second site on the enzyme, the consequence of this binding being increased reactivity. Mn^{2+} and Co^{2+} can also stimulate enzyme activity; the K_a for Mn^{2+} is 0.7 mM. ATP concentrations in excess of divalent cation strongly inhibit the cardiac enzyme. Ca^{2+} inhibits the enzyme in a manner which appears competitive with Mg^{2+}; the K_i for Ca^{2+} is about 0.3 mM.

The K_m of the skeletal muscle sarcolemmal enzyme for ATP is 0.3–0.5 mM.[4] The K_a for Mg^{2+} is 3–5 mM and for Mn^{2+} and Co^{2+}, 1–2 mM. As for the cardiac enzyme, binding of cation to a presumed second site increased V_{max}.[4] ATP in excess of divalent cation strongly inhibits the sarcolemmal enzyme. Pyrophosphate, a product of the adenylate cyclase reaction, inhibits the sarcolemmal enzyme in a manner competitive with ATP (K_i, 0.45 mM).

Stimulation by Fluoride. The ability of F⁻ to stimulate adenylate cyclase from various sources has been known for some years.[7,8] This anion markedly increases basal activity in washed particulate preparations of hearts from clam, crab, frog, turtle, pigeon, guinea pig, rabbit, rat, dog, etc.[2] The degree of stimulation in guinea pig and rabbit heart washed particle, and lithium bromide extracted preparations is usually 5- to 8-fold. The skeletal muscle enzyme from a variety of species is similarly stimulated.[4] Maximal stimulation occurs at 8–12 mM F⁻; higher concentrations cause reduced stimulation. No other anion possesses this action and organic fluorine compounds are inactive. When examined kinetically, it was found that F⁻ increased reaction velocity of the cardiac (LiBr extracted preparation) and skeletal muscle sarcolemmal enzyme at all Mg²⁺ concentrations, even those above saturation.[3,4] The prominent kinetic effect was to increase V_{max}; there was no appreciable effect on K_m for substrate or on K_a for Mg²⁺. Fluoride stimulation of the skeletal muscle enzyme is partially irreversible; i.e., it can be only partially removed by dialysis.

Effect of pH. Activity of the cardiac enzyme is measured between pH 6 and 9.5 with a maximum between pH 7.0 and 8. The pH optimum of the skeletal muscle sarcolemmal enzyme is slightly more alkaline—pH 8.5. Fluoride stimulates enzyme activity from both sources at all pH values.

Stimulation by Hormones. Several investigators have examined the effect of hormones on the cardiac enzyme since it was first recognized that adenylate cyclase in a variety of tissues is under hormonal control.[7,8] The table summarizes data which show that the enzyme in washed particle preparations of hearts of several species is stimulated by a variety of hormones and neurohormones in low concentrations—catecholamines, glucagon, histamine, thyroxine, and prostaglandins. For catecholamines, the order of potency is isopropylnorepinephrine > epinephrine > norepinephrine. The stimulatory action of these agents is blocked by β-adrenergic blocking agents such as dichloroisoproterenol and propranolol.[3,8,12–16] These facts are consistent with adenylate cyclase being associated with the β-adrenergic receptor. Adrenergic blocking agents do not prevent the stimulatory action of glucagon,[12,13] histamine,[16] or prosta-

[12] F. Murad and M. Vaughan, *Biochem. Pharmacol.* **18**, 1053 (1969).
[13] G. S. Levey and S. E. Epstein, *Circ. Res.* **24**, 151 (1969).
[14] I. Klein and G. S. Levey, *Metabolism* **20**, 890 (1971).
[15] S. E. Mayer, *J. Pharmacol. Expl. Ther.* **181**, 116 (1972).
[16] I. Klein and G. S. Levey, *J. Clin. Invest.* **50**, 1012 (1971).

EFFECT OF HORMONES ON ADENYLATE CYCLASE IN WASHED PARTICULATE PREPARATIONS OF HEARTS OF VARIOUS SPECIES

Hormone or neurohormone	Species	Concentration for ½ maximal stimulation (μM)	Concentration for maximal effect (μM)	Stimulation (percent over basal)	References
Norepinephrine	Rat	5	100	260	a
	Cat	5	50	250	b
	Guinea pig	8	50	200	c
	Guinea pig	5	50	200	d
	Dog	8	60	125	e
	Dog	8	100	165	f
	Human	—	50	160	g
Epinephrine	Dog	3	100	175	f
	Dog	7	20	150	e
	Rat	0.5	10	125	h
Isopropylnorepinephrine	Dog	0.7	100	195	f
	Dog	0.8	5	125	e
Glucagon	Rat	0.5	10	125	h
	Cat	1.0	10	180	g
	Human		10	165	g
Histamine	Guinea pig	9	10	300	d
	Cat	—	300	160	d
	Human	—	300	190	d
Thyroxine	Cat	0.5	5	170	i
Triiodothyronine	Cat	0.6	1	165	i
Prostaglandin (PGE₁)	Guinea pig	0.0004	0.001	170	c
Prostaglandin (PGE₂)	Guinea pig	—	100	150	c

[a] P. J. Laraia and W. J. Reddy, *Biochim. Biophys. Acta* **177**, 189 (1969).
[b] G. S. Levey, C. L. Skelton, and S. E. Epstein, *Endocrinology* **85**, 1004 (1969).
[c] I. Klein and G. S. Levey, *Metabolism* **20**, 890 (1971).
[d] I. Klein and G. S. Levey, *J. Clin. Invest.* **50**, 1012 (1971).
[e] S. E. Mayer, *J. Pharmacol. Exp. Ther.* **181**, 116 (1972).
[f] F. Murad, Y.-M. Chi, T. W. Rall, and E. W. Sutherland, *J. Biol. Chem.* **237**, 1233 (1962).
[g] G. S. Levey and S. Epstein, *Circ. Res.* **24**, 151 (1969).
[h] F. Murad and M. Vaughan, *Biochem. Pharmacol.* **18**, 1053 (1969).
[i] G. S. Levey and S. E. Epstein, *Biochem. Biophys. Res. Commun.* **33**, 990 (1968).

glandin (PGE₁).[14] The antihistaminic diphenhydramine specifically blocks histamine stimulation,[16] but not that produced by norepinephrine. These facts are consistent with the generally accepted idea that separate and specific receptors (discriminators) exist for each hormone. In all cases, stimulation of cardiac adenylate cyclase by hormone is relatively modest (see the table), most producing less than a 2-fold increase in velocity. Such stimulation is significantly less than that produced by F^-.

Adenylate cyclase present in lithium bromide extracted preparations of guinea pig heart and sarcolemma from rabbit skeletal muscle is stimulated 2- to 3-fold by epinephrine. Concentrations of isopropylnorepinephrine, epinephrine, and norepinephrine required to produce half-maximal stimulation of the sarcolemmal enzyme were 0.15, 0.5, and 10 μM, respectively.[4] These responses are blocked by propranolol. When examined kinetically, epinephrine, like F^-, was found to stimulate the cardiac and skeletal muscle enzyme at all Mg^{2+} concentrations. The effect was predominantly on V_{max}; there was no pronounced effect on the K_m for substrate or on the K_a for Mg^{2+}. Thus the stimulatory effect of the neurohormone seems primarily due to increased reactivity of the catalytic site for substrate.

Solubilization of Adenylate Cyclase from Heart

In their early studies, Sutherland et al.[7] reported solubilization of the enzyme from heart and skeletal muscle with Triton X-100, but the recovery of activity was low. Recently Levey[17,18] has solubilized the enzyme from canine ventricle using the nonionic detergent Lubrol-PX. The solubilized enzyme retains its sensitivity to F^-, but is not sensitive to norepinephrine, glucagon, or thyroxine at concentrations which fully stimulate the particulate enzyme. Addition of phosphatidylserine to the assay mixture (3 mg/ml) of a solubilized preparation (freed of detergent by chromatography on DEAE-cellulose) completely restored glucagon responsiveness.[19] Norepinephrine responsiveness was not restored by phosphatidylserine. However sensitivity to this neurohormone was restored by the addition of phosphatidylinositol at concentrations of 0.4 to 4 μg/ml to the assay.[20] In the presence of this phospholipid, half-maximal stimulation was achieved by norepinephrine at 0.08 μM,[20] a concentration only 1% that required for the particulate enzyme (see the table). Stimulation by norepinephrine under these conditions was blocked by propranolol. These observations clearly emphasize a possible role of phospholipids in hormonal regulation of adenylate cyclase and are described in detail by Levey[21] in this volume.

[17] G. S. Levey, Biochem. Biophys. Res. Commun. **38**, 86 (1970).
[18] G. S. Levey, Ann. N.Y. Acad. Sci. **185**, 449 (1971).
[19] G. S. Levey, Biochem. Biophys. Res. Commun. **43**, 108 (1971).
[20] G. S. Levey, J. Biol. Chem. **246**, 7405 (1971).
[21] G. S. Levey, this volume [24].

[19] Adenylate Cyclase from Kidney and Bone

By S. J. Marx and G. D. Aurbach

Preparation of Enzyme

Crude Enzyme

Male Sprague-Dawley rats, 150–200 g, are killed by decapitation. The kidneys are removed, chilled in 0.25 M sucrose, and the hilar structures are dissected away. If desired, the kidneys can be sectioned further into cortical, corticomedullary, and medullary zones. The three zones contain distinct receptors for parathyroid hormone, calcitonin, and vasopressin.[1,2] The tissue is transferred to 8 volumes (w/v) of Tris·HCl 0.05 M, pH 7.4 containing 10% (v/v) DMSO and homogenized in glass with a motor-driven Teflon pestle. The particulate fraction is obtained by centrifugation at 2000 g for 10 minutes; the pellet is resuspended in the original volume of Tris·HCl–DMSO and centrifuged again. The pellet is resuspended in Tris·HCl–DMSO to give a protein concentration of 15 mg/ml. This preparation is stored in small aliquots in liquid nitrogen. The enzyme in this form is stable indefinitely and has been used for *in vitro* bioassays for parathyroid hormone[3] or calcitonin.[4]

Purification by Density Gradient Centrifugation[5]

Materials

Dounce homogenizer (8 ml tube with large-clearance pestle), Kontes
Male Sprague-Dawley rats (150–200 g)
Swinging-bucket ultracentrifuge (Beckman L2-65B SW 40 rotor)
Buffers: (A) 0.25 M sucrose, 10 mM Tris, 1 mM Na$_2$EDTA, pH 7.5
(B) 10 mM Tris, 1 mM Na$_2$EDTA, pH 7.5

The kidneys from male Sprague-Dawley rats (150–200 g) are removed and chilled in buffer A. Solutions are kept on ice throughout. The renal capsules and perirenal fat pads are removed, and the hilar structures are cut out with a wedge-shaped incision that includes a small amount of the perihilar renal parenchyma. This removes the intrarenal portion

---

[1] L. R. Chase and G. D. Aurbach, *Science* **159**, 545 (1968).
[2] S. J. Marx, C. J. Woodard, and G. D. Aurbach, *Science* **178**, 999 (1972).
[3] R. Marcus and G. D. Aurbach, *Endocrinology* **85**, 801 (1969).
[4] J. Heersche, R. Marcus, and G. D. Aurbach. *Endocrinology* **94**, 241 (1974).
[5] S. J. Marx, S. A. Fedak, and G. D. Aurbach, *J. Biol. Chem.* **247**, 6913 (1972).

of the renal arteries that are difficult to process in the homogenizer. The kidneys are then bisected in a plane through the hilum and both poles. Hydronephrotic kidneys are discarded. Two volumes of buffer A are added to the renal tissue, and it is homogenized with 10 strokes in a Dounce homogenizer. This crude homogenate is centrifuged in the SS-34 rotor of a Sorvall RC-2B centrifuge. As soon as the rotor has reached 4500 rpm (2200 g), approximately 30 seconds, the rotor drive switch is turned off with the brake on. The pellet is discarded, and the supernatant solution is centrifuged twice again, the precipitates being discarded. Then the remaining supernatant fluid is centrifuged at 4500 rpm for 15 minutes at 4°. The reddish supernatant fluid is gently poured off and discarded, and the upper portion of the resulting double-layered pellet is resuspended in a small excess of buffer A by gentle swirling and then mixing with 10 strokes of the Dounce homogenizer. This material is diluted to 20–40 ml with buffer A and again centrifuged at 4500 rpm for 15 minutes. The resulting pellet is resuspended in a minimal volume of buffer A giving a thick preparation with protein concentration of 15–25 mg/ml. When tissue from a few animals is being processed suspension is accomplished by aspirating several times through a Pasteur pipette. With larger amounts the Dounce homogenizer is used. This partially purified membrane fraction is overlayered onto continuous gradients of 32% to 42% sucrose (w/w) made up in buffer B. Gradients of 11–12 ml are prepared and samples of 0.5 to 1.5 ml are applied. Material from up to 40 animals can be processed in a single ultracentrifuge run. The gradients should be approximately linear. Centrifugation is carried out at 100,000 g (23,000 rpm in the SW 40 rotor of the Beckman L2-65B ultracentrifuge) at 4° with the brake on. After centrifugation, two major bands of turbidity as well as a large pellet are obtained. The upper band (corresponding to a density of 38% sucrose) is aspirated with a Pasteur pipette, diluted with 3 volumes of buffer B, and centrifuged 15 minutes at 2200 g. This pellet is the purified membrane fraction and can be resuspended in appropriate buffers for adenylate cyclase assays.

The characteristics of the purified membrane fraction have been described in detail.[5] On electron microscopy, it is seen to be composed of vesicles and large sacs, some containing mitochondria. No brush border elements are detectable in the preparation. It is enriched with respect to both adenylate cyclase and Na,K-ATPase but depleted of brush border enzymes such as alkaline phosphatase and 5′-nucleotidase. Thus, this preparation of adenylate cyclase represents membranes not derived from the brush border; it is activated by parathyroid hormone, calcitonin, vasopressin, or isoproterenol. Where indicated, the kidneys can be dissected into gross anatomical regions prior to processing. This allows segre-

ADENYLATE CYCLASE ACTIVITY[a]

Species	None	Addition		
		PTH	SCT	AVP
Rat	2.6 ± 0.0	15.2 ± 2	24.1 ± 0.3	25.7 ± 0.1
Cow	3.8 ± 0.2	8.6 ± 0.5	3.4 ± 0.2	9.3 ± 0.2
Man	2.3 ± 0.1	12.1 ± 0.3	2.9 ± 0.1	2.6 ± 0.1

[a] Enzyme was prepared as described from renal corticomedullary junction of the indicated species and tested with maximally effective amounts of parathyroid hormone, salmon calcitonin, or arginine vasopressin. In the bovine and human preparations, similar results were obtained using bovine or human calcitonins, respectively. Results are the mean ± SE of three determinations, expressed as nanomoles of cyclic AMP generated per milligram of protein in 30 minutes at 22°. The abbreviations used are PTH, bovine parathyroid hormone; SCT, salmon calcitonin; AVP, arginine vasopressin.

gation into regions specifically sensitive to particular hormones; the cortical adenylate cyclase is most responsive to parathyroid hormone, the medullary junction most responsive to vasopressin, and that from the corticomedullary junction most responsive to calcitonin or vasopressin.[1,2] The same method has been used to prepare enzyme from kidneys of the rat, cow, and man, in which a similar distribution of parathyroid hormone and vasopressin-sensitive enzyme has been found. The bovine and human kidneys, however, contain much less calcitonin-sensitive enzyme (see the table).

In contrast to the crude enzyme preparation, the stability of the purified (38% sucrose fraction) adenylate cyclase on storage has been variable whether the material has been stored in Tris buffer A, or Tris with 10% DMSO. Thus, it is best to prepare the membranes freshly for use in the adenylate cyclase system. On the other hand, the preparation seems stable (stored in aliquots as pellets) for radioligand studies. Specific binding activity for ^{125}I-labeled salmon calcitonin was not significantly altered after storage in liquid nitrogen for many months.

Preparation of Enzyme from Neonatal Bone

Materials

Neonatal rats (0–24 hours old)
Stainless steel mortar and pestle
Electrically driven rotating Teflon pestle (Tri-R S63)
Buffer C: 10 mM Tris, 20 mM Na$_2$EDTA, pH 7.5 in 10% DMSO

Solutions are kept on ice throughout the procedure. Calvaria from 20-40 rat fetuses are dissected and suspended in buffer C at 4°. They are then transferred to a stainless steel mortar (prechilled on dry ice) and powdered while frozen using a prechilled stainless steel pestle. The preparation is then suspended in 6 volumes of buffer C and homogenized for 30 seconds with an electrically driven rotating Teflon pestle. The homogenate is then filtered once through glass wool and centrifuged 15 minutes at 2200 g at 4°. This pellet can then be resuspended in appropriate buffers for adenylate cyclase assays.

The adenylate cyclase in this preparation is sensitive to parathyroid hormone and to calcitonin. The calvarial preparation contains calcitonin-binding receptors similar to those in the kidney.[2] Earlier experiments showed no effect of porcine calcitonin on adenylate cyclase in fetal calvaria.[6] Recent studies, however, using the buffer (C) containing DMSO show marked stimulation of the enzyme by the highly potent molecule salmon calcitonin.[4] In these latter studies, salmon calcitonin in concentrations as low as 1 ng/ml caused activation of the skeletal enzyme. Porcine calcitonin was also active, but at a higher dose range.

[6] L. R. Chase, S. A. Fedak, and G. D. Aurbach, *Endocrinology* **84**, 761 (1969).

[20] Preparation of Vertebrate Photoreceptor Membranes for Study of Adenylate Cyclase, Guanylate Cyclase, and Cyclic Nucleotide Phosphodiesterase

By J. J. Keirns, N. Miki, and M. W. Bitensky

Principle

Photoreceptor membranes (vertebrate rod outer segments) are purified by flotation on heavy sucrose.

Instructions for Isolation[1,2]

All procedures except deliberate light adaptation are carried out in the complete absence of visible light with the aid of infrared light sources (made from ordinary microscope or high intensity tungsten lamps fitted with a Corning CS No. 7-56 infrared filter) and infrared image converters

[1] M. W. Bitensky, R. E. Gorman, and W. H. Miller, *Proc. Nat. Acad. Sci. U.S.* **68**, 561 (1971).
[2] N. Miki, J. J. Keirns, F. R. Marcus, and M. W. Bitensky, *Exp. Eye Res.* **18** (1974) in press.

or "sniperscopes" (Metascope 9902E, Varo Inc., Garland, Texas). Frogs (*Rana pipiens* or *R. catesbeiana*) are dark adapted overnight and sacrificed by decapitation; the retina and pigment epithelium removed and placed in 47.6% sucrose at 4° (40 g of sucrose and 60 ml of water) in a half-full centrifuge tube, covered with Parafilm, and shaken vigorously 25 times by hand. The tubes are filled with 47.6% sucrose and centrifuged at 50,000 *g* (Spinco SW 50L) for 90 minutes. The rod outer segments form a thick paste at the top of the tube. The paste is harvested with a spatula and diluted with 40 µl of water per retina. On examination by phase contrast and electron microscopy the paste (suspended in isotonic sucrose) is found to be primarily composed of material readily recognizable as intact rod outer segments or stacks of discs. Following suspension in water the preparation consists of discs and vesicles smaller than discs. Outer segment membranes are prepared in a similar way from dark-adapted rats and from frozen cow retinas (Hormel, Austin, Minnesota). These mammalian outer segment membranes give results qualitatively the same as those described below for the frog membranes.

Measurements of Adenylate and Guanylate Cyclase and Cyclic Nucleotide Phosphodiesterase

Adenylate cyclase and guanylate cyclase are assayed using radioactive substrate ([^{3}H]ATP or [^{3}H]GTP) and isolating the cAMP or GMP by thin-layer chromatography.[3] The assay mixtures for the cyclases contain a regenerating system (creatine phosphokinase and phosphocreatine) and an inhibitor of phosphodiesterase (isobutyl methylxanthine). Phosphodiesterase is assayed using [^{3}H]cAMP or [^{3}H]cGMP and isolating the remaining substrate by thin-layer chromatography. The standard reaction mixture for phosphodiesterase contains only substrate, buffer, and magnesium chloride. Frog rod outer segments isolated and assayed in the dark exhibit an adenylate cyclase activity (V_{max}) of 0.14 nmoles/minute/mg protein, guanylate cyclase 0.3 nmoles/minute/mg protein, and cAMP phosphodiesterase 0.5 µmoles/minute/mg/protein.[4] If the phosphodiesterase is assayed in the presence of ATP (or another nucleoside triphosphate),[5] the activity in the dark is the same, but a 5- to 10-fold higher activity is found in membranes exposed to visible light sufficient to bleach the rhodopsin. The activation of phosphodiesterase

---

[3] J. J. Keirns and M. W. Bitensky, *Anal. Biochem.* (1974) in press.

[4] M. W. Bitensky, W. H. Miller, R. E. Gorman, A. H. Neufeld, and R. Robison, *Advan. Cyclic Nucleotide Res.* **1**, 317 (1972).

[5] M. W. Bitensky, N. Miki, J. J. Keirns, and F. R. Marcus, *Miami Winter Symp.*, **18**, (1974) in press.

is not proportional to bleaching of rhodopsin, but is complete with about a 3% bleach of the rhodopsin (whether this bleach is achieved with a flash of light or by mixing three parts of bleached material with 97 parts of dark material). The cyclic nucleotide phosphodiesterase has a substantial preference for cyclic GMP (K_m = 0.15 mM) over cAMP (8 mM) but approximately the same V_{max} for each substrate (4 μmoles/minute/mg protein, in the light in the presence of 1 mM ATP or CTP).

Certain deviations from the above procedure can reduce or abolish light regulation of cyclic nucleotide levels.[2,6] Use of "dim red" light during dissection and assay produces a substantial part of the 3% bleach necessary to observe "light" levels of enzyme activity. Glass on glass homogenization produces an activated phosphodiesterase which is not influenced further by light and ATP. Detergents (eg., 0.5% Triton X-100) prevent activation of the phosphodiesterase by light and ATP. Finally, a substantial loss of light sensitivity occurs on storage for more than 1 hour at 25° or 2 days at 4° (part of this loss can be prevented by dithiothreitol).

[6] N. Miki, J. J. Keirns, F. R. Marcus, J. Freeman, and M. W. Bitensky, *Proc. Nat. Acad. Sci. U.S.* **70**, 3820 (1973).

[21] Preparation and Properties of Adenylate Cyclase from *Escherichia coli*[1]

By Mariano Tao

Adenylate cyclase catalyzes the formation of adenosine 3′,5′-cyclic monophosphate (cAMP) from ATP and occurs widely both in animal tissues and in bacteria.[2] The bacterial enzyme differs considerably from that of the animal in its ease of solubilization and insensitivity toward hormones. However, the activity of the *E. coli* adenylate cyclase may be regulated by some metabolites since Makman and Sutherland[3] have found lower intracellular level of cAMP when cells are grown in the presence of glucose or other carbon sources.

[1] This work was supported in part by grants from the National Science Foundation (GB-27435 A#1) and from the American Cancer Society (BC-65 and BC-65A). The author is an Established Investigator of the American Heart Association.
[2] G. A. Robison, R. W. Butcher, and E. W. Sutherland, "Cyclic AMP." Academic Press, New York, 1971.
[3] R. S. Makman and E. W. Sutherland, *J. Biol. Chem.* **240**, 1309 (1965).

Assay Method

Principle. Adenylate cyclase activity is determined by measuring the amount of cyclic [^{32}P]AMP or cyclic [^{3}H]AMP formed from [α-^{32}P]ATP or [^{3}H]ATP, respectively. Thin-layer chromatography is used to separate the cyclic nucleotide from ATP and other breakdown products such as adenosine, ADP, and AMP.[4] The procedure of Krishna et al.[5] may also be employed for this purpose.

Reagents

Buffer A: 0.25 M Tris·HCl, pH 9.0; 0.1 M MgSO$_4$
Buffer B: 20 mM Tris·HCl, pH 8.0; 10 mM MgSO$_4$; 0.25 mM EDTA; 1 mM dithiothreitol
Buffer C: 10 mM Tris·HCl, pH 8.0; 1 mM dithiothreitol
[α-^{32}P]ATP (from International Chemical and Nuclear Corp.) or [^{3}H]ATP (from Schwarz BioResearch), 2 mM, pH 8–9, specific activity 50–100 cpm/picomole
cAMP, 5 mM
Ammonium acetate, 1.0 M
Ethanol, 95%
Cellulose thin-layer sheet (20 × 20 cm) with fluorescent indicator (Eastman Chromagram Sheet 6065)

Procedure.[6] The reaction mixture contains 10 μl of buffer A, 10 μl of 2 mM [α-^{32}P]ATP or [^{3}H]ATP, enzyme protein, and sufficient water to bring to a final volume of 50 μl. The incubation is carried out in a 2-ml micro-conical centrifuge tube for 30 minutes at 34°. The reaction is terminated by heating in a boiling water bath for 3 minutes after the addition of 5 μl of carrier cAMP (5 mM). The blank contains the same reaction composition except that the incubation period is omitted. The protein precipitate is removed by centrifugation and 25 μl of the supernatant is transferred to a 2-ml micro-conical centrifuge tube. Using a capillary tube, the total sample is applied to a cellulose thin-layer sheet at the origin, which is marked by drawing a line with a pencil at 2 cm from the edge. Following each application, the spot is dried under a jet of cool air. For best resolution, the spot should be no larger than 5 mm in diameter. Six to eight samples may be applied to each thin-layer sheet. The sheet is then placed between two glass plates (Eastman Chromagram Developing Apparatus 6071) and developed at room temperature using 1.0 M ammonium acetate–95% ethanol (30:75, v/v) as the solvent

---

[4] M. Tao and F. Lipmann, *Proc. Nat. Acad. Sci. U.S.* **63**, 86 (1969).
[5] G. Krishna, B. Weiss, and B. B. Brodie, *J. Pharmacol. Exp. Ther.* **163**, 379 (1968).
[6] M. Tao and A. Huberman, *Arch. Biochem. Biophys.* **141**, 236 (1970).

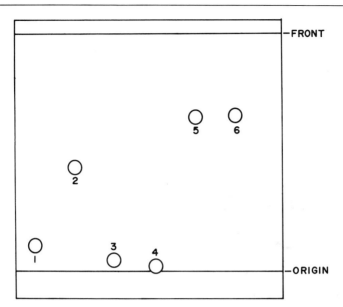

FIG. 1. Cellulose thin-layer chromatography of adenine and its derivatives. 1, 5′-AMP; 2, cyclic AMP; 3, ADP; 4, ATP; 5, adenosine; 6, adenine.

system. The time required for the solvent front to migrate to the top (18 cm from origin) of the thin-layer sheet is about 3 hours. The chromatogram is then removed and dried under a jet of cool air. Figure 1 illustrates the separation of adenine and its derivatives by the system described. The spot corresponding to cAMP is located under incident short wave ultraviolet light, cut out, and placed in a liquid scintillation. vial. When the radioisotope ^{32}P is used, it may be counted directly with 5 ml of Bray's solution.[7] In order to obtain higher counting efficiency for ^{3}H, the strips are soaked in 1 ml of water and counted in 10 ml of Bray's solution. The specific activity of the isotope used is determined by spotting a known amount of the isotope on a thin-layer strip about the size of the cAMP spot and counted under the same conditions.

Evaluation of the Assay Method. The thin-layer assay system is relatively simple and does not require excessive manipulation. The recovery of cAMP is almost quantitative. The system may also be used to assay for the formation of other cyclic nucleotides, such as uridine 3′,5′-cyclic monophosphate (cUMP), cytidine 3′,5′-cyclic monophosphate (cCMP), and guanosine 3′,5′-cyclic monophosphate (cGMP). The R_f values of these compounds and their derivatives are given in the table. One dis-

[7] G. A. Bray, *Anal. Biochem.* 1, 279 (1960).

R_f Values of Bases, Nucleosides, and Nucleotides on Cellulose Thin Layer

Compound	R_f
Adenine, adenosine	0.63
Guanosine	0.58
Uridine	0.73
Cytidine	0.68
ATP, GTP, CTP, UTP	0.02
ADP, GDP, UDP, CDP	0.04
5'-AMP	0.11
5'-GMP	0.06
5'-UMP	0.13
5'-CMP	0.09
Cyclic AMP	0.44
Cyclic GMP	0.40
Cyclic UMP	0.54
Cyclic CMP	0.47
Orthophosphate	0.06

advantage of the system is the amount of time required for developing the chromatogram. Furthermore, the use of [³H]ATP as the substrate for the determination of cyclase activity in a crude preparation may lead to an overestimate of the amount of cAMP formed. This could be attributed to either adenosine or adenine formed by contaminating enzyme(s), trailing into the region of cAMP. However, this may be avoided by employing [α-^{32}P]ATP as substrate.

Growth of E. coli Cells. In this study, Crooke's strain of *E. coli* (ATCC 8739) is used. However, several other strains of *E. coli* such as Q13, K12, and B are also found to contain adenylate cyclase activity. The cells are grown in a medium containing 14 g of K_2HPO_4, 6 g of KH_2PO_4, 2 g of $(NH_4)_2SO_4$, 0.2 g of $MgSO_4$, 10 g of glucose, 4 g of nutrient broth (Difco 0003-01), and 2.5 g of yeast extract (Difco 0127-01) per liter,[6] to an optical density of 400–700 units as measured with a 540-nm filter in a Klett-Summerson colorimeter, and harvested by centrifugation. The cells are washed with a buffer containing 10 mM Tris·HCl, pH 8.0, and 10 mM $MgCl_2$, and stored frozen at $-20°$ until used.

Although the intracellular level of cAMP is dependent on the presence of glucose in the growth medium,[3] extracts of cells appears to exhibit cyclase activity relatively independent of the glucose concentration in the medium.[8]

[8] A. Peterkofsky and C. Gazdar, *Proc. Nat. Acad. Sci. U.S.* **68**, 2794 (1971).

Purification of Adenylate Cyclase[6]

All operations are carried out at 0–4°. About 200 g of the frozen cells (wet weight) are resuspended in 200 ml of buffer B containing 1 mg of DNase and disrupted by passage through a French pressure cell at 10,000 psi. The disrupted cells are allowed to stand for 30 minutes to permit digestion by DNase and centrifuged at 30,000 g for 40 minutes. The cell debris is reextracted with 100 ml of buffer B by resuspension with a Teflon homogenizer and recentrifugation. To the combined supernatants, streptomycin sulfate is added to a final concentration of 1% (w/v). After 30 minutes, the precipitate is removed by centrifugation; and solid ammonium sulfate is added to the supernatant to 35% saturation. The solution is allowed to stand for 30 minutes; and the precipitate collected by centrifugation. The precipitate is washed twice with a total volume of 500 ml of buffer B containing 28% saturated ammonium sulfate. The final pellet is dissolved in about 50 ml of buffer C and centrifuged at 150,000 g for 2.5 hours, and the sediment is washed once more with about 30 ml of buffer C. The supernatants are combined, precipitated with 40% saturated ammonium sulfate, and the precipitate dissolved in buffer C. At this stage, the enzyme may be stored in liquid nitrogen or processed immediately as follows.

The protein solution is applied to a DEAE-cellulose (Gallard-Schlesinger Chemical Mfg. Corp.) column (2 × 6 cm), which has been previously equilibrated with buffer C. The column is then washed with buffer C, followed by buffer C containing 0.18 M KCl. During each wash, the eluate is monitored at 280 nm until the absorption reaches that of the buffer. Adenylate cyclase is eluted with a salt concentration of 0.3 M KCl and collected in 5-ml fractions. The elution of the protein peak is monitored by absorption at 280 nm. The protein fractions are pooled and concentrated to 5–10 ml by Diaflo ultrafiltration using a UM-10 membrane (Amicon Corp.) and stored frozen in liquid nitrogen. It is not necessary to remove the salt from the enzyme preparation since KCl has no effect on the enzyme activity. About 5 mg of enzyme protein is obtained from 200 g of wet cells. About a hundredfold purification is achieved by this procedure.

Properties

pH Optimum. The pH optimum for adenylate cyclase in 50 mM Tris·HCl buffer is 9.0.

Stability. When stored in liquid nitrogen, the enzyme is stable over a period of several months. Overnight storage at 0–4° leads to a loss of greater than 50% of activity.

Specificity. The enzyme preparation also catalyzes the formation of cGMP and cyclic dAMP. However, GTP and dATP are about 10% and 30% as active as ATP, respectively. The K_m for ATP is $4.5 \times 10^{-4}\ M$. 5'-D-Arabinosyladenine triphosphate has also been shown to be a substrate for the *E. coli* adenylate cyclase.[9]

Metal Ion Requirement. Mg^{2+} is required for enzymatic activity. Mn^{2+}, at low concentration, is more effective than Mg^{2+}. At higher concentration, however, Mn^{2+} is inhibitory.[10] Other divalent cations, such as Ca^{2+}, Co^{2+}, and Zn^{2+}, are found to inhibit the cyclase reaction. Monovalent cations, such as K^+, NH_4^+, Li^+, and Na^+, have no effect on the reaction.

Inhibitors. Many compounds are capable of completely or partially inhibiting the adenylate cyclase reaction; among them are the ribonucleoside triphosphates, pyridoxal phosphate, inorganic pyrophosphate, fluoride ion, and mercuric acetate.

Sedimentation Properties. Sedimentation in a 5–20% sucrose gradient using beef liver catalase (244,000 MW) and rabbit muscle lactate dehydrogenase (140,000 MW) as markers indicates that the molecular weight of *E. coli* adenylate cyclase is approximately 110,000.

[9] P. J. Ortiz, *Biochem. Biophys. Res. Commun.* **46**, 1728 (1972).
[10] M. Ide, *Arch. Biochem. Biophys.* **144**, 262 (1971).

[22] Adenylate Cyclase from *Brevibacterium liquefaciens*[1]

By Katsuji Takai, Yoshikazu Kurashina, and Osamu Hayaishi

$$\text{ATP} \xrightleftharpoons[Mg^{2+},\ \text{pyruvate}]{} 3',5'\text{-cyclic AMP} + PP_i$$

Adenylate cyclase from *Brevibacterium liquefaciens* catalyzes the conversion of ATP to 3',5'-cyclic AMP (cAMP) and PP_i (*forward reaction*) as well as the formation of ATP from 3',5'-cAMP and PP_i (*reverse reaction*). The enzyme requires a divalent metal ion, such as Mg^{2+}, and an α-keto acid, such as pyruvate, for maximum activity.[2,3] Therefore, the enzyme assay for both forward and reverse directions is conducted in the presence of Mg^{2+} and pyruvate.

[1] K. Takai, Y. Kurashina, C. Suzuki-Hori, H. Okamoto, and O. Hayaishi, *J. Biol. Chem.* **249**, 1965 (1974).
[2] M. Hirata and O. Hayaishi, *Biochem. Biophys. Res. Commun.* **21**, 361 (1965).
[3] M. Hirata and O. Hayaishi, *Biochim. Biophys. Acta* **149**, 1 (1967).

Assay of Forward Reaction

Principle. The assay is based on the radiometric determination of 3′,5′-cAMP produced from radioactive ATP.

Reagents

[8-^{14}C]ATP (0.375 µCi/µmole), 20 mM
Sodium pyruvate, 0.25 M
Bovine serum albumin (BSA), 1 mg/ml
Dithiothreitol (DTT), 10 mM
Tris·Cl buffer, pH 9.0, 0.2 M, containing 60 mM MgSO$_4$; the pH of the solution adjusted at 33°
Enzyme, diluted if necessary with 50 mM Tris·Cl, pH 8 containing 0.1 M KCl, 0.1 mg of BSA per milliliter, and 1 mM DTT and stored at 0°
EDTA, 0.15 M, containing 70 mM 3′,5′-cAMP (a carrier solution)
Ammonium acetate, 1 M:ethanol (2:5, v/v)

Procedure. The reaction mixture (50 µl) contains Tris buffer–Mg solution, 25 µl; [^{14}C]ATP, 5 µl; pyruvate 2 µl; BSA, 5 µl; and DTT, 5 µl. The reaction is initiated by the addition of the enzyme, and incubation is carried out at 33° for 30 minutes. The reaction is terminated by immersing the tube in a boiling water bath for 2 minutes. After the addition of 5 µl of an EDTA–cAMP mixture as carrier, the denatured protein is removed by centrifugation. An aliquot of the supernatant solution is applied on a 0.5 mm-thick cellulose thin-layer plate or a filter paper under a stream of hot air and is developed with ammonium acetate–ethanol as solvent. The area corresponding to cAMP is visualized under an ultraviolet lamp, scraped off or cut into pieces, and transferred into a counting vial. After extraction for at least 15 minutes with 1 ml of H$_2$O, the radioactivity is determined by a liquid scintillation spectrometer with 8 ml of scintillator solution which consists of 333 ml of Triton X-100, 667 ml of toluene, 5.5 g of 2,5-diphenyloxazole, and 0.1 g of 1,4-bis-2-(4-methyl-5-phenyloxazolyl)benzene.

Definition of Unit and Specific Activity. One unit of enzyme is defined as the amount which catalyzes the formation of 1 nmole of cAMP per minute under the standard assay conditions. Specific activity is expressed as units per milligram protein. The protein concentration is determined by the procedure of Lowry et al.[4] using bovine serum albumin as a standard.

---

[4] O. H. Lowry, N. J. Rosebrough, A. L. Farr, and R. J. Randall, *J. Biol. Chem.* **193**, 265 (1951); see also Vol. 3 this series, [73].

Assay of Reverse Reaction

Principle. The assay is based on the enzymatic determination of ATP formed from cAMP and PP_i using a fluorometric method.[5,6] This assay is not suitable for crude fractions which are not yet highly purified and contain a considerable amount of enzymes interfering with the reverse reaction, such as inorganic pyrophosphatase, ATPase, and adenylate kinase. However, it can be used after the ammonium sulfate step and is employed routinely for monitoring the enzyme activity in the column eluates, because it is much easier and quicker than the forward procedure.

Reagents

N-2-Hydroxyethylpiperazine-N'-2-ethanesulfonic acid (HEPES), 0.1 M, titrated to pH 7.3 with KOH at 33°
Sodium cAMP, 0.25 M
Tetrasodium pyrophosphate, 60 mM, titrated to pH 7.8 with HCl
$MgSO_4$, 0.1 M
Sodium pyruvate, 0.25 M
Bovine serum albumin (BSA), 5 mg/ml
Dithiothreitol (DTT), 50 mM
Enzyme, diluted if necessary with the same mixture as used for the forward reaction.
$MgSO_4$, 1 M
Tris·Cl buffer, 0.1 M, pH 8, containing 5 mM EDTA, 0.2 mM NADP, and 1 mM glucose (ATP-assay mixture)
Crystalline yeast hexokinase and crystalline yeast glucose-6-phosphate dehydrogenase (Boehringer Mannheim Corp.), diluted appropriately with the ATP-assay mixture before use.

Procedure. The reaction mixture (100 μl) contains HEPES-buffer, 50 μl; cAMP, 10 μl; 0.1 M $MgSO_4$, 5 μl; BSA, 2 μl; DTT, 2 μl; pyruvate, 4 μl; and pyrophosphate, 5 μl. Because of the precipitability of the Mg-PP_i complex, the components of the reaction mixture are added in the sequence described above and stored in an ice bath. The reaction is initiated by the addition of the enzyme, and incubation is carried out for 30 minutes at 33°. The reaction is terminated by immersing the tube

---

[5] Formed ATP yields a stoichiometric amount of NADPH through the reaction catalyzed by hexokinase and glucose-6-phosphate dehydrogenase. The amount of NADPH is estimated from the intensity of fluorescence emitted at 470 nm on excitation at 365 nm. See H. U. Bergmeyer, "Methods of Enzymatic Analysis," Academic Press, New York (1963); see also, S. Udenfriend, "Fluorescence Assay in Biology and Medicine," Vol. I, Academic Press, New York (1962).

[6] K. Takai, Y. Kurashina, C. Suzuki, H. Okamoto, A. Ueki, and O. Hayaishi, *J. Biol. Chem.* **246**, 5843 (1971).

in a boiling water bath for 2 minutes. To the heated reaction mixture are then added 1.5 ml of the ATP-assay mixture, 15 µl of 1 M $MgSO_4$, 3 µg of hexokinase, and 0.3 µg of glucose-6-phosphate dehydrogenase. The formation of NADPH is determined by the increment in fluorescence intensity. The amount of ATP formed is estimated using authentic ATP as an internal standard.

The reverse reaction proceeds at a rate of about $\frac{1}{25}$ of the forward reaction under these standard assay conditions.

Purification Procedure

Growth of Bacterial Cells. Brevibacterium liquefaciens (ATCC 14929) is grown in a medium containing 7 g of K_2HPO_4, 3 g of KH_2PO_4, 20 g of DL-alanine, 20 g of glucose, and 0.2 g of $MgSO_4 \cdot 7H_2O$ per liter of distilled water.[7] An inoculum (300–600 ml) is prepared by overnight shaking at 25° and is added to a large-scale culture medium (60 liters) which is maintained at 25° in a 100-liter fermenter. The medium is then aerated at a rate of 60 liters per minute and stirred at 160 rpm. After 40–45 hours the cells are harvested at their prestationary phase by the use of a Sharples centrifuge. The pellet is washed once with cold deionized water and stored frozen at −20°. The yield of wet packed cells is about 40 g per liter of medium.

Step 1. Preparation of Cell Extracts. Frozen cells, about 500 g, are suspended in 2.5 liters of 50 mM Tris·Cl, pH 7.9, containing 5 mM EDTA and dispersed with the aid of an electric mixer. The suspension of cells is then incubated with 400 mg of crystalline egg white lysozyme (6 times recrystallized, Seikagaku Kogyo Ltd., Japan) at 20–25°. The incubation is carried out with vigorous stirring by the use of a large stirring bar (1 × 5 cm) which is driven by a powerful magnetic stirrer. The digestion is finished after 30–60 minutes.[8] The viscous suspension is sub-

[7] *Brevibacterium liquefaciens* (ATCC 14929), identified by T. Okabayashi, excretes a large amount of cAMP into the culture medium, especially when cultured in the presence of glucose and DL-alanine. T. Okabayashi, *J. Bacteriol.* **84,** 1 (1962); T. Okabayashi, M. Ide, and A. Yoshimoto, *Arch. Biochem. Biophys.* **100,** 158 (1963).

[8] The conditions for full extraction of the enzyme differ from batch to batch of the bacterial cells. Therefore, pilot extraction on a small scale is necessary to determine the time required for the extraction and the proper amount of lysozyme to be added. Bacterial cells (about 25 g) are suspended in 150 ml of 50 mM Tris·Cl, pH 7.9, containing 5 mM EDTA, dispersed finely and incubated with a tentative amount (20 mg) of lysozyme at 20–25°. At time intervals of 10–15 minutes, an aliquot of the bacterial suspension is removed, cooled in an ice bath, and centrifuged at 15,000 g for 15 minutes at 0°. Supernatants thus obtained are assayed for the enzyme activity and protein concentrations as de-

jected as promptly as possible to centrifugation at 10,000 g for 60 minutes at 0° yielding 1.6–1.9 liters of a clear supernatant solution. At this stage the supernatant solution can be stored at −20° in the presence of 5 mM β-mercaptoethanol for at least several days without loss of activity.

All subsequent manipulations are carried out at 0–4°.

Step 2. Concentration by Acid. The pH of the supernatant is adjusted to 4.8 by a slow addition of 1 N HCl and after stirring for an additional hour the turbid solution is centrifuged at 10,000 g for 30 minutes. The resultant precipitate can be stored in an ice bath for at least several hours as a pellet.

Step 3. Streptomycin Sulfate Treatment. Five batches of the acid concentrate from step 2 are prepared sequentially, combined, and suspended in 1 liter of 0.1 M Tris·Cl, pH 6.8, containing 5 mM β-mercaptoethanol by the use of a Potter-Elvehjem-type homogenizer. The pH of the suspension is adjusted to 6.8 with 1 N KOH and 500 ml of cold 5% streptomycin sulfate are added slowly to the slightly turbid solution. The solution is stirred for an additional hour and centrifuged at 10,000 g for 30 minutes.

Step 4. pH Fractionation. The supernatant from step 3 is titrated to pH 4.8 with 1 N HCl and stirred for 1 hour. The precipitate formed is collected by centrifugation and suspended in a small volume (about 60 ml) of 0.1 M Tris·Cl, pH 6.8, containing 5 mM β mercaptoethanol by the use of a homogenizer as described above. The suspension is carefully titrated to pH 6.3 to 6.4 with 1 N KOH. The massive insoluble material is removed by centrifugation at 20,000 g for 30 minutes and the pH of the supernatant is then brought to 8.0 with 1 N KOH. The colored supernatant solution is stored at −20° until further purification.

Step 5. Ammonium Sulfate Fractionation. The protein concentration of the supernatant from step 4 is adjusted to about 7 mg/ml with 0.1 M Tris·Cl, pH 8.0, containing 5 mM β-mercaptoethanol. Saturated ammonium sulfate solution containing 2 mM EDTA, pH 7, is added slowly to 30% saturation and after 15 minutes of stirring the precipitate is removed by centrifugation at 20,000 g for 15 minutes. The supernatant is then brought to 42.5% saturation and stirred for 1 hour. The precipitate is recovered by centrifugation and dissolved in 5 ml of 0.1 M Tris·Cl, pH 8.0, containing 1 mM DDT and stored at −20° or preferably at

---

scribed above. The optimum time for incubation is estimated from the plots of the enzyme activity, protein concentration, and specific activity of extracts vs incubation time. Usually, once the enzyme activity has reached its maximum, further incubation results in loss of activity. Excess digestion also makes the suspension too gelatinous for recovery of the supernatant. When the time required for maximum extraction of the enzyme activity exceeds 60 minutes, the amount of lysozyme is increased.

−80°. The ammonium sulfate fractions derived from ten batches of crude extract are pooled for the next step.

Step 6. Hydroxyapatite Column Chromatography. A hydroxyapatite column (2 × 50 cm) is equilibrated with 5 mM potassium phosphate, pH 7.0, containing 50 mM KCl, 5% glycerol, and 5 mM β-mercaptoethanol. The combined ammonium sulfate fraction is desalted by passing it through a Sephadex G-25 column (2.8 × 50 cm) which has been equilibrated with the same buffer used for equilibration of the hydroxyapatite column. The desalted fraction is applied on the hydroxyapatite column, and a linear gradient elution is carried out between 400 ml each of 5 mM and 100 mM potassium phosphate, pH 7.0. Both solutions contain 50 mM KCl, 5% glycerol, and 5mM β-mercaptoethanol. Ten-milliliter fractions are collected at a flow rate of about 30 ml per hour. The enzyme activity in the eluate is assayed for the reverse reaction and absorbance at 280 nm is followed to monitor the protein concentration. Fractions with high specific activity are eluted at 50 to 80 mM potassium phosphate concentrations.[9] These active fractions (100–150 ml) are combined and passed through a Sephadex G-50 column (5 × 60 cm) which has been equilibrated with 0.1 M Tris·Cl, pH 8.3 (at 4°), containing 0.1 M KCl and 5 mM β-mercaptoethanol. Eluates are monitored for protein concentration by absorbance at 280 nm and the combined peak fraction is stored at 4°.

Step 7. DEAE-Sephadex A-50 Column Chromatography. The enzyme preparation from step 6 is applied on a DEAE-Sephadex A-50 column (2 × 40 cm) which has been equilibrated with 0.1 M Tris·Cl, pH 8.3 (at 4°), containing 0.1 M KCl and 0.2 mM DTT. After the column is washed with 50 ml of the same buffer, a linear gradient elution is applied between 350 ml each of 0.1 M and 0.35 M KCl. Both solutions contain 0.1 M Tris·Cl, pH 8.3 (at 4°), and 0.2 mM DTT. The flow rate is regulated to 10 ml per hour and 10-ml fractions are collected. The enzyme activity and protein concentrations are monitored as above. Fractions with high specific activity are eluted at about 0.2 M KCl concentration. They are combined and concentrated to less than 2 ml with the aid of a collodion bag (Sartorius Membranfilter GmbH, Germany).[10]

---

[9] Sometimes the enzyme activity is eluted as two peaks from the hydroxyapatite column. These two peak fractions have been purified separately through next two steps and shown to be indistinguishable from each other on polyacrylamide gel electrophoresis and gel filtration on Sephadex G-200. Therefore, these two peak fractions are routinely combined and subjected to the next purification step.

[10] Concentration of the enzyme in a collodion bag is carried out under reduced pressure using an apparatus (bag holder) available from Sartorius Membranfilter GmbH. The rate of concentration is about 5 ml per hour. Since the enzyme

PURIFICATION OF ADENYLATE CYCLASE FROM *Brevibacterium liquefaciens*[a]

Step	Total volume (ml)	Total protein (mg)	Total activity (units)[b]	Specific activity (units/mg)	Yield (%)
Crude extracts	25,885	146,689	947,516	6.5	100
pH fraction	248	4,480	540,305	121	58
Ammonium sulfate	34	1,083	325,200	300	34
Hydroxyapatite	280	191	188,333	934	20
DEAE-Sephadex A-50	1.5	9	123,750	13,600	13
Sephadex G-200	1.0	3.2	96,260	30,030	10

[a] From 6.4 kg (wet weight) of bacteria.
[b] Units as defined for the forward reaction.

Step 8. Gel Filtration on Sephedex G-200 Column. The concentrated fraction from step 7 is applied on a Sephadex G-200 column (2 × 88 cm) which has been equilibrated with 50 mM Tris·Cl, pH 8.3, containing 0.1 M KCl and 0.2 mM DTT. Ascending elution is carried out at a flow rate of less than 10 ml per hour and the enzyme activity and protein concentrations are monitored as above. Fractions with nearly constant high specific activity are combined, concentrated with the aid of a collodion bag as above[10] and stored in the same buffer containing 1 mM DTT at −80°.

A representative result of the purification is presented in the table.[11]

Crystallization. Crystallization is carried out according to the method of Jakoby.[12] The purified protein (about 1.5 mg) is precipitated in a small conical test tube by the addition of an equal volume of saturated ammonium sulfate solution. The precipitate is successively extracted at 0°

tends to be adsorbed by the surface of the bag, the concentrated solution in the bag is repeatedly agitated with a small Pasteur pipette; after removal of the solution, the bag is rinsed with an additional small volume of an appropriate buffer. These manipulations liberate the enzyme from the bag, resulting in a high recovery (over 90%) of the enzyme. A successful concentration is carried out alternatively by the following procedure, especially when the enzyme preparation to be concentrated is too low in protein concentration or too large in volume. The enzyme preparation is made up to 5% in glycerol and concentrated in the same manner as above except that the apparatus for concentration is filled (about 500 ml) with 30% polyethylene glycol 6000 containing 0.1 M KCl, 50 mM Tris·Cl, pH 8, 2 mM EDTA, and 5 mM β-mercaptoethanol to cover the outer surface of the collodion bag and promote concentration.

[11] The specific activity of various final preparations ranges from 30,000 to 48,000 nmoles per minute per milligram of protein, although these preparations behave as apparently homogeneous protein upon polyacrylamide gel electrophoresis.

[12] W. B. Jakoby, *Anal. Biochem.* **26**, 295 (1968); see also this series, Vol. 22 [23].

with 50 µl each of 45%, 42%, 39%, 36%, 33%, 30%, 27%, and 24% saturated ammonium sulfate solutions containing 50 mM Tris-maleate, pH 7.0, 1 mM EDTA, and 0.2 mM DTT. The extracts are allowed to stand at about 20°. Crystallization takes place predominantly in the 30 and 33% saturated ammonium sulfate solutions after about 20 hours. Crystals appear to be polyhedral in shape and grow to 10–15 µm in size.

Alternative Purification Procedure. The successful purification according to the procedure described above relies predominantly upon the high specific activity of the crude extract.[13] Sometimes, a final preparation with a low specific activity and some protein impurities is obtained when starting from a crude extract with low specific activity. Such a preparation can be further purified to near homogeneity by the following procedure: The enzyme solution is diluted with an equal volume of 50 mM Tris·Cl, pH 8.3 (at 4°), containing 1 mM DTT and 0.4% Triton X-100 and is applied on a column of DEAE-cellulose (Whatman DE 52) which has been equilibrated with 50 mM Tris·Cl, pH 8.3 (at 4°), containing 50 mM KCl, 0.1 mM DTT, and 0.2% Triton X-100. A linear gradient elution is applied between 50 mM and 0.3 M KCl each containing 50 mM Tris·Cl, pH 8.3 (at 4°), 0.1 mM DTT, and 0.2% Triton X-100. Although the size of the column and the mixing chamber for making the salt gradient depends upon an amount of the protein to be purified, about 10 mg of protein can be purified on about 4 ml of DEAE-cellulose using a mixing chamber of 50-ml volume. The enzyme activity is eluted at about 0.1 M KCl concentration. Fractions with the enzyme activity are combined, diluted to an appropriate ionic strength and adsorbed on to a small column (about 1 ml) of DEAE-cellulose which has been equilibrated as described above except without Triton X-100. The column is then extensively washed with the same buffer and the washings are monitored for the detergent remaining by absorbance at 280 nm. After elimination of the detergent, the enzyme is eluted with 50 mM Tris·Cl, pH 8.3 (at 4°), containing 0.2 M KCl and 0.5 mM DTT. By these procedures several batches of the impure enzyme have been successfully purified to near homogeneity.

Properties

Purity. The final preparation is free from ATPase, inorganic pyrophosphatase, and cAMP phosphodiesterase activities and is apparently homogeneous as judged by the following polyacrylamide gel electro-

---

[13] The specific activity of the crude extract obtained by ultrasonic disruption of the cells is 2.3 nmoles per minute per milligram of protein,[3] whereas it is 5–7 nmoles per minute per milligram by digestion with lysozyme.

phoreses. The native enzyme migrates as a single band on electrophoresis in polyacrylamide of different gel concentrations and pH's. The enzyme protein also migrates as a single band on a gel containing 8 M urea or 0.1% SDS.

Stability. The supernatant obtained after streptomycin sulfate treatment is extremely unstable. Overnight storage of the preparation at 4° or −20° causes a considerable loss of activity. The ammonium sulfate fraction is quite stable over wide range of pH; even at 2 or 13 the enzyme can be stored at 0° for at least several hours.

At a concentration of 1–3 mg/ml, the final preparation is stable for several months on storage at −80° and at least 2 weeks at 4°. Repeated freezing and thawing are detrimental. Dilution inactivation is observed and is prevented by the addition of bovine serum albumin or 0.2% Triton X-100. Heat treatment of the final preparation at 45° for 10 minutes causes about 50% loss of activity. The heat inactivation is not prevented by activators or substrates of the enzyme.

pH Optima. In the absence of pyruvate, the enzyme exhibits a broad pH optimum for the forward reaction between 7 and 9. In the presence of pyruvate, the maximum activity is obtained at pH 9–10 for the forward reaction and at pH 7.3 for the reverse reaction.

Molecular Properties. The molecular weight of the native enzyme is approximately 175,000 as determined by gel filtration on Sephadex G-200 according to the method of Andrews.[14] The minimum weight average molecular weight of the native enzyme is calculated to be 92,400 from the low speed sedimentation equilibrium experiment when a partial specific volume of 0.728 ml/g is used. The sedimentation coefficient estimated from the sucrose density gradient ultracentrifugation is 4.9 S. The molecular weight of the subunit is 46,000 as determined by SDS polyacrylamide gel electrophoresis.[15]

Equilibrium Constant. The adenylate cyclase reaction is readily reversible.[6,16,17] The equilibrium constant of the reaction at pH 7.0 and 25° is obtained as shown in equation below.

$$K'_{eq} = \frac{[cAMP][PP_i]}{[ATP]} = 0.065 \text{ mole per liter}$$

Thus, the standard free energy change ($\Delta G^{\circ\prime}$) of cAMP formation is estimated as +1.6 kcal/mole indicating that the reaction is ender-

---

[14] P. Andrews, *Biochem. J.* **96**, 595 (1965).
[15] K. Weber and M. Osborn, *J. Biol. Chem.* **244**, 4406 (1969).
[16] P. Greengard, O. Hayaishi, and S. P. Colowick, *Fed. Proc., Fed. Amer. Soc. Exp. Biol.* **28**, 467 (1969).
[17] O. Hayaishi, P. Greengard, and S. P. Colowick, *J. Biol. Chem.* **246**, 5840 (1971).

gonic.[6,16,17] Using this value and the value of the standard free energy change of hydrolysis of ATP between α and β position (-10.3 kcal/mole, pH 7),[18] the standard free energy change of hydrolysis is 3′-phosphodiester bond of cAMP is calculated to be -11.9 kcal/mole at pH 7.[16,17]

Substrate Specificity. The reaction is highly specific for adenine nucleotides. In the forward reaction, ATP and 2′-deoxy-ATP serve as substrates. The K_m value for ATP is 0.4 mM under the standard assay conditions. The synthetic ATP analogs, β, γ-methylene ATP (AMPPCP) and adenylate imidodiphosphate (AMPPNP) also serve as substrates. In the reverse reaction, cAMP and 2′-deoxy-cAMP serve as substrates. The K_m values are 10 mM for cyclic AMP and 1.9 mM for PP$_i$.

Metals. The reaction requires Mg^{2+} as an essential component. This requirement is substituted to various extents by several divalent cations, such as Mn^{2+}, Zn^{2+}, Fe^{2+}, Cd^{2+}, and Co^{2+}. The K_a value for free Mg^{2+} is 1 mM in the presence of 10 mM pyruvate.

Activators. Several α-keto acids strongly activate the enzyme for both the forward and the reverse reactions. The relative enzyme activity in the presence of various α-keto acids at 5 mM is as follows, as assayed for the forward reaction: pyruvate, 100; α-ketobutyrate, 80; α-ketovalerate, 35; and without activator, 1–1.5. The K_a value for pyruvate is 0.24 mM. At high concentrations, several monocarboxylic acids are also effective in activating the enzyme. NaF does not activate the enzyme at any concentration either in the presence or in the absence of pyruvate.

Inhibitors. Oxaloacetate inhibits the enzyme in a competitive manner with respect to ATP. The K_i value for oxaloacetate is 0.1 mM. Adenosine inhibits the enzyme competitively with respect to cAMP and uncompetitively with respect to ATP. The K_i value for adenosine is 10 μM. The enzyme activity is depressed to about 50% by p-chloromercuribenzoate at 0.1 mM, although no cysteine residues have so far been detected upon amino acid analysis of the enzyme. The inhibition is reversed by excess thiol compounds.

---

[18] R. A. Alberty, *J. Biol. Chem.* **244**, 3290 (1969).

[23] Preparation of Adenylate Cyclase from Frog Erythrocytes

By JACK ERLICHMAN and ORA M. ROSEN

$$\text{ATP} \xrightarrow[\text{Mg}^{2+}]{\text{F}^- \text{ or catecholamine}} \text{cyclic 3',5'-AMP} + \text{PP}_i$$

Assay Method

Principle. Adenylate cyclase activity is assayed by measuring the [^{14}C]cAMP formed from [^{14}C]ATP in the presence of Mg^{2+}, dithiothreitol, and either NaF or a catecholamine (e.g., epinephrine, isoproterenol).[1] The rate of cAMP synthesis is directly proportional to the concentration of erythrocyte membranes (5–50 μg) and hormone (1–50 μM).

Reagents for Assay

Tris·HCl buffer, 1.0 M, pH 8.1
Dithiothreitol, 1.0 M
MgSO$_4$, 0.10 M
[^{14}C]ATP, 20 mM (specific activity 900–1000 cpm/nmole)
Cyclic 3',5'-AMP, 20 mM
NaF, 1.0 M
Catecholamine (epinephrine, isoproterenol), 1.0 mM
Particulate adenylate cyclase (5–10 mg/ml)
8% ZnSO$_4$
7.2% Ba(OH)$_2$

Procedure. The reaction is carried out in 3-ml centrifuge tubes containing 50 mM Tris·HCl buffer (pH 8.1), 3.0 mM MgSO$_4$, 20 mM dithiothreitol, 1 mM [^{14}C]ATP, 10–20 μg of enzyme and either 0.01 M NaF or 10 μM catecholamine, in a final volume of 0.2 ml. The tubes are shaken at 35° for 15 minutes in a water bath and then rapidly chilled in ice. To each tube is added 5 μl of the nonradioactive carrier cAMP, followed by 50 μl of the ZnSO$_4$ solution and 50 μl of the suspension of Ba(OH)$_2$. The contents of the tubes are thoroughly mixed, and the heavy white precipitate (containing most of the [^{14}C]ATP as well as the [^{14}C]ADP and [^{14}C]AMP formed during the reaction) is removed by centrifugation at 10,000 g for 5 minutes. The supernatant fluid is then streaked (0.5 × 2.5 cm) on paper and chromatographed for 16–20 hours in descending fashion using a solvent system consisting of 1 M ammonium acetate:ethanol, 30:70 (v/v). After the papers are air-dried, the areas

---

[1] O. M. Rosen and S. M. Rosen, *Biochem. Biophys. Res. Commun.* **31**, 82 (1968).

corresponding to cAMP are detected by their absorption of UV light, cut into strips (2 × 1 cm) and counted in a liquid scintillation spectrometer in 10 ml of fluid containing 4 g of Omnifluor per liter of toluene. A reaction mixture containing boiled enzyme is assayed in parallel to determine the background radioactivity.

A useful check on the validity of the assay is to carry out a duplicate assay in the presence of added cyclic nucleotide phosphodiesterase (commercially available for Sigma Chemical Co.). The radioactivity in the area corresponding to cAMP is then reduced to background levels.

Definition of Unit. One unit of activity is defined as the amount of enzyme necessary to catalyze the formation of 1 nmole of cAMP per minute at 35°. Specific activity is expressed as units per milligram of protein.[2]

Application of Assay Method to Crude Tissue Preparation. Assays of unfractionated erythrocyte lysates yield only approximate values since such lysates contain cyclic nucleotide phosphodiesterase activity. In order to demonstrate full activity, diesterase inhibitors, such as theophylline (10 mM), should be added to the reaction mixture.

Purification of Adenylate Cyclase[3]

Reagents

> Amphibian Ringer's solution: NaCl, 6.5 g; KCl, 0.14 g; CaCl$_2$, 0.12 g; NaHCO$_3$, 0.2 g; H$_2$O to 1 liter
> Heparin, 2.5 mg/100 ml of Amphibian Ringer's solution
> Buffer A: 10 mM Tris·HCl buffer, pH 8.1 containing 1 mM dithiothreitol
> Deoxyribonuclease: (2000 units/mg), 5 mg/ml H$_2$O
> Buffer B: 10 mM potassium phosphate buffer pH 6.9 containing 0.1 mM dithiothreitol
> Carboxymethyl Sephadex equilibrated with buffer B
> Buffer C: 20 mM Tris·HCl buffer, pH 8.1, containing 0.1 mM dithiothreitol

Procedure. Frogs which may be stored at 4° are brought to room temperature for at least 30 minutes, then pithed and bled directly by cardiac puncture. The blood is collected in cold amphibian Ringer's solution containing heparin. The cell suspension is centrifuged at 5000 g for 5 minutes.

---

[2] O. H. Lowry, N. J. Rosebrough, A. L. Farr, and R. J. Randall, *J. Biol. Chem.* **193**, 265 (1951).

[3] O. M. Rosen and S. M. Rosen, *Arch. Biochem. Biophys.* **131**, 449 (1969).

The supernatant fluid is removed, and the packed cells are resuspended in 10 volumes of amphibian Ringer's solution. The cell suspension is then centrifuged and washed again. After each centrifugation, the superficial layer of white blood bells is removed with a Pasteur pipette. DNase (40 μg/ml of final lysate) is then added to the packed cells followed by 20 volumes of buffer A. The cells are permitted to lyse at room temperature for 15–30 minutes (lysate, see the table). All subsequent steps are performed at 0–4°. After completion of lysis (detectable by a visible clearing of the cell suspension), the lysate is spun in 2–3-ml aliquots at 5000 g for 30 minutes. The 5000 g supernatant fluid is removed with a Pasteur pipette and applied directly to a column of CM Sephadex (15 ml bed volume/1 ml of lysate), which has been equilibrated with buffer B. The column is washed with the equilibrating buffer. The resin retains hemoglobin while the adenylate cyclase activity is eluted with the solvent front in 1.5–2 times its original volume (CM Sephadex eluate). The yellow and slightly opalescent fractions containing adenylate cyclase activity are pooled and centrifuged at 17,000 g for 10 minutes. The sedimented material contains all the adenylate cyclase activity. The particles are then repeatedly washed with buffer C until the absorbance of the supernatant fluid is negligible at 280 nm (17,000 g fraction). The washed residue is resuspended in a small volume (1–2 ml) of buffer A and quickly frozen in liquid N_2. It is then rapidly thawed at 30° and centrifuged at 17,000 g for 10 minutes. This sequence of freezing and thawing releases a yellow pigment and some additional protein into the supernatant fluid. After two additional washes with buffer A, the pellet is resuspended in an equal volume of buffer C and stored in 0.5-ml aliquots in liquid N_2 (residue after freezing and thawing).

The enzyme preparation retains 80–90% of both its catalytic activity and sensitivity to catecholamine stimulation for several months when stored in liquid nitrogen. A summary of the purification procedure is given in the table.

PURIFICATION OF ADENYLATE CYCLASE

Fraction	Volume (ml)	Protein (mg/ml)	Specific activity (units/mg protein)	Recovery (%)
Lysate	20.0	13.0	0.20	100
CM Sephadex eluate	30.0	0.72	2.53	70
17,000 g fraction	2.0	0.50	35.41	68
Residue after freezing and thawing	2.0	0.43	38.0	63

Properties

Specificity. The purified enzyme preparation is free of cyclic nucleotide phosphodiesterase, alkaline phosphatase, and inorganic pyrophosphatase activities but contains ATPase activity. The enzyme utilizes only ATP or dATP as substrate for cyclic nucleotide formation.

Activators and Inhibitors. Activity is dependent upon the presence of either sodium fluoride (10 mM) or catecholamines (see below). The purified enzyme also exhibits an absolute requirement for a sulfhydryl containing compound such as dithiothreitol. In the absence of fluoride, the addition of catecholamines activates the adenylate cyclase of adult frog erythrocyte membranes. Maximal activation occurs at 10^{-5} M isoproterenol and is approximately 40–50% of that inducible by optimal concentrations of fluoride. The order of potency of catecholamines is isoproterenol > epinephrine > norepinephrine. The stimulation exerted by adrenergic agonists can be completely blocked by the simultaneous addition of equimolar concentrations of β-adrenergic blockers such as propranolol but not by α-adrenergic blockers, such as dibenzyline. In general, the hormone sensitivity of the adenylate cyclase derived from frog erythrocytes behaves as predicted for a β-adrenergic receptor system.[4] Glucagon, insulin, vasopressin, acetylcholine, thyroxine, triiodothyronine, or prostaglandin (E_1, E_2, A_1, $F_{1\alpha}$) do not influence its activity. Unlike the adenylate cyclase activity present in frog erythrocytes, the adenylate cyclase in tadpole erythrocytes does not respond significantly to the addition of either catecholamines or other hormones, although it is activated by sodium fluoride.[1]

pH Optimum and Metal Requirements. In Tris buffer, enzymatic activity is optimal at pH 8.0 and falls off sharply on either side of this pH. The enzyme requires either Mg^{2+} (3–5 mM) or Mn^{2+} (1 mM) for activity. Neither Zn^{2+} nor Ca^{2+} can satisfy this requirement. Enzymatic activity in the presence of Mg^{2+} is inhibited by the addition of either Zn^{2+} or Ca^{2+}.

Kinetic Constants. The K_m for ATP is 0.16 mM in the presence of fluoride and 1.3 mM in the presence of 50 μM isoproterenol (in the absence of fluoride). Reversal of the adenylate cyclase reaction could not be demonstrated.

---

[4] O. M. Rosen, J. Erlichman, and S. M. Rosen, *Mol. Pharmacol.* **6**, 524 (1970).

[24] Method for Solubilization of Myocardial Adenylate Cyclase and Assessment of the Role of Phospholipids in Hormone Stimulation

By GERALD S. LEVEY

Sutherland *et al.*[1] have postulated that adenylate cyclase may be a lipoprotein whose functional activity *in vivo* may be dependent on the presence of membrane lipid. In recent years several observations have been reported demonstrating the critical role of phospholipids in hormone-responsive adenylate cyclase systems.[1-7] Solubilization of membrane-bound adenylate cyclase abolishes its ability to be activated by hormones. Treatment of thyroid slices[8] and isolated liver plasma membranes[9] with phospholipases results in a loss of hormone responsiveness to thyroid-stimulating hormone and glucagon, respectively.

This section describes the preparation of soluble cat myocardial adenylate cyclase and the use of specific phospholipids in restoring hormone-activation of the soluble enzyme.

Methods

Solubilization of Myocardial Adenylate Cyclase

Table I outlines the solubilization procedure utilized for heart muscle.[10] Lubrol-PX is a nonionic detergent, an ethylene oxide condensate of dodecanol. The quantity of Lubrol-PX used for the solubilization is first dissolved in a small amount of 95% ETOH (10–20 mg per 0.14 cc of ETOH) and then added to the solubilization mixture. Upon homogenization of cat heart muscle in the presence of Lubrol-PX, almost all adenylate cyclase activity is located in the 12,000 g supernatant and no

---

[1] E. W. Sutherland, T. W. Rall, and T. Menon, *J. Biol. Chem.* **237**, 1220 (1962).
[2] S. L. Pohl, H. M. J. Krans, V. Kozyreff, L. Birnbaumer, and M. Rodbell, *J. Biol. Chem.* **246**, 447 (1971).
[3] G. S. Levey, *Biochem. Biophys. Res. Commun.* **38**, 86 (1970).
[4] G. S. Levey, *Ann. N.Y. Acad. Sci.* **185**, 449 (1971).
[5] G. S. Levey, *Biochem. Biophys. Res. Commun.* **43**, 108 (1971).
[6] G. S. Levey, *J. Biol. Chem.* **246**, 7405 (1971).
[7] G. S. Levey and I. Klein, *J. Clin. Invest.* **51**, 1578 (1972).
[8] V. Macchia and I. Pastan, *J. Biol. Chem.* **242**, 1864 (1967).
[9] M. Rodbell, H. M. J. Krans, S. L. Pohl, and L. Birnbaumer. *J. Biol. Chem.* **246**, 1857 (1971).
[10] G. S. Levey, *Recent Progr. Horm. Res.* **29**, Chap. 9 (1973).

TABLE I
PROCEDURE FOR SOLUBILIZING MYOCARDIAL ADENYLATE CYCLASE[a]

Step	Procedure
1	Cat anesthetized with pentobarbital, 25–35 mg/kg intraperitoneally
2	Heart excised
3	Left ventricle dissected free of epicardium and endocardium
4	Muscle, 300–350 mg, homogenized at 1° in 4.5 ml of a solution containing: a. 0.25 M sucrose; b. 10 mM Tris · HCl, pH 7.7; c. 1 mM EDTA-Mg^{2+}; d. 20 mM Lubrol-PX
5	Centrifuged at 12,000 g, for 10 minutes at 4°
6	Precipitate discarded

[a] From G. S. Levey, *Recent Progr. Horm. Res.* **29**, Chap. 9 (1973).

significant activity is found in the precipitate (Fig. 1). In contrast, adenylate cyclase activity in homogenates of cat heart, prepared in the absence of detergent, is almost totally precipitated when centrifuged at 12,000 g for 10 minutes (Fig. 1). This is a simple, one-step method resulting in solubilization of 90–100% of adenylate cyclase in heart muscle.

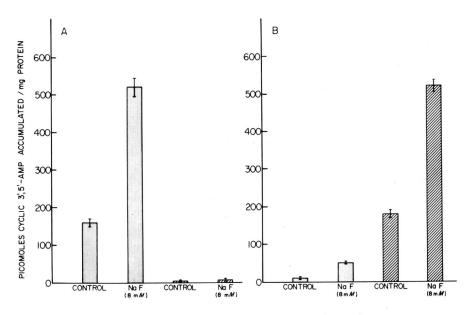

FIG. 1. The effect of Lubrol-PX on the precipitation of myocardial adenylate cyclase. (From G. S. Levey, *Ann. N.Y. Acad. Sci.* **185**, 449 (1971). (A) 12,000 g precipitate; (B) 12,000 g supernatant; stippled; Lubrol absent; hatched, Lubrol present.

The adenylate cyclase is soluble as defined by several criteria.[3,4] (1) Adenylate cyclase activity quantitatively passes through a 0.22 µm Millipore filter. (2) The solubilized enzyme present in the 12,000 g supernatant is not precipitated when centrifuged at speeds as great as 250,000 g for 2 hours. (3) No significant particulate material is observed with electron microscopy in this supernatant fraction. (4) The solubilized adenylate cyclase has an estimated molecular weight of 100,000 to 200,000 based upon Sephadex chromatography.

The soluble adenylate cyclase in the presence of detergent is not activated by the hormones which activate the membrane-bound enzyme, the catecholamines, glucagon, histamine, and thyroxine and triiodothyronine. In order to further define the hormone unresponsiveness, specifically the role of phospholipids, the detergent must be removed. This can be achieved utilizing DEAE-cellulose chromatography, since the nonionic detergent is not adsorbed to the resin whereas the enzyme is adsorbed. A representative experiment is shown in Table II. Approximately 1.3 ml of the 12,000 g supernatant (protein concentration 3–5 mg/ml) is applied to a DEAE-cellulose column (1.0 × 12.0 cm) equilibrated at 4° with 10 mM Tris·HCl, pH 7.7. The column is washed with 20 volumes of 10 mM Tris·HCl, pH 7.7 at a flow rate of 0.2 ml per minute. After removal of the detergent, adenylate cyclase can be eluted with 1 M Tris·HCl, pH 7.7 (Table II). The enzyme, freed of detergent, remains in a soluble state as defined by the criteria noted previously and continues to be unresponsive to hormone activation. The chromatography results in a loss of 50–75% of the total activity of the solubilized adenylate cyclase. The addition of phospholipids, especially phosphatidylserine restores most of this lost activity and phosphatidylethanolamine and sphingomyelin are

TABLE II
PREPARATION OF DETERGENT-FREE SOLUBILIZED ADENYLATE CYCLASE BY DEAE-CELLULOSE CHROMATOGRAPHY[a]

Tris · HCl, pH 7.7 (M)	[^{14}C]Lubrol-PX[b]		Cyclic 3′,5′-AMP accumulated (pmoles/5 min/mg protein)	
	Applied	Recovered	Applied	Recovered
0.01	13,874	—	1330	—
	—	13,930	—	0
1.0	—	—	—	1400

[a] From G. S. Levey, Ann. N.Y. Acad. Sci. **185**, 449 (1971).
[b] Lubrol-PX was labeled with ^{14}C in the ethylene oxide moiety. Each value represents the mean of duplicate samples.

partially effective. Cardiolipin, lecithin, and lysolecithin are without effect.

Phospholipids and Hormone Activation of Adenylate Cyclase

Pohl et al.[2] demonstrated that the addition of pure phosphatidylserine partially restored glucagon-activation of adenylate cyclase from digitonin-treated liver plasma membranes. The addition of specific phospholipids also restores hormone responsiveness of the solubilized myocardial adenylate cyclase.[5-7] The phospholipids utilized for the studies with the myocardial enzyme are prepared from bovine brain[6] and are chromatographically pure as determined in two solvent systems $CHCl_3:CH_3OH:CH_3COOH:H_2O$, 100:60:16:8 and $CHCl_3:CH_3OH:H_2O$, 65:25:4. Phosphatidylserine is obtained as a 25-mg/ml solution in $CHCl_3$ and phosphatidylinositol as a 10-mg/ml solution in $CHCl_3$. The appropriate aliquot (0.25 ml for phosphatidylinositol and 0.17 ml for phosphatidylserine) is placed in a 10×75 mm glass test tube and the $CHCl_3$ removed by evaporation with a stream of nitrogen. One milliliter of Tris·HCl, 10 mM, pH 7.7 is added to the residue and the lipid dispersed by sonication at 1°C until no further change in clarity of the solution is observed, usually 0.5 to 1 minute. The dispersed phospholipid is utilized immediately thereafter and not stored for future use since its effectiveness markedly diminishes after a period of several hours at 1°C. In addition, the phospholipids should be chromatographed every few weeks since their stability, even in chloroform, is limited. Oxidation of the fatty acids is particularly troublesome.

Adenylate cyclase is assayed by the method of Krishna et al.[11] The incubation mixture contains 1.6 mM [α-^{32}P]ATP, 8 mM theophylline, 2 mM $MgCl_2$, 21 mM Tris·HCl, pH 7.7, 0.8 mg/ml human serum albumin, and 25–50 µg of enzyme protein. The assay is initiated by adding an enzyme–phospholipid–hormone mixture prepared at 1° to the other components which are at 25°. After 5 minutes, the incubations are stopped by the addition of a solution containing 4 µmoles of ATP, 1.25 µmole of cyclic 3′,5′-AMP, and 0.15 µCi of cyclic [^{3}H]3′,5′-AMP followed by boiling for 3 minutes. After boiling, 0.4 ml of water is added, the precipitate is removed by centrifugation, and the supernatant is applied to a 0.5×2.0 cm Dowex-50 column. The column is washed with water, and the elute, between 3.0 and 6.0 ml, is collected and precipitated twice with 0.17 M $ZnSO_4$ and 0.15 M $Ba(OH)_2$. The cyclic [^{32}P]3′,5′-AMP and cyclic [^{3}H]3′,5′-AMP, which are in the supernatant, are counted in a liquid scintillation spectrometer.

[11] G. Krishna, B. Weiss, and B. B. Brodie, *J. Pharmacol. Exp. Ther.* **163**, 379 (1968).

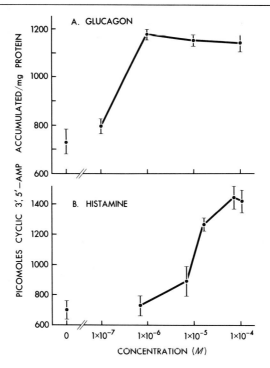

Fig. 2. Concentration-response curves for glucagon (A) and histamine (B) with solubilized adenylate cyclase and phosphatidylserine, 128 μg per milliliter of incubation. From G. S. Levey, "The Role of Membranes in Metabolic Regulation" (M. A. Mehlman and R. W. Hanson, eds.), p. 249. Academic Press, New York, 1972.

Figure 2 shows the effect of the addition of phosphatidylserine on the responsiveness of the solubilized adenylate cyclase to glucagon and histamine. In the presence of phosphatidylserine, 128 μg per milliliter of incubation mixture, both hormones activate the solubilized adenylate cyclase over concentration ranges almost identical to that observed with the membrane-bound enzyme.[7] Glucagon produces a 60% increase and histamine a 100 percent increase; half-maximal activation of the enzyme is produced by concentrations of glucagon and histamine of 0.5 μM and 9 μM, respectively. The values for maximal stimulation are also in close agreement with those found with the membrane-bound enzyme. In addition, receptor specificity is maintained in this reconstituted system. The antihistamine, diphenhydramine, Benadryl, at 80 μM abolishes the activation of solubilized adenylate cyclase produced by histamine, 80 μM, but not that due to glucagon, 10 μM, similar to the findings reported for the particulate enzyme.

Another acidic phospholipid, phosphatidylinositol, at 80 μg per milliliter of incubation mixture, restores norepinephrine responsiveness (Fig. 3). However, the sensitivity of the solublized adenylate cyclase to norepinephrine is increased approximately 100-fold as compared to the membrane-bound enzyme. Half-maximal activation was produced by a concentration of norepinephrine, 0.8 mM, as compared to 9 μM with the particulate enzyme.[6] It appears that homogenization alone alters the phosphatidylinositol–enzyme relationship resulting in decreased sensitivity to norepinephrine, a suggestion supported by the finding that the addition of phosphatidylinositol directly to particulate fractions from heart homogenates increases the sensitivity of the adenylate cyclase to norepinephrine.[10] Receptor specificity is also maintained in this system since the beta-adrenergic blocking drug, DL-propranolol, abolishes the activation of

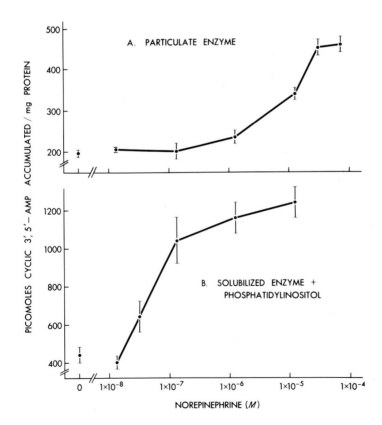

FIG. 3. The effect of increasing concentrations of norepinephrine on the activation of myocardial adenylate cyclase. From G. S. Levey, *J. Biol. Chem.* **246**, 7405 (1971).

solubilized adenylate cyclase by norepinephrine in the presence of phosphatidylinositol.[6]

The minimally effective concentrations of phospholipid in these studies were 64 µg per milliliter of incubation mixture, for phosphatidylserine and 0.8 µg per milliliter of incubation mixture for phosphatidylinositol. A degree of specificity is found with the phospholipids as demonstrated by the observation that phosphatidylserine restores the glucagon and histamine activation, but not that of norepinephrine, whereas phosphatidylinositol restores the norepinephrine activation, but not that of glucagon and histamine. Phosphatidylethanolamine, cardiolipin, and phosphatidylcholine do not restore reponsiveness to any of these hormones.

Acknowledgments

These studies were supported in part by NIH grant 1 R01 HE 13715-01, 02 and by the Heart Association of Broward County, a chapter of the Florida Heart Association.

[25] Production of Glass Bead-Immobilized Catecholamines

By J. Craig Venter and Jack E. Dixon

Porous glass beads were used originally in 1969[1] for the insolubilization of enzymes. Since that time numerous enzymes, proteins and cofactors have been immobilized on this solid support.[2]

The catecholamines, epinephrine, norepinephrine, and isoproterenol when bound to glass beads have been shown to be biologically active on beating chick embryo heart cells in culture, whole chicken embryo hearts, hearts in open-chested anesthetized dogs, cat papillary muscles, and isolated perfused guinea pig hearts, as well as stimulating cAMP formation in glial tumor cells grown *in vitro*.[3,4] The immobilized catecholamines have also been used as an affinity system to partially purify the β-adrenergic-receptor adenylate cyclase complex.[5] The purpose of this

---

[1] H. H. Weetall, *Science* **166**, 615 (1969).

[2] R. Goldman, L. Goldstein, and E. Katchalski, *in* "Biochemical Aspects of Reactions on Solid Supports" (G. R. Stark, ed.). Academic Press (1971).

[3] J. C. Venter, J. E. Dixon, P. R. Maroko, and N. O. Kaplan, *Proc. Nat. Acad. Sci. U.S.* **69**, 1141 (1972).

[4] J. C. Venter, J. Ross, Jr., J. E. Dixon, S. E. Mayer, and N. O. Kaplan, *Proc. Nat. Acad. Sci. U.S.* **70**, 1214 (1973).

[5] J. C. Venter and N. O. Kaplan this volume [26].

article is to describe the methods of derivatizing glass beads and the procedures for coupling biogenic amines to porous glass beads. Some of the procedures outlined here for derivatizing the glass are modifications of previous procedures.[1,6,7]

Alkylamine glass can be obtained from the Pierce Chemical Co., GAO or MAO series; however, our laboratory has found it more satisfactory to synthesize the alkylamine derivatives from unreacted glass beads, GZO or MZO series (Pierce Chemical Co.). The GZO 550 Å pore glass was used for the reported studies.

Synthesis of Alkyl Amine Glass

Reagents

GZO 550 Å porous glass
Nitric acid solution 5% (reagent grade)
α-Aminopropyltriethoxysilane,[8] 10% (by weight) solution with water adjust pH to 3.45 with HNO_3)
Picrylsulfonic acid sodium salt dihydrate, 3% solution
Saturated borate solution

Procedure. Boil the glass in the 5% nitric acid solution 5 ml per gram of glass for 5 hours, decant acid, and rinse glass with 2 volumes of distilled water, then dry the glass in an oven at 100° for 4 hours. Place the glass in 10% (by weight) of α-aminopropyltriethoxysilane solution pH 3.5, 5 ml per gram of glass, and mix the glass for 22 hours at 75°.[9] Decant the solution and wash the glass with distilled water, then allow it to dry overnight in an oven at 100°C.

The product formed is alkylamine glass (Fig. 1). The glass is washed extensively with distilled water.[10]

The alkylamine glass should give a positive reaction when tested with picrylsulfonic acid.[11,12] This is a test for primary amines; however, secondary amines will react. Take 1 ml of a saturated borate solution and 20–50 mg of glass in a test tube. Add three drops of a 3% picrylsulfonic

---

[6] M. K. Weibel, H. H. Weetall, and H. J. Bright, *Biochem. Biophys. Res. Commun.* **44**, 347 (1971).
[7] H. H. Weetall and L. S. Hersh, *Biochim. Biophys. Acta* **185**, 464 (1969).
[8] Aldrich 3-amino propyltriethoxysilane or Union Carbide Corp. A-1100 solution.
[9] Avoid the use of magnetic stirring bars, as they will grind the glass. A suspended stirring bar, overhead stirrer (at low speed), or a shaking water bath will provide adequate mixing.
[10] Coarse sintered-glass funnels, with vacuum, have proved to be best for washing the glass beads.
[11] J. K. Inman and H. M. Dintzis, *Biochemistry* **8**, 4074 (1969).
[12] P. Cuatrecasas, *J. Biol. Chem.* **245**, 3049 (1970).

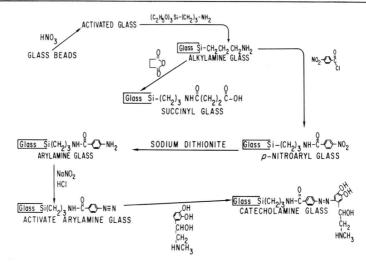

FIG. 1. Scheme for the synthesis of catecholamine glass beads.

acid solution to each tube. Let the reaction proceed for 30 minutes and then compare color of reacted glass with starting glass material. The alkylamine glass should be yellow to orange (a positive test) and the unreacted glass colorless (a negative test).

The alkylamine glass can be used at this point for direct coupling of enzymes and hormones by using dicyclohexylcarbodiimide or glutaraldehyde.[13,14]

The alkylamine glass can be further derivatized to make: (a) arylamine glass, (b) succinyl glass, (c) long-chain alkylamine, and (d) long-chain arylamine glass.

Production of Arylamine Glass

Reagents

Chloroform (reagent grade)
p-Nitrobenzyl chloride[15]
Triethylamine
Sodium dithionite 1% solution
Alkyl amine glass

---

[13] J. E. Dixon, F. E. Stolzenbach, J. A. Berenson, and N. O. Kaplan, *Biochem. Biophys. Res. Commun.* **52**, 905 (1973).

[14] J. E. Dixon, F. E. Stolzenbach, C.-L. T. Lee, and N. O. Kaplan, *Isr. J. Chem.* **12**, in press (1974).

[15] p-Nitrobenzoylchloride (Eastman Chemicals) was recrystallized from carbon tetrachloride.

Procedure. To 1 g of alkylamine glass, add 10 ml of a chloroform solution containing 100 mg of *p*-nitrobenzoyl chloride and 50 mg of triethylamine. Reflux the reaction mixture for 1 hour. Decant the solution and wash the glass three times with chloroform. The glass can be air dried or heated for 30 minutes at 80° to remove the chloroform. This is *p*-nitroaryl glass (Fig. 1). Test this glass with the picrylsulfonic acid color test; use the alkyl amine glass for comparison. The *p*-nitroaryl glass should give a negative test. The *p*-nitroaryl glass is reduced by adding 10 ml of a 1% aqueous sodium dithionite solution and refluxing for 30 minutes or by sitting 6 hours at room temperature. The reaction solution is decanted and the aromatic amine glass washed with distilled water. This is arylamine glass (Fig. 1) and should give a positive picrylsulfonic acid color test.

Synthesis of Succinyl Glass and of Long-Chain Alkylamine Glass

Reagents

Alkylamine glass
Succinic anhydride
Triethylamine
Chloroform
Dimethyl formamide or ether
Diamine
Dicyclohexylcarbodiimide

Procedure. SUCCINYL GLASS. Alkylamine glass, 1.5 g, is placed in a 100-ml round-bottom flask with 2 g of succinic anhydride. Exactly 3 ml of triethylamine and 50 ml of chloroform are put in the reaction vessel and shaken for 3 hours. The solution is originally a milky white but ultimately turns rather clear, thus indicating the completion of the reaction. The triethylamine and succinic acid serve as a buffer for the reaction. The product, succinyl glass (Fig. 1) is washed with chloroform and water. The succinyl glass should give a negative color test with picrylsulfonic acid.

LONG-CHAIN ALKYLAMINE GLASS. This reaction can be run in either dimethylformamide or ether. One gram of the succinyl glass is placed in 60 ml of the appropriate solvent along with equimolar amounts of the diamine and dicyclohexylcarbodiimide at room temperature until a cloudy solution forms. The reaction is then cooled to 4° for 12 hours. This reaction mixture must be kept dry and free of water until the coupling reaction is complete. Wash the glass with distilled water. The product is long-chain alkylamine glass and will give a positive color test.

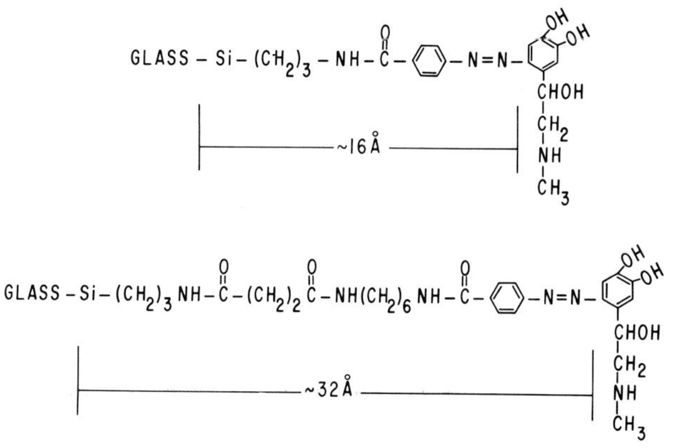

$$\text{GLASS} - \text{Si} - (CH_2)_3 - NH - \overset{O}{\underset{\|}{C}} - \langle \rangle - N=N- \text{[catechol]}$$

$$\vdash\!\!—\!\!\sim\!16\text{Å}\!\!—\!\!\dashv$$

with side chain: CHOH–CH$_2$–NH–CH$_3$

$$\text{GLASS} - \text{Si} - (CH_2)_3 NH - \overset{O}{\underset{\|}{C}} - (CH_2)_2 \overset{O}{\underset{\|}{C}} - NH(CH_2)_6 NH - \overset{O}{\underset{\|}{C}} - \langle \rangle - N=N- \text{[catechol]}$$

$$\vdash\!\!—\!\!\sim\!32\text{Å}\!\!—\!\!\dashv$$

with side chain: CHOH–CH$_2$–NH–CH$_3$

FIG. 2. Structure of short- and long-chain epinephrine glass beads.

The long-chain alkylamine glass (Fig. 2) can be converted to the arylamine derivative by the procedure for the short-chain glass.

Production of Catecholamine Glass

Materials

Hydrochloric acid 2 N, 10 ml
Arylamine glass (long or short chain)
Sodium nitrite, 250 mg per gram of glass
Sulfamic acid, 1% aqueous solution
Epinephrine HCl or appropriate compound
Hydrochloric acid 0.1 N, 12 liters
Phosphate buffer, 0.1 M, pH 7.5

Procedure. Add 10 ml of 2 N HCl to 1 g of arylamine glass and cool to 0° in an ice bath. Add 250 mg of sodium nitrite and let stand at 0° under vacuum for 20–30 minutes. The activated glass (Fig. 1) is then rapidly washed with 60 ml of ice cold distilled water followed by 60 ml of an ice cold 1% sulfamic acid solution and then twice more with cold water.

The glass is rapidly added as a wet cake to at least 50 mg of catecholamine in 10 ml of phosphate buffer, 0.1 M pH 7.5.[16] The reaction with catecholamine is immediate and the glass should turn a dark reddish brown. With less reactive compounds, such as histamine, the reaction can be continued overnight at 0°.

---

[16] This catecholamine solution should not be exposed to light and should be kept at 0–4°C.

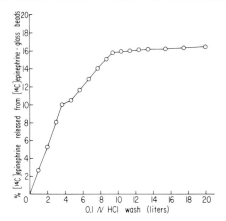

FIG. 3. HCl, 0.1 N, wash profile of [^{14}C]epinephrine-glass beads. One gram of [^{14}C]epinephrine-glass beads was packed into a small column and washed continuously for 6 days with 20 liters of 0.1 N HCl: aliquots were counted in Aquasol, and the cumulative percent release was calculated.

Washing Procedure and Comments

The importance of following proper washing procedures cannot be overstressed. Venter and Kaplan[17,18] have shown the importance of acid washing prior to chemical or biological utilization of the catecholamine glass. The methods described that the bead-bound catecholamines were extensively washed with several liters of 0.1 N HCl to remove any noncovalently bound catecholamines. In addition, the beads were washed extensively with 0.8% NaCl or distilled water immediately prior to biological assay, to remove the acid and any free catecholamines that might have accumulated since the acid wash. The acid wash profile of 1 g of [^{14}C]epinephrine-glass beads is shown in Fig. 3. The [^{14}C]epinephrine-glass beads were washed slowly over a 6-day period and the cumulative percentage of catecholamine per liter of 0.1 N HCl is calculated.

Failure to ensure complete absence of unbound amines can lead to erroneous results, such as those reported recently.[19]

Procedure. When the coupling reaction is completed, pack the beads into a small column.[20] Wash the beads at room temperature by continuous flow over the column with at least 12 liters of 0.1 N HCl at a flow rate of 3 liters per day, followed by at least 3 liters of 0.8% sodium

---

[17] J. C. Venter and N. O. Kaplan, *Science* in press.
[18] J. C. Venter, L. J. Arnold, Jr., and N. O. Kaplan, manuscript submitted.
[19] M. S. Yong, *Science* **182**, 157 (1973).
[20] The barrel of a 5-ml disposable syringe is very satisfactory for this purpose.

chloride at the same flow rate. If the beads are not going to be used immediately, they should be stored in a small volume of 0.1 N HCl at 4°. Do not freeze the glass, as it will crack. After storage and immediately prior to use, wash the catecholamine glass with at least 1 liter of 0.1 N HCl followed by 3 liters of 0.8% sodium chloride.[10]

In order to quantitate the amounts of amine bound to the glass, or to obtain a catecholamine derivative with a new amine group at the point of original diazo substitution for structural[18] or for biological activity determinations,[21] the following procedures should be used.

Hydrogen Fluoride Method. This method should be used when working with radioactively labeled catecholamines, to determine the total amount of radioactivity coupled to the glass beads. Ten milligrams dry weight of glass beads are placed into a plastic scintillation vial, and 0.1 ml of 48% hydrogen fluoride is added. After the glass beads have totally dissolved (usually within 5 minutes), 9.9 ml of distilled water is added and an aliquot is counted in 10 ml of Aquasol (New England Nuclear). Less distilled water can be added if the specific activity on the glass is low.

Dithionite Reduction Method. This method utilizes the fact that dithionite readily cleaves azo linkages to amino groups. As a result the determination of the point of substitution of the new amino group on the catecholamines gives the site of the diazo substitution.[18] This procedure also gives a substituted catecholamine analog which can be tested for potential biological activity.[21] Dithionite 0.5% solution is added to catecholamine glass beads in a sintered-glass funnel for 10 minutes. The glass is then vacuumed dry and washed twice with 0.01% HCl. The amino catecholamines can then be collected, lyophilized, and further purified.[18] The structure of epinephrine when coupled to long- and short-chain aryl amine glass, as determined by NMR spectroscopy following dithionite reduction,[18] is shown in Fig. 2.

Acknowledgment

The authors would like to acknowledge the helpful guidance of Dr. N. O. Kaplan, in whose laboratory this work was performed, by grants from the American Cancer Society (IN-93C) and from the National Institutes of Health (CA 11683-05).

[26] A Partial Purification of the β-Adrenergic Receptor Adenylate Cyclase Complex by Affinity Chromatography to Glass Bead-Immobilized Isoproterenol

By J. CRAIG VENTER and NATHAN O. KAPLAN

Hormones, drugs, and other ligands immobilized on solid supports have been used for the affinity isolation of whole cells[1,2] and the isolation and purification of cell membrane receptors and subcellular drug and hormone binding proteins from many sources.[3-13] These techniques have been applied to the β adrenergic receptor[14]; but, possibly owing to the method of ligand attachment to the solid support, it is unclear whether the functional receptor has been isolated. For ease of identification of the β-adrenergic receptor and to enable the study of the relationship of the receptor to the membrane enzyme adenylate cyclase, a known component of the β adrenergic function,[15] we attempted to purify the β-receptor–adenylate cyclase complex from crude membrane preparations.[16] This chapter will deal with the specific binding of erythrocyte membrane fragments, with known β-receptor activity[17,18] to catecholamines covalently

[1] G. M. Edelman, V. Rutishauser, and C. F. Milette, *Proc. Nat. Acad. Sci. U.S.* **68**, 2153 (1971).
[2] D. D. Soderman, J. Germershausen, and H. M. Katzen, *Proc. Nat. Acad. Sci. U.S.* **70**, 792 (1973).
[3] P. Cuatrecasas, *Proc. Nat. Acad. Sci. U.S.* **68**, 1264 (1971); **69**, 318 (1972).
[4] H. I. Yamamura, D. W. Reichard, T. L. Gardner, J. D. Horrisett, and C. A. Broomfield, *Biochim. Biophys. Acta* **302**, 305 (1973).
[5] J. D. Berman, *Biochemistry* **12**, 1710 (1973).
[6] M. A. Rafferty, *Arch. Biochem. Biophys.* **154**, 270 (1973).
[7] J. Schmidt and M. A. Rafferty, *Biochemistry* **12**, 852 (1973).
[8] V. K. Jansons and M. M. Burger, *Biochim. Biophys. Acta* **291**, 127 (1973).
[9] J. H. Ludens, J. R. Devries, and D. D. Fanestil, *J. Steroid Biochem.* **3**, 193 (1972).
[10] P. Cuatrecasas and G. P. Tell, *Proc. Nat. Acad. Sci. U.S.* **70**, 485 (1973).
[11] R. W. Olsen, J. C. Meunier, and S. P. Changeux, *FEBS Lett.* **28**, 96 (1973).
[12] E. Karlsson, E. Heilbronn, and L. Widlund, *FEBS Lett.* **28**, 107 (1972).
[13] P. M. Blumberg and J. L. Strominger, *Proc. Nat. Acad. Sci. U.S.* **69**, 3751 (1972).
[14] R. J. Lefkowitz, E. Haber, and D. O'Hara, *Proc. Nat. Acad. Sci. U.S.* **69**, 2828 (1972).
[15] G. A. Robinson, R. W. Butcher, and E. W. Sutherland, *in* "Cyclic AMP." Academic Press, New York, 1971.
[16] J. C. Venter, G. A. Weiland, J. Ross, Jr., and N. O. Kaplan, manuscript in preparation.
[17] I. Oye and E. W. Sutherland, *Biochim. Biophys. Acta* **127**, 347 (1966).
[18] M. Schramm, H. Feinstein, E. Maim, M. Lang, and M. Lasser, *Proc. Nat. Acad. Sci. U.S.* **69**, 523 (1973).

coupled to glass beads[16,19-21] and the resultant partial purification of the β-receptor adenylate cyclase membrane complex.

Preparation of Hormone-Sensitive Adenylate Cyclase

Erythrocyte ghosts and cell membrane fragments were prepared from turkey erythrocytes by the procedure of Oye and Sutherland.[17]

Buffers

A: Citric acid 75 mM, glucose 75 mM, pH 5.0 with NaOH
B: 0.145 M NaCl
C: 0.145 M NaCl, 10 mM Tris, 1 mM EDTA, pH 7.4 (buffered saline)

Erythrocyte Preparation

Blood was collected in heparinized 50-ml syringes from the wing veins of domestic turkeys (*Melleagus gallopairo*), diluted with 0.25 volume of buffer A and cooled to 3°. For cell preparations the blood was diluted further with equal volumes of cold 0.145 M NaCl and centrifuged at 800 g for 10 minutes at 4°. The supernatant and white buffy coat were aspirated and the cells washed twice in 4 volumes of ice cold buffered saline. The cells collected at 800 g were mixed rapidly with 10 volumes of ice cold distilled water and centrifuged at 1000 g for 15 minutes. The supernatant was aspirated and the process repeated four times. The hemolyzed cells were collected at 1000 g and suspended in 2 volumes of buffered saline. These cells are referred to as fraction I and, as Fig. 1C shows, are membrane ghosts one half cell size or larger.

Preparation of an Epinephrine-Sensitive Membrane Fraction

Hemolyzed cells (fraction I in 2 volumes of buffered saline) were treated for two 20-second periods (at high speed) in a Waring blender precooled to 0°. The homogenate was then centrifuged at 5000 g for 15 minutes, yielding a 2-layered pellet consisting of a pink fluffy layer on top of a gelatinous material. The supernatant was gently removed, then the fluffy layer was resuspended by adding a small amount of cold distilled water and swirling gently. This suspension was then removed and

[19] J. C. Venter, J. E. Dixon, P. R. Maroko, and N. O. Kaplan, *Proc. Nat. Acad. Sci. U.S.* **69**, 1141 (1972).
[20] J. C. Venter, J. Ross, Jr., J. E. Dixon, S. E. Mayer, and N. O. Kaplan, *Proc. Nat. Acad. Sci. U.S.* **70**, 1214 (1973).
[21] J. C. Venter, L. A. Arnold, Jr., and N. O. Kaplan, manuscript submitted.

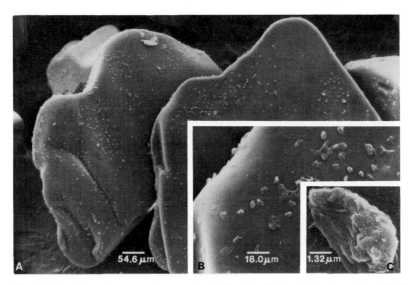

FIG. 1. Scanning electron micrographs of fraction I bound to isoproterenol-glass beads.

diluted with 5 volumes of buffered saline; the centrifugation procedure was repeated twice more. This procedure yields cell membrane fragments less than 2 μm in diameter as determined by scanning electron microscopy.

Binding of Hormone-Sensitive Adenylate Cyclase Membrane Fragments to Isoproterenol on Glass Beads

Porous glass beads were obtained from the Pierce Chemical Co. and derivatized to the aryl glass, then reacted with 50 mg of DL-isoproterenol (Sigma) by the procedure outlined by Venter and Dixon (this volume [25]). After extensive washings, 50 mg of isoproterenol on glass beads (dry weight) were added to each of fractions I and II of the erythrocyte membranes and incubated for 10 minutes with gentle mixing. The glass beads were allowed to settle out and the supernatant enzyme removed. Ten milliliters of ice cold buffered saline were added to the glass beads, and the mixture was vortexed at high speed for 10 seconds. The beads were then allowed to settle, and the supernatant was removed. This washing procedure was repeated 6 times with each sample. The same procedure was followed for unreacted arylamine glass to check for nonspecific absorption to the glass. Figure 1 shows a scanning electron micrograph of fraction I bound to isoproterenol on glass beads.

Adenyl Cyclase Assay of Membrane and Bead-Bound Enzyme

Assay Mixture

Tris·HCl, 40 mM
Theophylline, 9 mM
KCl, 5.5 mM
Phosphenol pyruvate, 20 mM
MgSO$_4$, 15 mM
ATP, 0.4 mM
Pyruvate kinase, 0.13 µg
NaF, 8 mM, or 0.1 mM isoproterenol, as indicated to stimulate adenylate cyclase
Fraction I or II, 30 µl, or 2.0 mg (dry weight) isoproterenol glass bead bound adenylate cyclase in a total volume of 200 µl

All solutions were adjusted to pH 7.4 prior to addition. After the addition of all the components except enzyme, the assay tubes were incubated at 37° for 4 minutes. The reaction was started by the addition of the 30 µl of enzyme or 2.0 mg of glass-bound enzyme, and continued for 10 minutes at 37°, with continuous mixing. Mixing was attained by sitting a number of tubes in the well of a Vortex mixer running at slow speed. The reaction was stopped by boiling the samples for 3 minutes, then centrifuging at 9000 g for 5 minutes. The supernatant was removed and assayed for cAMP as described.[22] The precipitate was assayed for protein by the method of Lowry et al.[23] Protein determinations were performed on duplicate samples of glass bead enzyme by the micro-Kjeldahl method.[24]

Properties and Comments

The table shows the enzymatic activity of the adenylate cyclase fractions prior to and after binding to isoproterenol on glass beads. When unreacted aryl amine glass was used, no adenylate cyclase activity could be measured even in the presence of NaF. The adenylate cyclase without NaF or isoproterenol present had no enzymatic activity, whereas the fragments bound to the isoproterenol on glass beads were maximally stimulated. As the table shows, the specific binding of the hormone sensitive adenylate cyclase to the isoproterenol on glass results in as much as a 39-fold purification of the hormone-sensitive activity. Attempts to

[22] W. B. Wastila, J. T. Stull, S. E. Mayer, and D. A. Walsh, *J. Biol. Chem.* **246**, 1996 (1971).
[23] O. H. Lowry, N. J. Rosebrough, A. L. Farr, and R. J. Randall, *J. Biol. Chem.* **193**, 265 (1951).
[24] R. Ballentine, *Methods Enzymol.* **3**, 984 (1957).

TURKEY ERYTHROCYTE ADENYLATE CYCLASE PURIFICATION SUMMARY

Enzyme system	Units (10^{-5} μmoles cAMP/min/mg protein)	Purification (-fold)
Fraction I (hemolyzed cells)		
Activator		
a. None	0	0
b. 0.1 mM isoproterenol	1.45	0
c. 8 mM NaF	6.28	0
Fraction II (membrane fragment)		
Activator		
a. None	0	
b. 0.1 mM Isoproterenol	5.32	3.6
c. 8 mM NaF	12.30	1.9
Fraction I bound to Isoproterenol on Glass Beads		
Activator		
a. Isoproterenol bound to glass beads	30.9	21
b. (a) + 0.1 mM isoproterenol	29.0	
c. (a) + 8 mM NaF	31.7	5
Fraction II bound to Isoproterenol on Glass Beads		
Activator		
a. Isoproterenol bound to glass beads	207	142
b. (a) + 0.1 mM isoproterenol	188	
c. (a) + 8 mM NaF	215	34

chemically remove the activity from the glass beads have thus far been only partially successful.

The β-adrenergic competitive blocking agent, propranolol, completely inhibited the unbound membrane fragments response to isoproterenol and their binding to the isoproterenol glass beads, whereas there was no inhibition by the blocking agent of the adenylate cyclase while bound to the isoproterenol on glass beads. Techniques other than competitive agents were not attempted for removal of the enzyme from the glass beads. These affinity procedures may allow further fractionation of the membrane particles while bound to the glass beads.

Acknowledgment

This work was supported by grants from the American Cancer Society (IN-93C) and from the National Institutes of Health (CA 11683-05).

[27] Preparation and Characterization of Guanylate Cyclase from Bovine Lung

By ARNOLD A. WHITE and TERRY V. ZENSER

$$\text{GTP} \longrightarrow \text{cGMP} + \text{PP}_i$$

Guanylate cyclase activity was detected in three laboratories nearly simultaneously.[1-4] The reaction appears to be analogous to that catalyzed by adenylate cyclase. Rat or bovine lung has a large amount of readily soluble guanylate cyclase activity. Although the specific activity of a rat lung extract is about four times that of bovine lung, we have preferred to work with the latter because of its ready availability.

Assay Method

The enzyme converts $[\alpha\text{-}^{32}\text{P}]\text{GTP}$ to $[^{32}\text{P}]\text{cGMP}$, and the product is purified from the reaction mixture on an aluminum oxide column and counted in a scintillation spectrometer.[5] Activity is expressed in microunits, which is the amount of enzyme catalyzing the transformation of 1 picomole of substrate per minute at 30°. Protein is determined by the method of Lowry et al.[6]

Enzyme Purification

Beef lungs were obtained from animals freshly killed at a local slaughterhouse and placed in crushed ice. In the experiment described they were minced with commercial machinery into pieces slightly larger than 0.5 cm in diameter, before being brought to the laboratory. All further steps were performed at 4°.

The tissue was homogenized in a 1-gallon capacity stainless steel Waring blender. Nine hundred grams of minced lung were placed in 2700 ml of 50 mM Tris · HCl buffer, pH 7.6 containing 10 mM 2-mer-

---

[1] A. A. White, G. D. Aurbach, and S. F. Carlson, *Fed. Proc., Fed. Amer. Soc. Exp. Biol.* **28**, 473 (1969).
[2] A. A. White and G. D. Aurbach, *Biochim. Biophys. Acta* **191**, 686 (1969).
[3] J. G. Hardman and E. W. Sutherland, *J. Biol. Chem.* **244**, 6363 (1969).
[4] G. Schultz, E. Böhme, and K. Munske, *Life Sci.* **8**, 1323 (1969).
[5] This volume [6].
[6] O. H. Lowry, N. J. Rosebrough, A. L. Farr, and R. J. Randall, *J. Biol. Chem.* **193**, 265 (1951).

captoethanol and 0.5 mM EDTA (TEM buffer) and homogenized at the "Hi" setting for 1 minute. After a 1-minute pause, the homogenization was repeated for 30 seconds. The homogenate was centrifuged at 14,000 g for 30 minutes at 3° (Sorvall RC-2 centrifuge with a GSA rotor), after which the supernatant solution was filtered through glass wool. This process was repeated until 6600 ml of filtered supernatant was accumulated.

Ammonium Sulfate Fractionation. While this supernatant solution was mixed rapidly on a magnetic stirrer, solid ammonium sulfate was added, 10.6 g for each 100 ml. Care was taken to prevent an accumulation of the solid salt during its addition. The resulting solution, which was 20% of saturation with respect to ammonium sulfate, was stirred for 30 minutes. The precipitated protein was then removed by centrifuging as before. The clear supernatants were pooled, and ammonium sulfate was added to 40% of saturation (11.3 g for each 100 ml). The stirring for 30 minutes was repeated, while the pH was maintained above 7 by the addition of Tris base. The precipitate obtained was harvested by centrifugation. This pellet was resuspended in TEM buffer to one-half of the original homogenate supernatant volume and stirred for 1 hour in order to completely dissolve it. Ammonium sulfate precipitation was then repeated, first at 20% and then at 40% of saturation. This second 40% precipitate was redissolved as before in TEM buffer at one-fourth of the original volume. The solution obtained (1650 ml) was dialyzed for 12 hours against 16 liters of TEM buffer, then for 12 hours against 32 liters of buffer, and finally for 8 hours against another 32 liters of buffer. A flocculent precipitate appeared in the dialysis sacks at the end of this time, which was removed by centrifugation. The clear solution was divided into several portions and quick-frozen in liquid nitrogen preliminary to storage at $-20°$. The enzyme was stable at this step. An 8-fold purification of guanylate cyclase was obtained with a 108% recovery of activity. It is important to note that guanylate cyclase is strongly inhibited by ammonium sulfate. Therefore, the apparent purification can vary with the efficiency of sulfate removal. The amount of protein precipitating during dialysis also obviously affects purification. In this experiment, the precipitate contained 19.2% of the total protein. The precipitate must also contain an inhibitor, since when the undialyzed solution was desalted on a Sephadex G-25 column, the specific activity was 40% of that found after dialysis.

DEAE-Cellulose Chromatography. Guanylate cyclase activity binds to DEAE-cellulose when applied in TEM buffer. If a dialyzed ammonium sulfate fraction is used, a large peak of protein and nucleic acid is not retained, and may be washed through the column with the buffer. A linear

NaCl gradient will then elute the enzyme, activity appearing at approximately 80 mM NaCl. We applied 92 ml of a dialyzed ammonium sulfate preparation (966 mg protein) to a 2.5 × 27 cm DE-52 column. The column was washed with one column volume of TEM buffer, and a linear gradient of NaCl was started. This consisted of 850 ml of 0.5 M NaCl in TEM in the reservoir and 850 ml of TEM in the mixing chamber. Five milliliter fractions were collected at a flow rate of 45 ml per hour. The enzyme peak was collected in fractions 93 to 133, which contained 16.9% of the applied protein and 127.6% of the applied activity. This increase in total activity after DEAE-cellulose chromatography has not been consistently obtained and apparently depends upon the previous treatment of the preparation. It probably involves the removal of an inhibitor. We have usually obtained a 6- to 7.5-fold purification in this step. The fractions containing the enzyme were pooled and concentrated to 13 ml on an Amicon UM20E ultrafilter, during which the protein content decreased from 163 to 130 mg. The total activity, however, increased 1.56 times, so that an approximate doubling of specific activity occurred. Since we cannot account for the increase in specific activity on the basis of protein loss, we must conclude that ultrafiltration removed an inhibitor or that the activity of the enzyme was concentration dependent. As we shall see a similar phenomenon occurred during ultrafiltration of the enzyme following gel filtration. At this step in the purification the specific activity was 1776 μU/mg, or 136.6 times the original tissue extract.

Gel Filtration. Guanylate cyclase may be fractionated on a Sephadex G-200 column. A small peak of activity appears in the void volume (peak I), while a larger peak of activity elutes with a K_d of 0.2 (peak II). The same pattern of activity is obtained with either a dialyzed ammonium sulfate preparation or after the preparation is further purified on a DEAE-cellulose column. This suggests that peak I is not a complex of peak II with other macromolecules appearing in the void volume, but rather is another form of the enzyme.

In a typical experiment we applied 4.5 ml, containing 45 mg protein, of the concentrated DEAE preparation to a 2.5 × 83 cm Sephadex G-200 column. The column was eluted with TEM buffer at a flow rate of 17 ml per minute, and 3.3 ml fractions were collected. Peak I appeared from fraction 40 to 48 with a maximum activity at tube 42. Activity increased again from tube 52, peaking at tube 62. We pooled tubes 56 to 70 as peak 2. The specific activity of peak I was 109 μU/mg or 6.1% of the applied sample. That of peak II was 1719 μU/mg or essentially unchanged from the previous step. The two peaks were separately concentrated on an Amicon XM300 ultrafilter. Ultrafiltration of peak I until the volume was reduced 2.5 times did not change the protein content or the specific activity. Ultrafiltration of Peak II until the volume was reduced four times, removed 37% of the protein and increased the specific activity 3.4-fold to 5783 μU/mg. This is the most active preparation that we have yet obtained. Calculated recovery was 63%, but this is of little significance because of the marked increases in recovery produced by prolonged dialysis, DEAE-cellulose chromatography and ultrafiltration. The preparation contained detectable cyclic nucleotide phosphodiesterase activity, but no GTPase.

Properties

Molecular Weight. The beef lung enzyme has a molecular weight of about 300,000[7] as determined by gel filtration on a calibrated Sephadex G-200 column.

pH Optimum. The beef lung enzyme has an optimum at pH 7.6[2] and liver guanylate cyclase at 7.4.[3]

Activators and Inhibitors. Enzyme activity is dependent upon the presence of Mn^{2+}, Mg^{2+}, being much less effective.[2,3] Ca^{2+} is even less effective than Mg^{2+}, however, it will stimulate the Mn^{2+}-dependent activity.[8] Zn^{2+}, Cd^{2+}, and Hg^{2+} inhibit the enzyme,[2,3] and this inhibition is prevented by reduced glutathione or dithiothreitol.[3] The apparent requirement for a free enzyme sulfhydryl group is supported by the demonstration that p-chloromercuriphenyl sulfonate inhibited liver guanylate cyclase, and this inhibition was reversed by dithiothreitol.[9] Nucleotides inhibit activity, in particular, the nucleoside di- and triphosphates,[2,3] as does oxaloacetate[3] and phosphoenolpyruvate.[3,10]

Stability. The requirement for free sulfhydryl groups is met by including 1 mM dithiothreitol or 10 mM 2-mercaptoethanol. The purified enzyme will inactivate on freezing unless 10–20% sucrose or glycerol is included.

Kinetics. Apparent K_m values for GTP are 0.02 to 0.1 mM for activity from rat lung,[3] 0.3 mM for beef lung,[2] 0.04 mM for rat liver,[11] and 0.4 mM for rat kidney.[12]

[7] A. A. White and T. V. Zenser, unpublished experiments (1971).
[8] J. G. Hardman, J. A. Beavo, J. P. Gray, T. D. Chrisman, W. D. Patterson, and E. W. Sutherland, *Ann. N.Y. Acad. Sci.* **185**, 27 (1971).
[9] W. J. Thompson, R. H. Williams, and S. A. Little, *Biochim. Biophys. Acta* **302**, 329 (1973).
[10] A. A. White, S. J. Northup, and T. V. Zenser, *In* "Methods in Cyclic Nucleotide Research" (M. Chasin, ed.), p. 126. Dekker, New York, 1972.
[11] W. J. Thompson, R. H. Williams, and S. A. Little, *Arch. Biochem. Biophys.* **159**, 206 (1973).
[12] E. Böhme, *Eur. J. Biochem.* **14**, 422 (1970).

[28] Guanylate Cyclase from Sperm of the Sea Urchin, *Strongylocentrotus purpuratus*

By DAVID L. GARBERS and J. P. GRAY

$$\text{GTP} \xrightarrow{\text{Mn}^{2+}} \text{cyclic-3',5'-GMP} + (\text{PP}_i?)$$

Assay Method

Principle. Guanylate cyclase activity is measured by estimating radioactive cGMP[1] formation from [³H]GTP. The nature of the phosphate product formed in the reaction has not been reported, although it is known that the α-phosphate of GTP is incorporated into cGMP.[2]

Reagents

Triethanolamine (TEA), 50 mM pH 7.8; 1 mM dithiothreitol (DTT) (prepared weekly)
Manganous chloride solution, 1–100 mM
Sodium azide solution, 100 mM
Theophylline solution, 100 mM
Guanosine triphosphate solution, 0.075–10 mM, in H₂O, pH adjusted to 6.9 with NaOH
Tritiated guanosine triphosphate, about 5 Ci/mmole, 1 mCi/ml in 50% ethanol (New England Nuclear)

Procedure. All reagents are stored at −20°. [³H]GTP is added to the TEA-DTT buffer to give 2–5 × 10⁵ cpm per 0.4 ml buffer. Each assay tube contains 0.4 ml of this mixture plus 50 µl of each of the following: MnCl₂ solution (to give 1–8 mM Mn²⁺ in excess of MnGTP); sodium azide solution; theophylline solution; GTP solution (6–800 µM final concentration), and suspension or dispersion of sonicated sea urchin sperm containing 5–150 µg of protein. With the above conditions at 30°, linear progress curves are obtained for at least 10 minutes at 6 µM Mn-GTP and for at least 60 minutes at 500 µM Mn-GTP. Sodium azide is included in the reaction mixture to inhibit nucleoside triphosphatase activity.

At the end of an incubation, 0.25 ml of 0.2 M Zn(acetate)₂·2H₂O

---

[1] Abbreviations used are as follows: cGMP, cyclic 3',5'-guanosine monophosphate; deoxy-cGMP, 2'-deoxy cyclic 3',5'-guanosine monophosphate; GTP or ATP represent the respective nucleoside triphosphates, and dGTP or dATP the respective 2'-deoxynucleoside triphosphate; ADP, adenosine diphosphate.

[2] J. G. Hardman, J. A. Beavo, J. P. Gray, T. D. Chrisman, W. D. Patterson, and E. W. Sutherland, "Cyclic AMP and Cell Function." *Ann. N. Y. Acad. Sci.* **185**, 27 (1971).

is added to stop the reaction. Unlabeled cGMP (approximately 1 mM) is included in the Zn(acetate)$_2$ solution to monitor cGMP recoveries after purification procedures; 0.25 ml of 0.2 M Na$_2$CO$_3$ is then added, and most of the GMP, GDP, and GTP is coprecipitated with ZnCO$_3$. The precipitate is removed by centrifugation, the resultant supernatant fluid is applied to polyethyleneimine-cellulose columns, and the cGMP is eluted as described by Schultz et al.[3] Part of the cGMP fraction collected from the column is counted for [^{3}H]cGMP in a liquid scintillation counter. The optical density at 252 nm is estimated using the remaining part of the collected sample. cGMP recoveries are usually about 40%.

Enzyme Source. Guanylate cyclase is found in much higher activity in sea urchin sperm than in mammalian tissues.[4] The activity in homogenates of sea urchin (*Strongylocentrotus purpuratus*) sperm ranges from 15 to 60 nmoles of cGMP formed min^{-1} mg protein^{-1} at 37° at pH 7.8 (0.3 mM GTP, 5 mM Mn^{2+}). When measured under similar conditions, the activity in rat lung homogenates ranges from about 0.1 to 0.4 nmole of cGMP formed min^{-1} mg protein^{-1}. Although guanylate cyclase is largely soluble in most mammalian tissues,[5] the sea urchin sperm enzyme is found in particulate fractions. The enzyme is primarily of flagellar origin and appears to be membrane associated.[6]

Sperm from other invertebrates (tube worms, clams, scallops, and abalone) also contain very high guanylate cyclase activity, but in sperm from vertebrates (human, dog, bull, salmon, and herring) activity is either very low or not detectable.[6,7]

Enzyme Preparation. Live specimens of the sea urchin, *S. purpuratus*, are purchased during the months from November to June from Pacific Bio-Marine Supply Co., Venice, California and are maintained at 15° to 17° in artificial sea water (454 mM NaCl; 24.9 mM MgCl$_2$; 9.2 mM KCl; 9.6 mM CaCl$_2$; 27.1 mM MgSO$_4$; and 4.4 mM NaHCO$_3$ at pH 7.8–8.0). The sperm yields are extremely variable among different shipments of urchins and among individual urchins from a given shipment. An average yield is about 30 g of packed sperm per 100 male urchins.

The shedding of gametes is induced by injection of approximately 0.5 ml of 0.5 M KCl into the sea urchin body cavity.[8] So-called "dry

[3] G. Schultz, E. Böhme, and J. G. Hardman, this volume [2].
[4] J. P. Gray, J. G. Hardman, T. Bibring, and E. W. Sutherland, *Fed. Proc., Fed. Amer. Soc. Exp. Biol.* **29**, 608 (1970).
[5] J. G. Hardman and E. W. Sutherland, *J. Biol. Chem.* **244**, 6363 (1969).
[6] J. P. Gray and G. I. Drummond, *Proc. Can. Fed. Biol. Soc.* **16**, 79 (1973).
[7] J. P. Gray, Ph.D. Thesis, Vanderbilt University Library (1971).
[8] E. B. Harvey, "The American Arbacia and Other Sea Urchins." Princeton Univ. Press, Princeton, New Jersey, 1956.

sperm" are obtained by inverting the male urchins over small Syracuse watchglasses. The dry samples are combined and strained through bolting cloth (silk 35 × 35 μm pore size). Samples may be contaminated with pigmented particles which are removed by a low speed centrifugation in a clinical centrifuge. The supernatant fluid containing the sperm is decanted and recentrifuged at about 2000 g for 10 minutes. The supernatant fluid is discarded, and the sperm are resuspended in artificial sea water and centrifuged one more time at about 10,000 g for 15 minutes.

The final sperm pellet is resuspended in about 40 volumes (wet weight/volume) of a solution containing 50 mM TEA, 1.0 mM DTT, and 5 mM disodium ethylenediaminetetraacetate (EDTA) at pH 7.0. The suspension is sonicated briefly and centrifuged at 39,000 g for 20 minutes. The supernatant fluid is discarded, and the above procedure is repeated once. Final enzyme activity is about the same when samples are homogenized with a Ten Broeck homogenizer instead of being sonicated. The pellet is then washed twice with 40 volumes of 50 mM TEA, 10 mM DTT at pH 7.0. Finally, 1 g of sperm particles is suspended in 50 ml of the TEA-DTT buffer without EDTA. This dilution results in approximately 2–3 mg of protein per milliliter. Assuming EDTA does not bind in significant quantities to particulate protein, the final concentration will be less than 10^{-8} M.

Guanylate cyclase activity in the particulate fraction of sea urchin sperm sonicates can be increased 2- to 10-fold by addition of 1% Triton X-100. Stimulation of guanylate cyclase activity in whole homogenates by Triton X-100 is paralleled by appearance of activity in the supernatant fluid of samples centrifuged at high speed (100,000 g for 60 minutes).[6] Dispersed guanylate cyclase can be prepared by adding Triton to sonicated sperm suspensions to a final concentration of 1%. The mixture is again sonicated briefly and then centrifuged at 100,000 g for 1–2 hours. The supernatant fluid contains more than 90% of the enzyme activity. The specific activity ranges from 100 to 500 nmoles of cGMP formed min^{-1} mg protein^{-1} at 37° at pH 7.8.

Properties

Metal Requirement. Sea urchin sperm guanylate cyclase requires free Mn^{2+} for maximum catalytic activity in addition to the chelate, Mn-GTP. Up to 10% of the activity obtained with Mn^{2+} is observed with Fe^{2+}, but detectable activity is not seen using Ni^{2+}, Mg^{2+}, Ca^{2+}, or Sr^{2+}.[9] Activities less than 0.5% of the rate with Mn^{2+} would not be detected.

[9] D. L. Garbers and J. G. Hardman, unpublished observations.

pH optimum. The pH optimum for the reaction is distinctly alkaline (pH 7.6–8.0).[6,9]

Stability. The enzyme is labile to either heat or sulfhydryl reagents, particularly in the presence of Triton X-100. In the absence of substrate or free metal, activity is rapidly lost by incubation at 37°. Heavy metals, N-ethylmaleimide, or p-hydroxymercuribenzoate rapidly inactivate the enzyme.[9] Particulate enzyme preparations are stable at $-70°$ for at least 3 months at pH 7.0 but should be frozen and thawed only one time, since enzyme affinity for Mn^{2+} as well as stimulation by Triton X-100 decrease with successive freeze-thaws.[7,9] Enzyme preparations stored at $-70°$ in the presence of Triton X-100 also show decreases in affinity for Mn^{2+} with increased storage time.[9]

Kinetics. When Mn-GTP is varied from 6 to 800 μM with free Mn^{2+} fixed, reciprocal (Lineweaver-Burk) plots are concave upward. Hill plots have a slope greater than 1.0. When Mn-GTP concentration is fixed, and free Mn^{2+} is varied from 400 to 3000 μM, reciprocal plots are linear.[9]

Inhibitors. At concentrations of Mn-GTP greater than 0.1 mM, dATP, ATP, dGTP, and ADP are inhibitors of the enzyme. With Mn-GTP varied, reciprocal plots change from concave upward with no inhibitor, to linear at high concentrations of these inhibitors.[9]

Activators. At low concentrations of Mn-GTP (less than 10 μM) dATP, ATP dGTP, and ADP can activate the enzyme. Ca^{2+} can also activate the enzyme, but the mechanism of activation is obscure.[9]

Alternate Substrates. When Mn-dGTP is used as a substrate, deoxy-cGMP is formed. Although it has not been proved that guanylate cyclase catalyzes this reaction, GTP and ATP are inhibitors, and the enzyme involved in deoxy-cGMP formation is denatured by heat at the same rate as the enzyme catalyzing cGMP formation.[10]

[10] D. L. Garbers, J. L. Suddath, and J. G. Hardman, unpublished observations.

[29] Guanylate Cyclase in Human Platelets

By EYCKE BÖHME, REGINE JUNG, and ILSE MECHLER

Determination of Guanylate Cyclase Activity

Guanylate cyclase activity was determined by incubation for 15 minutes or longer at 37° with 0.5 mM [8-^{3}H]GTP (0.25–1 μCi), 3 mM $MnCl_2$, 2 mM unlabeled cGMP, and 100 mM N-Tris(hydroxymethyl)methyl-2-aminoethane sulfonate (TES)–NaOH buffer, pH 7.4, containing 50 mM NaCl, if not otherwise indicated. The product formed and the substrate remaining at the end of the incubation period were isolated by column chromatography on polyethyleneimine cellulose.[1]

Distribution of Guanylate Cyclase in Human Blood Cells

Human blood cells were separated by differential centrifugation and by use of Dextran.[2] In plasma and homogenates of erythrocytes, no guanylate cyclase activity was detectable. In homogenates of mixed leukocytes (containing some platelets), the guanylate cyclase activity was about 50 pmoles min^{-1} mg $protein^{-1}$. The enzyme activity in platelets varied between 0.2 and 2.0 nmoles min^{-1} mg $protein^{-1}$; this is substantially (up to an order of magnitude) higher than in any other mammalian tissue so far studied except outer segments of retinal rods.

Preparation of Platelet Homogenates

Platelets were isolated from human blood by differential centrifugation at 4° with 4 mM EDTA added as an anticoagulant. The blood was centrifugated at 400 g for 15 minutes, and platelet-rich plasma was obtained. The platelet-rich plasma was centrifugated at 1500 g for 10 minutes, and the platelets were suspended in 100 mM TES-NaOH-buffer, pH 7.4, containing 90 mM NaCl. The platelet suspension was then centrifugated at 600 g for 10 minutes, and the platelets were resuspended in fresh buffer.

The guanylate cyclase activity measured in the platelet homogenates was not significantly affected if EDTA was replaced by 15 mM sodium citrate. The enzyme activity was slightly higher if the platelet isolation procedure was performed at room temperature (20–25°).

Suspensions of platelets were usually disintegrated ultrasonically using a MSE 60-W Ultrasonic Disintegrator for 15 seconds. The guanylate cyclase activity was not significantly different if the platelets were disintegrated by hypotonic shock in 5 mM TES-NaOH buffer, pH 7.4, or by repeated freezing and thawing. If the ultrasonic treatment was performed for more than 30 seconds or if the platelets were exposed to the hypotonic medium for more than 30 minutes or if freezing and thawing was repeated more than 3 times, the guanylate cyclase activity was decreased.

[1] G. Schultz, E. Böhme, and J. G. Hardman, this volume [2].
[2] W. A. Skoog and W. S. Beck, *Blood* **11**, 436 (1956).

Apparent Subcellular Distribution of Guanylate Cyclase

More than 90% of the total guanylate cyclase activity measured in subcellular fractions was found in the 250,000 g, 60-minute supernatant fluid. The apparent distribution of guanylate cyclase between particulate and soluble fractions was not affected by the mode of cell disintegration.

Unlike the particulate guanylate cyclase activity from a number of other sources, that in platelets was not increased by Triton X-100. The enzyme in platelet homogenates or high speed supernatant fractions was inhibited by 50% by about 1 mg/ml of Triton X-100 in the incubation medium.

Influence of Divalent Cations on Platelet Guanylate Cyclase

The formation of cGMP depends on the presence of a divalent cation. If the guanylate cyclase activity was measured with 1 mM GTP and 10 mM Me^{2+}, the activity with Mg^{2+} was 60%, with Fe^{2+} 35%, and with Ca^{2+} 5% of that observed with Mn^{2+}. The formation of cGMP measured in the presence of 1 mM GTP and 5 mM Mn^{2+} was completely inhibited by Hg^{2+}, Zn^{2+} or Cu^{2+} (0.1–10 mM) and partially inhibited by Co^{2+}, Pb^{2+}, Fe^{3+}, Fe^{2+}, or Ba^{2+} (1 or 10 mM). The addition of Mg^{2+} and especially of Ca^{2+}, however, caused an increase in the guanylate cyclase activity measured with 1 mM GTP and 5 mM Mn^{2+}. The effect of Ca^{2+} was more than additive while that of Mg^{2+} was less than additive. The effect of Ca^{2+} on guanylate cyclase activity highly depended on the respective concentrations of Ca^{2+}, Mn^{2+}, and GTP.

Apparent Activation of Platelet Guanylate Cyclase with Incubation Time

The formation of cGMP in assays using crude homogenates or 250,000 g, 30-minute supernatant fractions was not proportional to the incubation time; the enzyme activity increased with the length of the incubation time. A maximal activity was observed after about 60 minutes of incubation at 37°, then the activity slowly declined. The apparent activation of guanylate cyclase with time was also observed if the enzyme preparations were preincubated at 37° in the absence of Mn^{2+}, GTP, or cGMP (Fig. 1). Addition of EDTA (1 or 5 mM) to guanylate cyclase preparations preincubated for various lengths of time retarded but did not prevent the enzyme activation with time; in the presence of EDTA, the same maximal enzyme activity was observed as without addition of this agent.

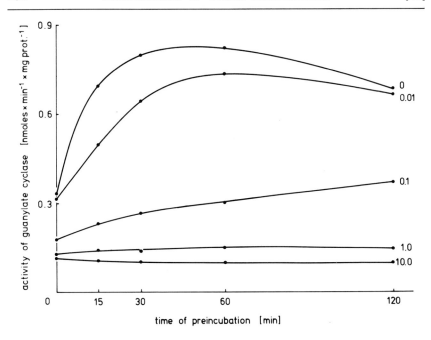

FIG. 1. Influence of dithiothreitol (DTT) on the apparent activation of human platelet guanylate cyclase with time. A platelet homogenate was preincubated for the indicated lengths of time at 37° in 100 mM TES-NaOH buffer, pH 7.4, containing 90 mM NaCl and DTT at the millimolar concentrations indicated in the figure. Guanylate cyclase activity was then measured by incubation for 15 minutes at 37° with the addition of 0.5 mM GTP, 3 mM MnCl$_2$, and 2 mM cGMP.

Addition of a sulfhydryl group protecting agent, e.g., dithiothreitol (DTT), reduced the apparent activation of the platelet guanylate cyclase with time (see Fig. 1). DTT in high concentrations (1 or 10 mM) completely abolished the effect of time on guanylate cyclase activity. Dithioerythritol, but not the oxidized form of DTT, also prevented the enzyme activation.

Acknowledgment

The authors' studies were supported by the Deutsche Forschungsgemeinschaft.

Section IV

Degradation of Cyclic Nucleotides

[30] Assay of Cyclic Nucleotide Phosphodiesterases with Radioactive Substrates

By W. Joseph Thompson, Gary Brooker, and M. Michael Appleman

Assay Method

Principle. The hydrolysis of the 3',5'-diester bond of 3',5'-cyclic nucleotides yields the 5'-mononucleotides and is catalyzed by cyclic nucleotide phosphodiesterases. Until the use of radioactive substrates to detect picomole quantities of product formation, phosphodiesterase enzymology was severely limited. Tritiated cyclic nucleotides are now widely used as substrates for the assay of cAMP phosphodiesterase[1-9] and cyclic GMP (cGMP) phosphodiesterase.[2,4,5]

The assay scheme for cAMP phosphodiesterase is shown in Fig. 1. The same scheme is used for other cyclic nucleotides. The separation of [^{3}H]5'-AMP from [^{3}H]3',5'-AMP formed by the phosphodiesterase reaction is simplified by enzymatically converting [^{3}H]5'-AMP formed by the phosphodiesterase to [^{3}H]adenosine with nucleotidase found in king cobra snake venom.

Adenosine can be easily separated from the charged cyclic nucleotide substrate by precipitation with an anion exchange resin and the resultant radioactivity of adenosine in the supernatant measured by liquid scintillation techniques.

These assays can be performed in a liquid scintillation vial wherein the phosphodiesterase and snake venom are incubated together and the reaction is terminated by the addition of the anion exchange resin. The anion exchange resin also absorbs unreacted cyclic [^{3}H]nucleotide and quenches its scintillation when liquid scintillation fluid is subsequently

---

[1] R. G. Kemp and M. M. Appleman, *Biochem. Biophys. Res. Commun.* **24**, 564 (1966).
[2] G. Brooker, L. J. Thomas, Jr., and M. M. Appleman, *Biochemistry* **7**, 4177 (1968).
[3] R. J. DeLange, R. G. Kemp, W. D. Riley, R. A. Cooper, and E. G. Krebs, *J. Biol. Chem.* **243**, 2200 (1968).
[4] W. J. Thompson and M. M. Appleman, *Biochemistry* **10**, 311 (1971).
[5] J. A. Beavo, J. G. Hardman, and E. W. Sutherland, *J. Biol. Chem.* **245**, 5649 (1970).
[6] O. M. Rosen, *Arch. Biochem. Biophys* **137**, 435 (1970).
[7] S. Hynie, G. Krishna, and B. B. Brodie, *J. Pharmacol. Exp. Ther.* **153**, 90 (1966).
[8] G. Pöch, *N. S. Arch. Pharmacol.* **268**, 272 (1971).
[9] P. S. Schonhofer, I. F. Skidmore, H. R. Bourne, and G. Krishna, *Pharmacology* **7**, 65 (1972).

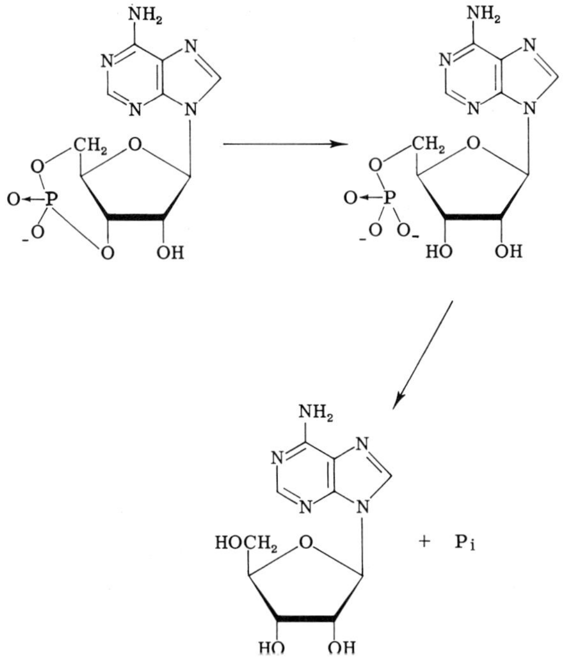

Cyclic $[^3H]3',5'$-AMP $\xrightarrow[\text{diesterase}]{\text{phospho-}}$ $[^3H]5'$-AMP $\xrightarrow{\text{snake venom}}$ $[^3H]$adenosine

FIG. 1. Assay scheme for cyclic AMP phosphodiesterase.

added. Even though kinetic analysis of these enzyme activities has been possible using this one-step system, a two-step method wherein the phosphodiesterase reaction is conducted in the absence of snake venom is theoretically preferable for kinetic analysis since the possible effects of snake venom on the phosphodiesterase are eliminated. Since both the one-step and two-step methods can be effectively used in studying cyclic nucleotide phosphodiesterases, both methods will be described.

Theory of Assay

The theory of cAMP phosphodiesterase assay and general procedural outline are shown in the table. The theory of all the assay references cited is basically identical and only the isolation of radioactive products

Assay Procedure for Cyclic Adenosine Monophosphate Phosphodiesterase

$$\text{Cyclic[}^3\text{H] 3',5'-AMP} \xrightarrow[1]{\text{phosphodiesterase}} \text{[}^3\text{H]5'-AMP} \xrightarrow[2]{\text{snake venom}} \text{[}^3\text{H]adenosine}$$

Step 1: Phosphodiesterase reaction

Cyclic [^{3}H]adenosine monophosphate (200,000 cpm)
Cyclic adenosine monophosphate (0.125 μM to 0.1 mM)
MgCl$_2$ 5 mM
Tris · Cl (pH 8.0, 40 mM)
2-Mercaptoethanol (3.75 mM) + enzyme in 0.4 ml final volume
10-minute incubation at 30°, stopped by boiling 2.5 minutes

Step 2: Snake venom nucleotidase

Contents step 1 + 0.1 ml of snake venom (*Ophiophagus hannah*) (1 mg/ml)
10-minute incubation at 30°, stopped by 1.0 ml of a 1:3 slurry Bio-Rad resin, AG1-X2, 200–400 mesh
Contents spun in clinical centrifuge and 0.5-ml aliquot counted by liquid scintillation

---

differ. [8-^{3}H]cAMP is hydrolyzed to [8-^{3}H]5'-AMP by cyclic nucleotide phosphodiesterase necessitating magnesium cofactor (step 1). All the [8-^{3}H]5'-AMP formed is converted to [8-^{3}H]adenosine by the nucleotidase action of snake venom (step 2). Labeled adenosine is then separated from charged nucleotides by precipitation with anion-exchange resin, and adenosine is measured by liquid scintillation techniques.

Since the only known catabolism of cAMP is cyclic nucleotide phosphodiesterase, the critical assumption of this assay method is that labeled products produced represent direct stoichiometry from cyclic nucleotide hydrolysis through 5'-AMP. Care must be taken to assure that this is indeed the case, particularly when crude homogenate enzyme preparations are used. This procedure has an added advantage as side reactions in crude homogenates, such as 5'-AMP deaminase, will not interefere with the assay stoichiometry of cAMP conversion to labeled uncharged nucleosides. This procedure also has advantages over other methods when handling multiple samples.

cGMP phosphodiesterase activity can be measured similarly. cGMP is, however, a general tritium-labeled isotopic compound. As such there are inherent problems of tritium exchange and substrate instability with cGMP phosphodiesterase assay that should be taken into account apart from those of cAMP phosphodiesterase assay.

Reagents and Assay Constituents

[8-^{3}H]cAMP and [G-^{3}H]cGMP. The highest specific activity cyclic nucleotides currently commercially available are 28.0 Ci/mmole for cAMP and ~4 Ci/mmole for cGMP. We have found it necessary to purify commercially available compounds, especially when exacting, detailed kinetic determinations are to be made and assay blank values as low as possible are desired. Several purification procedures for cGMP and cAMP are available. Thin-layer chromatography methods are those most often used.[4,10] All general precautions with reference to ^{3}H-labeled compounds should be observed, especially with [^{3}H]cGMP, as [^{3}H]guanine compounds are inherently much more unstable than [^{3}H]adenine compounds. Storage at $-20°$ in 50% ethanol has proved most satisfactory for stock solutions which generally need to be repurified every 3–5 weeks for best assay results.

cAMP and cGMP. Standard cyclic nucleotides solutions are prepared in 40 mM Tris-Cl, pH 8.0 as ~1 mM bacteria-free solutions and calculated to exact concentration by using extinction coefficients of 14.65 mM and 13.7 mM, respectively, at 259 nm pH 7.0. The desired concentrations for cyclic nucleotide phosphodiesterase assay are then prepared using accurate, near-sterile dilution procedures. The solutions should be aliquoted and stored at $-20°$ for routine use.

MgCl$_2$ is stored at 4° as 1 M solution and maintained bacteria-free.

Snake (*Ophiaphagus hannah*) venom is obtained from Sigma Chemical Co. and a solution adequate for 100–500 assays made as 1 mg/ml H$_2$O can be stored at 4° for several weeks. Lyophilized venom powder should be stored at $-20°$. Other venoms should not be used for this type of assay unless carefully analyzed for phosphodiesterase and proteolytic enzyme activity.

Tris·Cl, 40 mM, pH 8.0–3.75 mM mercaptoethanol (referred to as assay buffer) should be prepared fresh daily

Anion-exchange resin: Bio-Rad AG 1-X2, 200–400 mesh (Bio-Rad Laboratories) must be washed with 0.5 N HCl, 0.5 N NaOH 0.5 N HCl, and then repeatedly with H$_2$O to pH 5.0. If necessary, the pH can be adjusted with a small amount of Tris base to maintain the pH. The resin is then allowed to settle for at least 45 minutes, and 2 volumes H$_2$O added to 1 volume of settled resin.

Scintillation cocktail: naphthalene, 125 g; 2,5-diphenyloxazole, 7.5 g; 1,4-bis(2,5-phenyloxazolyl)benzene, 0.382 g; and 1,4-diethylene dioxide, 1 liter.

[10] G. Brooker, *J. Biol. Chem.* **246**, 7810 (1971).

Procedure for Two-Step Assay

The following procedures are currently in use and designed for maximum flexibility and convenience for several assay purposes, i.e., kinetic analyses, homogenate studies, and effector modification of activity, etc. Calculations and additions are given to assay low K_m cAMP phosphodiesterase activity at 62.5 nM cAMP and cGMP phosphodiesterase at 10 μM cGMP. Assay volumes and concentrations can be modified to fit any need and the amounts of radioactivity adjusted to budgetary needs and usage.

Step 1A. Reagents: 0.1 ml buffer or cAMP; 0.1 ml [^{3}H]cAMP: Mg^{2+}: buffer; 0.2 ml of enzyme preparation.

Equilibrate a shaking water bath at 30° and prepare a dry ice–acetone and a boiling water bath all capable of holding reaction vessels (12 × 75 mm culture tubes, preferably disposable). All pipettings are critical and should be made with both accurate and repeatable instruments, such as Hamilton syringes or calibrated and leak-tested Eppendorf pipettes. Pipette 0.1 ml of assay buffer in our example case or for other purposes desired cAMP to the test tubes kept at 4°. This pipetting is easily modified for experiments involving other test substances. Next, pipette 0.1 ml radioactive cAMP, Mg^{2+}, and buffer mixture. This addition contains 25 pmoles of cAMP and ~220,000 cpm. Mixture enough for 10 assays is made by adding 0.1 ml of 1.36 μM cAMP, 0.02 ml of 1 M $MgCl_2$, 0.01 ml purified stock solution [^{3}H]cAMP, and 0.87 ml of assay buffer. [Purified stock 50% ethanol [^{3}H]cAMP solution (28.0 Ci/mmole; 1 mCi/2 ml) is adjusted so 0.010 ml $\cong 2.2 \times 10^6$ cpm (33% efficiency) when quenched with 0.5 ml of assay buffer.] Enzyme reactions are initiated by addition of 0.20 ml of enzyme preparation (4°), gently vortexed, and incubated in the shaking water bath at constant 30°.

Or Step 1B. Reagents: 0.1 ml cGMP; 0.1 ml [^{3}H]cGMP: Mg^{2+}: buffer; 0.2 ml of enzyme preparation.

Cyclic GMP phosphodiesterase assay is identical to step 1A except that aliquot 0.1-ml of standardized 40 μM cGMP to the reaction vessels. Next, aliquot 0.1-ml radioactive substrate containing Mg^{2+}. Mixture of radioactive substrate enough for 10 assays is made by adding 0.01 ml stock, purified [^{3}H]cGMP [purified stock 50% ethanol solution (3.96 Ci/mmole, 1 mCi/2 ml) is adjusted so 0.01 ml $\cong 2.7 \times 10^6$ cpm (33% counting efficiency) when quenched with 0.5 ml of assay buffer], 0.020 ml 1 M $MgCl_2$, and 0.970 ml assay buffer. Initiate reactions as in step 1A.

We recommend termination of both the cAMP and cGMP phosphodiesterase reactions by immersing the reaction tubes into a dry ice–acetone bath until frozen (12 seconds) followed by subsequent immersion in a

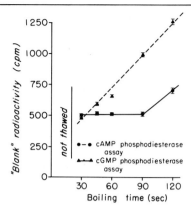

Fig. 2. Effects of boiling on "blank" values for assay. ●——●, cAMP phosphodiesterase assay; ▲——▲, cGMP assay.

boiling water bath for exactly 45 seconds. However, reactions may be terminated by boiling directly, but blank values (particularly cGMP) may be higher (Fig. 2). Termination of phosphodiesterase activity by dilution, pH variance, or metal poisoning has proved to be unsatisfactory.

Step 2. The reaction vessels of step 1A or B are again immersed in an ice bath until cooled to 4°. 5'-Nucleotides are then hydrolyzed to nucleosides by addition of 0.1 ml of snake venom (*Ophiaphagus hannah*) solution (1 mg/ml). The tubes are thoroughly vortexed, and incubated for 10 minutes at 30° in a shaking water bath. At the end of the incubation period the tubes are immersed in an ice bath, 1.0 ml of stirred resin slurry is added (Cornwall repeating syringe fitted with 2.5-ml glass syringe works well if washed clean of resin when not in use), thoroughly vortexed, and allowed to equilibrate for 15 minutes at 4°. The tubes are then revortexed, and centrifuged in a clinical centrifuge for 10 minutes to form a resin pellet at the bottom of the tube. With the use of an automatic dilutor (Lab Industries), a 0.5-ml aliquot is carefully pipetted out of the tube and washed into a scintillation vial with 10 ml of the scintillant mixture. The samples are then counted to 2% statistical accuracy.

Procedure for the One-Step Assay

The complete assay incubation including [^{3}H]cAMP, phosphodiesterase, and snake venom and liquid scintillation counting of the reaction products is conducted in a single liquid scintillation vial. The nucleoside products of the reaction are separated by addition of an anion-exchange resin slurry. Liquid scintillation fluid is added, and the unreacted

[³H]cAMP substrate counting efficiency is reduced about 90% while the product of the reaction [³H]adenosine is not quenched by the resin.

The assay volume of the one-step assay has been reduced to 35 μl.[11] Into a glass or plastic scintillation vial is placed: 5 μl of 100 mM Tris·HCl, pH 8.0, containing 0.1 μCi [³H]cAMP; 1.2 μmoles 5'-AMP (optional), and 1.2 μmoles $MgCl_2$; 20 μl of cAMP solutions calculated to alter the total assay cAMP concentrations from 0.1 μM to 0.1 mM.

The reaction is then started by addition of 5 μl of king cobra venom solution and 5 μl of the phosphodiesterase preparation. The vials are incubated for 10 minutes at 30° and stopped by addition of 0.8 ml of the resin slurry. After 10 minutes' equilibration, the vials are counted by liquid scintillation using 10 ml of dioxane-naphthalene scintillant.

Comments

Assay Limits. The theoretical limits of substrate concentration of this assay method is a function of the specific activity of labeled cyclic nucleotide. The practical, convenient limit of the method as presented here is 10 nM for cAMP phosphodiesterase and 70 nM for cGMP phosphodiesterase. The reproducibility of an assay is primarily a function of pipetting and liquid scintillation counting errors.

Velocity Calculations. Reaction velocity is calculated according to the following equation:

$$v = \frac{\text{cpm (measured)} - \text{blank}}{\text{cpm (maximum)} - \text{blank}} \times \frac{\text{picomoles substrate}}{0.4 \text{ ml}} \div \text{time of incubation}$$

Velocity is then expressed in microinternational units per assay and subsequent specific activity can be determined.

Determination of Michaelis-Menten Constants. Determination of the apparent Michaelis-Menten constant of any phosphodiesterase activity is at the present time the most useful and meaningful identification of enzyme activity of an extremely complex nature. Based on current enzyme data we suggest the following cAMP substrate concentrations for differentiation of low K_m enzyme activity versus the mixture of cyclic nucleotide phosphodiesterase activities present in most tissues[12]: 100, 50, 32.5, 25, 18, 16.25, 14.375, 12.375, 10, 7.875, 6.5, 5.0, 3.94, 3.25, 2.5, 2.0, 1.625, 1.25, 1, 0.875, and 0.8125 μM cAMP with constant [³H]cAMP adjusted to 0.125 μM. For determination of the apparent Michaelis-Menten

---

[11] G. Brooker, *in* "Advances in Cyclic Nucleotide Research" (P. Greengard and A. Robison, eds.), p. 111. Raven, New York, 1972.

[12] W. J. Thompson and M. M. Appleman, *J. Biol. Chem.* **246**, 3145 (1971).

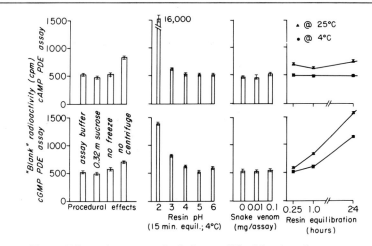

FIG. 3. Effect of assay manipulation on "blank" values for assay.

constant for cyclic GMP phosphodiesterase, we suggest the following concentrations of unlabeled cGMP: 200, 100, 65, 50, 30, 25, 20, 14.28, 12.5, 10, 8, 5, and 0.1 μM with constant [H^3]cGMP adjusted to 0.25 μM. All values given indicate final assay concentrations and, although seemingly inconvenient, provide the best graphic presentation for both enzyme activities when plotted according to Lineweaver-Burk.[13] Solutions can be diluted, stored at −20°C indefinitely, and used repeatedly. One should expect diminution of the hydrolysis of cyclic [^{3}H]nucleotide substrate with nonlabeled cyclic nucleotide addition[14] and should adjust enzyme concentrations accordingly. Maximum hydrolysis, however, should never exceed 25% saturation. This should be strictly observed using crude enzyme preparations.

Assay "Blanks." Assay "blank" values either determined with assay buffer substituted as enzyme or with boiled enzyme preparations should be 1% or less of maximum cyclic [^{3}H]nucleotide added. These blank values are linear with respect to labeled cyclic nucleotide added and are probably a function of ^{3}H exchange and substrate instability at assay temperatures. The effect of boiling on blank values is shown in Fig. 2. The effect of other manipulations on assay procedure performed identically as described is shown in Fig. 3. Abscissas express actual counts per minute one should expect as blank values.

---

[13] H. Lineweaver and D. J. Burk, *J. Amer. Chem. Soc.* **56**, 658 (1934).
[14] G. Brooker and M. M. Appleman, *Biochemistry* **7**, 4182 (1968).

[31] Assay of Cyclic Nucleotide Phosphodiesterase by a Continuous Titrimetric Technique[1]

By WAI YIU CHEUNG

Most methods available to monitor phosphodiesterase either make use of auxiliary enzymes, and are thereby subject to inherent limitations, or fail to monitor continuously the enzymatic reaction. The advantage of the titrimetric method to be described here is that it is simple, direct, and offers initial velocity measurement. The disadvantage is that it is rather insensitive and would not be applicable to tissue homogenates that are low in phosphodiesterase activity; nor would it be sensitive enough to monitor the "low K_m" enzyme at micromolar substrate concentration.

Principle

The method is based on the fact that the phosphate group of cyclic AMP (cAMP) has one titratable species, whereas that of 5'-AMP has two. Hydrolysis of cAMP by phosphodiesterase to 5'-AMP makes available an additional titratable species as follows:

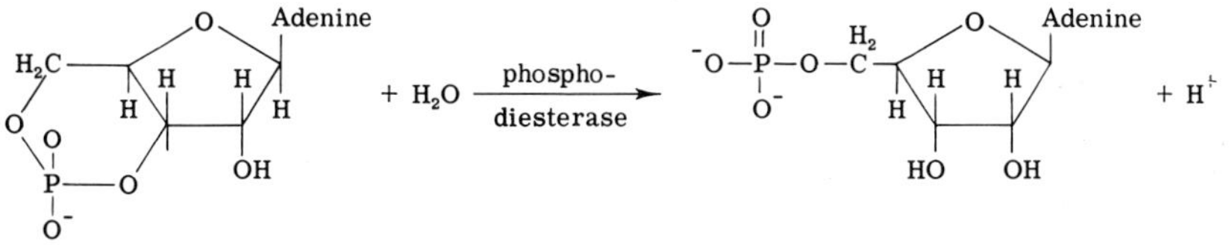

cAMP + H$_2$O $\xrightarrow{\text{phosphodiesterase}}$ 5'-AMP + H$^+$

As hydrolysis of cAMP to 5'-AMP proceeds, a stoichiometric quantity of protons is generated, and the reaction mixture becomes acidic. The rate of addition of alkali required to maintain a constant pH will be related quantitatively to the rate of cAMP hydrolysis.[2]

Reagents

cAMP, 0.1 M, adjusted to pH 8.0

CO_2-free water. Boil 4 liters of glass-distilled water for 10–20 minutes, cool under a stream of nitrogen, transfer to a 4-liter aspirator bottle, and store under nitrogen or a CO_2-free atmosphere. The

---

[1] This work was supported by ALSAC, by Grants NS-08059, CA-13537, and CA-08480, and by a Career Development Award (1-K-4-NS42576) from the U.S. Public Health Service.

[2] W. Y. Cheung, *Anal. Biochem.* **28**, 182 (1969).

latter is easily achieved by connecting the mouth of the bottle to a plastic tube filled with ascarite.

NaCl, 40 mM. Prepare a 1 M stock solution from which a 40 mM solution is made daily with CO_2-free water.

$MnCl_2$, 1 mM

NaOH, 0.5 to 2.0 mM, CO_2-free. Make a saturated solution with NaOH pellets and let it stand at room temperature for a few days until the supernatant solution becomes clear. Carefully remove some of the solution and dilute it with CO_2-free water. Standardize it against a potassium hydrogen phthalate solution. A saturated NaOH solution is about 12 M. Prepare a 0.1 M stock solution from which an appropriate concentration of the titrant is made each day. A commercial standardized CO_2-free NaOH solution may also be used. Keep all NaOH solutions under nitrogen.

Apparatus and Experimental Conditions

A Metrohm Combi-Titrator 3D equipped with a 0.2-ml syringe and a combination electrode is used. Other pH-stat models adaptable to a micro-version may be used also. The reaction mixture is stirred magnetically and thermostatted at 30°. Standard assay is carried out at pH 8.0 in a final volume of 1.5 ml containing 40 mM NaCl, 0.1 mM $MnCl_2$, phosphodiesterase, and 1 mM cAMP; the latter is usually added last to start the reaction. The reaction mixture is adjusted with the titrant to about 0.05 pH unit above the set point, which is the pH at which the reaction is to be maintained. cAMP (15 μl of 0.1 M, pH 8.0) is then introduced to initiate the reaction, and the rate of cAMP hydrolysis is followed by that of the consumption of NaOH. The concentration of the titrant varies from 0.5 mM to 2 mM, depending upon the activity of phosphodiesterase. The pH of the reaction system is easily maintained within 0.02 unit of the set point.

Titration involves low levels of protons (400 nmoles or less per 1.5 ml reaction mixture) at an alkaline pH; it is imperative to protect the reaction mixture from atmospheric CO_2 by a stream of N_2. In fact, when a reaction is carried out under air instead of nitrogen, a background titration is easily discernible.

Selection of pH for Titrimetric Assay

Hydrolysis of cAMP to 5'-AMP involves not only generation of protons, but an additional pK; the pH at which titration is carried out deserves consideration. Figure 1 shows the titration curves of cAMP and

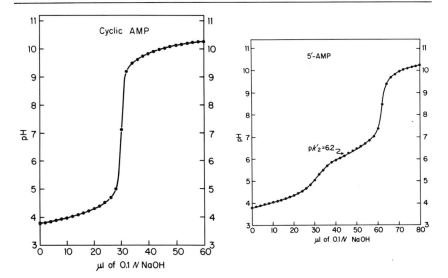

Fig. 1. Titration curves of cAMP and 5'-AMP. cAMP or 5'-AMP (3.0 μmoles in 1.5 ml of water) was manually titrated in the pH stat with 0.1 M NaOH. An aliquot of 2 μl of titrant was added stepwise, and the resultant pH was recorded. The abrupt change in pH indicated an end point. Note the absence of buffer capacity between pH 7.5 and pH 9.0 in both curves.

5'-AMP with a dilute HCl solution. In the former there is virtually no buffer capacity from pH 5 to 9, whereas in the latter, buffer capacity exists between pH 6 and 7, with pK_2 = 6.2. In following the hydrolysis of cAMP to 5'-AMP by the pH stat, one should carry out the titration as far removed from pK_2 as practical so that possible interference by this buffer region may be minimized. As neither cAMP nor 5'-AMP exhibits significant buffer capacity between pH 7.5 and 9, the activity of phosphodiesterase may be followed best within this range. The reaction mixture in the experiments described here is kept at pH 8.0 because it is optimal for the activity of the brain enzyme.

Sensitivity of the Technique

Figure 2 reproduces a set of titration curves obtained with different amounts of phosphodiesterase. Numbers in parentheses indicate titration rates in nanomoles per minute. Titration proceeded from left to right with a full-scale equivalent to 0.2 ml of a titrant. In this experiment the titrant was 0.5 mM NaOH, and a full chart scale represented 100 nmoles of NaOH. As shown in the figure, a reaction giving 2 nmoles per minute can be easily measured.

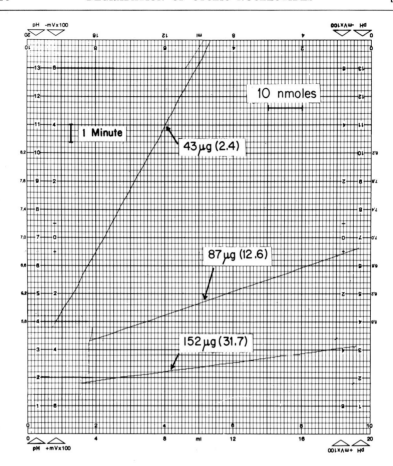

Fig. 2. Reproduction of titration curves. The reaction mixture was of standard composition, and the amount of phosphodiesterase was an indicated. The titrant was 0.5 mM NaOH. A full scale of the recorder chart was equivalent to 100 nmoles of NaOH. The chart speed was 12 mm/min. Numbers in parentheses indicate corresponding rates expressed as nanomoles of cyclic AMP (cAMP) hydrolyzed per minute. Specific activity of phosphodiesterase as calculated from the figure is 210 units (1 unit = 1 nmole of cAMP hydrolyzed per milligram of protein per minute at 30°).

Dynamic Aspect of the Titrimetric Assay

The chief advantage of this technique is that it affords initial velocity. Figure 3 illutrates this point with Mn^{2+}, which is required for the bovine brain enzyme to express full activity. In the absence of Mn^{2+}, the rate of cAMP hydrolysis was 41 nmoles per minute. Upon addition of 0.1 mM Mn^{2+}, the rate increased to 102 nmoles per minute after a lag of

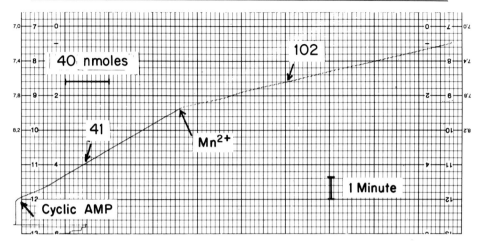

FIG. 3. Effect of Mn^{2+} on the activity of phosphodiesterase. Mn^{2+} was omitted in the initial part of the curve and added later (0.1 mM $MnCl_2$) as indicated. The rates of cyclic AMP hydrolysis were 41 and 102 nmoles/minute in the absence and in the presence of Mn^{2+}, respectively.

a couple of seconds. This lag represents the response time of the titrator to the change of acidity in the reaction mixture.

Concluding Remarks

Results obtained by the titrimetric technique have been found to correlate well with those by other methods used in the author's laboratory. A reaction mixture that catalyzes the hydrolysis of cAMP to 5'-AMP at a rate of 2 nmoles per minute or greater under standard conditions is measured accurately. The technique has been used to assay phosphodiesterase activity in the partially purified state as well as in the crude homogenates of bovine brain and *Serratia marcescens*.[3] Other tissue homogenates, such as rat liver, muscle, and kidney, that have low phosphodiesterase activity may not be accurately detected by the pH stat. Although one can increase the protein concentration to give a higher rate, it is usually not practical, as proteins are amphoteric and usually exhibit buffer capacity in the neutral pH range, and therefore may interfere with the assay.

[32] Cyclic 3′,5′-Nucleotide Phosphodiesterase from Bovine Heart

By R. W. Butcher

Cyclic AMP + $H_2O \to$ 5′-AMP
(Cyclic GMP + $H_2O \to$ 5′-GMP)

Assay Method[1]

Principle. The assay used in the purification procedure described herein utilizes the 5′-nucleotidase of *Crotalus atrox* venom as a means for hydrolyzing the phosphate of the reaction product, 5′-AMP, with the subsequent measurement of P_i by a suitable method (e.g., Fiske and SubbaRow[2] or Buell *et al.*[3]). Apparent zero-order kinetics are maintained so long as the substrate concentration is well above the higher K_m (around 0.1 mM). However, it should be noted that this assay system is relatively insensitive and is of limited value in that it cannot be used at low substrate concentrations.

Reagents

Cyclic AMP (cAMP) 1.2 mM
$MgSO_4$, 12.0 mM
Tris buffer, 240.0 mM, pH 7.5
Crotalus atrox venom, 0.1%, in 10.0 mM Tris, pH 7.5
Trichloroacetic acid, 55%
Reagents for inorganic phosphate by the method of Fiske and SubbaRow, except that the ammonium molybdate is prepared as 2.5% in 5.0 N H_2SO_4.

Procedure. Place 0.3 ml of cAMP, 0.15 ml of $MgSO_4$, and 0.15 ml of Tris in small culture tubes and cool in an ice bath. Add 0.3 ml of suitably diluted enzyme and incubate for 30 minutes at 30°. After the first 20 minutes of incubation, 0.1 ml of *C. atrox* venom is added. The reaction is terminated by the addition of 0.1 ml of 55% trichloroacetic acid. The precipitates are removed by centrifugation, and the inorganic phosphate in 0.5 ml of the supernatants is measured by the Fiske–SubbaRow method with volumes modified to 0.1 ml of ammonium molybdate, 0.35 ml of distilled H_2O, and 0.05 ml of reducer.

[1] R. W. Butcher and E. W. Sutherland, *J. Biol. Chem.* **237**, 1244 (1962).
[2] C. H. Fiske and Y. SubbaRow, *J. Biol. Chem.* **66**, 375 (1925).
[3] M. V. Buell, O. H. Lowry, N. R. Roberts, M. L. W. Chang, and J. I. Kapphahn, *J. Biol. Chem.* **232**, 979 (1958).

It is important to include reagent blanks, and, in the case of crude enzyme preparations, to determine the amount of inorganic phosphate therein.

Definition of Unit and Specific Activity. A unit of phosphodiesterase activity is defined as that amount causing the transformation of 1 μmole of cAMP to 5'-AMP per 30 minutes at 30°.[1] Specific activity refers to units per milligram of protein as determined by the method of Lowry.

Purification Procedure

Step 1. Homogenization and Centrifugation. Hearts are obtained at the abattoir and packed in ice for transportation to the cold-room. The hearts are rinsed with cold 0.33 M sucrose, and are trimmed of fat and large vessels with scissors. The myocardial tissue is then macerated in a meat grinder and homogenized for 2 minutes in 2 volumes of 0.33 M sucrose and 1 mM Tris (pH 7.4) in a Waring commercial blender at high speed. The homogenate is centrifuged (usually at 16,000 g for 15 minutes at 4°), and the resulting supernatant fluid (which provides the starting material for the purification) is decanted through glass wool prewashed in sucrose.

Step 2. Ammonium Sulfate Precipitations, Dialysis, and Centrifugation. The supernatant fluid is adjusted to 0.5 saturation with $(NH_4)_2SO_4$, neutrality being maintained by the addition of 1 N KOH as required. After standing for at least 30 minutes with stirring (the preparation may be allowed to stir as long as overnight at 4–8° without apparent loss of activity), the precipitate is collected by centrifugation for 20 minutes at 16,000 g at 4°. The precipitate is taken up in 15% of the original extract volume in a solution containing 1 mM MgSO$_4$ and 1 mM imidazole (pH 7.5). Neutralized saturated $(NH_4)_2SO_4$ solution is added to a final concentration of 0.45 saturation, and the resulting mixture is stirred from 30 minutes to 12 hours, as convenient. The precipitate is collected by centrifugation as above and taken up in 5% of the extract volume in 1 mM MgSO$_4$ and 1 mM imidazole. The resuspended precipitate is dialyzed at 4° against 10–30 volumes of 1 mM MgSO$_4$ and 1 mM imidazole (pH 7.4) with constant agitation for 12 to 24 hours, the dialyzate being changed two or three times. After dialysis, the preparation is frozen at −20°C and thawed, and a heavy flocculent precipitate, which appears during dialysis and freezing, is collected by centrifugation for 20 minutes at 20,000 g and discarded.

Step 3. Fractionation on DEAE Cellulose. This is the most effective but the least reproducible of the steps in this preparation. The DEAE-cellulose originally used was purchased from the Brown Co., type 20, reagent grade, with a capacity of 0.57 mEq/g and is now offered by Carl Schleicher and Schuell of Keene, New Hampshire. It is readied for use by suspending 15 g of the resin in 1 liter of distilled water and pouring off the finer particles several times. The slurry is taken to a volume of 1 liter and adjusted to approximately 10 mM imidazole (pH 7.0) and 80 mM KCl. The DEAE-cellulose is put into a chromatographic tube with a sintered-glass disk [0.6–0.8 g of resin (dry weight) per gram of protein to be applied] at 4°. The column is washed with several volumes of 20 mM imidazole (pH 7.0)–80 mM KCl. The enzyme solution is also adjusted to pH 7.0 in 10 mM imidazole and 80 mM KCl. The column (1.5 cm in diameter) is packed under air pressure so that the flow rate is between 1 and 2 ml per minute. The phosphodiesterase preparation is added and allowed to run through the column. As the resin becomes heavily loaded with protein, the flow rate may tend to fall, and sometimes pressure (either air or nitrogen pressure) must be applied to keep the flow rates at reasonable levels. After the adsorption step is complete, 2 bed volumes of 10 mM imidazole (pH 7.0) and 80 mM KCl are added to the column as a washing step. The phosphodiesterase is then eluted with 20 mM imidazole (pH 6.0) containing 0.4 M KCl. The elution may be monitored by observing the appearance of a dark yellow band which forms as the eluting mixture moves down the column. The phosphodiesterase activity invariably accompanies this intense yellow color in the eluate, and collection is continued until the color disappears. The fractions containing the highest specific activity are dialyzed as described above.

Certain problems have been encountered in reproducing the DEAE-cellulose chromatographic step. It is an unusual sort of protein chromatograph, in that the resin is heavily overloaded with protein, but this is required in order to recover any activity. Early attempts at fractionation on DEAE were unsuccessful, for no activity could be recovered as long as more reasonable amounts of protein were applied to the resin. More recently, the capacities of the DEAE-celluloses (for phosphodiesterase activity) have been far less than expected. Nonetheless, the elution patterns have been essentially unchanged, and despite the difficulties, it seems that the step can be established following preliminary experiments on a smaller scale.

Step 4. Heat Step and Differential Isoelectric Precipitation. A volume of 20 mM glycine buffer (pH 10.0) and 10 mM imidazole (pH 7.0) equal to the volume of the dialyzed DEAE eluate is added to that fraction and the pH adjusted to 9.6. The mixture is brought to 45° rapidly with shaking in 60° bath, and the temperature is maintained at 45° for 16 minutes; the sample is chilled rapidly. The preparation is stirred in an ice water bath and 0.1 N acetic acid is added dropwise. The end point of the acid step is the appearance of a heavy white haze at approximately pH 5.8 (the pH is monitored continuously with a glass electrode). Caution should be taken to avoid adding too much acetic acid during the acid precipitation, as pH's below 5.8 have resulted in loss of supernatant phosphodiesterase activity, which was not recovered from the precipitate, even when the pH was restored to neutrality. The precipitate is collected by centrifugation for 30 minutes at 29,000 g at 4°, and the clear supernatant solution containing the phosphodiesterase is dialyzed as above.

Step 5. Adsorption on and Elution from Calcium Phosphate Gel. Calcium phosphate is added dropwise to the dialyzed supernatant fraction until a gel–protein ratio of 1.8 is reached. The mixture is stirred for at least 30 minutes, and the gel is collected by centrifugation at 16,300 g for 15 minutes. The supernatant is discarded, and the gel is washed in a volume equal to the original dialyzed supernatant volume with 100 mM imidazole, pH 6.0. It is stirred for 15 minutes, and then centrifuged again at 16,300 g for 15 minutes. The washing is repeated in the same volume of 10 mM imidazole pH 7.5, and the phosphodiesterase is eluted from the gel with a volume of 10% saturated ammonium sulfate in 100 mM glycine, pH 9.8, equal to the original supernatant volume. The preparation is stirred in the presence of this mixture for 30 minutes and then the gel is removed by centrifugation at 7000 g for 15 minutes at 4°. The supernatant solution is dialyzed as described above, and, stored at $-20°$, represents the final purified preparation.

A summary of the procedure is presented in the table.

Summary of Purification and Yield in Preparation of Cyclic Nucleotide Phosphodiesterase from Beef Heart[a,b]

Fraction	Total units	Specific activity	Original activity (%)
Extract	3150	0.2	100
First ammonium sulfate precipitate	2720	0.7	85.6
Second ammonium sulfate precipitate	2530	1.2	79.7
20,000 g supernatant fraction	1895	1.3	59.7
Dialyzed DEAE-cellulose eluate	769	7.2	24.2
Dialyzed acid supernatant fraction	461	11.9	14.5
Dialyzed calcium phosphate gel eluate	231	25.6	7.3

[a] From R. W. Butcher and E. W. Sutherland, *J. Biol. Chem.* **237**, 1244 (1962).
[b] The extract was taken as the starting material, but it should be noted that a very high (67.1) percentage of the phosphodiesterase activity in the whole homogenate was found in the particulate fraction.

General Comments and Properties

This preparation was developed as an enzymatic reagent which is capable of destroying cAMP in biological samples, thus providing an additional parameter of specificity to the measurement of the nucleotide.[1] Since that time, the preparation has been adapted as a component of a number of assay systems involving the conversion of cAMP or cGMP to their corresponding 5' forms with subsequent formation of the di- and trinucleotides. For this use, it is imperative that the preparation be free of 5'-nucleotidase activity. Such is the case after the calcium phosphate gel step; however, considerable 5'-nucleotidase activity is present in dialyzed DEAE-cellulose eluates.[4]

The soluble bovine heart phosphodiesterase has been purified more extensively by Hrapchak and Rasmussen[5] and Goren and Rosen.[6] While both preparations are of a higher specific activity than that obtainable by the method described here, and hence are more desirable for detailed enzymology, neither would seem to present any particular advantage as a reagent enzyme for cyclic nucleotide hydrolysis.

Teo and Wang[7] have shown that the soluble, high K_m beef heart phosphodiesterase can be greatly activated by an endogenous heat-stable protein factor and that the activation is Ca^{2+} dependent. It seems possible that the purification of the enzyme may be complicated by separation from the activator, and that the yields might be much improved by attention to this factor.

Substrate Specificity. The enzyme preparation has a K_m for cAMP reported to be from around 45 μM[8] to as high as 500 μM.[9] A figure of 200 μM was determined by the methods described herein. cGMP is also rapidly hydrolyzed, and the K_m for this nucleotide is lower[8] (1–3 μM) than for cAMP. Beavo et al.[8] found competitive inhibition between cAMP and cGMP, and suggested that only a single phosphodiesterase was present. The partially purified enzyme has little activity against cyclic 3',5' pyrimidine mononucleotides[10] and does not hydrolyze adenosine 2',3'-monophosphate.[1]

---

[4] J. G. Hardman, personal communication (1973).
[5] R. J. Hrapchak and H. Rasmussen, *Biochemistry* 11, 4458 (1972).
[6] E. N. Goren and O. M. Rosen, *Arch. Biochem. Biophys.* 153, 384 (1972).
[7] T. S. Teo and J. H. Wang, *J. Biol. Chem.* 248, 5950 (1973).
[8] J. A. Beavo, J. G. Hardman, and E. W. Sutherland, *J. Biol. Chem.* 245, 5649 (1970).
[9] M. M. Appleman, W. J. Thompson, and T. R. Russell, in "Advances in Cyclic Nucleotide Research" (P. Greengard and G. A. Robison, eds.), Vol. 3, p. 65. Raven, New York, 1973.
[10] J. G. Hardman and E. W. Sutherland, *J. Biol. Chem.* 240, 3704 (1965).

Activators and Inhibitors. In addition to the protein activator described by Teo and Wang,[7] imidazole, at pH's between 7.0 and 8.0 and at high substrate concentrations, is an effective stimulator of phosphodiesterase activity.[1] The methylxanthines are competitive inhibitors of the soluble beef heart enzyme.[1] More recently a whole spectrum of related and unrelated pharmacological agents have been reported to be inhibitors, many of which are more potent than the methylxanthines.[9]

Mg^{2+} is absolutely essential for the activity of the enzyme,[1] and Ca^{2+} is apparently required for the activation by the heat-stable protein factor.[7]

Stability. The partially purified preparation is stable over many months at $-70°$ in plastic vessels.

[33] Purification and Characterization of Cyclic 3',5'-Nucleotide Phosphodiesterase from Bovine Brain[1]

By Wai Yiu Cheung and Ying Ming Lin

Cyclic 3',5'-nucleotide phosphodiesterase was discovered by Sutherland and Rall[2] and partially purified from bovine heart by Butcher and Sutherland.[3] The enzyme partially purified from bovine brain catalyzes the hydrolysis of cyclic AMP (cAMP) and other cyclic 3',5'-nucleotides with a purine base to their corresponding 5'-nucleoside monophosphates.

The purification and characterization of phosphodiesterase from bovine brain are described herein. It also describes purification conditions under which phosphodiesterase is prepared in a partially inactive form, due to the dissociation of a protein activator from the enzyme.[4] Full activity of the partially inactive enzyme is restored by the addition of the activator. The purification and characterization of the activator is described separately in this volume [39].

Reagents. Prepare all reagents in glass-distilled water, and store them at 0 or $-20°$.

cAMP, 20 mM, adjusted to pH 8.0 with 1 N NaOH
Tris·HCl, 0.4 M, pH 8.0, at 22° (buffer A)

---

[1] The work was supported by ALSAC and by Grants NS08059, CA13537, CA08480, and Career Development Award NS42576 (W. Y. C.) from the U.S. Public Health Service.
[2] E. W. Sutherland and T. W. Rall, *J. Biol. Chem.* **232**, 1077 (1958).
[3] R. W. Butcher and E. W. Sutherland, *J. Biol. Chem.* **237**, 1244 (1962).
[4] W. Y. Cheung, *Biochim. Biophys. Acta* **191**, 303 (1969).

Tris·HCl, 20 mM, pH 7.5, at 4° (buffer B)
(NH₄)₂SO₄, 0.5 M in buffer B
NaCl, 0.1 M in buffer B
MnCl₂, 1 mM
Snake venom (*Crotalus atrox* for 5′-nucleotidase activity), 1 mg/ml in buffer B
TCA (trichloroacetic acid), 55%
Ammonium molybdate, 2.5% in 5 N H₂SO₄
Fiske–SubbaRow reagent: Mix 2 g of 1-amino-2-naphthol-4-sulfonic acid, 12 g of sodium bisulfite, and 12 g of sodium sulfite. Dissolve 2.5 g of this mixture in 100 ml of water
Acetic acid, 6 N
NH₄OH, 1 N
NaOH, 1 N
Calcium phosphate gel, 38 mg dry weight/ml in water prepared according to the procedure of Keilin and Hartree.[5]
Diethylaminoethyl cellulose (Sigma, coarse mesh with a capacity of 0.9 meq/g). Wash successively with 0.5 N NaOH, H₂O, 0.5 N HCl, and buffer B, until the final pH is 7.5
Sepharose 4B (Pharmacia)
Acrylamide, 30 g/100 ml (solution A)
N,N′-Methylenebisacrylamide, 0.8 g/100 ml (solution B)
Gel buffer (Tris, 2.06 g/100 ml; glycine 1.39 g/100 ml)
N,N,N′,N′-Tetramethylenediamine, 0.15 ml/100 ml (solution C)
Ammonium persulfate, 0.8 g/100 ml (solution D)
Upper buffer (Tris, 5.16 g/liter; glycine, 3.48 g/liter) pH 8.9 at 22°
Lower buffer (Tris, 14.5 g/liter, 1 N HCl 60 ml/liter) pH 8.1 at 22°
Elution buffer, same as lower buffer
Membrane holder buffer, 5 times concentration of the lower buffer
Bovine brain phosphodiesterase activator, 0.14 mg/ml (purified through the stage of DEAE-cellulose chromatography with a specific activity of 200 units (this volume [39]).

Assay Methods

Principle. Phosphodiesterase may be assayed by one of several methods: (1) by measuring inorganic phosphate released by the coupled phosphodiesterase and 5′-nucleotidease reactions (see this volume [32]), (2) by using radioactive substrates (see this volume [30]); (3) by titrating the protons released from hydrolysis of cAMP (see this volume

[5] D. Keilin and E. F. Hartree, *Proc. Roy. Soc. Ser. B* **124**, 397 (1938).

[31]). Other techniques may be used, depending upon the sensitivity required of the assay.

Procedure. The procedure of Butcher and Sutherland[3] employing 5'-nucleotidease of snake venom as an auxiliary enzyme is used with a slight modification for routine assays. The modified procedure involves a two-stage incubation.[6] The incubation is performed in two stages because snake venom contains active proteolytic enzymes which stimulate the phosphodiesterase purified through a DEAE-cellulose column chromatography.[7] To perform the first stage of incubation, prepare a reaction mixture containing 0.05 ml of buffer A, 0.05 ml of $MnCl_2$, and an appropriate concentration of enzyme; add water to make this volume 0.45 ml. Initiate the reaction with 0.05 ml of 20 mM cAMP. After a 10-minute incubation at 30°, terminate the first stage by transferring the tube containing the reaction mixture to a boiling water bath for 2 minutes. After thermal equilibration to 30°, add 0.05 ml of snake venom and incubate for another 10 minutes. Stop this reaction by adding 0.05 ml of TCA solution; then add 0.70 ml of water and 0.15 ml ammonium molybdate. Remove denatured proteins by centrifugation and decant the clear supernatant fluid into clean tubes containing 0.05 ml of the Fiske-SubbaRow reagent. After 10 minutes at room temperature, measure the blue color at 660 nm. A word of technical interest should be added. The protein sediment removed by centrifugation is yellowish, and the subsequent color loss is proportional to the amount of protein removed. In experiments involving addition of more proteins to some tubes than others, compensation for color loss due to unequal protein concentrations is necessary. After the second stage reaction has been terminated, add sample protein as needed until all tubes have the same final protein concentration. Perform all assays in duplicate under conditions in which the reaction is linear with the time of incubation and the protein concentration. Correct all data for a control that contains all components except phosphodiesterase.

Unit of Activity. One unit of activity is that amount of protein which catalyzes the hydrolysis of 1 μmole of cAMP per minute at 30° under the conditions described in the text. Specific activity is expressed as units per milligram of protein.

Purification Procedure

Transport fresh bovine brain from a local slaughterhouse to the laboratory in packed ice. Clean the cortices and rinse them with water. Use

[6] W. Y. Cheung, *J. Biol. Chem.* **246**, 2859 (1971).
[7] W. Y. Cheung, *Biochem. Biophys. Res. Commun.* **29**, 478 (1967).

the cortices fresh or store them at $-20°$ until use. No appreciable difference has been noted in using either fresh or frozen tissue. When frozen tissues are used, thaw them first at room temperature. Perform subsequent operations at 0–4°. Figure 1 presents the scheme of purification of the brain enzyme.

Step 1. Homogenization. Process 2 kg of brain tissue in 500-g batches. Homogenize each 500 g of tissue in 1.5 liters of glass-distilled water in

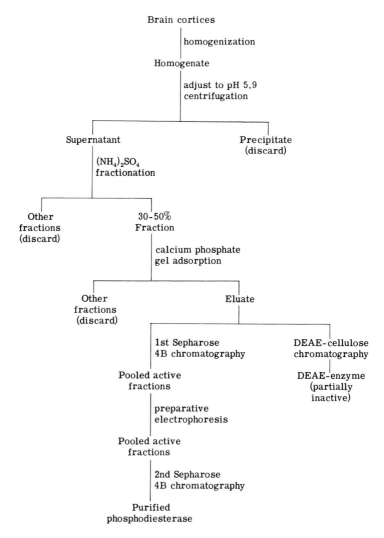

FIG. 1. Scheme of purification of cyclic 3′,5′-nucleotide phosphodiesterase from bovine brain.

a 1-gallon Waring commercial blender at low speed for 30 seconds and then at medium for 1 minute. Filter the homogenate (pH 6.4) through a double layer of cheesecloth to remove tissue debris. Adjust the filtrate to pH 5.9 with 6 N acetic acid, and centrifuge it at 12,000 g for 30 minutes. Filter the supernatant fluid through glass wool to remove lipid flocculants.

Step 2. Ammonium Sulfate Fractionation. Add, with stirring, solid ammonium sulfate to the supernatant fluid to 30% saturation (176 g/liter) and maintain the solution at pH 7 by adding 1 N NH$_4$OH. After stirring for 30 minutes, remove the precipitate by centrifugation (12,000 g, 20 minutes). Add more ammonium sulfate to 50% saturation (127 g/liter), and maintain the pH at neutrality as before. Stir the solution for 30 minutes and collect the precipitate by centrifugation (12,000 g, 20 minutes). Dissolve the 30–50% fraction in a minimal volume of buffer B, and dialyze the sample against the same buffer with several changes. Remove turbidity by centrifugation (39,000 g, 10 minutes). Adjust the protein concentration to 10 mg/ml with buffer B.

Step 3. Calcium Phosphate Gel Adsorption. Add calcium phosphate gel slowly to the protein solution, with stirring, to reach a ratio of 0.8 g of gel to 1 g of protein. After stirring for 2 hours, collect the gel by centrifugation (16,000 g, 10 minutes) and discard the supernatant fluid. Transfer the gel to a tissue grinder fitted with a Teflon pestle (Thomas Tissue Grinder with a chamber volume of 55 ml). Wash the gel by homogenization to a fine suspension in a volume of buffer B equivalent to the volume of supernatant fluid discarded. Washing is done in 10–12 batches with about 500 mg of gel each time. Combine the suspensions, stir them for 30 minutes, and collect the gel by centrifugation (16,000 g, 10 minutes). Discard the supernatant fluid. Elute the gel with a volume of 0.5 M, (NH$_4$)$_2$SO$_4$, pH 7.5, which is equivalent to the supernatant fluid discarded. Elution is done batchwise by homogenization of the gel in a manner identical to the washing procedure with buffer B. Stir the suspension for 30 minutes, centrifuge, and save the eluant. Repeat the process of elution and combine the two eluents. Add solid (NH$_4$)$_2$SO$_4$ to the eluents to 100% saturation (about 700 g/liter) and stir overnight. Collect the precipitate by centrifugation (12,000 g, 20 minutes), dissolve it in buffer B, and dialyze it against the same buffer with two changes.

Step 4. First Sepharose 4B Chromatography. Suspend Sepharose 4B in buffer B and pack the gel into a 2.5 × 100 cm column (Pharmacia K25/100, equipped with flow adapters and column jackets) at a hydrostatic pressure of 5 cm. Raise the pressue to 22 cm when the packed gel reaches the desired height. Pass buffer B through the column until the bed height is stabilized (93 cm). Equilibrate the column overnight with buffer B by upward flow, and maintain upward flow for subsequent operations.

Apply the sample solution from step 3 (120–150 mg/protein) in 10–12 ml) to the column. Collect fractions of 4 ml at 20-minute intervals. Pool the active fractions and concentrate the solution by ultrafiltration with UM-10 membrane (Amicon) at a pressure of 50 psi. Remove turbidity by centrifugation (39,000 g, 20 minutes) if necessary. Figure 2 shows the protein and the activity profiles of the Sepharose 4B column. Pool active fractions from 2 to 3 columns for the next step.

Step 5. Preparative Acrylamide Gel Electrophoresis. EQUIPMENT. Buchler's "Poly-prep" electrophoresis apparatus was used.

PREPARATION OF THE SAMPLE. Dialyze the pooled fractions from step 4 (70–130 mg in 6–8 ml) against 2 liters of 1:10 dilution of upper buffer for 4–6 hours. Add sucrose (100 mg/ml) to increase the density of the solution.

PREPARATION OF THE GEL COLUMN. To make a 7.5% gel with respect to acrylamide, measure 25 ml each of solutions A, B, C, and D into 250-ml Erlenmeyer flask and remove dissolved air under negative pressure. Introduce the solution into the electrophoresis column, and cover the solution with a layer of water about 0.5 cm deep. Allow the gel to polymerize in 1 hour. The length of the polymerized gel is 5 cm. Maintain the temperature at 4° by circulating a refrigerated coolant through the column jacket.

ELECTROPHORESIS. Perform a preliminary electrophoresis at a current

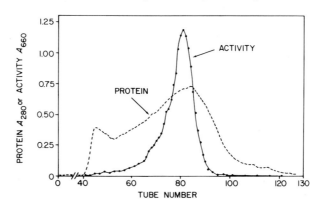

FIG. 2. First Sepharose 4B chromatography. The gel column, 2.5 × 93 cm, was previously equilibrated with buffer B. Twelve milliliters of a calcium phosphate gel eluate containing 122 mg of protein was applied to the column by upward flow. Fractions of 4 ml were collected over 20 minutes. Tubes 73–87 were pooled and concentrated. The pooled activity represented 58% of the activity applied to the column.

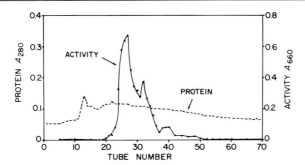

FIG. 3. Preparative polyacrylamide gel electrophoresis. A 7.5% acrylamide gel column, 5 cm long, was run overnight to remove residual persulfate embedded in the gel. Seven milliliters of a sample containing 70 mg of protein was applied to the column. Fractions of 12 ml were collected at a rate of 0.6 ml per minute maintained by a peristaltic pump. Tubes 24–30, which accounted for 43% of the initial activity, were pooled and concentrated.

of 30 mA for 10 hours to remove any residual persulfate embedded in the gel. Carefully layer the sample onto the top of the gel, which is immersed in the upper buffer. Start electrophoresis with a current of 23 mA at 160 V. After the protein has entered the gel in about 1 hour, increase the current to 30 mA and maintain it constant throughout the course of electrophoresis (about 24 hours). The voltage increases from the initial 160 V to about 300 V at the end of electrophoresis. Collect fractions of about 12 ml at a rate of 0.6 ml per minute maintained by a peristaltic pump. Figure 3 shows the protein and the activity profiles of the preparative electrophoresis. Pool the active fractions and concentrate them by ultrafiltration with a UM-10 membrane as in step 4. Dialyze the concentrated solution against buffer B. Clarify the solution by centrifugation.

Step 6. Second Sepharose 4B Chromatography Apply the dialyzed concentrated solution (5–9 mg protein in 6–10 ml) from step 5 into a column (2.5 × 93 cm) which has been equilibrated with buffer B by upward flow as described in step 4. Collect fractions of 4 ml at 20-minute intervals. Figure 4 shows the protein and activity profiles of the filtration column. Pool the active fractions and concentrate them by ultrafiltration through a UM-10 membrane. The enzyme at this stage is not dependent on an exogenous protein activator for maximal activity.

The overall purification is about 180-fold from the crude homogenate with a yield of 1% of the total activity (table). The purified enzyme catalyzes the hydrolysis of cAMP at a rate of 9 μmoles per milligram of protein per minute under standard conditions, and does not require an exogenous activator for maximal activity.

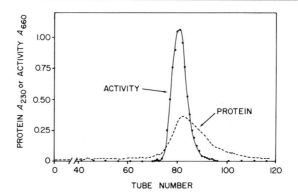

Fig. 4. Second Sepharose 4B chromatography. Seven milliliters of a sample from step 5 were applied by upward flow to a gel column, 2.5 × 93 cm, which had been equilibrated with buffer B. Fractions of 4 ml were collected over 20 minutes. Tubes 76–84, representing 56% of the initial activity, were pooled and concentrated.

Preparation of a Partially Inactive Phosphodiesterase. Apply 240 mg of the protein sample from step 3 to a column packed with DEAE-cellulose (2.5 cm × 42 cm), which has been equilibrated with buffer B. Elute the column with an exponential gradient from a lower reservoir contain-

PURIFICATION OF CYCLIC 3′,5′-NUCLEOTIDE PHOSPHODIESTERASE FROM BOVINE BRAIN[a]

Step	Fraction	Protein[b] (mg)	Activity (units)	Specific activity (units/mg)	Purification (x-fold)	% Recovery
1	Homogenate	86,800	4383	0.05	1	100
2	pH 5.9 Supernatant	23,800	2570	0.11	2	59
3	30–50% $(NH_4)_2SO_4$	3,690	963	0.26	5	22
4	Calcium phosphate gel eluate	800	434	0.54	11	10
5	1st Sepharose 4B chromatography	228	193	0.85	17	4.4
6	Preparative electrophoresis	19.8	83.3	4.23	84	1.9
7	2nd Sepharose 4B chromatography	5.4	48.4	8.97	178	1.1

[a] The data presented in the table are based on 1 kg of brain cortices.
[b] Protein in the homogenate was determined by the biuret reagent and in subsequent steps by the spectrophotometric technique O. Warburg and W. Christian, Biochem. Z., *310*, 384 (1941).

ing 130 ml of buffer B and an upper reservoir containing 1000 ml of 0.5 M $(NH_4)_2SO_4$ in buffer B. Alternatively, elute the column with a linear gradient produced by 500 ml of buffer B and 500 ml of 0.5 M $(NH_4)_2SO_4$ in buffer B. Collect fractions of 10 ml per tube. Phosphodiesterase activity emerges as a single peak. Figures 5 and 6 depict the protein and enzyme profiles of the elution pattern by the two different gradients, respectively. Concentrate active fractions by ultrafiltration through a UM-10 membrane at a pressure of 50 psi. Dialyze the concentrated solution against buffer B, with two changes of the buffer. The preparation is partially inactive due to dissociation of a protein activator from the enzyme on the column. The activator may be eluted from this column free of

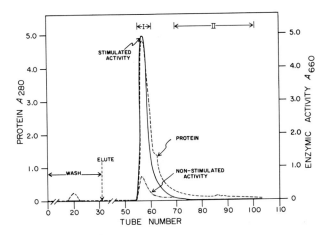

Fig. 5. Resolution of phosphosdiesterase into a partially inactive enzyme and its activator on a DEAE-cellulose column by an exponential gradient. The column, 2.5 × 42 cm, was equilibrated with buffer B and then charged with 240 mg of protein of a calcium phosphate gel eluate from step 3. Each tube collected 10 ml. Approximately 300 ml of buffer B was passed through the column as a wash. A small protein peak came out with the wash, which had no phosphodiesterase activity. After the A_{280} in the washing returned to a base line level, the column was eluted with an exponential gradient generated from a lower reservoir containing 130 ml of buffer B and an upper reservoir containing 1000 ml of 0.5 M $(NH_4)_2SO_4$ in buffer B. A single protein peak emerged after about 240 ml of the eluate had been collected. Phosphodiesterase activity coincided with this peak. Phosphodiesterase was assayed by the two-stage incubation, once with snake venom in the first stage of incubation to give the stimulated activity, and again without venom to give the nonstimulated activity. Tubes were pooled as follows: fraction I, tubes 56–63; fraction II, tubes 70–100. Fraction I was concentrated in 60% sucrose solution and then dialyzed extensively against buffer B. Phosphodiesterase in this fraction was mostly inactive. Fraction II was lyophilized and then dialyzed against buffer B. Fraction II contained the activator and was free of phosphodiesterase activity. From W. Y. Cheung, *J. Biol. Chem.* **246**, 2859 (1971).

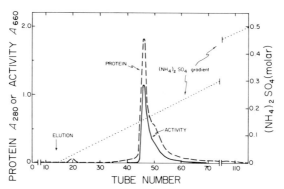

FIG. 6. Resolution of phosphodiesterase into a partially inactive enzyme and its activator on DEAE-cellulose column by a linear gradient of $(NH_4)_2SO_4$. The column, 2.5×42 cm, was equilibrated with buffer B, and charged with 100 mg of protein of a calcium phosphate gel eluate from step 3. The linear gradient was generated from 500 ml of buffer B and 500 ml of 0.5 M $(NH_4)_2SO_4$ in buffer B. Elution was started at tube 14. Phosphodiesterase activity emerged at 0.1 M $(NH_4)_2SO_4$. Fractions of 10 ml were collected at 10-minute intervals. Tubes 45–48 were pooled. The enzyme in the pooled tubes were mostly inactive, owing to the dissociation of a protein activator, which emerged after the enzyme peak, tube 55 and thereafter.

phosphodiesterase activity after the enzyme peak. The specific activity of this preparation is about 0.1 unit/mg and 0.5 unit/mg in the absence and presence of an exogenous activator, respectively.

The use of a coarse DEAE-cellulose with a capacity of 0.9 meg/g from Sigma was necessary to achieve dissociation of the activator from phosphodiesterase on the column. For unknown reasons, Whatman DE-52 DEAE-cellulose, for example, did not dissociate the activator from the enzyme under comparable conditions.

Properties

pH Optimum. Phosphodiesterase from most tissues exhibits maximal activity between pH 7.5 and 8. The optimal pH of rat and bovine brain phosphodiesterase is 8.[8,9] The enzymes from dog heart[10] and fish brain[11] appear to function best between pH 8.5 and pH 9, whereas those from plants are more active around pH 4.[12]

[8] W. Y. Cheung, *Biochemistry* **6**, 1079 (1967).
[9] W. Y. Cheung, *Biochim. Biophys. Acta* **243**, 395 (1971).
[10] K. G. Nair, *Biochemistry* **5**, 150 (1966).
[11] M. Yamamoto and K. L. Massay, *Comp. Biochem. Physiol.* **30**, 941 (1969).
[12] P. P.-C. Liu and J. E. Varner, *Biochim. Biophys. Acta* **276**, 454 (1972).

Effect of Divalent Cations. The earlier postulation that phosphodiesterase is a metalloenzyme[8] appears compatible with recent findings. Phosphodiesterase of bovine brain cortex purified through the stage of calcium phosphate gel contained approximately 1 mole each of calcium and zinc per 250,000 g of protein (see also this volume [39]). Extensive dialysis of the enzyme against an EDTA solution reduced the metals to about half their levels before dialysis.[9]

The enzyme of brain and other mammalian tissues requires a divalent cation for optimal activity. Mg^{2+} was slightly stimulatory, whereas, Cu^{2+} was inhibitory. Ca^{2+}, Mn^{2+}, Co^{2+}, and Zn^{2+} were stimulatory at low (0.1 mM) but inhibitory at high concentrations (> 1 mM). The terms "stimulatory" and "inhibitory" are used arbitrarily, since a cation may be either stimulatory or inhibitory, depending on its concentration. A pair of stimulatory cations increased activity; however, this increase was invariably less than the sum of individual cation increases. A stimulatory cation and an inhibitory cation together had an effect comparable to that of the stimulatory cation alone.

Kakiuchi *et al.*[13] noted that 1 μM to 10 μM Ca^{2+} stimulated rat brain phosphodiesterase activity in the presence of 3 mM Mg^{2+}, and they postulated the existence of a Ca^{2+} plus Mg^{2+}-dependent enzyme (see also this volume [39]).

Kinetic analysis indicated that a stimulatory cation (0.1 mM Mn^{2+}) increased the $K_{m_{app}}$ and V_{max} of phosphodiesterase, whereas an inhibitory cation (1 mM Ca^{2+}) decreased them. The simplest mechanism appears to be that the metal ions principally affect the rate of dissociation of the enzyme–substrate complex to its product and free enzyme.[9]

The optimal pH of phosphodiesterase varied with an exogenous divalent cation. Maximum activity was observed at pH 8.3 in the absence of added cation, at pH 8 in 0.1 mM Mn^{2+}, and at pH 7.6 in 1 mM Ca^{2+}. Likewise, the effectiveness of a divalent cation varied with pH. Whereas the stimulation of 0.1 mM Mn^{2+} was pronounced at a neutral pH, the inhibition by 1 mM Ca^{2+} was severe at an alkaline pH.[9]

EDTA and ATP were potent inhibitors of phosphodiesterase.[8] In the absence of an exogenous cation, 7 μM EDTA or 30 μM ATP caused 50% inhibition. Mn^{2+} completely reversed EDTA inhibition of phosphodiesterase, but only partially reversed ATP inhibition. However, Kakiuchi *et al.*[13] noted that Mg^{2+}, together with trace amounts of Ca^{2+}, could reverse the inhibition by ATP.

---

[13] S. Kakiuchi, R. Yamazaki, and Y. Teshima, *in* "Advances in Cyclic Nucleotide Research" (P. Greengard, G. A. Robison, and R. Paoletti, eds.), Vol. 1, p. 455, Raven, New York, 1972.

Phosphodiesterase was fully active in the crude state, and was relatively inactive when purified through DEAE-cellulose chromatography unless supplemented with a specific protein activator.[6] Mn^{2+} (0.1 mM) increased the activity of the nonstimulated enzyme as well as the enzyme after full stimulation by the activator.[9]

The brain enzyme also catalyzed the hydrolysis of cGMP, the other naturally occurring cyclic 3',5'-nucleotide. The effectiveness of divalent cations on hydrolysis of cAMP and cGMP appeared to vary with the substrate. Mn^{2+} was more effective for the hydrolysis of cAMP, whereas Mg^{2+} was more effective for that of cGMP.[9]

Molecular Weight. Gel filtration studies indicated that the molecular weight of phosphodiesterase varied over a wide range. The soluble enzyme was partially excluded from the matrix of a Sephadex G-200 column, indicating a molecular weight of 200,000 or greater. On a Sepharose 4B column the enzyme was resolved into two active peaks. The first peak was eluted close to the exclusion limit; the second peak was broad, trailing after the first. It was the dominant component and usually had a specific activity slightly higher than the first. The first peak appeared to increase upon storage of the enzyme at $-20°$. Overnight incubation of the stored preparation in the presence of β-mercaptoethanol prior to gel filtration converted the enzyme eluted in the first peak to that in the second (Fig. 7).

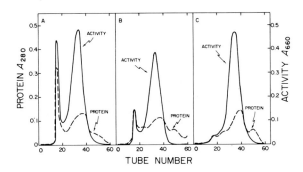

Fig. 7. The change of apparent molecular weights of bovine brain phosphodiesterase after treatment with β-mercaptoethanol. Two hundred microliters of a calcium phosphate gel eluate from step 3 containing 3.7 mg protein and 10 mg of sucrose was applied to a Sepharose 4B column, 0.9 × 29 cm, which had been previously equilibrated with buffer B. The enzyme was stored at $-20°$ for about 24 months. The column was eluted with buffer B at $22°$. Fractions of 1 ml were collected. In A, no β-mercaptoethanol was added; in B, the sample and buffer contained 10 mM β-mercaptoethanol; in C, the sample was incubated with 10 mM β-mercaptoethanol overnight at $0°$ and then applied to the column at $22°$. Buffer B also contained 10 mM β-mercaptoethanol.

Phosphodiesterase was inhibited by p-hydroxymercuribenzoate, and the inhibition was reversible by β-mercaptoethanol. This indicates that the enzyme possesses sulfhydryl groups and that at least some of them are necessary for enzymatic activity.[14]

The two molecular species of phosphodiesterase might be manifestations of different oxidation states of the sulfhydryl groups on the enzyme moiety. During storage, some of the sulfhydryl groups might have been oxidized to form intermolecular disulfide linkages, giving rise to an aggregate with a larger molecular weight. This molecular form could represent the first peak on the Sepharose 4B column. Exposure of the stored enzyme to β-mercaptoethanol might reduce these disulfide bonds to sulfhydryl groups. The reduced enzyme could represent the smaller molecular species.[14]

The molecular weight of phosphodiesterase may vary in another manner. In a Sephadex G-200 gel filtration column, phosphodiesterase purified through a DEAE-cellulose column emerged broad and heterogeneous. After trypsin activation, the activity pattern was accompanied by a marked decrease in the molecular weight (Fig. 8).

During the purification of phosphodiesterase, proteolytic enzymes inherent in the tissue extracts might act on the enzyme and change its molecular weight. It is not established whether the different molecular weights reported for phosphodiesterase represent native forms.

Increasing evidence indicates that phosphodiesterase may exist in different molecular forms.[14-20] Until the molecular components of the enzyme are better characterized, the molecular weights cited above should be taken as tentative figures.

Substrate Specificity. Phosphodiesterase exhibits a preference for cyclic nucleotides with a purine rather than a pyrimidine base. cAMP was the best substrate at millimolar concentrations, followed by cIMP and cGMP. However, cGMP was a better substrate than cAMP at micromolar concentrations. N^6,O^2-Dibutyryl cAMP was not hydrolyzed, nor did it affect the hydrolysis of cAMP. The hydrolysis of cUMP was barely discernible, and no hydrolysis of cCMP and cTMP was detected.[14]

[14] W. Y. Cheung, *in* "Advances in Biochemical Psychopharmacology" (P. Greengard and E. Costa, eds.), Vol. 3, p. 51. Raven, New York, 1970.
[15] W. J. Thompson and M. M. Appleman, *Biochemistry* **10**, 311 (1971).
[16] S. Jard and M. Bernard, *Biochem. Biophys. Res. Commun.* **41**, 781 (1970).
[17] E. Monn and R. O. Christiansen, *Science* **173**, 540 (1971).
[18] E. N. Goren, A. H. Hirsch, and O. M. Rosen, *Anal. Biochem.* **43**, 156 (1971).
[19] P. Uzunov and B. Weiss, *Biochim. Biophys. Acta* **284**, 220 (1972).
[20] S. Kakiuchi, R. Yamazaki, and Y. Teshima, *Biochem. Biophys. Res. Commun.* **42**, 968 (1971).

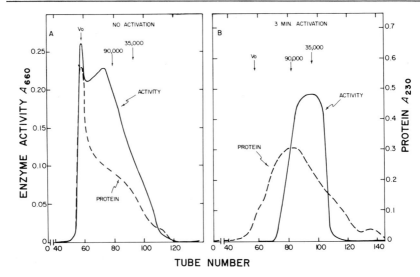

Fig. 8. The change of apparent molecular weights of bovine brain phosphodiesterase after activation by trypsin. A Sephadex G-200 column, 2.5 × 200 cm, with internal support of siliconized glass beads [D. H. Sachs and E. Painter, *Science* **175,** 78 (1972)] was equilibrated at 0° with 0.1 M NaCl in buffer B at a pressure head of 30 cm. Two milliliters of phosphodiesterase (14 mg of protein) purified through DEAE-cellulose chromatography was applied to the column, which was eluted with the NaCl-Tris buffer. Fractions of 2.8 ml were collected. In A, 280 µg of a soybean trypsin inhibitor was added to the enzyme prior to the addition of 140 µg of trypsin. In B, the enzyme was incubated with 140 µg of trypsin for 3 minutes before the addition of 280 µg of soybean trypsin inhibitor. The column was calibrated according to P. Andrews [*Biochem. J.* **91,** 222 (1964)] with thyroglobulin, lactate dehydrogenase, bovine serum albumin, horseradish peroxidase, and soybean trypsin inhibitor. The figures above the arrows in the figure refer to the elution position of proteins having those molecular weights. V_0 refers to the void volumn of the column.

Brooker et al.[21] noted that the enzyme from the rat brain had two K_m's, one of 0.1 mM and another of 5 µM. The high-affinity species was associated with a molecular weight of 200,000, and the low-affinity species of 400,000. The former did not catalyze the hydrolysis of cGMP and is believed to be derived from membranous fractions.

Enzyme Stability. Phosphodiesterase is stable; storage at −20° for several months caused no appreciable change in activity. A purified enzyme at a concentration of 1.5 mg protein per milliliter of buffer B was kept at 0–4° for 3 weeks. During this time enzymatic activity was checked periodically. No significant loss of activity was observed.[14]

[21] G. Brooker, L. J. Thomas, Jr., and M. M. Appleman, *Biochemistry* **7,** 4177 (1968).

Kinetic Properties. The K_m of bovine brain phosphodiesterase is about 0.1 mM for cAMP and about 80 μM for cGMP. The presence of cAMP in the reaction mixture inhibits the hydrolysis of cGMP, and vice versa. The K_i of one as an inhibitor of the other is comparable to its K_m as a substrate, in a manner predictable from kinetic considerations that the two nucleotides compete for a single substrate site (see also footnotes 21–23).

At low concentration of cAMP (e.g., 2 μM) 2–20 μM of cGMP caused a slight stimulation of cAMP hydrolysis catalyzed by bovine brain phosphodiesterase. Beavo et al.[24] reported marked stimulation of cAMP hydrolysis by cGMP catalyzed by a crude rat liver phosphodiesterase. The purity of the two preparations was not comparable. Tissue specificity of the two enzymes may account for the difference in the degree of stimulation.

Regional and Subcellular Distribution of Phosphodiesterase. Phosphodiesterase is widely distributed, and brain shows the highest activity.[3]

Weiss and Costa[25] found that the regional distribution of phosphodiesterase in the brain varied considerably. Enzymatic activity was high in the cerebral cortex, hippocampus, caudate nucleus, and septal nuclei, and low in the spinal cord, medulla, and pons. In general, those areas containing gray matter were higher in phosphodiesterase activity than those consisting of white matter. Williams et al.[26] obtained similar results.

Breckenridge and Johnson[27] confirmed and extended these findings. They noted that phosphodiesterase activity was not uniquely associated with the particular layers of the cerebral cortex or with specialized synaptic structures in the olfactory bulb. Correlation with cellular density, monoamine content, or other enzymes was also not apparent.

Subcellular fractionations of rat cerebra indicated that the enzyme was partly soluble and partly particulate. A considerable portion of the activity was concentrated inside nerve endings.[28,29] Treatment with a

[22] O. M. Rosen, *Arch. Biochem. Biophys.* **139**, 447 (1970).
[23] J. A. Beavo, J. G. Hardman, and E. W. Sutherland, *J. Biol. Chem.* **245**, 5649 (1970).
[24] J. A. Beavo, J. G. Hardman, and E. W. Sutherland, *J. Biol. Chem.* **246**, 3841 (1971).
[25] B. Weiss and E. Costa, *Biochem. Pharmacol.* **17**, 2107 (1968).
[26] R. H. Williams, S. A. Little, and J. W. Ensinck, *Amer. J. Med. Sci.* **258**, 190 (1969).
[27] B. M. Breckenridge and R. E. Johnston, *J. Histochem. Cytochem.* **8**, 505 (1969).
[28] W. Y. Cheung and L. Salganicoff, *Nature (London)* **214**, 5083 (1967).
[29] E. De Robertis, G. R. D. L. Arnaiz, A. Alberici, R. W. Butcher, and E. W. Sutherland, *J. Biol. Chem.* **242**, 3487 (1967).

nonionic detergent revealed substantial latent activity in a microsomal fraction.[28]

Most of the work on the properties of phosphodiesterase has been done with the soluble enzyme. Detailed investigation of the enzyme associated with particulate fractions has not been performed, and extrapolation of the literature data on the soluble to the particulate enzyme may be unwarranted.

The Extent of Reversibility of the Reaction Catalyzed by Phosphodiesterase. The reversibility of the reaction catalyzed by phosphodiesterase was studied from both the forward and the reverse directions using labeled cAMP and 5′-AMP, respectively. After a long incubation, cAMP was quantitatively converted to 5′-AMP, but no synthesis of radioactive cAMP was detected from isotopically labeled 5′-AMP, even in the presence of added nonlabeled cAMP. Similarly, cGMP was quantitatively hydrolyzed to 5′-GMP, and no 5′-GMP was synthesized from cGMP. These results indicated that the reverse reaction was too sluggish to be detected with isotopic techniques under our assay conditions and that the reaction could be considered essentially irreversible.[30]

The experiments were performed with a purified phosphodiesterase from the bacterium *Serratia marcesens*.[31] The bacterial enzyme was used because of its high specific activity and absence of interfering enzymes. Phosphodiesterase prepared from mammalian tissues, including the brain, is usually not homogeneous; contamination with other enzyme activities would make the results of such experiments equivocal.

Nevertheless, comparable experiments using a partially purified phosphodiesterase from bovine brain lead to similar conclusions, i.e., that the reaction catalyzed by phosphodiesterase is essentially irreversible.

The Enthalpy of Hydrolysis of cAMP. The enthalpy hydrolysis of cAMP was measured with a microcalorimeter and found to be -13.3 kcal/mole.[30] Greengard *et al.*[32] reported a value of -14.1 kcal/mole. On the basis of established thermodynamic relationships, they calculated the standard free energy change ($\Delta F°$) for the phosphodiesterase reaction to be -11.9 kcal/mole. From the values of $\Delta F°$ (-11.9 kcal/mole) and ΔH (13.3 kcal/mole) for the hydrolysis of the 3′-bond of cAMP and the thermodynamic relationship

$$\Delta F = \Delta H - T\Delta S$$

[30] W. Y. Cheung, M. H. Chiang, and S. N. Pennington, *Fed. Proc., Fed. Amer. Soc. Exp. Biol.* **30**, 1098 (1971).
[31] T. Okabayashi and M. Ide, *Biochim. Biophys. Acta* **220**, 116 (1970).
[32] P. Greengard, S. A. Rudolph, and J. M. Sturtevant, *J. Biol. Chem.* **244**, 4798 (1969).

we computed a change in entropy terms of −1.4 kcal, corresponding to an entrophy decrease of about 5 units. This indicates that the large decrease in free energy associated with the hydrolysis of cAMP is mainly due to a change in enthalpy rather than a change in entropy (see also Greengard and Kuo[33]).

From the standard free energy change (ΔF°), we can estimate the free energy change under physiological conditions by the following equation[34]:

ΔF Physiological conditions

$$= \Delta F^\circ + RT \ln \frac{(5'\text{-AMP})}{(\text{cAMP})}$$
$$\cong -11{,}900 + 1.9 \times 300 \times 2.3 \log[(10^{-4})/(10^{-7})]$$
$$\cong -11{,}900 + 3900$$
$$\cong -8{,}000 \text{ cal/mole}$$

The large negative free energy change suggests that under physiological conditions the reaction greatly favors the hydrolysis of cAMP to 5'-AMP.

We interpreted the high energy content as providing a thermodynamic barrier against a reversal of the phosphodiesterase reaction, thereby serving as an irreversible off-switch for the effective termination of the action of the cyclic nucleotide. With respect to regulation, this thermodynamic property appears to be functionally significant.[34]

Nonsusceptibility of Bound Cyclic AMP. cAMP bound to a cAMP-binding protein was not susceptible to phosphodiesterase. It was degraded only when dissociated from the protein, and the rate of hydrolysis was governed by its rate of dissociation from the protein. Possible physiological significance of this has been discussed elsewhere.[35]

---

[33] P. Greengard and J. F. Kuo, *in* "Advances in Biochemical Psychopharmacology" (P. Greengard and E. Costa, eds.), Vol. 3, p. 289. Raven, New York, 1970.
[34] W. Y. Cheung, *Perspect. Biol. Med.* **15,** 221 (1972).
[35] W. Y. Cheung. *Biochem. Biophys. Res. Commun.* **46,** 99 (1972).

[34] Purification and Characterization of Cyclic Nucleotide Phosphodiesterase from Skeletal Muscle

By ROBERT G. KEMP and YUNG-CHEN HUANG

Assay[1]

Principle. The assay is based upon the production of [^{3}H]nucleoside from cyclic [^{3}H]nucleotide by the combined action of the diesterase and snake venom nucleotidase. The unreacted nucleotide can be removed from the nucleoside by the use of a anion-exchange resin.

Reagents

Tris·Cl, 0.1 M, pH 7.5
MgCl$_2$, 50 mM
Cyclic [^{3}H]3′,5′-nucleotide, 1 mM, 500,000 cpm/ml
Crotalus atrox venom, 2 mg/ml

Dilution. The phosphodiesterase is diluted to the desired concentration in a medium consisting of 50 mM Tris·Cl, 2 mM EDTA, 0.1 mM dithiothreitol (all at pH 7.5).

Procedure. A reaction mixture of 0.9 ml is prepared by mixing 0.5 ml of Tris·Cl, 0.1 ml of MgCl$_2$, 0.1 ml of cyclic [^{3}H]nucleotide, and 0.2 ml of H$_2$O. The reaction mixture is placed in a 30° water bath, and the reaction is started by the addition of 0.1 ml of diluted phosphodiesterase. The reaction is usually allowed to proceed for 20 minutes; the reaction tube is then placed in a boiling water bath for exactly 1 minute. The mixture is quickly cooled in an ice bath and then placed again in a 30° bath. The *Crotalus atrox* venom is added (0.1 ml) and the mixture is further incubated for 10 minutes. The reaction mixture is then put on a column of DEAE-Sephadex A-25 (0.5 × 7 cm) previously equilibrated with 50 mM Tris·Cl (pH 7.5). After the sample has entered the gel, the column is washed with 50 mM Tris·Cl (pH 7.5) until the total volume of the initial effluent and wash is 5 ml. A sample of this effluent is counted in a liquid scintillation spectrometer. The radioactivity in a blank (without diesterase) is subtracted from the sample radioactivity.

Unit. Activity can be calculated by determining the micromoles of cyclic nucleotide hydrolyzed, based on the specific activity of the nucleotide in the reaction mixture. One unit of enzyme activity represents the

---

[1] Y.-C. Huang and R. G. Kemp, *Biochemistry* **10**, 2278 (1971).

amount of enzyme required to catalyze the hydrolysis of 1 µmole of cyclic nucleotide in 1 minute at pH 7.5 and 30°.

Purification Procedure[1]

Step 1. Extraction. A female New Zealand white rabbit is killed with an overdose of Nembutal and bled by cutting the blood vessels in the neck. The hind leg and back muscles are removed, chilled immediately in ice, and passed through a meat grinder. The ground muscle is homogenized in a Waring blender for 30 seconds in the presence of 3 volumes of a buffer containing 50 mM imidazole, 4 mM EDTA, and 0.1 mM dithiothreitol (pH 8.5). This and all subsequent steps are performed at 2–4°. The homogenate is centrifuged for 40 minutes at 10,000 g, and the supernatant is retained.

Step 2. Ammonium Sulfate. After the supernatant is passed through glass wool to remove lipid material, it is adjusted to 0.60 saturation of $(NH_4)_2SO_4$ by the addition of solid $(NH_4)_2SO_4$. The mixture is stirred for 30 minutes, and the sediment is collected by centrifugation at 10,000 g for 40 minutes and dissolved in a volume of 10 mM Tris·HCl, 100 mM NaCl, 2 mM EDTA, 0.1 mM dithiothreitol (pH 8.0) equal to one-tenth of the volume of the original extract. The Tris·NaCl buffer will be subsequently referred to as buffer A.

Step 3. First DEAE-Sephadex. The enzyme solution is dialyzed against two changes of this same buffer overnight and then applied to a column of DEAE-Sephadex A-50 (5 × 35 cm) that has been previously equilibrated with buffer A. After the protein sample has entered the column bed, the column is washed with the buffer A until the optical density of the effluent at 280 nm drops below 0.1. The enzyme is eluted with buffer A containing 0.8 M NaCl at a flow rate of 40 ml per hour. Those fractions containing activity are pooled and then dialyzed overnight against buffer A.

Step 4. Second DEAE-Sephadex. The dialyzed enzyme is applied to column of DEAE Sephadex A-50 (3.5 × 40 cm) equilibrated with buffer A. After the column is washed with 300 ml of buffer A, the enzyme is eluted with a linear gradient consisting of 1.5 liter of buffer A containing NaCl at a final concentration of 0.6 M flowing into buffer A (1.5 liters) in an open mixing chamber. The enzyme elutes from the column in fractions that contain 0.2–0.4 M NaCl. Fractions with a specific activity greater than 0.01 are concentrated by precipitation with ammonium sulfate at 0.6 saturation. The sediment is dissolved in buffer A containing 0.8 M NaCl.

Step 5. First Gel Filtration. The solution is layered on a column of Sephadex G-200 (5 × 90 cm) equilibrated with buffer A containing 0.8 M NaCl. The column is washed with this buffer at a flow rate of 30 ml per hour. A large peak of inactive protein emerges in the void volume. The phosphodiesterase activity elutes with the trailing shoulder of this peak.

Steps 6 and 7. Further purification involves repeating the steps of gradient elution from Sephadex A-50 (step 4) and gel filtration on Sephadex G-200 (step 5). Concentration of the enzyme solution both preceding and following step 7 is accomplished with an Amicon Ultrafiltration cell fitted with a UM-10 membrane.

Comments on the Purification Procedure. During the elution of the enzyme from the DEAE-Sephadex in step 4 a small peak of enzyme activity precedes the main activity peak. The specific activity of the leading peak is usually very low, and it is not included with the principal peak. Although it was not further purified, in several experiments the substrate specificity of the leading peak was compared with that of the principal component. It was found to hydrolyze cyclic 3′,5′-GMP more rapidly than cyclic 3′,5′-AMP at both high (100 μM) and low (2 μM) substrate concentrations. As will be noted below the principal phosphodiesterase component hydrolyzes low concentrations of cyclic GMP very slowly. When the activities of both phosphodiesterase peaks were measured at varying substrate concentrations and were analyzed with double reciprocal plots of velocity versus substrate concentration, both components gave nonlinear plots that indicated either the presence of two enzymes or negative cooperativity. The minor component contained a significant amount of phosphodiesterase with a high affinity for cyclic 3′,5′-GMP. Generally a small peak of activity with high affinity for cyclic 3′,5′-GMP precedes the major phosphodiesterase component in the gel filtration step also.

TABLE I
PURIFICATION PROCEDURE FOR MUSCLE PHOSPHODIESTERASE

Fraction	Volume (ml)	Total units	Specific activity (units/mg)	Yield (%)
1. 10,000 g supernatant	1500	12	0.0005	100
2. 60% (NH$_4$)$_2$SO$_4$ sediment	180	11	0.001	91
3. First DEAE–A-50 eluate	350	7.0	0.004	58
4. Second DEAE–A-50 eluate	30	6.0	0.020	50
5. Gel filtration	80	3.9	0.080	33
6. Third DEAE–A-50 eluate	5	2.5	0.12	21
7. Second gel filtration	5	0.96	0.16	8

After the major component is purified further through steps 5, 6, and 7, it exhibits the normal hyperbolic saturation properties of a single enzyme.

The yield at each step of a typical preparation of cAMP phosphodiesterase from 500 g of muscle is shown in Table I. The procedure has been repeated ten times with yields of 4–10% and purification factors of 300–1000-fold. The preparation is not homogeneous as indicated by large variations in specific activity across the activity peak in the final G-200 elution. On polyacrylamide gel electrophoresis, however, only two bands of protein are observed and phosphodiesterase activity is associated with one of the bands.

Properties[1]

Stability. In solutions of low ionic strength the enzyme loses about 50% of its activity in 1 month. Reasonable stability is achieved in the presence of high concentrations of NaCl (0.1–0.5 M) or of $MgCl_2$ (0.1–0.4 M). The presence of EDTA with these salts provides additional stability on prolonged storage. The enzyme can be stored for 2 months with better than 80% retention of activity either frozen or at 4° in the presence of 10 mM Tris·HCl, 2 mM EDTA, and 0.1 M $MgCl_2$ (pH 8.0).

Cation Requirement. Although the purified phosphodiesterase that is passed through Sephadex G-25 to remove metal ions displays about one-fifth of maximal activity in the absence of added metal ion, this residual activity is abolished by the addition of chelating agents. Magnesium, manganous, and cobalt ions satisfy the metal ion requirement whereas calcium, cupric, ferrous, zinc, and nickle ions do not when present at 2 mM. The enzyme *in vivo* probably employs magnesium ion for which it has a K_a of about 24 μM.

Kinetic Parameters. The affinity of the enzyme for cyclic 3′,5′-AMP and cyclic 3′,5′-GMP is shown in Table II. The two substrates are compet-

TABLE II
KINETIC PARAMETERS OF MUSCLE PHOSPHODIESTERASE

Substrate	pH	K_m (μM)	Relative V_{max}
Cyclic AMP	7.0	4	0.6
Cyclic AMP	7.5	6	1.0
Cyclic AMP	8.0	10	1.2
Cyclic AMP	9.0	26	2.2
Cyclic GMP	7.5	2700	0.9

itive with one another, indicating a common catalytic site. There is no evidence for cooperativity in substrate binding with the purified enzyme.

Activators. As with other phosphodiesterase preparations that have been described, imidazole is an activator of the muscle phosphodiesterase. The effects of imidazole appear to be complex; the pH optimum, which is around pH 9.0, is shifted to lower values in the presence of imidazole, the maximal velocity is increased, but the affinity for cyclic 3',5'-AMP is decreased. As a result of the opposing effects of imidazole on V_{max} and affinity, the amount of imidazole stimulation at low concentrations of cyclic 3',5'-AMP is relatively small. Beavo et al.[2] reported that low concentrations of cGMP activate the hydrolysis of cAMP in several tissues. The phosphodiesterase preparation described herein is not activated by cGMP when this effect is examined under a great variety of experimental conditions.

Inhibitors. As with other cyclic nucleotide phosphodiesterases, the methylxanthines are competitive inhibitors of the muscle enzyme. Theophylline is the most potent with a K_i of 0.32 mM at pH 7.5. 6-Dimethylaminopurine is also inhibitory and is more effective than caffeine and theobromine. Other inhibitors that are somewhat less potent are adenine, adenosine, and 6-hydroxypurine. The inhibitor action of ATP and citrate can be completely accounted for by the chelation of the required divalent metal ion.

The muscle phosphodiesterase is also inhibited by p-chloromercuribenzoate and 5,5'-dithiobis(2-nitrobenzoic acid). The inhibition is reversed by dithiothreitol.

[2] J. A. Beavo, J. G. Hardman, and E. W. Sutherland, *J. Biol. Chem.* **245**, 5649 (1970).

[35] Purification and Properties of Cyclic Nucleotide Phosphodiesterase from *Dictyostelium discoideum* (Extracellular)

By BRUCE M. CHASSY and E. VICTORIA PORTER

Nucleoside 3',5'-cyclic phosphate + H_2O → nucleoside 5'-phosphate

The cellular slime mold *Dictyostelium discoideum* produces both an extracellular and a cell-bound cyclic nucleotide phosphodiesterase.[1,2] The

[1] Y. Y. Chang, *Science* **160**, 57 (1968).
[2] R. G. Pannbacker and L. J. Bravard, *Science* **175**, 1014 (1972).

enzyme is an essential part of the cyclic AMP (cAMP)-mediated aggregation and differentiation of this organism.[3,4]

Assay

Principle. [^{3}H(G)]5′-AMP, enzymatically produced from [^{3}H(G)]cAMP, is isolated by paper chromatography and radioactivity measured by scintillation counting. Spectrophotometric and "phosphate-release" assay methods (described elsewhere in this volume) that are used with other cyclic nucleotide phosphodiesterases are not applicable. Not only are interfering substances present, but highly variable results are obtained with these techniques (even with purified fractions).

Procedure. Incubation mixtures (100 µl) are prepared to have a final concentration of 100 µM cAMP (0.1–0.3 µCi [^{3}H]cAMP), 50 mM Tris·HCl pH 7.4, 10 mM MgCl$_2$, and 0.0001–0.001 unit (IU) of phosphodiesterase. Incubations are carried out for 30 minutes at 30° and terminated with 100 µl of 0.12 N HCl in 95% ethanol. Aliquots are then subjected to paper chromatography.

Protein Determination. Protein is determined according to Lowry *et al.*[5] with bovine serum albumin as the reference.

Purification Procedure

Step 1. Preparation of Extracellular Supernatant. The contents of 6 flasks, each 1-liter flask containing about 400 ml of a suspension of *D. discoideum* grown for 3 days on autoclaved *Escherichia coli*,[6] are pooled and centrifuged for 10 minutes at 800 g (use of higher forces will rupture the amoeba, releasing large quantities of proteolytic enzymes). This, and all subsequent operations, are done at 0–4° unless otherwise noted. The supernatant is decanted and recentrifuged at 27,500 g for 1 hour to complete clarification. Two liters of supernatant results.

Step 2. 90% Ammonium Sulfate Precipitation. Over the course of 30 minutes, finely powdered ammonium sulfate (1260 g) is added with continuous stirring to the 2-liter supernatant. The solution is centrifuged for 30 minutes at 27,500 g. Both a precipitate and a flotage result. Careful

---

[3] B. M. Chassy, L. L. Love, and M. I. Krichevsky, *Proc. Nat. Acad. Sci. U.S.* **64**, 296 (1969).
[4] E. A. Goidl, B. M. Chassy, L. L. Love, and M. I. Krichevsky, *Proc. Nat. Acad. Sci. U.S.* **69**, 1128 (1972).
[5] O. H. Lowry, N. J. Rosebrough, A. L. Farr, and R. J. Randall, *J. Biol. Chem.* **193**, 265 (1951).
[6] M. I. Krichevsky and L. L. Love, this series, Vol. 39 [42].

SUMMARY OF PURIFICATION

Step	Volume (ml)	Units (IU)	Protein (mg/ml)	Specific activity (IU/mg)	Yield (%)	Purification (x-fold)
1	2000	178	1.0	0.09	100	1
2	240	147	1.3	0.48	82	5
3	25	184	2.2	3.20	103	36
4a	4	73	0.3	61.00	41	685
4b	9	186	2.5	8.28	104	93

decantation leaves the flotage adhering to the walls of the centrifuge tube. After carefully dissolving both in a minimum volume of 0.1 M Tris·HCl pH 7.4, 200 ml of a turbid brown suspension is obtained. The suspension is dialyzed overnight against one change of 6 liters of 10 mM Tris·HCl pH 7.4. The dialyzed preparation is clarified by centrifugation at 27,500 g for 30 minutes and pooled to give a final volume of 240 ml.

Step 3. Dithiothreitol (DTT)-$MgCl_2$ Activation. The following step removes some extraneous protein and nucleic acids and also activates the phosphodiesterase resulting in a net increase in total units. The supernatant from step 2 (240 ml) is adjusted to 0.1 M $MgCl_2$ by the addition of 26 ml of 1 M $MgCl_2$. Solid DTT (208 mg) is added to make the solution 5 mM with respect to DTT. The solution is stirred for 1 hour at 23° and then freed of precipitate by centrifugation at 41,000 g for 30 minutes. The supernatant is adjusted to 90% saturation of ammonium sulfate by slowly (15 minutes) adding 164 g of finely powdered ammonium sulfate with constant stirring. The suspension is stirred an additional 15 minutes. The precipitate is collected by centrifugation at 41,000 g for 30 minutes. The pellet is dissolved in 20 ml 0.1 M Tris·HCl pH 8.5–0.2 mM DTT and dialyzed overnight against 6 liters of 10 mM Tris·HCl pH 8.5–0.2 mM DTT (one change).

Step 4a,b. Chromatography on DEAE-Cellulose. A 40 × 2.5 cm chromatography column is filled (flow rate = 2 ml per minute) with Whatman DE-52 (a microgranular DEAE-cellulose) that is preequilibrated with 10 mM Tris·HCl pH 8.5–0.2 mM DTT. The phosphodiesterase preparation from step 3 (25.4 ml) is applied to the column at a flow rate of 2 ml per minute while the collection of 10-ml fractions is begun. After the sample is adsorbed onto the column, the column is washed with 100 ml 10 mM Tris·HCl, pH 8.5–0.2 mM DTT, followed by a linear gradient elution. The 10 mM Tris·HCl, pH 8.5–0.2 mM DTT starts the gradient, and the gradient ends at 0.1 M Tris·HCl pH 7.4–0.2 mM DTT–0.5 M KCl (2 liters total volume). Approximately 40% of the activity applied to the column emerges in fractions 10–20 (only slightly retained). The phosphodiesterase comprises 15–25% of the protein found associated with this sharp activity peak (purity judged by disc gel electrophoresis). Fractions 10–20 are pooled and concentrated to 4 ml in an Amicon pressure ultrafiltration cell using a PM-30 membrane. The remaining fractions are pooled and concentrated to 9 ml in the same fashion. This pooled fraction (step 4b) lacks the high specific activity of fraction 4a but contains the remainder of the activity applied to the column; it is not eluted as a peak but is "smeared" over the entire linear gradient. The omission of the DTT-$MgCl_2$ step and of DTT from the buffers gives rise to markedly different elution patterns. The proportion of activity eluted in fractions 10–20 falls and a second sharp peak centered around fractions 45–55 appears. These two fractions have distinct kinetic properties.[7]

Comments on Further Purification. Gel filtration on BioGel A, 0.5 mm, 200–400 mesh spherical agarose; preparative disc gel electrophoresis at pH 9.3; or isoelectric focusing all give rise to essentially pure phosphodiesterase preparations (disc gel electrophoresis) when applied to the fraction of step 4a. However, large losses of activity occur during these steps, and the resulting preparations are quite unstable. The purification, as described, is quite reproducible. The same methods may be applied to cultures of the slime mold grown in axenic media[6]; however, the final specific activity is lowered by the presence of extraneous proteins from the rich growth medium. There are several factors that may contribute to the net gain in units observed during the purification: (1) the DTT activation, (2) other interconversions of multiple forms, (3) the removal of a possible natural inhibitor,[8] and (4) the variability of the assay.

Properties

Stability. The material from step 4a,b (or earlier steps) is stable for at least 6 months at $-20°$ when stored in 0.1 M Tris·HCl, pH 7.4–0.2 mM DTT. After 1 month at 0–4°, 90% of the activity remains. More highly purified preparations lose activity rapidly under these conditions. Before assay, phosphodiesterase preparations that have been subjected to cold storage should remain at room temperature for 1 hour. This allows a reversible reactivation of unknown mechanism to take place.

Optimum pH and Buffers. Fraction 4a has a broad pH optimum asymmetrically centered around pH 7.4. Only 50% as much activity is observed at pH 6.2, while 85% of the original activity remains at pH

---

[7] B. M. Chassy, *Science* **175**, 1016 (1972).
[8] D. Malchow, B. Nagele, H. Schwarz, and G. Gerisch, *Eur. J. Biochem.* **28**, 136 (1972).

9.2. Identical pH-rate profiles are observed in the following 0.1 M buffers: MES (morpholinoethanesulfonic acid), MOPS (morpholinopropane sulfonic acid), PIPES (piperazineethanesulfonic acid), Tris·HCl, sodium borate, sodium glycinate, and sodium glycylglycinate.

Substrate Specificity. Besides cAMP, these preparations (taken as a mixture of forms) are capable of hydrolyzing cCMP, cUMP, cTMP and to a lesser extent cGMP, cIMP, and N^6-monobutyryl-cAMP, but not dibutyryl-cAMP.

K_m Values. Fraction 4a has a K_m of 15 μM and fraction 4b displays mixed kinetics with a predominant K_m of 2 mM.

Metal ions. While fraction 4a does not seem to require metal ions for activity, it can be totally deactivated by gel filtration in a column preequilibrated with 1 mM 1,10-phenanthroline. The addition of 1 mM $ZnCl_2$ allows an 11% restoration of activity. Divalent metals are not required for activity, but 1 mM $MgCl_2$ or $ZnCl_2$ stimulate 50%. The same concentrations of $CaCl_2$ ($MnCl_2$, $CoCl_2$, $NiCl_2$, and $BaCl_2$ are without effect. $CuCl_2$ reduces activity to 68% at 1 mM and 37% at 10 mM. EDTA stimulates as much as 50% at 5 mM, while the same concentration of 1,10-phenanthroline causes the activity to fall to about half that observed without any addition.

Activators and Inhibitors. ATP, GTP, GMP, imidazole and pretreatment with rattlesnake venom all activate fraction 4a. TPP is an effective inhibitor causing 70% inhibition at 1 mM and 90% inhibition at 10 mM. Concentrations of DTT greater than 0.1 mM are inhibitory in assay mixtures, even though these concentrations activate the enzyme.

Comment on Multiple Forms. There are at least two forms of slime mold phosphodiesterase, one having a K_m of 15 μM (higher V_{max} and a molecular weight of 65,000 by gel filtration) and the other having a K_m of 2 mM (lower V_{max} and a molecular weight of 120,000 by gel filtration). These are interconvertible; DTT causes the appearance of the low K_m form and aging converts the low K_m form into the high K_m form.[7] The chemistry underlying this process is not understood. The existence of other forms, perhaps inactivated by disc gel electrophoresis, appears likely. The possibility of multiple subunit forms, or interaction with the reported inhibitor protein[8] should be considered.

[36] Cyclic AMP Phosphodiesterase of *Escherichia coli*[1]

By LARRY D. NIELSEN and H. V. RICKENBERG

$$\text{Cyclic 3',5'-adenosine monophosphate} \xrightarrow{H_2O} \text{5'-AMP} + H^+$$

Braná and Chytil[2] first reported the occurrence of a cyclic 3',5'-adenosine monophosphate (cAMP) phosphodiesterase (EC 3.1.4.17) in *E. coli*. A role for this enzyme in the regulation of bacterial energy metabolism has been suggested[3,4]; mutants defective in cAMP phosphodiesterase are viable but are resistant to catabolite repression.

Assay Method

Principle. The enzyme is assayed conveniently by a modification of the procedure reported by Thompson and Appleman.[5] The principle of the method is based on the fact that the anion-exchange resin, Dowex 2, binds cAMP, but not adenosine. The assay is carried out in two steps. In step 1, ^{3}H-labeled cAMP is hydrolyzed by the cAMP phosphodiesterase to 5'-AMP. In step 2 an excess of 5'-nucleotidase (or snake venom) hydrolyzes the 5'-AMP formed in step 1 to adenosine. Unhydrolyzed cAMP, bound to the anion exchange resin, is removed by sedimentation and the radioactivity in the supernatant fluid, which represents adenosine (and possibly adenine), and hence is a measure of the cAMP hydrolyzed, is determined.

Reagents

Tris·HCl buffer, 0.5 M, pH 7 (37°)
$MgCl_2$, 20 mM in H_2O
Dithiothreitol, 25 mM in H_2O
^{3}H-Labeled cAMP, 2 mM, 5×10^5 to 10^6 cpm/ml, in H_2O
5'-Nucleotidase, 2.5 mg/ml in H_2O (or snake venom, 2.5 mg/ml in H_2O)

---

[1] Supported in part by grants from the National Institute of Arthritis and Metabolic Diseases and the National Science Foundation.
[2] H. Braná and F. Chytil, *Folia Microbiol (Prague)* **11**, 43 (1965).
[3] D. Monard, J. Janeček, and H. V. Rickenberg, *Biochem. Biophys. Res. Commun.* **35**, 584 (1969).
[4] D. Monard, J. Janeček, and H. V. Rickenberg, *in* "The Lactose Operon" (J. R. Beckwith and D. Zipser, eds.), p. 393. Cold Spring Harbor Laboratory of Quantitative Biology, Cold Spring Harbor, New York, 1970.
[5] W. J. Thompson and M. M. Appleman. *Biochemistry* **10**, 311 (1971).

17% (w/v) ethanolic slurry of Dowex AG 2-X8, 200–400 mesh, formate form

Scintillation fluid (POPOP, 100 mg/liter; PPO, 4 g/liter in toluene)

Procedure. A mixture containing the following is preincubated in tubes (suitable for low-speed centrifugation) at 37° for 3 minutes; Tris buffer, 0.02 ml (50 mM final); MgCl$_2$, 0.02 ml (2 mM final); dithiothreitol, 0.02 ml (2.5 mM final); preparation of proteinaceous activator (see below), 0.02 ml, when used; 0.02 ml of bacterial extract or of purified cAMP phosphodiesterase. The volume of the mixture is brought to 0.15 ml with water, and the reaction is started by the addition of 0.05 ml of ^{3}H-labeled cAMP (0.5 mM final) prewarmed at 37°. Incubation is at 37° and step 1 of the reaction is terminated by the immersion of the tubes containing the reaction mixtures in boiling water for 3 minutes. After cooling to 37°, 50 μg (0.02 ml) of 5'-nucleotidase (or of snake venom) are added to each tube, and incubation at 37° is continued for 15 minutes. Then 2.2 ml of the ethanolic slurry of Dowex are added, and the mixture is kept at room temperature for 30 minutes with occasional shaking. The resin is sedimented by low-speed centrifugation, and 1.0-ml aliquots of the supernatant fluids are removed to scintillation vials containing 2.0 ml of absolute ethanol and 7.0 ml of the scintillation fluid. The radioactivity is determined by liquid scintillation counting.

The rate of the reaction is followed by measuring the hydrolysis of cAMP at three time intervals. Duplicate samples are assayed for each time point. Values are corrected by the subtraction of counts of radioactivity obtained in reagent blanks treated in a manner identical with that of the experimental samples but containing either no, or heated, cAMP phosphodiesterase. The activity of a given preparation is calculated from the slope of the line representing the time course of the reaction.

Units. One unit of cAMP phosphodiesterase activity is that amount of enzyme which hydrolyzes 1 pmole of cAMP per minute under the conditions described above. Specific activity is expressed as units per milligram of protein.

Culture of Bacteria

Escherichia coli strain K12 is used for the preparation of the cAMP phosphodiesterase. Conditions of growth, e.g., the use of either mineral salts-based or complex media, do not appear to affect the specific cyclic AMP phosphodiesterase activity significantly. Cells may be stored frozen at −20° or below for prolonged periods of time. *E. coli* strain Crookes (American Type Culture Collection 8739) when grown on a mineral salts medium is devoid of detectable cAMP phosphodiesterase activity and serves as source of the activator. The medium contains per liter: 13.6 g of KH_2PO_4; 2.0 g of $(NH_4)_2SO_4$; 0.01 g of $CaCl_2$; 0.5 mg of $FeSO_4 \cdot 7H_2O$, and 0.2 g of $MgSO_4 \cdot 7H_2O$. The pH of the medium is adjusted to 7 with KOH; glucose at a concentration of 50 mM serves as source of carbon. The bacteria are grown with forced aeration at 37°. The generation time is approximately 50 minutes under these conditions and the yield is 6–8 g of bacteria wet weight per liter. Following sedimentation, the bacteria to be used for the purification of either cAMP phosphodiesterase, in the case of strain K12, or for the preparation of the activator, in the case of strain Crookes, are washed with 10 mM Tris·HCl, pH 7.3 (0°), buffer (buffer A), and stored at −20° or below.

Preparation of Extracts

The bacteria of both strains K12 and Crookes are extracted by any one of three equally effective methods. Extraction and all following operations are carried out at 0–4°. The bacteria are suspended in a volume (ml) of buffer A equal to their wet weight in grams. They are then disrupted either by passage through a French press at 5000 psi, or by treatment in a Raytheon 10-kcycles/sec magnetostrictive oscillator for two consecutive 15-minute periods, or by grinding with an approximately equal weight of levigated alumina. The preparation of the crude bacterial extract constitutes step 1 of the procedure of purification. DNase, to a final concentration of 1 μg/ml, is added just prior to the disruption of the bacteria. Extracts are centrifuged at 40,000 g for 20 minutes, and the centrifugation is repeated, if required for the clarification of the extracts. Both cAMP phosphodiesterase and the proteinaceous activator are in the supernatant fluid.

Purification

A problem encountered during the purification of the cAMP phosphodiesterase is the progressive loss of activity with purification. This is not due primarily to an inherent instability of the enzyme, but appears to reflect its separation from a proteinaceous effector which, without prejudice as to its mode of action, we term the "activator." The activator requires the simultaneous presence of a reduced thiol for it to exert its activating effect on purified preparations of the phosphodiesterase. Of the thiols tested (reduced glutathione, β-mercaptoethanol, reduced lipoic acid, and cysteine) dithiothreitol was the most effective and is used routinely at a concentration of 2.5 mM. A crude extract of the Crookes strain of *E. coli* (cAMP phosphodiesterase-negative) may be substituted for the activator, if employed at a sufficiently high concentration. The purification of the cAMP phosphodiesterase and the preparation of the activator are described separately. In the procedure of purification of the cAMP phosphodiesterase as discussed below, either a crude extract of strain Crookes at 2 mg per assay or a preparation of activator at 200 μg per assay is employed in monitoring cAMP phosphodiesterase activity.

A. Cyclic AMP Phosphodiesterase

Two procedures may be employed for the removal of the bulk of the protein from crude extracts. Fractionation with acetone is employed in method A and a combination of precipitation with streptomycin sulfate, heating, and fractionation with acetic acid in method B. Occasional variability with respect to the concentration of acetone at which the cAMP phosphodiesterase is precipitated is encountered, particularly in the case of commercial preparations of strain K12. The table summarizes the results of typical experiments in which the two methods were employed. Procedural steps, such as manner of disruption of the bacteria or elution from DEAE, are independent of the use of method A or B.

Step 2. Method A. Acetone. Acetone is precooled to $-10°$; a volume equal to that of the crude bacterial extract (step 1) is added slowly with constant stirring to the extract. After 30 minutes the large precipitate is removed by sedimentation and discarded. A second volume of cold acetone, equal to the first, is added slowly with stirring to the supernatant fluid obtained from the first treatment with acetone. After 30–60 minutes the preparation is centrifuged, and the supernatant fluid is discarded. The precipitate is dissolved in a volume of buffer A corresponding to approximately one-tenth of the volume of the original crude extract. The preparation is then dialyzed exhaustively against buffer A to remove residual acetone.

Step 2. Method B. Streptomycin sulfate, heat, acid. (a) Streptomycin sulfate (10% in water) is added to a final concentration of 1% to the crude bacterial extract (step 1). After several hours in the cold the precipitate is removed by centrifugation and discarded. (b) The supernatant solution from the preceding step is heated rapidly to 60° in an 80° water-bath and then cooled immediately in an ice bath. The precipitate is removed by centrifugation and discarded. (c) Normal acetic acid is added to the supernatant solution from the preceding step to a final concentration of 0.1 N (pH of approximately 3.5). The acidified preparation is kept on ice for 1 hour, the precipitate is collected by centrifugation at 10,000 g for 15 minutes and dissolved in buffer A. Solid $NaHCO_3$ is used

PURIFICATION OF *Escherichia coli* CYCLIC AMP PHOSPHODIESTERASE[a]

Step	Total protein (mg)		Total units[b]		Specific activity (units/mg)		Yield (%)		Purification	
	A	B	A	B	A	B	A	B	A	B
1. Crude extract[c]	13,800	23,000	505×10^6	551×10^6	3.7×10^4	2.4×10^4	100	100	1	1
2. Precipitation	1,160[d]	4,500[e]	448×10^6	460×10^6	4.0×10^5	1.02×10^5	89	83	11	4.2
3. Sephadex G-100	201	246	183×10^6	78×10^6	9.1×10^5	3.2×10^5	36	14	25	10.3
4. DEAE[f]	55	25	113×10^6	52×10^6	2.06×10^6	2.12×10^6	22	9.5	56	89

[a] From L. D. Nielsen, D. Monard and H. V. Rickenberg, *J. Bacteriol.* **116**, 857 (1973) by permission of the American Society for Microbiology.
[b] One unit = 1 pmole of cyclic AMP hydrolyzed per minute as assayed in the presence of activator.
[c] Crude extract: Starting material: 250 g (frozen wet weight) *E. coli* K12. Crude extract for method A was prepared in a French press. Crude extract for method B was prepared by treatment of the bacteria in a Raytheon 10 kcycles/sec magnetostrictive oscillator.
[d] Method A: Acetone.
[e] Method B: Streptomycin sulfate, heat, acid.
[f] Elution of the protein prepared by method A was by the application of successive portions of buffer containing increasing concentrations of NaCl, whereas in the case of the protein prepared by method B a gradient of NaCl was employed.

to adjust the pH to 7. The volume of the preparation at this point is between one-tenth and one-fifth that of the crude extract.

Step 3. Filtration through Sephadex G-100. The redissolved, dialyzed acetone precipitate or the redissolved acid precipitate, respectively, are applied in a volume equal to 2–3% of the bed volume to a column of Sephadex G-100 equilibrated with buffer A. Good separation of the enzyme is obtained, if the total protein in milligrams approximates the bed volume of the column in milliliters. The sample is eluted with buffer A, protein is monitored by UV absorbance, and fractions are assayed for cAMP phosphodiesterase activity. Fractions showing high activity are pooled and employed for the next step of purification.

Step 4. Chromatography on DEAE-Cellulose. Anion exchange columns are prepared with Whatman DE-52 which is equilibrated with buffer A and degassed using a water aspirator. The cAMP phosphodiesterase obtained from Sephadex G-100 (step 3) is applied to the column at a rate of 1–2 mg of protein per milliliter of bed volume of the exchanger. Protein may be eluted by either one of two methods: (a) the column is washed sequentially with portions of buffer A containing 0.1 M, 0.2 M, 0.3 M, and 0.5 M NaCl, respectively. Each portion of buffer corresponds to approximately twice the bed volume of the column. The enzyme is eluted in the 0.3 M NaCl wash. (b) The column is washed with a volume of buffer A containing 0.15 M NaCl and corresponding to twice the bed volume of the column. Then a gradient of NaCl ranging from 0.15 M to 0.3 M NaCl is applied to a volume three times that of the bed volume. An additional bed volume of 0.3 NaCl is applied at the end of this gradient in order to ensure recovery of all the material that can be eluted at 0.3 M NaCl. Protein is monitored by UV absorbance and fractions are assayed for cAMP phosphodiesterase activity. The enzyme is eluted as a single peak at an NaCl concentration of about 0.23 M. Method 4 (a) has the advantage of rapidity whereas 4 (b) yields cAMP phosphodiesterase of greater purity.

Fractions obtained by either method from the DEAE column and showing high phosphodiesterase activity are pooled, concentrated by ultrafiltration under N_2 pressure, and dialyzed against buffer A to remove the NaCl.

B. Preparation of the Activator

Step 1. Digestion with RNase. The extract of strain Crookes at a concentration of approximately 60 mg/ml is incubated at 37° for 1 hour with RNAse at a final concentration of 10 μg/ml. This is followed by dialysis against buffer A.

Step 2. Heat Treatment. The dialyzed extract is kept for 3 minutes in a boiling water bath. The precipitate is removed by centrifugation and discarded.

Step 3. Acid Precipitation. The pH of the supernatant fluid from step 2 is brought to about 3.5 by the addition of 1 N acetic acid to a final concentration of 0.1 N acid. The precipitate is collected by centrifugation at 10,000 g for 10 minutes and is dissolved in a volume of buffer A corresponding to approximately one-tenth the volume of the original supernatant. The pH is adjusted to 7 with solid $NaHCO_3$. Precipitation with acetic acid may be repeated several times for the removal of residual acid-soluble material. The protein in the final acid precipitate corresponds to approximately 1% of that of the crude extract. The material may be stored at $-20°$ or it may be freeze-dried without loss of activity. Treatment of the activator with proteolytic, but not with nucleolytic, enzymes destroys its activity. Filtration of activator, prepared as described, through Sephadex G-100 indicates a molecular weight of 75,000. However, occasionally, activator is also detected in material with a molecular weight of 150,000 or greater (in void volume).

Properties of the cAMP Phosphodiesterase

Requirement for Activator. The dependence of the *E. coli* cAMP phosphodiesterase for activity on the simultaneous presence of a proteinaceous activator and a reduced thiol is notable. The degree of activation is a function of both the concentration of protein at which the cAMP phosphodiesterase is assayed and of the extent of its purification. The activity of crude extracts when assayed at 100 µg per 0.2 ml of assay mixture is stimulated approximately 10-fold by the addition of dithiothreitol to a final concentration of 2.5 mM. At this concentration of protein, the addition of activator enhances cAMP phosphodiesterase activity less than 2-fold in crude extracts. When the same crude extract, however, is assayed at a concentration of 10 µg of protein per 0.2 ml in the presence of dithiothreitol, no cAMP phosphodiesterase activity is detected unless activator, at 200 µg/0.2 ml, or Crookes crude extract, at 2 mg/0.2 ml, is added. When an approximately 80-fold purified preparation of the enzyme is assayed at less than 2 µg of protein per 0.2 ml final volume, complete dependence on the presence of activator (or Crookes extract) is also observed. Neither bovine serum albumin nor a variety of other proteins, including fractions of *E. coli* protein devoid of activator, can substitute for the activator.

It is noteworthy that in either crude extracts or in purified preparations, supplemented with *unheated* Crookes extract, reduced pyridine nucleotides are as effective as dithiothreitol in stimulating the activity of the cAMP phosphodiesterase.

Physical and Catalytic Properties of cAMP Phosphodiesterase. The molecular weight of the cAMP phosphodiesterase is approximately 28,000; this is an estimate based on the behavior of either a crude extract or a partially purified preparation on a calibrated Sephadex G-100 column. The pH optimum of the enzyme is 7; its K_m for cAMP is 0.2 mM. The cAMP phosphodiesterase shows normal Michaelis-Menten kinetics in the presence or in the absence of activator; neither pH optimum nor K_m appear to be affected by the activator. The activator, however, increases significantly the velocity of the hydrolysis of cAMP (see above).

Phosphate ion inhibits the cAMP phosphodiesterase of *E. coli* severely, succinate inhibits slightly. The product of the reaction, 5'-AMP, at 0.5 mM inhibits by approximately 40% when the substrate, cAMP, is present at the same concentration. Theophylline at concentrations of up to 10 mM does not inhibit the enzyme from *E. coli*.

Omission of Mg^{2+} from the assay does not reduce cAMP phosphodiesterase activity, i.e., the addition of Mg^{2+} to the assay mixture may be unnecessary. Similarly, Mn^{2+} and Ca^{2+} at 2 mM do not stimulate; Co^{2+} and Zn^{2+} inhibit. Chelators of metal ions, specifically EDTA (0.1 mM final), *o*-phenanthroline (1 mM final), and α,α'-dipyridyl (1 mM final) inhibit the activity of the enzyme to varying degrees. The possibility that these agents, particularly those with avidity for the Fe^{2+} ion, affect the system of activation rather than the cAMP phosphodiesterase proper, is not excluded.

cGMP is hydrolyzed by neither the partially purified cAMP phosphodiesterase nor by crude extracts of *E. coli* under the conditions of assay described. Furthermore, cGMP does not affect the hydrolysis of cAMP by *E. coli* cAMP phosphodiesterase when employed at a concentration equimolar with that of cAMP.

The cAMP phosphodiesterase of *E. coli* resembles the cAMP phosphodiesterase of the related enteric bacterium *Serratia marcescens*[6] in its K_m and its lack of a requirement for Mg^{2+}. The enzyme from *Serratia* differs from that of *E. coli* in its sensitivity to theophylline, its resistance to inhibition by EDTA, and its ability to hydrolyze a number of cyclic nucleotides. The most interesting difference, however, is the requirement for a proteinaceous activator for the hydrolysis of cAMP by the *E. coli* phosphodiesterase.

[6] T. Okabayashi and M. Ide, *Biochim. Biophys. Acta* **220**, 116 (1970).

[37] Preparation and Characterization of Multiple Forms of Cyclic Nucleotide Phosphodiesterase from Liver

By WESLEY L. TERASAKI, THOMAS R. RUSSELL and M. MICHAEL APPLEMAN

At least three separable and kinetically distinct forms of cyclic nucleotide phosphodiesterase exist in rat liver.[1] These forms are not all detectable in fresh liver extracts because of physical or kinetic masking,[2] but can be revealed in part by storage at 4° for 24 hours,[3] by trypsin treatment,[1] by ammonium sulfate fractionation,[4] or by DEAE-cellulose chromatography.[1] This latter method serves as the basis for the separation of the phosphodiesterase activities from liver into fractions suitable for kinetic characterization.

Assay Method

The assay used is described elsewhere in this volume.[5] Incubations are carried out at pH 8.0 unless noted otherwise.

Preparation of Tissue Extract

The starting material for this preparation may be either fresh or frozen liver from male Sprague-Dawley rats (200–300 g) fed *ad libitum* on laboratory chow. All procedures were carried out on ice or in a 4° cold room. Five grams of liver were minced and homogenized in 25 ml of cold deionized water for 1 minute using a high speed blender. The homogenate was transferred to a small beaker and sonicated for 15 minutes (0.5 minute per milliliter) using the large probe at a setting of 40 on a Bronwill Biosonik III or equivalent. The preparation was centrifuged at 30,000 g for 20 minutes, and the supernatant was filtered through glass wool to remove fat particles. Centrifugation without prior sonication can cause a large loss in phosphodiesterase activity. This preparation may be stored at 4° for a number of days without problems.

---

[1] T. R. Russell, W. L. Terasaki, and M. M. Appleman, *J. Biol. Chem.* **248**, 1334 (1973).
[2] W. J. Thompson and M. M. Appleman, *J. Biol. Chem.* **246**, 3145 (1971).
[3] J. G. Hemington and A. S. Dunn, unpublished results, 1971.
[4] J. A. Beavo, J. G. Hardman, and E. W. Sutherland, *J. Biol. Chem.* **246**, 3841 (1971).
[5] W. J. Thompson, G. Brooker, and M. M. Appleman, this volume [30].

DEAE-Cellulose Chromatography

A portion of the extract (about 8 ml) was applied to a 1.5 × 30 cm column of DEAE-cellulose (exchange capacity 0.66 mEq/g) equilibrated with 50 mM Tris·acetate, pH 6.0, containing 4 mM 2-mercaptoethanol. With the flow rate at 40 ml per hour, two column volumes (80 ml) of the same buffer were passed through the column without any phosphodiesterase activity appearing in the wash. A 300-ml total volume linear gradient from zero to 1.0 M sodium acetate in the above buffer was then passed through the column, and 3-ml fractions were collected. Fractions were assayed directly for cAMP and cGMP phosphodiesterase activity using substrate concentrations below 1 µM, and the profile shown in Fig. 1 was obtained. Three discrete activities identified as DI, DII, and DIII were pooled and concentrated by ultrafiltration on an Amicon apparatus with a UM-20E filter. Estimates from gel filtration on Bio-Gel A-5M in the chromatography buffer indicated that all three activities were of similar size with molecular weight close to 400,000.

Alternate Preparations

A method involving a subcellular fractionation prior to chromatography has been reported.[1] The phosphodiesterase activities identified as

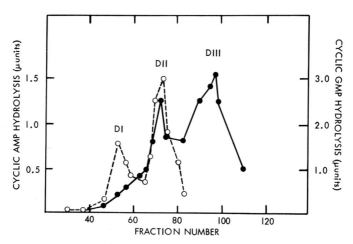

Fig. 1. Activity profile of rat liver phosphodiesterase on DEAE-cellulose. An extract (8.0 ml) was applied to the column, followed by an initial wash of 80 ml of buffer. A 300-ml linear gradient from zero to 1.0 M sodium acetate was applied from fraction 30 to fraction 130. Aliquots of 0.2 ml were assayed directly for phosphodiesterase activity at 0.25 µM cGMP (○-----○) or 0.125 µM cAMP (●——●). From T. R. Russell, W. L. Terasaki, and M. M. Appleman, *J. Biol. Chem.* **248**, 1334 (1973).

DI and DII appear to be soluble while DIII is derived from particulate or membranous material. A different fractionation of liver phosphodiesterase using ammonium sulfate has also been published.[4]

Properties of Liver Phosphodiesterases

DI. The phosphodiesterase activity eluted from the DEAE-cellulose column at a sodium acetate concentration of about 0.15 M is quite specific for the hydrolysis of cGMP and has a high affinity ($K_m = 6$ μM). This activity shows normal kinetic behavior.

DII. The fraction eluted at a salt concentration of about 0.3 M contains phosphodiesterase active against both cAMP and cGMP with approximately equal affinity ($K_m = 20$–40 μM). The activation of cAMP hydrolysis by cGMP, reported by others,[4] is observed with this fraction ($K_a = 0.2$ μM cGMP). Hydrolysis of both purine nucleotides is positively cooperative at pH 7.4 as indicated by an upward curvature of double reciprocal plots.

DIII. The last active fraction to elute from DEAE-cellulose (at about 0.6 M sodium acetate) has a high specificity and affinity for cAMP ($K_m = 6$ μM). This value is obtained by extrapolation of a downward curved double reciprocal plot indicative of negative cooperative kinetics.[6]

[6] T. R. Russell, W. J. Thompson, F. W. Schneider, and M. M. Appleman, *Proc. Nat. Acad. Sci. U.S.* **69**, 1791 (1972).

[38] Activity Stain for the Detection of Cyclic Nucleotide Phosphodiesterase in Polyacrylamide Gels

By Elihu N. Goren, Allen H. Hirsch, and Ora M. Rosen

Assay Method

Principle. The localization of cyclic nucleotide phosphodiesterase activity in acrylamide gels[1] is based upon methods developed for the detection of alkaline phosphatase activity.[2,3] A similar method has been util-

[1] E. N. Goren, A. H. Hirsch, and O. M. Rosen, *Anal. Biochem.* **43**, 156 (1971).
[2] G. Gomori, "Microscopic Histochemistry." Univ. of Chicago Press, Chicago, Illinois, 1952.
[3] J. M. Allen and G. Hyncik, *J. Histochem. Cytochem.* **11**, 169 (1963).

ized to localize cyclic nucleotide phosphodiesterase histochemically.[4,5] The reaction employed results in a white precipitate of lead phosphate during the first incubation period:

$$\text{Cyclic 3',5'-AMP} \xrightarrow{\text{phosphodiesterase}} \text{5'AMP} \xrightarrow[\text{Pb}^{2+}]{\text{alkaline phosphatase}} \text{adenosine} + \text{Pb}_3(\text{PO}_4)_2$$

This is then converted to a black precipitate of lead sulfide by the addition of ammonium sulfide.

$$\text{Pb}_3(\text{PO}_4)_2 \xrightarrow{(\text{NH}_4)_2\text{S}} \text{PbS}$$

Reagents

Cyclic 3',5'-AMP, 20 mM
Cyclic 3',5'-GMP, 20 mM
Tris-maleate buffer, 0.10 M, pH 7.0
MgSO$_4$, 1.0 M
Lead nitrate, 30 mM
5% Ammonium sulfide
Bacterial alkaline phosphatase (10 units/mg protein, commercially available from Sigma Chemical Co.)
Cyclic nucleotide phosphodiesterase in appropriate dilution

Definition of Unit. The unit of diesterase activity is defined as the amount of enzyme which converts 1 μmole of cyclic 3',5'-AMP to 5'-AMP per minute at 35° in a reaction mixture (0.2 ml) containing 0.1 M Tris·HCl buffer, pH 8.1, 10 mM MgSO$_4$ and 1.0 mM cyclic 3',5'-AMP.

Procedure for Electrophoresis

Standard polyacrylamide disc gel electrophoresis in 7.5% acrylamide gel is performed according to the method of Davis[6] at pH 9.0. Electrophoresis can also be performed at pH 7.0 and in various concentrations of polyacrylamide gel. Approximately 0.003 unit of cyclic nucleotide phosphodiesterase in 200 μl of 20% sucrose is layered on the stacking gel, and a constant current of 1.5 mA/gel is applied until the tracking dye (bromophenol blue) is about 2 mm from the end of the gel.

Procedure for Activity Stain

Upon completion of electrophoresis, the gels are removed from the glass tubes and the gel below the level of the dye marker is cut off and discarded. The gels are washed thoroughly in 0.1 M Tris-maleate buffer,

[4] T. R. Santa, W. D. Woods, M. B. Waitzman, and G. H. Bourne, *Histochemistry* **7**, 177 (1966).
[5] N. T. Florendo, R. J. Barnett, and P. Greengard, *Science* **173**, 745 (1971).
[6] B. J. Davis, "Disc Electrophoresis," *Ann. N.Y. Acad. Sci.* **121**, 404 (1964).

pH 7.0, and placed in a reaction mixture (4 ml) containing 12 μmoles of lead nitrate, 10 μmoles of $MgSO_4$, 320 μmoles of Tris-maleate buffer, pH 7.0, 2.5 units of alkaline phosphatase and 8 μmoles of cyclic 3',5'-AMP. Incubation is carried out at 25° for 30 minutes; during this time a white band of precipitate appears in the gel at the site of diesterase activity. The gels are washed for 1 hour in running water, then exposed to 5% ammonium sulfide for 2 minutes and a black band replaces the white band of precipitate. Excess ammonium sulfide is then removed, and the gel is washed in water for another 5 minutes and stored in water.

The black band within the gel matrix is stable for approximately 3 days at room temperature and for 4–5 days at 4°.

Permanent records of the activity stain can be obtained by scanning with a Gilford density scanning and recording device attached to a Beckman DU spectrophotometer set at 500 nm.

Specificity. The cyclic nucleotide substrates that can be used in the activity stain depend upon the substrate specificity of the particular cyclic nucleotide phosphodiesterase. The black bands of PbS do not appear in the gel when either cyclic 2',3'-AMP or poly(A) are added instead of the cyclic 3',5'-purine nucleotides, when alkaline phosphatase is omitted from the reaction mixture, or when 5'-AMP is substituted for cyclic 3',5'-AMP in the absence of alkaline phosphatase. Thus, the bands that develop are due to the cyclic nucleotide phosphodiesterase present in the gel, not to 5'-nucleotidase or nonspecific alkaline phosphatase activity. With the purified enzyme from bovine cardiac muscle the activity bands appear only after incubation with cyclic purine nucleotides and are in identical positions when either cyclic 3',5'-AMP or cyclic 3',5'-GMP are used as substrates.

Sensitivity. This procedure can be used to detect phosphodiesterase after electrophoresis of either crude homogenates or purified preparations of enzyme and is capable of detecting 0.0003 unit of diesterase activity. Sensitivity can be enhanced by change to increasing either the concentration of the cyclic nucleotide substrate or the concentration of alkaline phosphatase used during the first incubation.

Application. Using this method, multiple electrophoretically distinct bands of cyclic nucleotide phosphodiesterase activity can be detected in extracts from various tissues and in purified preparations of cyclic nucleotide phosphodiesterase from bovine cardiac muscle.[1,7] The activity stain may also be performed on a longitudinal slice of a preparative disc gel facilitating localization of the appropriate segment of the remaining gel for homogenization and elution of phosphodiesterase activity.

[7] E. N. Goren and O. M. Rosen, *Arch. Biochem. Biophys.* **153**, 384 (1972).

[39] Purification and Characterization of a Protein Activator of Cyclic Nucleotide Phosphodiesterase from Bovine Brain[1]

By YING MING LIN, YUNG PIN LIU, and WAI YIU CHEUNG

Cyclic nucleotide phosphodiesterase of bovine brain and other mammalian tissues loses activity upon purification under certain conditions, owing to the removal of a protein activator from the enzyme during purification.[2,3] The enzyme, purified through diethylaminoethyl cellulose chromatography and referred to as the DEAE enzyme, is partially inactive; its activity is fully restored by the addition of an exogenous activator. The activator has no effect on the activity of phosphodiesterase in a crude homogenate, which possesses an excess of activator.[4,5]

Reagents. Prepare all reagents in glass-distilled water, and store them at 0 to $-20°$.

Tris·HCl, 0.4 M, pH 8.0, at 22° (buffer A)
Tris·HCl, 20 mM, and MgSO$_4$, 1 mM, pH 7.5, at 4° (buffer B)
MnCl$_2$, 1 mM
(NH$_4$)$_2$SO$_4$, 0.15 M in buffer B (buffer C)
(NH$_4$)$_2$SO$_4$, 0.3 M in buffer B (buffer D).
Diethylaminoethyl cellulose (Sigma, coarse mesh with a capacity of 0.9 mEq/g). Wash successively with 0.5 N NaOH, H$_2$O, 0.5 N HCl, and buffer B until the final pH is 7.5.
Bovine brain phosphodiesterase, partially purified through DEAE-cellulose chromatography (DEAE enzyme, see this volume [33])
Snake venom (*Crotalus atrox* for 5'-nucleotidase activity), 1 mg/ml in buffer B
TCA (trichloroacetic acid), 55%
Ammonium molybdate, 2.5% in 5 N H$_2$SO$_4$
Fiske–SubbaRow reagent. Mix 2 g of 1-amino-2-naphthol-4-sulfonic acid, 12 g of sodium bisulfite, and 12 g of sodium sulfite. Dissolve 2.5 g of this mixture in 100 ml of water.

---

[1] We are grateful to Dr. Andrew Kang for amino acid analysis and to Fineberg Packing Company, Memphis, Tennessee, for a generous supply of bovine brains. This work has been supported by ALSAC, by Grants NS-08059, CA 13537, and CA-08480, and by a Career Development Award I-K4-NS-42576 (W.Y.C.) from the U.S. Public Health Service.
[2] W. Y. Cheung and Y. M. Lin, this volume [33].
[3] W. Y. Cheung, *Biochim. Biophys. Acta* **191**, 303 (1969).
[4] W. Y. Cheung, *Biochem. Biophys. Res. Commun.* **38**, 533 (1970).
[5] W. Y. Cheung, *J. Biol. Chem.* **246**, 2859 (1971).

Cyclic AMP (cAMP), 20 mM, adjust to pH 8 with 1 N NaOH
Acrylamide, 30 g/100 ml (solution A)
N,N'-Methylenebisacrylamide, 0.8 g/100 ml (solution B)
Gel buffer (Tris, 2.06 g/100 ml, glycine, 1.39 g/100 ml, N,N,N',N'-tetramethylenediamine, 0.15 ml/100 ml) (solution C)
Ammonium persulfate, 2.4 g/100 ml (solution D)
Upper buffer, pH 8.9 at 22° (Tris 5.16 g/liter; glycine 3.48 g/liter)
Lower buffer, pH 8.1 at 22° (Tris 14.5 g/liter; 1 N HCl, 60 ml/liter)
Elution buffer (same as lower buffer)
Membrane holder buffer (5 times concentration of lower buffer)

Assay Method

Principle. The bovine brain DEAE enzyme has been deprived of most of its activator, and its activity is enhanced by an exogenous activator. The activator is assayed by its ability to stimulate the DEAE enzyme as indicated below:

Activity of the activator
= (phosphodiesterase activity of DEAE enzyme plus activator)
− (phosphodiesterase activity of DEAE enzyme)
− (phosphodiesterase activity in the activator preparation)

For activator preparations that do not have phosphodiesterase activity, omit the last term for the above equation. Figure 1 shows the activity of the DEAE enzyme as a function of activator concentration. The activator is estimated using that part of the curve around half-maximal activation.

Unit of Activity. One unit of activator is defined as the amount of protein which, under standard conditions, stimulates hydrolysis of 1 µmole of cAMP per minute above the basal level of the DEAE enzyme activity and any phosphodiesterase activity which may be present in the activator preparation. Specific activity is expressed as units per milligram of protein.

Procedure. Phosphodiesterase activity of the DEAE enzyme may be determined by one of several methods available.[6-8] The procedure described below employs 5'-nucleotidase of snake venom (*Crotalus atrox*) as an auxiliary enzyme and is essentially that of Butcher and Sutherland.[9] The venom contains active proteolytic enzymes which stimulate the DEAE enzyme.[3] To prevent the proteolytic enzymes from acting on

[6] R. W. Butcher, this volume [32].
[7] W. J. Thompson, G. Brooker, and M. Appleman, this volume [30].
[8] W. Y. Cheung, this volume [31].
[9] R. W. Butcher and E. W. Sutherland, *J. Biol. Chem.* **237**, 1244 (1962).

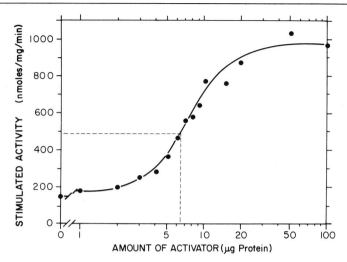

Fig. 1. Effect of activator concentration on phosphodiesterase activity. The stimulatory activity was assayed as described in the text. A boiled pH 5.9 supernatant fluid from step 2 was used as a source of the activator. Purified phosphodiesterase was 50 μg, and the amount of activator varied from 1 to 100 μg of protein. From W. Y. Cheung, *J. Biol. Chem.* **246**, 2859 (1971).

phosphodiesterase, separate the incubation into two stages and add venom to the reaction mixture at the end of the first stage of incubation. Prepare the first stage reaction mixture which contains 0.05 ml of buffer A, 0.05 ml of $MnCl_2$, about 50 μg of DEAE enzyme, an appropriate amount of the activator preparation, and water to bring the volume to 0.45 ml. Start the reaction by adding 0.05 ml cAMP. After a 10-minute incubation at 30°, stop the reaction by transferring the test tube into a boiling water bath for 2 minutes. Cool the tube to 30°, add 0.05 ml of snake venom, and incubate for another 10 minutes. Terminate the reaction with 0.05 ml of TCA. Add 0.15 ml of ammonium molybdate and 0.7 ml of H_2O. Remove denatured proteins by a low-speed centrifugation. Decant the clear supernatant into a clean tube containing 0.05 ml of Fiske–SubbaRow reagent.[10] Measure the blue color at 660 nm after 10 minutes at room temperature.

Purification Procedure

Transport fresh bovine brains from a local slaughterhouse to the laboratory in packed ice. Remove brain stems and large blood clots. Use the

[10] C. H. Fiske and Y. SubbaRow, *J. Biol. Chem.* **66**, 375 (1925).

cleaned tissue immediately or store it at $-20°$; no appreciable difference in activity has been noted. When frozen brains are used, first thaw them at room temperature. Carry out all purification steps at 0–4° unless stated otherwise.

Step 1. Homogenization. Use 6 kg of brain tissue in each preparation, and extract it batchwise. Homogenize 500 g of tissue in 1.5 liter of buffer B in a Waring commercial blender with a capacity of 1 gallon (Waring Products). Set the speed at low for 0.5 minute and then at medium for 1 minute. Remove tissue debris by filtering the homogenate through a double layer of cheesecloth. Centrifuge the homogenate at 12,000 g for 30 minutes and save the supernatant.

Step 2. Heat Treatment. Heat 1.5-liter fraction of the supernatant solution in a 4-liter stainless steel beaker immersed in a large boiling water bath. Bring the temperature of the solution quickly to 95°. After 5 minutes at 95°, transfer the beaker to an ice slurry. Stir the solution throughout the process of heating and cooling. Remove coagulated proteins by centrifugation (12,000 g, 20 minutes), and save the supernatant fluid.

Step 3. Diethylaminoethyl Cellulose Chromatography. Pack DEAE-cellulose into a column (5 × 56 cm) at a hydrostatic pressure of 200 cm. Wash the column with buffer B until the bed height is stabilized. After each run, regenerate the column by passing through it 4–6 liters of 0.1 N HCl, and then equilibrate it with buffer B.

Apply the solution from step 2 directly to the column at a rate of about 1 liter per hour. Discard the solution that passes through the column at this stage. Wash the column with 4 liters of buffer C. This washing contains the bulk of protein without activity. Elute the column with a linear gradient generated by 2 liters of buffer C and 2 liters of buffer D. At the end of the gradient, elute the column further with 1 liter of buffer D. Collect fractions of 24 ml at a rate of 3 ml per minute. The activator peak comes out at about 0.3 M $(NH_4)_2SO_4$ (Fig. 2). Combine active fractions and precipitate the protein by adding granular $(NH_4)_2SO_4$ to 100% saturation (about 700 g/liter). After stirring for 2 hours, collect the protein by centrifugation at 21,000 g for 30 minutes. Dissolve the sediment in a minimal volume of buffer B and dialyze it against the same buffer with two changes. Remove any turbidity by centrifugation and save the clear supernatant fluid.

Step 4. Preparative Acrylamide Gel Electrophoresis. EQUIPMENT. Buchler's "Poly-prep" electrophoresis apparatus has been employed.

PREPARATION OF THE GEL COLUMN. To make a 15% gel with respect to acrylamide, measure 120 ml of solution A, 40 ml of solution B, 60 ml of solution C, and 20 ml of solution D successively into a 500-ml

Erlenmeyer flask and remove dissolved air under negative pressure. Introduce the solution into the column of the electrophoresis apparatus, and cover the solution with a layer of water about 0.5 cm deep. Allow the gel to polymerize in 1 hour. Maintain the temperature around 4° by circulating the refrigerated coolant through the column jacket.

PREPARATION OF THE SAMPLE. Dialyze the concentrated protein solution from step 5 against 2 liters of 1:10 dilution of upper buffer for 6 hours. Add granular sucrose (100 mg/ml) to increase the density of the solution. Keep the sample volume less than 8 ml and the amount of protein less than 25 mg.

ELECTROPHORESIS. Do a preliminary electrophoresis at a current of 25 mA for 10 hours to remove any residual persulfate embedded in the gel. Carefully layer the sample onto the top of the gel which is immersed

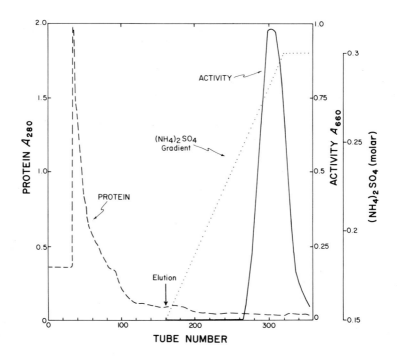

FIG. 2. DEAE-cellulose column chromatography. The boiled supernatant fluid from 6 kg of bovine brain (about 14 liters) was applied to a column (5 × 56 cm). The column was washed with 4 liters of buffer C and then eluted with a gradient generated by 21 liters of buffer C and 2 liters of buffer D. At the end of the gradient, the column was eluted further with 1 liter of buffer D. Fractions of 24 ml were collected. The solid line shows the activity, the dashed line the protein, and the dotted line the gradient of $(NH_4)_2SO_4$ concentration. Tubes 281–338 were pooled and concentrated by $(NH_4)_2SO_4$ precipitation as described in the text.

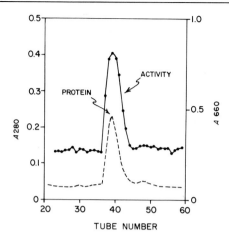

Fig. 3. Preparative polyacrylamide gel electrophoresis. Eleven milligrams of a concentrated sample from the DEAE-cellulose column was applied to the gel after a preliminary electrophoresis of 10 hours. Electrophoresis was started at a current of 25 mA. The current was increased to 40 mA after the protein had entered the gel in about 1 hour. Fractions of 10 ml were collected 5 hours after electrophoresis had started. No protein was eluted prior to this time since the tracking dye, bromophenol blue, came out after 5 hours. The solid line shows the activity profile (A_{660}); and the dashed line, the protein profile (A_{280}). Fractions 37–41 were pooled and concentrated.

in the upper buffer. Start electrophoresis with a constant current of 25 mA at 200 V. After the protein has entered the gel in about 1 hour, increase the current to 40 mA, and maintain it constant throughout the course of electrophoresis (about 15 hours). The voltage increases from the initial 200 V to about 450 V at the end of electrophoresis. Collect fractions of about 10 ml at a rate of 1 ml per minute maintained by a peristaltic pump. Figure 3 shows the activity and protein profiles of electrophoresis. Examine each fraction from the activity peak for purity by analytical acrylamide gel electrophoresis. Combine the fractions with good purity as revealed by Coomassie blue staining of the protein. Concentrate the solution by ultrafiltration through a UM2 DIAFLO membrane at a pressure of 50 psi. Dialyze the concentrated solution against buffer B.

The overall purification is 1700-fold with a yield of 7% (Table I).

Properties

Purity. The activator obtained at the final stage of purification appears to be homogeneous with respect to the following criteria: (a) ana-

TABLE I
PURIFICATION OF PHOSPHODIESTERASE ACTIVATOR FROM BOVINE BRAIN[a]

Fraction	Total protein[b] (mg)	Total activity (units)	Specific activity (units/mg)	Yield (%)	Purification (x-fold)
Homogenate	102,000	22,300	0.22	100	1
Supernatant	21,300	7,670	0.36	34.4	1.6
Boiled supernatant	1,100	4,030	3.67	18.1	16.7
DEAE-Cellulose chromatography	17.5	3,500	200	15.7	909
Preparative electrophoresis	4.2	1,590	383	7.1	1,740

[a] The data presented in the table are based on 1 kg of brain tissue.

[b] Protein in the homogenate was determined by the biuret reagent and in subsequent steps by the spectrophotometric technique of O. Warburg and W. Christian, *Biochem. Z.* **310,** 384 (1941).

lytical polyacrylamide gel electrophoresis in the presence or absence of sodium dodecyl sulfate; (b) polyacrylamide gel isoelectric focusing; (c) gel filtration in Sephadex G-100 chromatography; (d) sucrose density gradient centrifugation; (e) sedimentation velocity; and (f) sedimentation equilibrium.

Nature of Activator. Incubation of the activator with trypsin, but not with DNase or RNase, resulted in a loss of activity.[5] Staining of the activator after analytical gel electrophoresis with the periodate–Schiff reagent[11] did not reveal the presence of glycoprotein. Staining with pyronine Y,[12] a test for ribonucleic acid, was also negative.

Stability. The activator is remarkably stable. The activity survived exposure to pH 1.7, boiling, and boiling at pH 1.7. Exposure to pH 12.3 markedly reduced its potency, and boiling at this pH obliterated all activity. Incubation with 8 M urea did not affect its potency.[5] A heat-stable protein activator for phosphodiesterase has been reported in beef heart,[13,14] and in rat brain.[15]

Walsh *et al.*[16] described a heat-stable protein inhibitor of protein kinase with several properties similar to the activator of phosphodiesterase.

[11] J. T. Clarke, *Ann. N.Y. Acad. Sci.* **121,** 428 (1964).
[12] K. Marcinka, *Anal. Biochem.* **50,** 304 (1972).
[13] E. N. Goren and O. M. Rosen, *Arch. Biochem. Biophys.* **142,** 720 (1971).
[14] T. S. Teo, T. H. Wang, and J. H. Wang, *J. Biol. Chem.* **248,** 588 (1973).
[15] S. Kakiuchi and R. Yamazaki, *Biochem. Biophys. Res. Commun.* **41,** 1104 (1970).
[16] D. A. Walsh, C. D. Ashby, C. Gonzalez, D. Catkins, E. H. Fischer, and E. G. Krebs, *J. Biol. Chem.* **246,** 1977 (1971).

TABLE II
Determination of the Molecular Weight of Phosphodiesterase Activator from Bovine Brain

Method of determination	MW
Sedimentation velocity and diffusion	14,500[a]
Sedimentation equilibrium	15,000[a]
Sucrose density gradient centrifugation[b]	
Horseradish peroxidase as marker	15,000
Soybean trypsin inhibitor as marker	16,000
SDS polyacrylamide electrophoresis[c]	15,000
Amino acid analysis[d]	18,920
Gel filtration chromatography[e]	31,000

[a] Partial specific volume of the activator was calculated to be 0.72 cm³/g from the amino acid analysis. E. J. Cohn and J. T. Edsall, "Proteins, Amino Acids and Peptides as Ions and Dipolar Ions," p. 374. Reinhold, New York, 1943.
[b] R. G. Martin and B. N. Ames, *J. Biol. Chem.* **236**, 1372 (1961).
[c] K. Weber and M. Osborn, *J. Biol. Chem.* **244**, 4406 (1969).
[d] S. Moore and W. H. Stein, this series, Vol. 6, [117].
[e] P. Andrews, *Biochem. J.* **91**, 222 (1964).

Assays of exchanged purified samples did not show that the two entities substituted for each other.[17]

Molecular Weight. Several methods have been used to determine the molecular weight of the activator. Table II shows that value obtained from several methods is around 15,000. The molecular weight calculated from the amino acid composition is 18,920. For reasons not yet known, gel filtration chromatography with Sephadex G-100 gave a molecular weight of 31,000.

Other Physical Constants. The partial specific volume, $\bar{v}$, calculated from the data of amino acid analysis is 0.72 cm³/g. The sedimentation coefficient, $s_{20,w}$, determined at a protein concentration of 1 mg/ml from sedimentation velocity, is 1.85 S. The diffusion coefficient, $D_{20,w}$, obtained by the synthetic boundary method at 1 mg protein per milliliter, is 1.09×10^{-6} cm^{-2}/sec. The frictional ratio, $f:f_0$, is calculated to be 1.20, using a molecular weight of 15,000 obtained from sedimentation equilibrium. This ratio suggests that the activator is a globular protein. The isoelectric point, pI, was determined to be 4.3 by isoelectric focusing in polyacrylamide gel containing 8 M urea. Table III presents these constants.

Amino Acid Composition and End Groups. The amino acid composition of the purified activator presented in Table IV is based on 1 mole

[17] D. A. Walsh and C. D. Ashby, personal communication.

TABLE III
Physical Parameters of Phosphodiesterase Activator from Bovine Brain

Determination	Method used	Value
$s_{20,w}$ (sedimentation coefficient)	Sedimentation velocity[a]	1.85 S
$D_{20,w}$ (diffusion coefficient)	Analytical ultracentrifugation	1.09×10^{-6} cm^2/sec
$\bar{v}$ (partial specific volume)	Calculated from amino acid composition[b]	0.72 cm^2/g
$f:f_0$ (fractional ratio)	Calculated from $D_{20,w}$ and MW[c]	1.20
pI (isoelectric point)	Isoelectric focusing[d]	4.3

[a] H. K. Schachman, this series, Vol. 4 [2].
[b] E. J. Cohn and J. T. Edsall. "Proteins, Amino Acids and Peptides as Ions and Dipolar Ions," p. 374. Reinhold, New York, 1943.
[c] Using a molecular weight (MW) of 15,000 from sedimentation equilibrium. C. T. Svedberg and K. O. Pedersen, "The Ultracentrifuge," p. 38. Oxford Univ. Press (Clarendon), London and New York (1940).
[d] N. Catsimpoolas, Anal. Biochem. **26**, 480 (1968).

of histidine per mole of activator. One striking feature of the data is the preponderance of acidic amino acids, which make up one-third of the total amino acids. Another feature is the absence of tryptophan and half-cystine. The NH$_2$-terminal residues was identified by the method of dansylation[18] as valine.

The absence of half-cystine in amino acid analysis indicates that the activator does not possess a disulfide bridge. Because the molecular weight obtained from SDS polyacrylamide gel electrophoresis is the same as that obtained from analytical ultracentrifugation and because valine is the only NH$_2$-terminal group, the activator probably possesses only one polypeptide. A frictional ratio of 1.1 to 1.2 indicates a globular protein. The prevalence of aspartic and glutamic acids accounts for the isoelectric point of 4.3.[19]

Specificity and Distribution. The requirement of phosphodiesterase for the activator appears specific; none of a variety of proteins tested affected the activity of the DEAE enzyme. The proteins tested included cytochrome *c*, RNase, DNase, myoglobin, trypsin inhibitor, horseradish peroxidase, ovalbumin, bovine serum albumin, and thyroglobulin.[5]

The activator, however, lacks tissue specificity. Phosphodiesterase and its activator prepared from human brain, pork brain, and bovine brain and heart cross-activated each other effectively. Homogenates of various

[18] W. R. Gray, this series, Vol. 25 [8].
[19] Y. M. Lin, Y. P. Liu, and W. Y. Cheung, *J. Biol. Chem.* (in press) 1974.

TABLE IV
AMINO ACID COMPOSITION[a] OF PHOSPHODIESTERASE
ACTIVATOR FROM BOVINE BRAIN

Amino acid	Molar ratio
Aspartic acid	24
Threonine[b]	14
Serine[b]	5
Glutamic acid	30
Proline	2
Glycine	13
Alanine	13
Half-cystine[c]	0
Valine	8
Methionine	11
Isoleucine	9
Leucine	11
Tyrosine	2
Phenylalanine	10
Lysine	8
Histidine	1
Arginine	7
Tryptophan[d]	0

[a] Amino acid analyses were performed according to S. Moore and W. H. Stein, this series, Vol. 6 [117].

[b] Values were extrapolated to zero time, based upon data obtained after 24, 48, and 72 hours of hydrolysis assuming first-order kinetics.

[c] Determined as cysteic acid following performic acid oxidation (C. H. W. Hirs, this series, Vol. 11 [6]).

[d] Determined after alkaline hydrolysis [J. L. Stokes, M. Gunness, I. M. Dwyer, and M. C. Caswell, *J. Biol. Chem.* **160**, 35 (1945).]

rabbit tissues effectively augmented the activity of the DEAE enzyme from bovine brain. These tissues were: brain, testis, adrenal, spleen, liver, thyroid, kidney, lung, heart, skeletal muscle, duodenum, thymus, adipose, and blood. Homogenates of rat tissues also stimulated the DEAE enzyme from bovine brain; they were brain, blood, skeletal muscle, adrenal, heart, testis, liver, fat pad, thymus and kidney.[20]

Although the distribution of the activities of the activator and phosphodiesterase paralleled one another subcellularly, the ratios of their activities varied greatly from one tissue to another; for example, the activator was high in both rat brain and testis but low in liver, whereas the enzyme was high in brain and low in testis and liver.[21]

[20] W. Y. Cheung, unpublished experiment.
[21] J. A. Smoake, S. Y. Song, and W. Y. Cheung, *Biochim. Biophys. Acta* **341**, 402 (1974).

Development of Activator and Phosphodiesterase. The activator was present in excess of phosphodiesterase in brain, liver, and testis from the embryo through the adult. Although the activity of the activator did not change in the brain or liver from 8 days before birth to adulthood, phosphodiesterase activity increased 20-fold in brain and 2-fold in liver over the same period. In the testis, the activity of phosphodiesterase decreased 4-fold from birth to 25 days of age while that of the activator was unaltered. These results indicate that while the activity of phosphodiesterase may vary during development, that of the activator remains relatively constant, and that the activator and phosphodiesterase are probably under separate genetic control.[21]

Effect of Activator on Kinetics of Phosphodiesterase. The activation of phosphodiesterase was dependent on the concentration of the activator, not on the time of its exposure to the enzyme, indicating a stoichiometric process. This is in contrast to the action of trypsin, which was time-dependent and appeared to be a catalytic process. Although activation by the two agents appeared different mechanistically, a stoichiometric versus a catalytic process, the result was comparable: an accelerated rate of hydrolysis of cAMP catalyzed by phosphodiesterase.

The activator decreased the K_m of the DEAE enzyme of bovine brain for cAMP and increased the V_{max}. Likewise, trypsin decreased the K_m and increased the V_{max}. Either of the two agents caused maximal stimulation, and no synergism was noted.[5] The activator increased the rate of hydrolysis of cAMP and cGMP at both millimolar and micromolar concentrations.[19]

The activator stimulated the activity of the purified, but not the crude, enzyme. Likewise, trypsin stimulated the activity of the purified, but not the crude, enzyme. Although a prolonged incubation of phosphodiesterase with trypsin led to a loss of phosphodiesterase activity, no such loss was seen with the activator, which exhibited no proteolytic activity.

Theophylline, an inhibitor of phosphodiesterase, has been used extensively to potentiate the effects of hormones suspected to mediate their effects through the adenylate cyclase and cAMP system. Although the activator stimulates phosphodiesterase, it does not counteract the inhibition of phosphodiesterase by theophylline.[22]

Mode of Action of Activator. Atomic absorption spectrophotometry revealed that the activator binds Ca^{2+} specifically. A Scatchard plot of data obtained from equilibrium binding indicated 4 Ca^{2+} binding sites per molecule of activator; their dissociation constants ranged from 4 to 18 μM. Stimulation of phosphodiesterase by the activator required a low con-

[22] W. Y. Cheung, *in* "Advances in Psychopharmacology" (P. Greengard and E. Costa, eds.), Vol. 3, p. 51. Raven, New York, 1970.

centration of Ca^{2+}, and chelation of Ca^{2+} by EGTA rendered the activator inactive. The concentration of Ca^{2+} needed to give half-maximum activation of phosphodiesterase was approximately 3 μM. Ca^{2+} alone did not stimulate phosphodiesterase. Since the activator binds Ca^{2+}, the active form of the activator appears to be a Ca^{2+}-activator complex.[19]

The activator is known to be dissociated from phosphodiesterase by a salt gradient on a DEAE–cellulose column. Dissociation of the two proteins reduced phosphodiesterase activity to its basal level. EGTA, which chelated Ca^{2+} in the activator, also reduced phosphodiesterase activity to its basal level, indicating that EGTA probably caused the dissociation of the two proteins. This notion was supported by a filtration experiment in which a mixture of phosphodiesterase and activator was passed through a Sephadex G-200 column that had been equilibrated with EGTA. Under this condition, the two proteins were separated and the activator was eluted after the enzyme. On the other hand, in a separate column which had been equilibrated with Ca^{2+} instead of EGTA, the two proteins were eluted together, indicating that Ca promoted the formation of the enzyme-activator complex.[23]

Since the activator is usually in excess of the enzyme, phosphodiesterase activity may be regulated by the cellular flux of Ca^{2+}, as illustrated below:

$$Ca^{2+} + \text{activator} \rightleftharpoons Ca^{2+}-\text{activator}$$
$$[\text{Enzyme}]_{\text{inactive}} + Ca^{2+}-\text{activator} \rightleftharpoons [\text{enzyme}-Ca^{2+}-\text{activator}]_{\text{active}}$$

Although the notion depicted above appears physiologically significant, it awaits confirmation by *in vivo* experiments.

[23] W. Y. Cheung, Y. M. Lin, Y. P. Liu, and J. A. Smoake, in "Cyclic Nucleotides in Disease" (B. Weiss, ed.), University Park Press, Baltimore (in press).

[40] Techniques for the Formation, Partial Purification, and Assay of a Cyclic AMP Inhibitor

By Ferid Murad

The intracellular concentration of adenosine 3′,5′-monophosphate (cAMP) may be modified by altering its rate of formation, breakdown, and/or excretion from cells. With preparations from numerous tissues, a variety of hormones have been found to enhance the activity of adenylate cyclase and the formation of cAMP from ATP. This is discussed elsewhere in this volume.

The action or effects of cAMP in tissues may be determined as much by the concentrations of other substances, e.g., ions or other nucleotides, as they are by the concentration of cAMP itself. It has been reported that tissue extracts[1,2] and heated extracts of incubations of adenylate cyclase preparations[2-5] contain materials that inhibit the effect of cAMP in the phosphorylase activation assay system.

This section describes the methods for the formation, partial purification, and assay of a specific inhibitor of cAMP. The properties of this, as yet unidentified, inhibitory material in many ways resemble those of cAMP.[2,4] Studies in this laboratory support the view that this inhibitor is a nucleotide with cyclic 3′,5′-monophosphate and a structure resembling cAMP. Although several other possibilities exist, the simplest interpretation is that the inhibitor is a metabolite of cAMP. Others have also characterized a protein in tissue extracts which inhibits the activation of cAMP-dependent protein kinases.[6,7] This is described elsewhere in this volume. The properties of the latter are distinctly different from those of the inhibitor described here. Both materials are found naturally in tissue extracts and/or incubation mixtures and may prove to be useful tools in future studies with cyclic nucleotides. The physiological significance of either of these inhibitory materials, however, is unknown.

Formation of the Inhibitor in Cell-Free Incubations

Homogenates and washed particulate preparations of adenylate cyclase from dog liver and from dog and rat heart are prepared in 0.25 M sucrose as previously described.[2,3] Enzyme preparations from other sources, such as rat kidney[8] and testis,[5] have also been used successfully

Enzyme preparations are incubated in duplicate in the presence of 40 mM Tris buffer (pH 7.4), 6.6 mM MgSO$_4$, 4 mM ATP, 16 mM caffeine, with and without hormones and/or adrenergic blocking agents as previ-

[1] R. W. Butcher, E. W. Sutherland, and T. W. Rall, *Pharmacologist* **2**, 66 (1960).
[2] F. Murad, T. W. Rall, and M. Vaughan, *Biochim. Biophys. Acta* **192**, 430 (1969).
[3] F. Murad, T. W. Rall, and E. W. Sutherland, *J. Biol. Chem.* **237**, 1233 (1962).
[4] F. Murad, *Fed. Proc., Fed. Amer. Soc. Exp. Biol.* **24**, 150 (1965).
[5] F. Murad, B. S. Strauch, and M. Vaughan, *Biochim. Biophys. Acta* **177**, 591 (1969).
[6] J. B. Posner, K. E. Hammermeister, G. E. Bratvold, and E. G. Krebs, *Biochemistry* **3**, 1040 (1964).
[7] D. A. Walsh, C. D. Ashby, C. Gonzalez, D. Calkins, E. H. Fischer, and E. G. Krebs, *J. Biol. Chem.* **246**, 1977 (1971).
[8] F. Murad, H. B. Brewer, and M. Vaughan, *Proc. Nat. Acad. Sci. U.S.* **65**, 446 (1970).

ously described.[2,3,5] Mixtures are incubated for various times at 30° with gentle shaking. Under these conditions of incubation the accumulation of cAMP and the inhibitor is linear for several minutes and plateaus at about 10–20 minutes. The incubations are terminated by heating at 100° for 3–5 minutes, and the mixtures are centrifuged. The supernatant fractions are assayed for apparent cAMP and inhibitory activity before and after purification as described below.

With most preparations the accumulation of the inhibitor in incubations parallels that of cAMP. Furthermore, the concentration of inhibitor in partially purified tissue extracts has been found to parallel that of cAMP; for example, tissue extracts of rat brain and lung have the highest concentrations of inhibitor and cAMP.[2] Agents which enhance cAMP formation in adenylate cyclase incubations such as NaF, catecholamines, glucagon, or other appropriate stimulatory hormones with other preparations also increase the accumulation of the inhibitor. The stimulatory effect of catecholamines can be prevented with β-adrenergic blocking agents. Conversely, choline esters decrease the formation of both cAMP and the inhibitor with enzyme preparations from dog heart.[2]

With the incubations performed as described the low speed (1000 g for 10 minutes) particulate fraction of heart homogenates contains most of the adenylate cyclase activity as well as the ability to synthesize the inhibitor.[2] Enzyme preparations which are quickly frozen in a dry ice-ethanol bath can be stored for several months at $-80°$ without apparent losses in adenylate cyclase or inhibitor-forming activities.

It has not been possible to substitute either UTP, GTP, or ITP for ATP in the incubations and observe the formation of the inhibitor or cAMP.[2]

The addition of exogenous cAMP to the incubations increases the accumulation of the inhibitor.[2] From preliminary studies the stimulatory effect of cAMP on the formation of the inhibitor appears to require both ATP and Mg^{2+}.[2] In addition the stimulatory effect of cAMP on formation of the inhibitor can be prevented in the presence of N^6-monobutyryl cAMP. The addition of N^6-$2'$-O-dibutyryl cAMP or $2'$-O-monobutyryl cAMP to incubations does not prevent the enhanced accumulation of the inhibitor with added cAMP. These studies and other properties of the inhibitor suggest that the inhibitor is a metabolite of cAMP.[2,4] Since the inhibitor as well as cAMP is hydrolyzed by the high K_m beef heart cyclic nucleotide phosphodiesterase, caffeine is included in the incubation to inhibit the latter and permit increased accumulation of the inhibitor.[2,4] With this phosphodiesterase preparation[9] the inhibitor is hydrolyzed at

[9] R. W. Butcher and E. W. Sutherland, *J. Biol. Chem.* **237**, 1244 (1962).

about two-thirds the rate of cAMP. The inhibitor and cAMP are resistant to hydrolysis by a variety of venoms and phosphatases as well as trypsin.[2] These studies distinguish this inhibitor from the protein inhibitor discussed elsewhere in this volume.

Purification of the Inhibitor from Incubation Mixtures and Tissue Extracts

Although the inhibitor and cAMP are stable to brief boiling at neutral pH, in the presence of 1 N HCl at 100° cAMP has a half-life of about 35 minutes compared to 7 minutes for the inhibitor.[2] With 0.05 N HCl at 100° the inhibitor is slowly degraded while cAMP is minimally altered.

The inhibitor and cAMP cochromatographed with the following supporting and solvent systems:

A. Column chromatography
  1. Dowex-2, chloride (AG 2-X8, 100–200 mesh) eluted with HCl (Fig. 1)

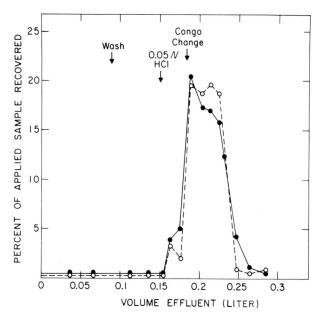

Fig. 1. Dowex-2 column chromatography. Ninety milliliters of a heated extract of an incubation mixture of dog myocardial particles which contained per milliliter 0.33 nmole of cAMP (●——●) and 6400 units of inhibitor (○---○) were applied to a Dowex-2 (Cl⁻) column (1 cm × 15 cm), washed with 70 ml of 10 mM Tris buffer (pH 7.4), and eluted with 50 mM HCl (as indicated by the arrows). From F. Murad, T. W. Rall, and M. Vaughan, *Biochim. Biophys. Acta* **192**, 430 (1969).

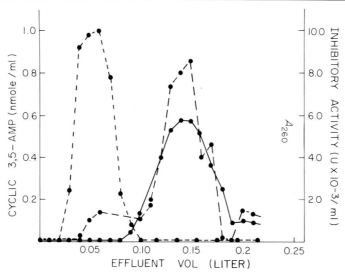

FIG. 2. Dowex-50 column chromatography. Selected Dowex-2 fractions from the experiment of Fig. 1 were applied to a Dowex-50 column (1 cm × 26 cm) which had been prewashed with 50 mM HCl. The sample was followed with 50 mM HCl. Aliquots of the effluent were used for A_{260} determinations (----) and after neutralization were assayed for cAMP (———) and inhibitory (— —) activity as described. The peak of A_{260} was identified as ADP with paper chromatography. The recoveries of cAMP and the inhibitor were 98 and 95%, respectively. From F. Murad, T. W. Rall, and M. Vaughan, *Biochim. Biophys. Acta* **192**, 430 (1969).

    2. Dowex-1, chloride (AG 1 X2, 50–100 mesh) eluted with HCl
    3. Dowex-50, H$^+$ form (AG 50W X8, 100–200 mesh) eluted with HCl (Fig. 2)
    4. DEAE-cellulose eluted with a gradient of ammonium acetate, pH 7.0 (Fig. 3)
  B. Cellulose thin-layer and paper chromatography
    1. Ethanol–0.5 M ammonium acetate (5:2, by volume)
    2. Ethanol–0.5 M ammonium acetate–1 M boric acid (5:2:0.2, by volume)
    3. Ethanol–acetone–acetic acid–5% NH$_4$OH–water (35:25:15:15:10, by volume)
    4. n-Butanol–acetone–acetic acid–5% NH$_4$OH–water (35:25:15:15:10, by volume)

Thin-layer or paper chromatography with isobutyric acid–1 M NH$_4$OH–0.1 M EDTA (100:60:1.6, by volume) or the upper phase of phenol–isopropanol–formic acid–water (40:5:5:50, by volume) or isopropanol–conc. NH$_4$OH–0.1 M boric acid (70:10:20, by volume) resulted in loss of the inhibitory activity. This is attributable to the lability of

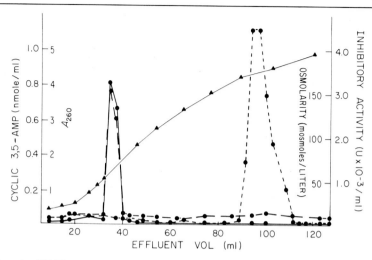

FIG. 3. DEAE-cellulose column chromatography. Selected Dowex-2 fractions were concentrated, neuralized and applied to a DEAE-cellulose column (0.5 cm × 15 cm) which had been prewashed with 10 mM ammonium acetate (pH 7.0). A gradient of ammonium acetate (pH 7.0) from 10 mM to 0.15 M (reservoir volume of 100 ml) was used to elute the column. Aliquots of the effluent were used for osmolarity (▲——▲) and A_{260} (- - -) determination and were assayed for cAMP (——) and inhibitory (— —) activity as described. The peak of A_{260} was identified as 5'-AMP, with thin-layer chromatography. The recoveries of cAMP and inhibitory activity were 105 and 108%, respectively. From F. Murad, T. W. Rall, and M. Vaughan, *Biochim. Biophys. Acta* **192**, 430 (1969).

the inhibitor. In addition, partially purified samples of inhibitor stored at −20° for several weeks frequently lose their inhibitory activity.

The inhibitor and cAMP were absorbed quantitatively onto Norit A in 0.1 N HCl. After the Norit is washed with distilled water and eluted with a mixture of 1% ammonia and 50% ethanol,[10] the inhibitor and cAMP are recovered equally in the eluate.[2]

The inhibitor remains in the supernatant fraction with cAMP[2] after mixtures are treated with $Ba(OH)_2$ and $ZnSO_4$ according to the method of Krishna et al.[11]

Assay of cAMP and the Inhibitor

The assay used[2] is a modification of that described by Butcher et al.[12] To disposable culture tubes, 10 mm × 75 mm, is added 0.1 ml of

[10] K. K. Tsuboi and T. D. Price, *Arch. Biochem. Biophys.* **81**, 223 (1959).
[11] G. Krishna, B. Weiss, and B. B. Brodie, *J. Pharmacol.* **163**, 379 (1968).
[12] R. W. Butcher, R. H. Ho, H. C. Meng, and E. W. Sutherland, *J. Biol. Chem.* **240**, 4515 (1965).

a mixture containing 1.64 μmoles of ATP, 2.05 μmoles of $MgSO_4$, 0.09 μmole of EDTA, and 1.09 μmoles of Tris buffer (pH 7.4). cAMP standards (1.25–15 pmoles), appropriately diluted unknowns or 10 mM Tris buffer (pH 7.4) are also added to make a total volume of 0.15 ml. For estimation of inhibitory activity, a known amount of commercial cAMP (generally 5 pmoles) is also included in some of the incubation mixtures. All standards and samples are assayed in duplicate. The vessels are chilled in an ice bath, and a mixture of 6–10 μl of a 20,000 g supernatant fraction from a dog liver or heart homogenate, 0.1–0.23 unit of purified inactive phosphorylase, 1 mg of reprecipitated oyster glycogen and 0.025 μmole of L-epinephrine in 0.1 ml 10 mM Tris buffer (pH 7.4) is added. Dephosphophosphorylase (inactive phosphorylase) is prepared from pig liver as described by Kakiuchi and Rall.[13]

After incubation for 20 minutes in an ice bath, the tubes are transferred to a 25° bath for 20 minutes. Phosphorylase activity is then assayed by adding 0.9 ml of a mixture containing 44.7 μmoles of glucose 1-phosphate, 4.53 mg of reprecipitated oyster glycogen, 0.114 mmole of NaF, and 2.0 μmoles of 5′-AMP at pH 6.1. The tubes are incubated at 37° for 30 minutes, after which glycogen content is determined by transferring 0.2 ml of the mixture to 1.0 ml of a solution containing 1.0–1.5 mg I_2, and 2–3 mg KI in 0.04 M HCl. These are diluted to 10 ml with distilled water, and A_{540} is determined. The absorbance of the mixtures which contain standard solutions of cAMP is used to obtain a standard curve with each assay. The apparent cAMP in unknowns in each assay can be estimated from such curves.

By assaying samples with and without added cAMP, it is also possible to determine the inhibitory activity in a sample. This is calculated as follows:

Inhibitory activity (units)

$$= 100 \left[1 - \frac{(\text{apparent cAMP with added cAMP}) - (\text{apparent cAMP})}{\text{amount of cAMP added}} \right]$$

Thus, 1 unit of inhibitory activity is defined as that amount of inhibitor which prevents the expression of 1% of the added cAMP in the assay. This method of quantifying the inhibitor provides a linear relationship between the amount of inhibitor in the assay and percentage inhibition of added cAMP between about 20 and 80% inhibition. The results from a typical assay for the apparent cAMP and inhibitory activity of a sample are illustrated in the table.

---

[13] S. Kakiuchi and T. W. Rall, *Mol. Pharmacol.* **4**, 367 (1968).

Quantification of Inhibitory Activity[a,b]

Volume of extract assayed (μl)	cAMP added (pmoles) (A)	Apparent cAMP		Inhibitory activity (units)	
		Without added cAMP (pmoles) (B)	With added cAMP (pmoles) (C)	Aliquot assayed $100 \times \left[1 - \dfrac{C - B}{A} \right]$	Per ml heated extract
25	—	8.5	—	—	—
25	5.0	—	9.4	82[c]	3280[c]
12.5	—	8.0	—	—	—
12.5	5.0	—	10.0	60	4800
8.3	—	6.0	—	—	—
8.3	2.5	—	7.7	32	3840
8.3	5.0	—	9.4	32	3840
8.3	7.5	—	10.6	39	4680

[a] From F. Murad, T. W. Rall, and M. Vaughan, *Biochim. Biophys. Acta* **192**, 430 (1969).

[b] A washed particle preparation from dog myocardium prepared in 0.25 M sucrose was incubated as described for 30 minutes. Each milliliter of incubation mixture contained particles derived from 46 mg of ventricular muscle. The heated extract from the incubation mixture was assayed at several concentrations in the absence and in the presence of different amounts of added commercial cyclic AMP (cAMP). The values reported are the means of duplicate assays.

[c] These values would not be considered usable, since inhibition in the assay was greater than 80%.

Several different commercial preparations of cAMP were found to contain no inhibitory activity either when assayed directly from the bottle, after incubation in the presence of denatured enzyme preparations, or after chromatography with Dowex-2 or paper as described above.

The inhibitor is equally effective when either dog liver or heart is used as a source of the supernatant fraction that supplies the phosphorylase activation system. When glycogen is added to liver extracts and samples are preincubated for 20 minutes at 2° in the assay, the apparent K_a for cAMP is about 20–30 nM, but when glycogen and a cold preincubation are omitted, the apparent K_a is about 100–200 nM. The use of glycogen and a cold preincubation does not increase the sensitivity of the assay to cAMP when heart extracts are used, but with liver extracts, this procedure increases the sensitivity of the system to both cAMP and the inhibitor.

The inhibitor is equally effective in the assay whether or not L-epinephrine is added. L-Epinephrine is routinely added to all assays because some supernatant fractions from homogenates of heart and liver contain small amounts of adenylate cyclase activity, and with these preparations L-epinephrine and glucagon produce a small stimulatory effect in the assay.

Inhibition could occur at one of several steps in the assay, e.g., activation of protein kinase, activation of phosphorylase kinase, activation of inactive phosphorylase, and/or phosphorylase activity. Samples containing inhibitor and cAMP at concentrations which markedly influenced the assay were found to have no effect on phosphorylase activity in the third step of the assay.[2] The last possibility, therefore, can be excluded. Furthermore, in recent preliminary studies the inhibitor prevented the activation of protein kinases from several tissues by cAMP. These studies suggest that phosphorylation of protein substrates by cAMP-dependent protein kinases would be a more convenient method of assaying the inhibitor. This latter method is described elsewhere in this volume [9] and [41].

With the phosphorylase activation assay method described, a variety of compounds have been found to mimic and/or inhibit the activity of cAMP.[2] Adenine, adenosine, 3′-AMP, and 5′-AMP are slightly inhibitory with preparations of supernatant fractions from heart or liver. Some of the nucleoside triphosphates produce slight inhibition with heart preparations. An inhibitory effect of glucose-1-P, glucose-6-P, and ADP has been previously reported.[1] Phenoxybenzamine (Dibenzyline) inhibits at relatively high concentrations. Some derivatives of cAMP and several other cyclic 3′,5′-nucleotides were also tested. With liver extracts, the dibutyryl, N^6-monobutyryl, and 2′-O-monobutyryl derivatives of cAMP had no inhibitory activity, but had 1.8, 48, and 6%, respectively, of the activity of cAMP. The cyclic 3′,5′-monophosphates of inosine, guanosine, cytidine, uridine, xanthosine, 2′-deoxythymidine, and 2′-deoxyadenosine (2–20 μM) likewise have no inhibitory activity but have 2.9, 0.6, 0.4, 0.3, 0.08, 0.02, and 0.1%, respectively, of the stimulatory activity of cAMP. Cyclic 2′,3′-AMP (200 μM) is without effect, stimulatory or inhibitory. Other agents that have been tested at a concentration of 20 μM and found to have no effect include glucosamine, glucosamine-6-P, N-acetylglucosamine, sorbitan-6-P, NAD^+, $NADP^+$, ergotamine, 1-(3,4-dichlorophenyl)-2-isopropylaminoethanol HCl (dichloroisoproterenol) and propranolol. Puromycin at 20 μM has no effect in the assay, but at a concentration of 300 μM it potentiates the effect of cAMP. At 1–5 mM it is stimulatory but has less than 0.01% of the activity of cAMP. The aminonucleoside derived from puromycin produces similar effects in the same range of concentrations even after passage through Dowex-1 resin at neutral pH.[2]

The mechanism of inhibition in the assay of some of these compounds is not known since it is possible that enzymatic alteration of some of these compounds may occur during incubation with the crude tissue fractions used in the assay.

Discussion

The experimental evidence for the role of adenylate cyclase and cAMP in mediating a variety of hormonal responses in many different tissues has been extensively reviewed. On the other hand, knowledge of other pathways of metabolism of cAMP and of conditions that may influence its activity in various enzymatic and physiological systems is relatively meager. Posner et al.[6] and Walsh et al.[7] described a heat-stable protein present in tissue extracts which inhibits the action of cAMP on protein kinases. This is described in this volume [48]. Unlike the compound described here, this material is destroyed with trypsin.

Cyclic nucleotide phosphodiesterase has been found in most tissues and in some cases has been studied fairly extensively. As mentioned earlier, the inhibition of this enzyme by methylxanthines, such as caffeine and theophylline, and the effects of these compounds in a variety of tissues has suggested that this is the predominant pathway for cAMP degradation. Other pathways for degradation of cAMP have not been ruled out, however. No direct evidence of another mechanism for degradation of cAMP has been obtained, although the data presented here and elsewhere[2,4] are compatible with such a possibility.

It is striking that the conditions which are optimal for accumulation of cAMP in cell-free preparations from heart and liver are very similar to the optimal conditions for accumulation of the inhibitor. Since the addition of cAMP increases the accumulation of inhibitor, it seems unlikely that the inhibitor is a direct product of the adenylate cyclase reaction. cAMP may increase the formation or decrease the degradation of the inhibitor. While the data to date cannot eliminate either of these possibilities, an attractive explanation for the observations with N^6-monobutyryl cAMP is that cAMP may be converted to the inhibitor, and the analog acts as a competitive substrate to give either no inhibitor formed or an ineffective inhibitor analog. This hypothesis is supported by the observation that the inhibitor is destroyed by cyclic nucleotide phosphodiesterase, which suggests that the inhibitor is a cyclic 3′,5′-nucleotide. The inability to separate the inhibitor and cAMP with numerous chromatographic procedures also suggests that the compounds are very similar. With other compounds that are inhibitory in the assay, relatively high concentrations are required, and these materials are removed with the purification procedures used. If the inhibitor is a metabolite of cAMP, it would appear that this is a result of a minor pathway in cAMP metabolism and that there is only a slight change in the structure of cAMP.

The physiological significance of the inhibitor remains to be determined. The inhibitor has been found in six different tissues, and its concentration parallels the concentration of cAMP.[2] Previous studies with fat cell[14] and toad bladder[15] preparations have shown that under some conditions the concentration of cAMP can be markedly elevated (2–20-fold) while its physiological or biochemical effect is prevented or attenuated. It is possible that the inhibitor described here could modify the physiological and biochemical responses produced by cAMP, hormones, and methylxanthines.

In intact tissue preparations, certain derivatives of cAMP are more effective than the parent nucleotide. To what extent this is due to more rapid entry into cells, less rapid destruction of the derivative as suggested by Posternak et al.,[16] or decreased cAMP hydrolysis[17] remains to be determined. If N^6-substituted analogs can depress accumulation of the inhibitor in tissues as N^6-monobutyryl cAMP does in the particulate systems, this action could also contribute to making it more effective than cAMP itself.

Acknowledgment

These studies were supported in part by USPHS Grant AM-15316 and a Virginia Heart Association grant. The author is the recipient of a USPHS Research Career Development Award AM-70456.

---

[14] F. Murad, V. Manganiello, and M. Vaughan, *J. Biol. Chem.* **245**, 3352 (1970).
[15] S. Mendoza, F. Murad, J. Handler, and J. Orloff, *Amer. J. Physiol.* **223**, 104 (1972).
[16] T. Posternak, E. W. Sutherland, and W. F. Henion, *Biochim. Biophys. Acta* **65**, 558 (1962).
[17] J. Heersche, S. A. Fedak, and G. D. Aurbach, *J. Biol. Chem.* **246**, 6770 (1971).

Section V

Cyclic Nucleotide-Dependent Protein Kinases and Binding Proteins

[41] Assay of Cyclic AMP-Dependent Protein Kinases

By JACKIE D. CORBIN *and* ERWIN M. REIMANN

$$\text{Protein} + \text{ATP} \rightarrow \text{protein-P} + \text{ADP}$$

Assay Method

Principle. Most tissues contain protein kinases which are stimulated several-fold by cyclic 3′,5′-adenosine monophosphate (cAMP) and catalyze the transfer of phosphate from ATP to several proteins.[1] Incorporation of phosphate into protein can be monitored by transfer of ^{32}P to protein from [γ-^{32}P]ATP in the presence of magnesium. The phosphorylated protein is separated from the labeled precursor by adsorption of the precipitated protein on filter paper disks and washing the disks as described by several investigators.[2-5] Several proteins may be used as substrates in the assay, including muscle phosphorylase kinase[6] and glycogen synthetase,[7,8] protamine,[9] adipose tissue lipase,[10] casein,[6] specific histones,[11] and histone mixtures. When either phosphorylase kinase, glycogen synthetase, or lipase are used, the phosphorylation causes enzymatic activity changes which may be developed into alternate methods for assay of protein kinases.[6-8,10] In practice, however, such assays are difficult to quantitate. A histone mixture is a suitable substrate for several reasons: (1) it is available from commercial sources, (2) there is little if any protein kinase contamination, (3) an adequate amount of phosphate is incorporated, (4) it is a stable and easily precipitable protein mixture, and (5) the degree of stimulation of histone phosphorylation by cAMP is usually relatively high.

---

[1] E. G. Krebs, *Curr. Top. Cell. Regul.* **5**, 99 (1972).
[2] F. J. Bollum, *J. Biol. Chem.* **234**, 2733 (1959).
[3] R. J. Mars and G. D. Novelli, *Arch. Biochem. Biophys.* **94**, 48 (1961).
[4] W. B. Wastila, J. T. Stull, S. E. Mayer, and D. A. Walsh, *J. Biol. Chem.* **246**, 1996 (1971).
[5] E. M. Reimann, D. A. Walsh, and E. G. Krebs, *J. Biol. Chem.* **246**, 1986 (1971).
[6] D. A. Walsh, J. P. Perkins, and E. G. Krebs, *J. Biol. Chem.* **243**, 3763 (1968).
[7] D. L. Friedman and J. Larner, *Biochemistry* **4**, 2261 (1965).
[8] T. R. Soderling, J. P. Hickenbottom, E. M. Reimann, F. L. Hunkeler, D. A. Walsh, and E. G. Krebs, *J. Biol. Chem.* **245**, 6317 (1970).
[9] B. Jergil and G. H. Dixon, *J. Biol. Chem.* **245**, 425 (1970).
[10] J. K. Huttunen, D. Steinberg, and S. E. Mayer, *Biochem. Biophys. Res. Commun.* **41**, 1350 (1970).
[11] T. A. Langan, *Science* **162**, 579 (1968).

Reagents

    Potassium phosphate, 50 mM, pH 6.8, with or without 6 μM cAMP
    Histone mixture (type II-A, Sigma Chemical Company), 30 mg/ml in H_2O
    Mg[γ-^{32}P]ATP (18 mM magnesium acetate, 1 mM [γ-^{32}P]ATP) in H_2O. Specific activity $\sim$100 cpm/pmole. The [γ-^{32}P]ATP can be obtained from commercial sources or prepared by modification of the method of Glynn and Chappell[12,13]
    Enzyme solution, 25–1000 units/ml
    Trichloroacetic acid, 10%
    Ethanol, 95%
    Ethyl ether

Procedure. Equal volumes of the potassium phosphate, histone, and Mg[^{32}P]ATP are combined. Two mixtures, one with and one without cAMP, are usually prepared. The mixtures are stable for several weeks with repeated freezing and thawing. To disposable glass test tubes (1.2 $\times$ 7.5 cm) are added 50 μl of the mixtures. The reactions are initiated by pipetting 20 μl of the appropriate buffer (blank) or enzyme solution into the mixtures, mixing, and placing the reaction tubes in a water bath at 30°. After incubation 10 minutes, 50 μl of the reaction mixtures are spotted either on filter paper squares (Whatman 31 ET, 2 $\times$ 2 cm) or disks (Whatman 3 MM, 2.3 cm in diameter) numbered with pencil lead. The disks are dropped immediately into a screen wire basket located in a beaker containing ice cold 10% trichloroacetic acid (see Fig. 1). The screen basket is constructed so that ample space is provided for a magnetic stirring bar underneath the basket. The beaker should contain approximately 5 ml of trichloroacetic acid per filter disk. The filter disks are washed for 15 minutes in an ice bath at the slowest possible rotating speed of the stirring bar. The screen basket containing the filter disks is then removed from the beaker with forceps, and the trichloroacetic acid is poured into an appropriate radioactive waste container and replaced with fresh 10% trichloroacetic acid. This wash procedure is repeated three times at room temperature (total = 4 washes) for 15 minutes each. The filter disks are then washed in 95% ethanol and then ethyl ether for 5 minutes each. The disks are dried for approximately 5 minutes with a hair dryer, and then placed into a toluene-based scintillant for counting. Since the radioactive protein remains adsorbed to the filter disks during counting, the disks can be removed and the scintillant reused fol-

---

[12] I. M. Glynn, and J. B. Chappell, *Biochem. J.* **90**, 147 (1964).
[13] D. A. Walsh, J. P. Perkins, C. O. Brostrom, E. S. Ho, and E. G. Krebs, *J. Biol. Chem.* **246**, 1968 (1971).

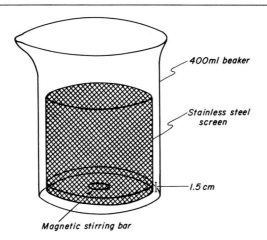

Fig. 1. Apparatus used for washing filter paper disks.

lowing counting. The last two trichloroacetic acid washes and the ethanol and ether washes can also be saved and reused. In addition to protecting the papers against damage by the stirring bar, the basket provides a convenient way of handling the papers during the washing and drying procedure. Stirring is normally included to ensure efficient washing, although we have found that low blanks can be obtained even without stirring. The blank is normally equivalent to 10 pmoles or less.

An alternate wash procedure to the filter disk method is that described by Walsh et al.[13] The filter disk method is preferable because it is less cumbersome, particularly when large numbers of samples (>25) are assayed.

The concentration of ATP can be reduced 10- to 100-fold, but this will reduce the reaction rate, increase substrate depletion by contaminating ATPases, and enhance the degree of inhibition by adenine nucleotides.[14]

Application of Assay Method to Crude Preparations. The procedure described above is suitable for assay of the enzyme in crude extracts of homogenates (2–30 ml/g). Dilute Tris or phosphate buffers containing 1 mM EDTA are appropriate homogenizing media. The ATPase activity of extracts of some tissues can be inhibitory. The inclusion of 40 mM NaF in the assay inhibits the ATPase activity in extracts of adipose tissue[15] but may not be necessary in other tissue extracts, such as heart

[14] H. Iwai, M. Inamasu, and S. Takeyama, *Biochem. Biophys. Res. Commun.* **46,** 824 (1972).
[15] J. D. Corbin, T. R. Sodering, and C. R. Park, *J. Biol. Chem.* **248,** 1813 (1973).

and skeletal muscle.[16] Because crude enzyme preparations may contain large amounts of endogenous substrates, there may be substantial phosphorylation without added substrate and it may be necessary to correct for this endogenous phosphorylation. If this is the case, two blanks [(1) no enzyme, (2) no substrate] are necessary.

Definition of Unit. One unit of protein kinase activity is that amount catalyzing transfer of 1 pmole of phosphate from [^{32}P]ATP to a histone mixture in 1 minute at 30°.

[16] T. R. Soderling, S. L. Keely, and J. D. Corbin, unpublished observations, 1972.

[42] Criteria for the Classification of Protein Kinases[1]

By J. A. TRAUGH, C. D. ASHBY, and D. A. WALSH

Protein kinases catalyze the transfer of the terminal phosphate moiety of ATP to a variety of different protein substrates. Initially[2] the classification of these enzymes was, in accord with standard enzymological nomenclature, based on specificity with respect to the protein substrate. More recently it has become recognized that cyclic AMP (cAMP) regulates the activity of some, but not all, of these enzymes,[3,4] and from experimentation to date it would appear that many cAMP-regulated protein kinases exhibit a broad protein substrate specificity.[5] Thus, classical nomenclature becomes inoperable at the experimental level, although in the future cAMP-regulated protein kinases of unique protein substrate specificity may be recognized. The activation of cAMP-regulated protein kinases occurs by a cyclic nucleotide-promoted dissociation of the holoenzyme (designated RC) to yield the active catalytic species (C) and a complex consisting of regulatory subunit (R) and cAMP.[6-9] This reaction is shown in Eq. (1).

$$RC + cAMP \rightleftharpoons R \cdot cAMP + C \qquad (1)$$

[1] Supported by research grants Am 13613 from the U.S. Public Health Service. J.A.T. is a recipient of a Public Health Service Fellowship (GM 505590). D.A.W. is an Established Investigator of the American Heart Association.

[2] M. Rabinowitz *in* "The Enzymes" (P. D. Boyer, H. Lardy, and K. Myrbäck, eds.) 2nd ed., Vol. 6, p. 119. Academic Press, New York, 1960.

[3] D. A. Walsh, J. P. Perkins, and E. G. Krebs, *J. Biol. Chem.* **243**, 376 (1968).

[4] J. F. Kuo and P. Greengard, *J. Biol. Chem.* **245**, 2493 (1970).

[5] For review see D. A. Walsh and E. G. Krebs *in* "The Enzymes" (P. D. Boyer, ed.), Vol. 8, p. 555. Academic Press, New York, 1973.

[6] M. A. Brostrom, E. M. Reimann, D. A. Walsh, and E. G. Krebs, *Advan. Enzyme Regul.* **8**, 191 (1970).

[7] M. Tao, M. L. Salas, and F. Lipmann, *Proc. Nat. Acad. Sci. U.S.* **67**, 408 (1970).

[8] G. N. Gill and L. D. Garren, *Biochem. Biophys. Res. Commun.* **39**, 335 (1970).

[9] A. Kumon, H. Yamamura, and Y. Nishizuka, *Biochem. Biophys. Res. Commun.* **41**, 1290 (1970).

Experimentally, the holoenzyme (RC) has been designated a cAMP-dependent protein kinase since the phosphotransferase activity is expressed only in the presence of cAMP.[10] From a physiological standpoint, free catalytic subunit (C) should also be considered as a cAMP-regulated protein kinase in that the activity in the cell is modulated in response to cAMP (by combination with R); nevertheless, the activity of the isolated enzyme (i.e., C) is expressed in the absence of the cyclic nucleotide.

In experimental situations, three categories of protein kinase can be recognized. Type *I*, holoenzyme (RC), is generally referred to as cAMP dependent protein kinase; type II is free catalytic subunit; and type III includes other protein kinases, the activity of which are not regulated by cAMP either *in vivo* or *in vitro*.

The purpose of this article is to present simple criteria that can be used experimentally to distinguish between the three categories of protein kinase. Of particular importance is the quantitative distinction between the enzymes designated types II and III, the activities of both of which are expressed in the absence of added cyclic nucleotide.

Physiologically, it is to be anticipated that free catalytic subunit will be formed in tissues in response to a hormonal stimulus, and that the amount present in a tissue homogenate will reflect the prior treatment of that tissue. Alternatively, free catalytic subunit may also occur by dissociation of the holoenzyme during the manipulations of extraction and purification. As studies of the effect of hormones on protein kinase dissociation *in vivo* become more extensive (see this volume [49]), it will become increasingly important to quantitate what amount of the activity expressed in the absence of cAMP represents free catalytic subunit. The criteria established here can be used quantitatively to distinguish the species of protein kinase in crude tissue extracts or, alternatively, to recognize the type of protein kinase in the purified enzyme preparations. Whereas these criteria cannot be conceived of as absolute, they are presented as working experimental guidelines.

Criteria for Classification

Four criteria may be used for the classification of various types of the protein kinases: stimulation of enzymatic activity by cAMP; binding of cAMP; inhibition by free regulatory subunit; and inhibition by the

---

[10] Although the holoenzyme has been designated cAMP dependent, it has not been demonstrated that this form is completely devoid of enzymatic activity in the absence of cAMP. Highly purified preparations of holoenzyme usually do exhibit some enzymatic activity in the absence of cyclic nucleotide, but it remains to be established whether this is due to holoenzyme per se or to free catalytic subunit produced by dissociation of holoenzyme at low protein concentration.

TABLE I
CRITERIA FOR THE CLASSIFICATION OF PROTEIN KINASES

Criteria	Type I (holoenzyme)	Type II (catalytic subunit)	Type III (other)
1. Stimulation by cAMP	+	−	−
2. Binding of cAMP	+	−	−
3. Inhibition by R[c]			
Assayed +cAMP	Complex[a]	Complex	−
Assayed −cAMP	N/A[b]	+	−
4. Inhibition by I[d]			
Assayed +cAMP	+	+	−
Assayed −cAMP	N/A	+	−

[a] See text.
[b] Not applicable; see text footnote 10.
[c] Free regulatory subunit.
[d] Inhibitor protein.

heat-stable inhibitor protein.[11] The role of these criteria in distinguishing the various types of protein kinases is summarized in Table I.

The cAMP-dependent holoenzyme (type I protein kinase) is stimulated by cAMP by the formation of a regulatory subunit·cAMP complex. The activity of the holoenzyme, as assayed in the presence of cAMP, is inhibited by addition of the inhibitor protein. The response obtained by adding regulatory subunit to the holoenzyme, as assayed in the presence of cAMP, is complex. At high concentrations of the cyclic nucleotide, addition of regulatory subunit is essentially of little effect on the position of equilibrium of Eq. (1) and thus of little consequence to the expressed enzymatic activity. However, at low concentrations of cAMP, addition of high levels of regulatory subunit are inhibitory through a reversal of the reaction of Eq. (1). The activity of holoenzyme in the absence of cAMP is currently not well defined in that it has not been conclusively established whether undissociated holoenzyme has enzymatic activity.[10]

Free catalytic subunit (type II protein kinase) is not stimulated by cAMP, nor does it bind the cyclic nucleotide. The enzymatic activity is inhibited by both regulatory subunit and inhibitor protein if the assay is performed in the absence of cAMP, and by the latter if the assay is performed in the presence of cAMP. As with the holoenzyme, the effect of regulatory subunit on catalytic subunit activity assayed in the presence of cAMP is a complex function of the concentrations of both proteins and of the nucleotide.

Type III protein kinases are characterized by negative responses to

[11] C. D. Ashby and D. A. Walsh, *J. Biol. Chem.* **247**, 6637 (1972); this volume [48].

all four criteria. Typical examples of this class of enzymes are phosphorylase b kinase,[12] histone kinase HK2,[13,14] phosvitin kinase,[14,15] casein kinase,[16] and some species of chromatin protein kinase.[14,17]

Generality of the Criteria

The potential utility of the criteria in the classification of mammalian protein kinases is enhanced by the observed lack of tissue specificity in the interaction of free regulatory subunit and of heat-stable inhibitor protein with the protein kinase. As depicted in Fig. 1A, the regulatory subunit purified from rabbit skeletal muscle blocks the activity of the catalytic subunit isolated from a wide range of rat tissues. Likewise, the cAMP-dependent protein kinase activity from a number of rabbit tissues is inhibited by the heat-stable inhibitor protein from rabbit skeletal muscle (Fig. 1B). The inhibitor protein is also effective in blocking the protein kinase activities from a number of different organisms. This lack of tissue and species specificity permits the general application of these criteria to a wide range of mammalian tissues. Furthermore, it has been shown that neither the interaction of the regulatory subunit with catalytic subunit nor the interaction of the inhibitor protein with the catalytic subunit is altered by the presence of undissociated holoenzyme (Fig. 2). Thus the presence in crude tissue extracts of undissociated holoenzyme (type I protein kinase) will not interfere with the use of either regulatory protein or inhibitor protein in the identification of free catalytic subunit (type II protein kinase).

Procedures for Classification

A number of different methods have been used for the assay of protein kinase and determination of cAMP binding to protein (see this volume). These various techniques are often readily adaptable for experimental classification. Examples of several assay techniques are presented in the legends of Figs. 1 and 2 and in Table III. Described below are the assay conditions that have been used for the classification of four different protein kinases isolated from rabbit reticulocytes.

[12] E. G. Krebs and E. H. Fischer, *Biochim. Biophys. Acta* **20**, 150 (1956).
[13] T. A. Langan, *Ann. N.Y. Acad. Sci.* **185**, 166 (1971).
[14] D. A. Walsh and C. D. Ashby, *Recent Progr. Horm. Res.* **29**, 329 (1973).
[15] R. Rodnight and B. E. Lavin, *Biochem. J.* **93**, 84 (1964).
[16] E. W. Bingham, H. M. Farrell, Jr., and J. J. Basch, *J. Biol. Chem.* **247**, 8193 (1972).
[17] M. Takeda, H. Yamamura, and Y. Ohga, *Biochem. Biophys. Res. Commun.* **42**, 103 (1971).

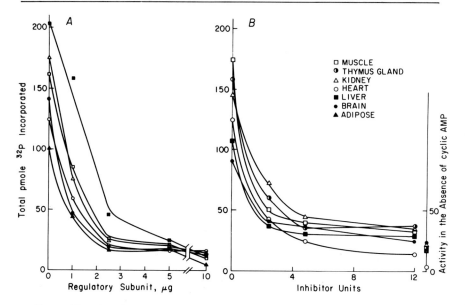

FIG. 1. The titration of protein kinase activity. (A) The effect of muscle regulatory subunit on type II enzymes derived from rat. (B) The effect of skeletal muscle heat-stable inhibitor protein on type I enzymes derived from rabbit. Isolation of type II enzyme (catalytic subunit) from various rat tissues. The tissues were homogenized in 2.5 volumes of 4 mM EDTA, pH 7.0. The homogenate was centrifuged first at 9000 g for 15 minutes and then at 105,000 g for 45 minutes. Cyclic AMP (cAMP) (1×10^{-4}) was added to the supernatant solutions which were then chromatographed on a CM 50 column (4×0.7 cm) equilibrated in 5 mM MES, pH 7.0; 1 mM EDTA. The catalytic subunit was eluted with the same buffer containing 0.25 M NaCl. All procedures were performed at 4°.

Isolation of type I protein kinase (the predominant form of cAMP-regulated enzymes) from various rabbit tissues. The tissues were excised and homogenized in 5 volumes of cold 10 mM Tris · HCl, pH 7.5, 2 mM EDTA. The homogenates were centrifuged at 9000 g for 20 minutes, and the resultant supernatant solutions, after filtration through glass wool, were centrifuged 1 hour at 105,000 g.

Protein kinase activity was determined in a reaction mixture containing: sodium glycerol-phosphate buffer, 5.3 μmoles; sodium fluoride, 2.0 μmoles; ethylene glycol bis(β-aminoethyl ether)-N,N'-tetraacetate, 0.03 μmole; aminophylline, 0.2 μmole; histone f$_2$b (0.2 mg/ml), cAMP, 0.4 nmole (added only in experiment B), magnesium acetate, 0.36 μmole; [γ-^{32}P]ATP, 0.12 μmole; and increasing concentrations of regulatory subunit or heat-stable protein in a total volume of 0.13 ml with a final pH of 6.0. The assays were initiated with the protein kinase afer appropriate dilution to give linear reaction rates. The incubation was performed at 30° for 20 minutes. The extent of ^{32}P incorporation into protein was determined as described elsewhere in this volume [48].

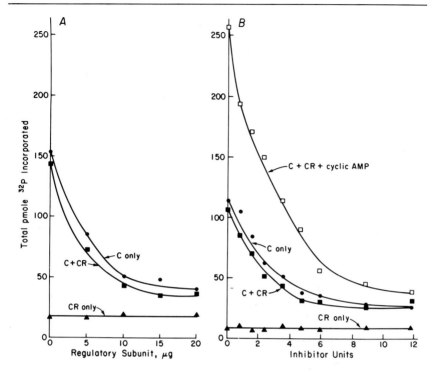

FIG. 2. (A) The titration of catalytic subunit activity by the regulatory subunit in the presence or in the absence of undissociated holoenzyme. The titration by regulatory subunit of protein kinase activity was examined in the absence of cAMP with histone f_2b as substrate. Incubation time was 20 minutes. Various concentrations of regulatory subunit were added to 0.8 µg of holoenzyme (▲——▲), 0.04 µg of catalytic subunit (●——●), and a combination of 0.8 µg of holoenzyme and 0.04 µg of catalytic subunit (■——■). All proteins were partially purified from rabbit skeletal muscle. The specific activities of the catalytic subunit and holoenzyme were 382 and 346 pmoles of ^{32}P incorporated into histone f_2b per microgram of protein, respectively.

(B) The titration of catalytic subunit activity by the inhibitor protein in the presence or absence of undissociated holoenzyme. The titration by inhibitor protein of casein phosphorylating activity was examined in the absence (filled symbols) or presence (open symbols) of cAMP. Various inhibitor concentrations were added to 0.04 µg catalytic subunit alone (●——●), 0.9 µg of cAMP-dependent holoenzyme alone (▲——▲) or to a combination of 0.04 µg of catalytic subunit plus 0.9 µg of holoenzyme (■——■, □——□). All other conditions were the same as in the experiment of Fig. 1 with the exception that for the experiment of Fig. 2B casein (6 mg/ml) was used as the substrate and the incubation time was 35 minutes.

Stimulation by cAMP. A reaction volume of 0.07 ml contained: 14 mM sodium phosphate or 50 mM Tris·HCl, pH, 7.0; 10 mM $MgCl_2$; 1.4 µM cAMP (where indicated); 0.14 mM radioactive ATP; and either

0.3 mg purified casein[18] or histone (type IIA, Sigma Chemical Co.). The reaction was initiated by addition of the enzyme fraction, and incubated for 15 minutes at 30°. Samples of 0.05 ml were pipetted onto 2-cm squares of Whatman ET 31 filter paper, placed in cold 10% TCA for 10 minutes, washed three times for 10 minutes each in 5% TCA, dried, and counted in liquid scintillation fluid. The reaction was initiated with enzyme since free catalytic subunit inactivates rapidly upon dilution. In addition, prior association of the holoenzyme with some substrates results in dissociation of the subunits prior to the addition of cAMP.[19,20] The amount of protein kinase was adjusted so that incorporation of ^{32}P into protein was linear over the 15-minute incubation period and linearly dependent on enzyme concentration.

Binding of cAMP. A reaction volume of 0.1 ml contained: 5 mM sodium acetate, pH 4.5; 1 mM $MgCl_2$; 15 nM cAMP (Schwarz-Mann, 28 Ci/mmole); 10 μg of bovine serum albumin; and enzyme. After incubation for 10 minutes at 30°, the reaction was terminated by addition of 5 ml of cold 10 mM Tris·HCl, pH 8.0, 40 mM $MgCl_2$. The R·cAMP complex was isolated by Millipore filtration,[21] washed three times with termination buffer, dried, and counted in liquid scintillation fluid.

Inhibition by Regulatory Subunit. Regulatory subunit of rabbit skeletal muscle cAMP-dependent protein kinase was prepared as described by Beavo *et al.*[22] Using the standard reaction conditions described above, the protein kinase fractions are assayed in the presence of varying concentrations of regulatory protein. As shown in Fig. 1A, the regulatory subunit interacts in approximately the same manner with catalytic subunit from all the tissues that have been examined.

Inhibition by the Heat-Stable Inhibitor Protein. The heat-stable inhibitor protein was isolated from rabbit skeletal muscle as described elsewhere in this volume.[23] Using the standard reaction conditions, the protein kinase fractions were assayed in the presence of varying concentrations of inhibitor protein. One unit of inhibitor protein has been defined,[23] and this value can be used to determine approximate inhibitor concentrations for use in the assay. Some quantitative differences are observed in the interaction between the skeletal muscle inhibitor protein and catalytic subunit obtained from various tissues. In consequence, the assay should be performed over a range of inhibitor protein concentrations.

[18] E. M. Riemann, D. A. Walsh, and E. G. Krebs, *J. Biol. Chem.* **246**, 1986 (1971).
[19] E. Miyamoto, G. L. Petzold, J. S. Harris, and P. Greengard, *Biochem. Biophys. Res. Commun.* **44**, 305 (1971).
[20] M. Tao, *Biochem. Biophys. Res. Commun.* **46**, 561 (1972).
[21] A. G. Gilman, *Proc. Nat. Acad. Sci. U.S.* **67**, 305 (1970).
[22] J. A. Beavo, P. J. Bechtel, and E. G. Krebs, this volume [43].
[23] C. D. Ashby and D. A. Walsh, this volume [48].

TABLE II
CLASSIFICATION OF PURIFIED PROTEIN KINASES[a]

Enzyme	Stim. by cAMP (x-fold)	[^{3}H]cAMP bound (pmoles/enzyme unit $\times 10^3$)	Inhibition by I	Inhibition by R	Protein kinase classification
Peak I	0	0	+	+	II
Peak II	6.5	10	+	−	I
Peak III$_H$	6.5	10	+	−	I
Peak III$_C$	0	0	−	−	III

[a] Conditions of the methods used are given in the text and full experimentation is presented elsewhere [J. A. Traugh and R. R. Traut, *J. Biol. Chem.* **249**, 1207 (1974)].

Example Uses of the Criteria

Classification of Resolved Enzymes. Recently four major protein kinase activities have been resolved and partially purified from rabbit reticulocytes by chromatography on DEAE-cellulose and phosphocellulose.[24] Each enzyme fraction was assayed in accordance with the four criteria, using the procedures described above. The enzymes were numbered in order of elution from DEAE-cellulose. Two enzymes, coincident as peak III were resolved by phosphocellulose into a fraction preferential for histone substrates, and one specific for casein, and designated III$_H$ and III$_C$, respectively. Table II shows that peak I activity was not stimulated by cAMP, did not bind cAMP and was inhibited by either free regulatory subunit or the heat-stable inhibitor protein. This enzyme fraction was thus characterized as free catalytic subunit (type II protein kinase). Peaks II and III$_H$ have similar properties. Both enzymes were stimulated greater than 6-fold by cAMP, each fraction bound cAMP, and the activity of each fraction was inhibited by the protein inhibitor. These enzymes were identified as cAMP-dependent protein kinases (type I enzyme). The enzyme, peak III$_C$, gave a negative response to each of the four criteria and was assigned the classification, type III.

Classification of Protein Kinases in Cell Extracts. The established criteria can also be used to characterize and quantify the protein kinase activities in crude tissue extracts as illustrated for six tissues from rabbit (Table III). For these determinations, the assays were performed within a minimal interval of time after the animal was sacrificed to reduce spurious dissociation of the cAMP-dependent protein kinase. Between 78% and 97% of the protein kinase activity in the cell extracts was de-

[24] J. A. Traugh and R. R. Traut, *J. Biol. Chem.* **249**, 1207 (1974).

TABLE III
CLASSIFICATION OF PROTEIN KINASE ACTIVITY IN CRUDE TISSUE EXTRACT[a]

Tissue	Total activity (pmoles ^{32}P incorp/min/ml extract $\times 10^{-2}$)	Stimulation by cAMP (% of total)	Inhibition of cAMP independent activity		Classification		
			By I (%)[b]	By R (%)[b]	Type I (%)[c]	Type II (%)	Type III (%)[c]
Thymus	32	88	0	0	88	0	12
Kidney	22	86	0	0	86	0	14
Liver	16	81	0	0	81	0	19
Skeletal muscle	26	88	0	0	88	0	12
Heart	25	96	50	50	96	2	2
Brain	18	78	0	0	78	0	22

[a] Rabbit tissues were excised and extracted as described in Fig. 1B. For the assays in the absence of cAMP, the dilutions were 1:3 for brain, 1:3 for thymus, and undiluted for the remainder. The dilutions for the assays performed in the presence of cAMP were 1:6 for muscle, kidney, and liver, and 1:8 for thymus, heart, and brain. The reaction mixtures contained: 2-(N-morpholino)ethanesulfonic acid buffer, pH 7.0, 1 μmole; sodium fluoride, 1 μmole; theophylline, 0.2 μmole; sodium glycerol-P, 0.2 μmole; histone F$_2$b, 0.02 mg/ml; cAMP, 0.4 nmole (where added); magnesium chloride, 0.36 μmole; [γ-^{32}P]ATP, 0.12 μmole; and aliquots of purified inhibitor and regulatory subunit in a total volume of 0.14 ml. Following a 20-minute incubation at 30°, a 100-μl aliquot was removed and pipetted on a paper square (Whatman No. 31 ET) and immediately placed in cold 10% trichloroacetic acid. The procedure for determining the amount of ^{32}P incorporated into the protein substrate and the method of preparation of regulatory subunit and inhibitor protein are described elsewhere in this volume [43] and [48].
[b] Percent of cAMP independent activity.
[c] Percent of total activity.

pendent on cAMP for activity. Thus, the majority of the protein kinase activity in the six tissues was identified as cAMP-dependent holoenzyme (greater than 90% of this activity was blocked by the addition of the inhibitor protein (not shown)). Five of the six tissues contained a significant amount of cAMP-independent protein kinase (12–23%). This activity was not inhibited by either free regulatory subunit or inhibitor protein; thus, the cAMP-independent activity was due to protein kinase, type III. In the sixth tissue (heart), the low level of cAMP-independent protein kinase activity appeared to be comprised of both type II and type III enzymes.

[43] Preparation of Homogeneous Cyclic AMP-Dependent Protein Kinase(s) and Its Subunits from Rabbit Skeletal Muscle

By J. A. BEAVO, P. J. BECHTEL, and E. G. KREBS

Cyclic AMP (cAMP)-dependent protein kinases catalyze the transfer of the terminal phosphate of ATP to serine or threonine residues of a variety of protein substrates. These enzymes are termed cAMP-dependent since most if not all of their activity is revealed only in the presence of low concentrations of cAMP. The enzymes are composed of two subunit types, one of which is catalytically active (C) and another (R) which regulates the activity of the catalytic subunit. The subunit R inhibits the activity of C until cAMP is bound to R at which time the subunits dissociate and the activity of the enzyme is expressed. This process is illustrated schematically by the following equation:

$$cAMP + R \cdot C \rightleftharpoons R \cdot cAMP + C$$

The evidence leading to the scheme outlined above has been reviewed recently.[1]

Currently, it appears that there is more than one cAMP dependent protein kinase present in muscle and liver.[2,3] In most of the cases investigated, at least two peaks of cAMP-dependent activity can be separated by chromatography on DEAE-cellulose. These fractions will be referred

[1] D. A. Walsh and E. G. Krebs, *in* "The Enzymes" (P. D. Boyer, ed.), Vol. 8, p. 555. Academic Press, New York, 1973.
[2] E. M. Reimann, D. A. Walsh, and E. G. Krebs, *J. Biol. Chem.* **246**, 1986 (1971).
[3] A. Kumon, K. Nishiyama, H. Yamamura, and Y. Nishizuka, *J. Biol. Chem.* **247**, 3726 (1972).

to as peak I and peak II protein kinase in this communication. To date, no clear-cut differences in the physiological roles of these enzyme forms have been demonstrated.

Section A of this article describes the procedures for preparation of homogeneous cyclic AMP-dependent protein kinase from peak I. Section B deals with the preparation of the catalytic subunit from either peak I or peak II protein kinase. Section C describes a simplified procedure for the preparation of nearly homogeneous catalytic subunit directly from an extract. An outline of the major fractionation steps described in the three sections is given below.

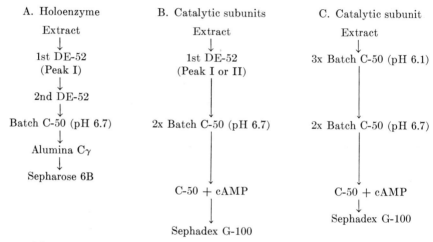

Schematic Outline of Fractionation Steps for Purification of Protein Kinases

Materials and Methods

DEAE-cellulose (Whatman DE-52) was purchased from Reeve Angel, Inc. and regenerated as indicated by the manufacturer. CM-Sephadex C-50, Sephadex G-100, and Sepharose 6B were obtained from Pharmacia Fine Chemicals. Electrophoresis equipment and chemicals were purchased from Bio-Rad Laboratories, and the larger chromatography columns used in several of the procedures were obtained from Glenco Scientific, Inc. Aged Alumina Cγ, lot **98B-3040**, and all other reagents were purchased from Sigma Chemical Co.

The various buffers used in the isolation procedures are listed below.

Buffer A: 5 mM 2-(N-morpholino)ethanesulfonic acid (MES), 9 mM NaCl, 15 mM β-mercaptoethanol, pH 6.5

Buffer B: 10 mM potassium phosphate, 0.1 mM ethylenediaminetetraacetic acid (EDTA), 15 mM β-mercaptoethanol, pH 6.7

Buffer C: 75 mM potassium phosphate, 15 mM β-mercaptoethanol, pH 7.0

Buffer D: 300 mM potassium phosphate, 15 mM β-mercaptoethanol, pH 7.0

Buffer E: 5 mM MES, 100 mM NaCl, 15 mM β-mercaptoethanol, 0.1 mM EDTA, pH 6.5

Buffer F: 30 mM potassium phosphate, 15 mM β-mercaptoethanol, 0.1 mM EDTA, pH 6.7

The gels and resins used in the procedures are equilibrated in the appropriate buffer before use. cAMP-dependent protein kinase activity was determined by the filter paper method essentially as described by Reimann et al.[2] The assays were carried out for 2 minutes in 80 μl of buffer containing 4 mM magnesium acetate, 0.25 mM ethyleneglycol-bis(β-aminoethyl ether)N,N'-tetraacetic acid (EGTA), 25 mM MES (pH 6.9), 0.125 mM [γ-^{32}P]ATP (10–100 cpm/pmole) and 75 mg of histone per milliliter (type II A mixture). Protein concentrations were determined by the method of Lowry[4] and by the biuret technique.[5]

A. Preparation of Homogeneous Holoenzyme (RC) from Peak I

Procedure

Extract. The hind leg and back muscles are removed from 15 mature female rabbits which have been anesthetized with sodium pentobarbital and bled from the jugular veins. After chilling on ice, the muscle (approximately 10 kg) is passed through the coarse disk of a meat grinder. Aliquots of the ground muscle are placed in 2.5 volumes of 4 mM EDTA (pH 7) containing 15 mM β-mercaptoethanol and homogenized for 1 minute at high speed in a Waring blender. The homogenate is centrifuged at 11,000 g_{max} for 20 minutes in a 3RA rotor of a Lourdes centrifuge. The resulting extract, approximately 25 liters, is decanted through glass wool.

First DE-52. The extract is diluted with about 2 volumes of cold deionized water until the conductivity is 0.75–0.80 mMHO (final volume approximately 75 liters). To the diluted extract are added 6 liters of packed DE-52 resin equilibrated in buffer A (see materials section for buffer compositions). The suspension is stirred for 15 minutes and then allowed to settle for an equal period of time. Most of the supernate above the settled gel can be siphoned off and the resin collected in large sintered-glass funnels. While in the funnels, the resin is washed with 10 liters of buffer A. The washed resin is resuspended in an equal volume

---

[4] O. H. Lowry, this series, Vol. 3, p. 448.
[5] T. E. Weichselbaum, *Amer. J. Clin. Pathol.* **16**, 40 (1946).

of the same buffer and poured into a 4-inch in diameter column. After packing, the column is washed with 3–5 liters of buffer A and then eluted with a 15-liter linear gradient of 0–0.50 M NaCl in buffer A. Usually the gradient can be started within 8 hours of homogenizing the muscle. The large gradient maker for the first step can be easily constructed by connecting the bottom portions of two plastic pipette washing containers such as the Nalgene pipette jar.

After assaying the gradient elution fractions to determine exactly where the two peaks of activity were eluted, the first peak is pooled as indicated in Fig. 1 and dialyzed overnight against 10 volumes of buffer A.

Second DE-52. One liter of DE-52 equilibrated in buffer A is then added to the dialyzed fraction from the previous step. The resin is gently stirred with the enzyme for 15 minutes and then poured into a column 2-inches in diameter. The column is washed with a column volume of buffer A and then eluted with a 5-liter linear gradient of 0–300 mM NaCl in buffer A. The gradient is assayed for protein kinase activity and pooled as indicated in Fig. 2. The pooled peak is dialyzed overnight against 10 volumes of buffer B.

Batch C-50. Packed CM Sephadex-C50 resin, 200 ml, equilibrated in buffer B, is then added to the dialyzed fraction and stirred for 15

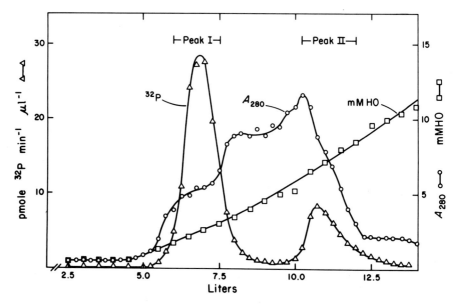

Fig. 1. First DE-52 column elution profile. Conditions and procedures are as described in the text. △——△, protein kinase activity in the presence of 1.25 μM cyclic AMP; ○——○, A_{280}; □——□, conductivity (mmho).

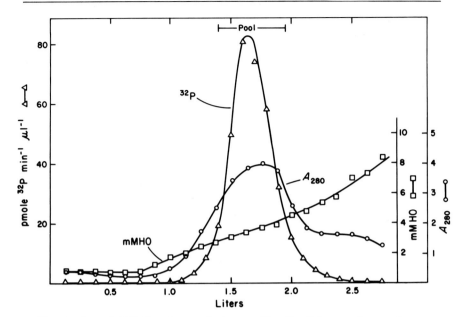

Fig. 2. Second DE-52 column elution profile. Conditions and procedures are as described in the text. Symbols are the same as Fig. 1.

minutes. The resin is separated by filtration and discarded. This step is then repeated with 100 ml of resin.

Alumina Cγ. To the filtrate is added 0.75 g dry weight of alumina Cγ for each gram of protein as estimated by spectrophotometric measurement at 280 nm. The gel is gently suspended and agitated slowly for 5 minutes at 0–4°. The gel is then collected by centrifugation at 400–1000 g for 2 minutes. Longer times and higher speeds must be used if the gel particles are fractured. The gel is resuspended and washed twice in a similar manner with 10 volumes of buffer C. The protein kinase is then eluted by washing the gel 3 times with 25 ml of buffer D. The authors have found considerable variation in the optimal ratios of gel to protein between different batches of alumina Cγ. Similarly, the concentration of phosphate in the washing and eluting buffers should be piloted for each new batch of alumina Cγ.

The enzyme fraction is concentrated by dialysis against 70% sucrose in buffer E. Alternatively, the sample can be concentrated by dialyzing or diluting the fraction until the conductivity is less than 1.0 mmho and then absorbing it to a small (2–4 ml) column of Whatman DE-52. The enzyme can then be eluted with 2 column volumes or less of 0.5 M NaCl in buffer A.

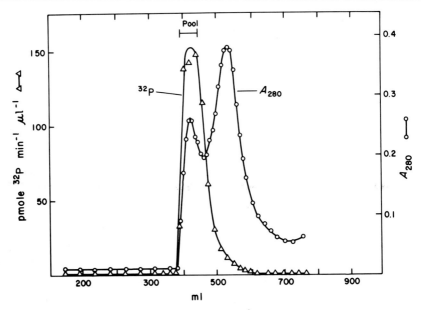

FIG. 3. Sepharose 6B elution profile. Conditions are as described in the text and the symbols are the same as in Fig. 1.

Sepharose 6B. The concentrated sample in sucrose is then applied to a 2-inch × 36-inch column of Sepharose 6B, equilibrated in buffer E. The sample volume must be less than 0.5% of the column volume for good separation. The protein kinase activity and absorption at 280 nm of the column fraction are determined and they are then pooled as indicated in Fig. 3. The pooled fractions contain essentially homogeneous protein kinase as determined by SDS gel electrophoresis. It is preferable to check the purity of each fraction of the first protein peak by SDS gel electrophoresis to determine which fractions should be pooled, as the protein profile varies slightly depending on the care taken in the previous steps. The pooled samples are concentrated to 1–2 mg/ml by dialysis against 70% sucrose in buffer E, then dialyzed against buffer E to remove the sucrose and stored at 0–5°. The shoulders of the activity peak may be pooled, concentrated, and passed again over the Sepharose 6B column to increase the yield of homogeneous enzyme.

Comments about the Preparation

A summary of the purification steps used in the preparation of peak I protein kinase is given in the table. This procedure will also yield a

SUMMARY OF PROCEDURE FOR PREPARATION OF PEAK I HOLOENZYME

Fraction	Total activity (μmoles min^{-1})	Specific activity (pmoles min^{-1} mg^{-1})	Protein (mg)	Purification (x-fold)	Yield (%)
Extract[a]	290	690	420,000	1	100
1st DE52	100	21,000	4,690	35	36
2nd DE52	90	75,000	1,200	100	30
Alumina Cγ	50	500,000	100	730	17
Sepharose 6B	10.2	1,600,000	6.5	2300	3.5

[a] Of the activity present at the extract stage, 10–15% does not require added cAMP, nor is it inhibited by the heat-stable inhibitor of cAMP-dependent protein kinase.

greatly purified cAMP-dependent protein kinase from peak II of the first DE-52 step. However, this enzyme preparation after the Sepharose 6B step still contains several other proteins as judged by SDS gel electrophoresis.

Properties of Peak I Holoenzyme

The enzyme is stable in terms of activity, activation by cAMP, binding capacity for cAMP and sedimentation characteristics for at least 1 month when stored at 0–5° in buffer E. The $s_{20,w}$ as determined by sucrose density centrifugation is 6.8, the same as most of the enzyme activity in the crude extract. The homogeneous enzyme is stimulated at least 20-fold by cAMP. Other investigators[6] have found that the enzyme after the second DE-52 step is suitable for the assay of cAMP by the Millipore filter method, since most of the cAMP phosphodiesterase activity has been removed by this step.

Preparation of the Subunits from cAMP-Dependent Protein Kinase

One method of preparing the regulatory and catalytic subunits from the holoenzyme depends on the separation of the subunits on a column of CM-Sephadex C-50 in the presence of cAMP. To a sample of holoenzyme (0.1–1 mg/ml equilibrated in buffer F), cAMP is added to a concentration of 100 μM. The enzyme is then passed through a small column of CM-Sephadex C-50 (0.5 ml of gel per milligram of enzyme) containing 100 μM cAMP in buffer F. The regulatory subunit is not retained by

[6] M. Castagna and D. A. Walsh, personal communication.

the column and can be concentrated from the "flow through" fraction. After washing the packed column with 5–10 ml of buffer F, the catalytic subunit can be eluted with 2 column volumes of buffer D. Other methods for separating the subunits of protein kinase including affinity chromatography and gel filtration in the presence of cAMP have been described.[3,7]

B. Preparation of Catalytic Subunit from Either Peak I or Peak II

The catalytic subunit from either peak I or peak II may be prepared by a simplified version of the procedure for obtaining holoenzyme. Starting with the pooled fractions of either peak I or peak II holoenzyme obtained from the first DE-52 step described previously, homogeneous catalytic subunit may be prepared as follows.[8]

Batch C-50. The pooled fraction is dialyzed overnight against 10 volumes of buffer F. To the dialyzed fraction is added 20 ml of packed CM-Sephadex C-50 for each gram of protein. The suspension is stirred gently for 15 minutes and then separated by filtration in a sintered-glass funnel. To the filtrate is added another 2–4 ml of CM-Sephadex C-50 per gram of protein. This suspension is stirred for an additional 15 minutes, after which it is separated by filtration.

C-50 Column. Enough cAMP is added to the filtrate to yield a final concentration of 10 μM. CM-Sephadex C-50 equilibrated in buffer F is then added at a ratio of 1 ml per gram of protein, and the suspension is stirred for 15 minutes. The resin is collected by filtration in a sintered-glass funnel, resuspended in an equal volume of buffer F and poured into a column of an appropriate size. After washing with 2–3 column volumes of buffer F, the catalytic subunit is eluted with an 8 column volume linear gradient of 30–300 mM potassium phosphate containing 0.1 mM EDTA and 15 mM β-mercaptoethanol (pH 6.7). The peak of protein corresponding to protein kinase activity elutes between 100 and 200 mM potassium phosphate. If the preceding stages are conducted carefully, the enzyme is essentially homogeneous at this stage as judged by SDS gel electrophoresis. However, in some cases it is necessary to remove a small number of a higher molecular weight contaminants by gel filtration of the sample on Sephadex G-100.

Sephadex G-100. The sample may be concentrated by dialysis against 70% sucrose in buffer F and applied to a 2.5 × 90 cm column of the

---

[7] E. M. Reimann, C. O. Brostrom, J. D. Corbin, C. A. King, and E. G. Krebs, *Biochem. Biophys. Res. Commun.* **42**, 187 (1971).

[8] The starting material for this preparation can also be obtained by eluting the holoenzyme from the washed DE-52 resin with 0.5 M NaCl in buffer A while the resin is still in the sintered-glass funnel.

Sephadex G-100 equilibrated in the same buffer. Alternately, the enzyme can be diluted or dialyzed until the conductivity is less than 2 mMHO and then concentrated by absorption to a small (1–2 ml) column of CM-Sephadex C-50 equilibrated in buffer F. The enzyme can then be eluted with 2–4 ml of 300 mM potassium phosphate in buffer F and applied to the G-100 column.

C. Preparation of Catalytic Subunit from the Extract

A modification of the above procedures can be used to obtain essentially pure catalytic subunit directly from an extract in a much shorter time. This modification allows the preparation of milligram quantities of the catalytic subunit in 3 days.

Batch C-50. An extract prepared as described in procedure A is adjusted to pH 6.1 with 1 N acetic acid and the precipitate which forms is removed by centrifugation. CM-Sephadex C-50, equilibrated in buffer F adjusted to pH 6.1, is added at a ratio of 500 ml of packed gel to 1000 ml of extract. The suspension is stirred for 15 minutes, and the gel is separated by filtration on a Büchner or sintered-glass funnel and washed with 500 ml of buffer F. This step is then repeated twice using the same amount of gel each time. After the third batch absorption step the extract is adjusted to pH 6.7 with 1 N potassium hydroxide. The same volume of gel equilibrated in buffer F adjusted to pH 6.7 is then added and the batch absorption step repeated. A last batch absorption at pH 6.7 is then carried out in the same manner.

CM-Sephadex C-50 Column. To the last filtrate cAMP is added to a concentration of 10 μM followed by the addition of 6 ml of packed CM-Sephadex C-50 for each liter of filtrate. After stirring for 15 minutes, the gel is collected in a sintered-glass funnel and washed with 10–12 volumes of buffer F. The gel is then resuspended in an equal volume of buffer F and poured into a column of an appropriate size. The column is washed with 2–3 column volumes of buffer F and eluted with an 8 column volume linear gradient of 30–300 mM potassium phosphate containing 0.1 mM EDTA and 15 mM β-mercaptoethanol (pH 6.7). The peak of protein eluting between 100–200 mM potassium phosphate contains the catalytic subunit. The enzyme at this stage is from 500- to 2000-fold purified depending on the care taken during the batch absorption steps.

Sephadex G-100. The enzyme can be further purified to near homogeneity by concentrating the sample and passing it through a 2.5 × 90 cm column of Sephadex G-100 equilibrated in either buffer E or F. The final product is 5000- to 6000-fold purified from the extract.

Comments about the Catalytic Subunit Preparations. Either of the procedures described for the preparation of the catalytic subunit of protein kinase may be easily scaled up to process large amounts of tissue which is often desirable, since the enzyme must be purified nearly 6000-fold to reach homogeneity. Accordingly, large quantities of muscle must be processed to obtain enough protein for many types of studies. The importance of using enough CM-Sephadex C-50 in the batch absorption steps and of repeating the step a sufficient number of times must be emphasized. Using too little gel or failing to repeat the procedure a sufficient number of times yields a product so contaminated with other proteins that they cannot be successfully separated from the enzyme by gel filtration on G-100. The thoroughness of the wash steps of the last batch absorptions are important for the same reasons. CM-Sephadex C-50 was chosen arbitrarily for this batch absorption and column steps. It is probable that other anion exchange resins having better handling characteristics could be substituted successfully.

Stability. The enzyme prepared by either of these procedures is stable in terms of activity and sedimentation characteristics for at least 1 month when stored at 0–5° in the final buffer at a protein concentration greater than 0.5 mg/ml. Attempts to freeze the preparations have met with mixed success. It is usually necessary to include 0.5 mg/ml bovine serum albumin in the diluent when the enzyme concentration is adjusted below 0.1 mg/ml for assaying.

[44] Cyclic AMP-Dependent Protein Kinase from Bovine Heart Muscle

By CHARLES S. RUBIN, JACK ERLICHMAN, and ORA M. ROSEN

$$\text{Protein} + \text{ATP} \xrightarrow{\text{Mg}^{2+}} \text{Protein-phosphate} + \text{ADP}$$

Assay Method

Principle. Cyclic AMP (cAMP)-dependent protein kinase catalyzes the transfer of the γ-phosphate of ATP to serine or threonine hydroxyl groups in various protein substrates. The addition of cAMP greatly enhances the velocity of the phosphotransferase reaction. The activity of the enzyme is determined by measuring the amount of ^{32}P transferred from [γ-^{32}P]ATP to protamine or histone. The labeled protein substrate is separated from the assay mixture by precipitation with trichloroacetic acid and collected on glass fiber filters. Assays are carried out in the presence and absence of cAMP to evaluate the activation of the enzyme by the cyclic nucleotide.

A supplementary measurement of the activity of cAMP-dependent protein kinase may be obtained by determining the cAMP-binding capacity of the enzyme. Binding assays are carried out according to the method of Gilman[1] with a single modification: 50 mM potassium phosphate buffer, pH 7.0, is substituted for acetate buffer, pH 4.0. Assays of cAMP-binding capacity and catalytic activity provide two independent methods for assessing the recovery of protein kinase during the course of purification[2] and also allow for the detection of cAMP-independent protein kinase and cAMP-binding protein resulting from the dissociation of the enzyme into its dissimilar subunits.[3]

Reagents

[γ-^{32}P]ATP, 2.0 mM
Potassium phosphate buffer, 1.0 M, pH 7.0
Magnesium sulfate, 1.0 M
Dithiothreitol, 1.0 M
cAMP, 0.20 mM
Bovine serum albumin, 100 mg per ml H$_2$O
Protamine sulfate, 25 mg per ml H$_2$O. Protamine is only partially soluble at the concentration indicated and often forms a clear gel after storage in solution at —20°. It is conveniently solubilized prior to use by incubation at 70° for 5 minutes with occasional stirring on a vortex mixer. This treatment provides a supersaturated stock solution of protamine which is stable for 1 hour at room temperature and, furthermore, has no effect on the phosphate-accepting capacity of protamine.
Trichloroacetic acid, 10% (w/w)
Trichloroacetic acid, 5% (w/w)
Sodium hydroxide, 1.0 M

Procedure. Protein kinase activity is measured by a modification of the method of DeLange et al.[4] The reaction mixture (0.2 ml) contains 50 mM potassium phosphate buffer, pH 7.0, 50 μM [γ-^{32}P]ATP (15,000

---

[1] A. G. Gilman, *Proc. Nat. Acad. Sci. U.S.* **67**, 305 (1970).
[2] Copurification of cAMP-binding and catalytic activities is observed only after interfering substances are removed in step 3 of the purification procedure.
[3] C. S. Rubin, J. Erlichman, and O. M. Rosen, *J. Biol. Chem.* **247**, 36 (1972).
[4] R. J. DeLange, R. G. Kemp, W. D. Riley, R. A. Cooper, and E. G. Krebs, *J. Biol. Chem.* **243**, 2200 (1968).

cpm per nmole), 10 mM MgSO$_4$, 10 mM dithiothreitol, 2.0 μM cAMP (when included), 0.25 mg of protamine, 0.5 mg of bovine serum albumin, and 0.02–0.2 unit of protein kinase. Incubations are performed at 30° for 5 minutes and the reactions are terminated by the addition of 0.5 ml of ice-cold 10% trichloroacetic acid. All further operations are carried out at 0°. After standing for 5 minutes the tubes are centrifuged at 8000 g for 5 minutes. The supernatant fluid is discarded and the protein precipitate is dissolved in 0.1 ml of 1 M NaOH and reprecipitated with 2 ml of 10% trichloroacetic acid. The protein precipitates are transferred onto 24-mm glass fiber filter disks (Whatman GF/C) in a suction apparatus and the assay tubes are rinsed 3 times with 5% trichloroacetic acid. Subsequently, the filter disks are washed with 20 ml of 5% trichloroacetic acid, transferred to aluminum planchets, and dried under a heat lamp. The radioactivity in the protein precipitates is determined by a gas-flow detector at an efficiency of 30%.

Units. One unit of protein kinase activity is defined as that amount of enzyme necessary to catalyze the transfer of 1 nmole of ^{32}P from [γ-^{32}P]ATP to protamine per minute at 30°. A unit of binding activity is equivalent to the binding of 1 nmole of cAMP. Specific activity is expressed as units per milligram of protein.[5]

Purification Procedure

Fresh beef hearts are purchased from a local slaughterhouse. After removal of the pericardium and fat tissue, the heart muscle is minced and stored at −20°.

Step 1. Homogenization. Frozen heart muscle is thawed overnight at 4°. All further operations are carried out at 4° and all buffer systems contain 4 mM 2-mercaptoethanol. Heart muscle (2 kg wet weight) is mixed with 4 liters of 40 mM potassium phosphate buffer, pH 6.1, containing 2 mM EDTA and homogenized in small batches at high speed for 1 minute in a Waring blender. The homogenate is then centrifuged at 10,000 g for 10 minutes and the supernatant fluid is collected and filtered through Whatman No. 54 paper. The pellets are extracted two more times with 1 liter of buffer, and the extracts are combined.

Step 2. Ammonium Sulfate Fractionation. The pooled extracts are brought to 55% saturation by the addition of solid (NH$_4$)$_2$SO$_4$ [320 g of (NH$_4$)$_2$SO$_4$ per liter], and the pH is maintained between pH 7 and 8 by the addition of NH$_4$OH (5.0 ml of concentrated NH$_4$OH per liter).

---

[5] O. H. Lowry, N. J. Rosebrough, A. L. Farr, and R. J. Randall, *J. Biol. Chem.* **193**, 265 (1951).

Protein is allowed to precipitate for 2.5 hours and is then collected by centrifugation at 10,000 g for 10 minutes. The supernatant fluid is discarded and the precipitate dissolved in 500 ml of 50 mM Tris buffer, pH 7.6, containing 10 mM NaCl. This solution is dialyzed overnight against two 4-liter volumes of the same buffer. All Tris buffers were adjusted to pH 7.6 by the addition of concentrated HCl.

Step 3. DEAE-Sephadex Batch Elution. The dialyzed enzyme preparation, containing 18 g of protein is stirred for 1 hour with 800 ml of DEAE-Sephadex that has been equilibrated with 0.05 M Tris buffer, pH 7.6, containing 10 mM NaCl. Under these conditions, the kinase is adsorbed by the resin and can be collected by filtration on a Büchner funnel with Whatman No. 54 filter paper. The resin is washed with approximately 3 liters of the Tris·NaCl buffer until the filtrate becomes colorless, suspended in 800 ml of 50 mM Tris buffer, pH 7.6, containing 0.30 M NaCl and stirred for 45 minutes. The DEAE-Sephadex is collected by filtration and washed twice with 400 ml of 50 mM Tris buffer, pH 7.6, containing 0.30 M NaCl. The pooled filtrates (1800 ml) containing the protein kinase activity, are brought to 35% saturation by the addition of solid $(NH_4)_2SO_4$ (199 g per liter), maintaining the pH of the solution between pH 7 and 8 as described under step 2. After 1 hour, the precipitate is collected by centrifugation at 10,000 g for 10 minutes and discarded. The supernatant fluid is then brought to 75% saturation by the addition of solid $(NH_4)_2SO_4$ (258 g per liter). The precipitate which forms after 1 hour is collected by centrifugation at 10,000 g for 10 minutes. The pellet is suspended in a minimal volume of 50 mM potassium phosphate buffer, pH 7.0, and dialyzed against 2 liters of the same buffer overnight. The combination of the DEAE batch elution and a second $(NH_4)_2SO_4$ fractionation serves to clarify the enzyme solution. Five individual preparations are purified through step 3 and pooled prior to step 4. Enzyme preparation from step 3 can be stored at $-20°$.

Step 4. Alumina $C\gamma$ Elution. The dialyzed $(NH_4)_2SO_4$ fraction (10.6 g of protein in 500 ml) is adsorbed to alumina $C\gamma$ gel by mixing the protein solution with gel (2.2 g of alumina $C\gamma$ per gram of protein) that had been previously washed with two 500-ml portions of 50 mM potassium phosphate buffer, pH 7.0. After stirring for 45 minutes, the gel is collected by centrifugation at 3000 g for 5 minutes and the supernatant fluid is discarded. The gel is then resuspended and washed two additional times with 250 ml of the same buffer. Protein kinase is eluted by suspending the alumina $C\gamma$ gel in 0.15 M potassium phosphate buffer, pH 7.0, stirring for 15 minutes and removing the gel by centrifugation. Elution is carried out once with 400 ml of buffer and twice with 250 ml of buffer. The supernatant fluids are pooled and concentrated by the addition of solid $(NH_4)_2SO_4$ to 85% saturation (569 g per liter). After centrifugation the pellet is suspended in 100 ml of 0.05 M Tris buffer, pH 7.6, containing 10 mM NaCl and dialyzed overnight against two 1-liter changes of the same buffer.

Step 5. Chromatography on DEAE-Cellulose. Concentrated protein kinase from the previous step (2 g of protein in 125 ml) is injected onto a column, 2.5 × 35 cm, of DEAE-cellulose previously equilibrated with 50 mM Tris buffer, pH 7.6, containing 10 mM NaCl. The column is then washed with 100 ml of 50 mM Tris buffer, pH 7.6, containing 60 mM NaCl. Protein kinase activity is eluted with a linear gradient of Cl⁻ generated by a reservoir containing 1 liter of 50 mM Tris buffer, pH 7.6, and 0.30 M NaCl (0.34 M Cl⁻), and a mixing chamber containing 1 liter of 50 mM Tris buffer, pH 7.6, and 60 mM NaCl (0.10 M Cl⁻). Ascending chromatography is carried out at a flow rate of 30 ml per hour and 15-ml fractions are collected. Protein kinase is eluted as a single peak of enzymatic activity between 0.13 and 0.15 M Cl⁻. The peak fractions are combined and concentrated by precipitation with $(NH_4)_2SO_4$ as described above, and the protein is redissolved and dialyzed against 50 mM potassium phosphate buffer, pH 7.0.

Step 6. Gel Filtration on Bio-Gel P-300. The concentrated protein kinase solution (510 mg in 12 ml) is next injected onto a column 2.5 × 95 cm, of Bio-Gel P-300 which had been previously washed with 50 mM potassium phosphate buffer, pH 7.0. Ascending gel filtration is performed with the same buffer at a flow rate of 5 ml per hour, and 3-ml fractions are collected. Protein kinase emerges from the column as a single peak of enzymatic activity which is coincident with the major peak of protein with K_{av}[6] = 0.167. Peak fractions are combined, concentrated with $(NH_4)_2SO_4$, and dialyzed against 50 mM potassium phosphate buffer, pH 7.0, as described earlier.

Step 7. Chromatography on Hydroxyapatite. The protein kinase preparation (111 mg in 5 ml) is injected onto a column (2.5 × 20 cm) of hydroxyapatite which had previously been equilibrated with 50 mM potassium phosphate buffer, pH 7.0. The column is washed in stepwise fashion with 200-ml aliquots of 50 mM, 0.10 M, and 0.15 M potassium phosphate buffer, pH 7.0. Protein kinase is then eluted with 0.25 M potassium phosphate buffer, pH 7.0.

Alternative Step 7. Chromatography on DEAE Cellulose. A second chromatographic separation on DEAE-cellulose can be substituted for chromatography on hydroxyapatite. The concentrated protein kinase preparation from step 6 is dialyzed against 50 mM Tris–10 mM NaCl

---

[6] K_{av} = (elution volume − void volume)/(total volume − void volume).

and then applied to a 0.9 × 20 cm column of DEAE-cellulose which had been previously equilibrated with the same buffer. The column is eluted with a linear gradient of Cl⁻ generated from a reservoir containing 150 ml of 50 mM Tris–0.21 M NaCl and a mixing chamber containing 150 ml of 50 mM Tris–10 mM NaCl. Descending chromatography is carried out at a flow rate of 20 ml per hour, and 2-ml fractions are collected. All the protein kinase activity is eluted as a symmetrical peak between 0.12 M Cl⁻ and 0.16 M Cl⁻.

A summary of the purification procedure is given in the table. Cyclic AMP-dependent protein kinase is purified approximately 1200-fold with a yield of 15% by this procedure. Similar results are obtained when alternate step 7 replaces hydroxyapatite chromatography.

PURIFICATION OF PROTEIN KINASE

Step[a]	Protein (mg)	Units		Specific activity	
		Catalytic (× 10⁻³)	Binding	Catalytic	Binding
1. Homogenate (10,000 g supernatant)	298,250	193.9	2326	0.65	0.0078
2. Ammonium sulfate	87,954	192.8	6245	2.2	0.071
3. DEAE-Sephadex	10,585	148.2	2752	14	0.26
4. Alumina Cγ	2,012	98.6	1730	49	0.86
5. DEAE-cellulose	510	57.1	1071	112	2.1
6. Bio-Gel P-300	111	31.9	488	287	4.4
7. Hydroxyapatite	37	29.9	333	807	9.0

[a] The data given in steps 1 to 3 represent the combined and averaged (where appropriate) values obtained from five preparations. Subsequent purification was carried out with the pooled protein kinase preparation.

Properties

Purity. Purified cAMP-dependent protein kinase is homogeneous as judged by sedimentation in the ultracentrifuge and electrophoresis in both standard polyacrylamide gels and polyacrylamide gels containing sodium dodecyl sulfate. Occasional preparations of protein kinase require a second cycle of steps 6 and 7 before homogeneity is achieved.

Specificity. Beef heart protein kinase catalyzes the phosphorylation of protamine, arginine-rich histone, and lysine-rich histone at relative rates of 1:0.5:0.2. Serum albumin and casein are poor substrates.

Nature of Product. Following incubation with protein kinase, protamine, and histones possess ^{32}P-labeled phosphoserine and/or phosphothreonine residues. These phosphoester bonds are stable to 5% trichloroacetic acid at 95° for 5 minutes, but are completely hydrolyzed by treatment with 1 M NaOH under the same conditions.

Activators. cAMP (2 μM) causes a 5-fold enhancement of the velocity of the phosphotransferase reaction. Other cyclic purine ribonucleotides (cIMP, cGMP) produce the same degree of stimulation at concentrations of 10–20 μM, while cyclic pyrimidine ribonucleotides (cUMP, cCMP) are effective at 100–200 μM.

The presence of a divalent cation is an absolute requirement for protein kinase activity. Mg^{2+} (10 mM) permits maximal enzymatic activity and optimal stimulation by cAMP. Co^{2+} can partially substitute for Mg^{2+}. No protein kinase activity is observed in the presence of Ca^{2+} (10 mM), and this cation also inhibits the enzyme in the presence of Mg^{2+}.

Reducing agents, such as dithiothreitol and 2-mercaptoethanol, are required in enzyme assays and all buffers in order to achieve maximal levels of enzymatic activity and stimulation by cAMP.

pH optimum. The enzyme exhibits a broad pH-activity profile with an optimum range between pH 7.0 and pH 7.8. Maximal stimulation by cAMP is observed at pH 7.0.

K_m Values. The apparent K_m values for ATP and Mg^{2+}, determined at a protamine concentration of 1.25 mg/ml, are 13 μM and 1.7 mM, respectively. Half maximal stimulation by cAMP occurs at a concentration of 60 nM.

Stability. Protein kinase activity is stable for 3 months at 4° in 50 mM phosphate buffer, pH 7.0 (protein concentration > 1.5 mg/ml), but the enzyme is converted into multiple molecular forms under these conditions.[3] This conversion can be minimized by storing the enzyme as a suspension in 3.3 M $(NH_4)_2SO_4$, pH 7.5, containing 2 mM EDTA.

Physicochemical Properties. cAMP-dependent protein kinase has an $s_{20,w}$ of 6.8, a Stokes radius of 60 Å, a partial specific volume of 0.735, and a molecular weight of 174,000.[7] It is composed of dissimilar catalytic and cAMP-binding subunits having molecular weights of approximately 40,000 and 50,000, respectively.[3]

Separation of Catalytic and cAMP-Binding Subunits. Active subunits may be prepared from purified protein kinase by the following procedure: Protein kinase (1 mg in 1 ml of 50 mM potassium phosphate buffer, pH 7.0) is applied to a column, 1 × 3 cm, of DEAE-cellulose which was previously equilibrated with the same buffer. After washing with 20 ml

---

[7] J. Erlichman, C. S. Rubin, and O. M. Rosen, *J. Biol. Chem.* **248**, 7607 (1973).

of buffer, cAMP-independent protein kinase (catalytic subunit) is eluted with 10 ml of buffer containing 10 μM cAMP. Approximately 85–90% of the original protein kinase activity is recovered in this fraction. Residual cAMP is then washed off the resin with 20 ml of 0.12 M potassium phosphate buffer, pH 7.0, and the cAMP-binding protein is eluted with 10 ml of 0.25 M potassium phosphate buffer, pH 7.0. Catalytic and binding components are separately dialyzed against 50 mM potassium phosphate buffer, pH 7.0, before further use.

[45] Preparation and Properties of Cyclic AMP-Dependent Protein Kinases from Rabbit Red Blood Cells[1]

By MARIANO TAO

The action of a wide variety of hormones has been shown by Sutherland and his associates[2] to be mediated by adenosine 3′,5′-cyclic monophosphate (cAMP). In rabbit skeletal muscle, a cAMP-dependent protein kinase which catalyzes the phosphorylation and activation of phosphorylase b kinase has been isolated by Walsh et al.[3] Thus one of the links in the sequence of reactions between hormonal stimulation of adenylate cyclase and increased glycogenolysis in muscle has been identified. Since the initial report, the occurrence of cAMP-dependent protein kinases has been shown to be widespread.[4–7] The action of these enzymes is found to be relatively nonspecific; several proteins, such as histone, protamine, casein, may also serve as substrates. The presence of multiple forms of protein kinase was first demonstrated in rabbit red blood cells[7] and, subsequently, was also found in other tissues.[8–10]

---

[1] This work was supported in part by grants from the American Cancer Society (BC-65 and BC-65A) and from the National Science Foundation (GB 27435 A#1). The author is an Established Investigator of the American Heart Association.

[2] G. A. Robison, R. W. Butcher, and E. W. Sutherland, "Cyclic AMP." Academic Press, New York, 1971.

[3] D. A. Walsh, J. P. Perkins, and E. G. Krebs, *J. Biol. Chem.* **243**, 3763 (1968).

[4] T. A. Langan, *Science* **162**, 579 (1968).

[5] J. F. Kuo and P. Greengard, *Proc. Nat. Acad. Sci. U.S.* **64**, 1349 (1969).

[6] J. D. Corbin and E. G. Krebs, *Biochem. Biophys. Res. Commun.* **36**, 328 (1969).

[7] M. Tao, M. L. Salas, and F. Lipmann, *Proc. Nat. Acad. Sci. U.S.* **67**, 408 (1970).

[8] E. M. Reimann, D. A. Walsh, and E. G. Krebs, *J. Biol. Chem.* **246**, 1986 (1971).

[9] A. Kumon, K. Nishiyama, H. Yamamura, and Y. Nishizuka. *J. Biol. Chem.* **247**, 3726 (1972).

[10] F. Lipmann, *Advan. Enzyme Regul.* **9**, 5 (1971).

The mechanism by which cAMP activates these enzymes is found to be similar for all the cAMP-dependent protein kinases studied.[10–15] These enzymes may exist as an inactive complex of two dissimilar functional subunits: a catalytic subunit and a regulatory subunit. cAMP causes the dissociation of the two subunits by binding to the regulatory moiety. It is in this dissociated form that the catalytic subunit functions as a phosphotransferase.

Assay Method

Principle. The phosphotransferase activity of cAMP-dependent protein kinases is determined by measuring the amount of ^{32}P incorporated into calf thymus histone with [γ-^{32}P]ATP as the phosphoryl donor. The cAMP binding activity is assayed by determining the amount of cyclic [^{3}H]AMP–enzyme complex retained on Millipore filters.

Reagents

    Buffer A: 1 M Tris·HCl, pH 8.5; 40 mM MgCl$_2$
    Buffer B: 20 mM Tris·HCl, pH 7.5; 4 mM MgCl$_2$
    Buffer C: 20 mM Tris·HCl, pH 7.5; 1 mM dithiothreitol
    Buffer D: 50 mM potassium phosphate, pH 6.8; 1 mM dithiothreitol
    Buffer E: 0.1 M Tris·HCl, pH 6.5; 1 mM dithiothreitol
    Calf thymus histone (Sigma Chemical Co.), 12 mg/ml, adjusted to pH 8.0 with KOH
    [γ-^{32}P]ATP (New England Nuclear), 2 mM, adjusted to pH 8.0 with KOH, specific activity 20–40 cpm/pmole
    Bovine serum albumin (Sigma Chemical Co.), 12 mg/ml
    Binding substrate: 1 μM cyclic [^{3}H]AMP, specific activity 3000–5000 cpm/picomole; 8 mM MgCl$_2$; 20 mM Tris-HCl, pH 7.5; 0.12 mg/ml protamine
    Trichloroacetic acid, 10% (w/v)
    Whatman GF/C glass-fiber filters, 2.4 cm diameter
    Millipore filters, HA 0.45 μm pore size, 2.4 cm diameter

Assay for Phosphotransferase Activity.[16] The reaction mixture contains 20 μl of buffer A, 20 μl of 2 mM [γ-^{32}P]ATP, 30 μl of 12 mg/ml

---

[11] G. N. Gill and L. D. Garren, *Biochem. Biophys. Res. Commun.* **39**, 335 (1970).
[12] M. Tao, *Ann. N.Y. Acad. Sci.* **185**, 227 (1971).
[13] A. Kumon, H. Yamamura, and Y. Nishizuka, *Biochem. Biophys. Res. Commun.* **41**, 1290 (1970).
[14] E. M. Reimann, C. O. Brostrom, J. D. Corbin, C. A. King, and E. G. Krebs, *Biochem. Biophys. Res. Commun.* **42**, 187 (1971).
[15] C. S. Rubin, J. Erlichman, and O. M. Rosen, *J. Biol. Chem.* **247**, 36 (1972).
[16] M. Tao, *Biochem. Biophys. Res. Commun.* **46**, 56 (1972).

of calf thymus histone, 10 µl of 20 µM cAMP, enzyme, and sufficient water to bring to a final volume of 0.2 ml. Incubation is performed at 37° for 5 minutes; and the reaction is terminated by the addition of 25 µl of 12 mg/ml of bovine serum albumin followed by 2 ml of 10% trichloroacetic acid. The precipitate is collected on Whatman GF/C glass-fiber filters, washed ten times with 2-ml portions of 10% trichloroacetic acid, and counted in 5 ml of Bray's solution.[17] One unit of enzyme activity is defined as that amount necessary to catalyze the incorporation of 1 pmole of ^{32}P into histone per minute. The enzyme specific activity is expressed in units per milligram of protein. Protein concentration is determined by the method of Lowry et al.[18] using bovine serum albumin as standard.

cAMP Binding Assay.[19] The binding mixture contains 50 µl of binding substrate and enzyme protein in a final volume of 100 µl. The mixture is incubated at 37° for 3 minutes and transferred to an ice-water bath. Under these conditions, maximum complex formation between cAMP and the enzyme is obtained. The mixture is then diluted with about 2 ml of cold buffer B and filtered through Millipore filter which has been presoaked in the same buffer. The filter is washed ten times with 2-ml portions of cold buffer B, dissolved in 5 ml of Bray's solution, and counted in a liquid scintillation spectrometer.

Preparation of Crude Cyclic AMP-Dependent Protein Kinases[20]

Preparation of Crude Lysate. Two liters of young (8–12 weeks) rabbit red blood cells (Pel-Freez, type 2, with heparin as anticoagulant) are added to 8 liters of 2.5 mM $MgCl_2$. The solution is stirred for 30 minutes and centrifuged at 15,000 g for 20 minutes. The supernatant is carefully separated from the loosely packed sediment and referred to as the "crude lysate."

Ammonium Sulfate Fractionation. To the crude lysate, solid ammonium sulfate is added until 50% saturation and let stand for 30 minutes. The protein precipitate is collected by centrifugation, dissolved in buffer C, and dialyzed overnight against 4 liters of the same buffer. After dialysis, the solution is centrifuged to remove any insoluble material. This ammonium sulfate step removes most of the hemoglobin and therefore gives rise to considerable enzyme purification.

---

[17] G. A. Bray, *Anal. Biochem.* 1, 279 (1960).
[18] O. H. Lowry, N. J. Rosebrough, A. L. Farr, and R. J. Randall, *J. Biol. Chem.* 193, 265 (1951).
[19] M. Tao, *Arch. Biochem. Biophys.* 143, 151 (1971).
[20] All operations are carried out at 0–4° unless otherwise specified.

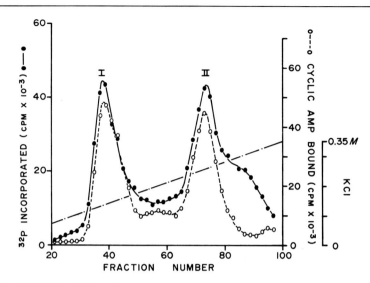

Fig. 1. DEAE-cellulose chromatography of the ammonium sulfate fraction. Experimental conditions are given in the text. The specific activities of [γ-^{32}P]ATP and cyclic [^{3}H]AMP are 28 cpm/pmole and 5.7×10^3 cpm/pmole, respectively.

DEAE-Cellulose Chromatography. The enzyme solution is applied to a DEAE-cellulose column (3.8 × 37 cm, Cellex D, Bio-Rad Laboratories) previously equilibrated with buffer C. The column is then washed with two column volumes of buffer C, and the enzymes are eluted with a linear gradient of 0–0.35 M KCl in buffer C, using a total volume of 2 liters and at a flow rate of about 80 ml per hour. Fractions of 21 ml each are collected and both kinase and cAMP binding activities are determined. As is shown in Fig. 1, two major kinase activity peaks with their corresponding cAMP binding activities are resolved on the column. The active fractions under each peak are pooled and designated as I (fractions 33–49) and II (fractions 65–80). Both fractions I and II are concentrated by precipitation with 50% saturated ammonium sulfate, dissolved in buffers D and E, respectively, and dialyzed overnight against 2 liters of their corresponding buffer.

Purification of Kinase I

Hydroxyapatite Column Chromatography. Fraction I is applied to a hydroxyapatite column (2.8 × 14 cm, Bio-Gel HTP, Bio-Rad Laboratories) which has been previously equilibrated with buffer D. The column is washed with two column volumes of buffer D and the enzyme eluted with a linear gradient of 50 mM to 0.3 M potassium phosphate buffer,

TABLE I
Purification of Cyclic AMP-Dependent Protein Kinase I

Fraction	Volume (ml)	Total protein (mg)	Specific activity[a] (units/mg)	Yield (%)	Purification (x-fold)
Crude lysate	8250	701,250	8.2	100	1
Ammonium sulfate	360	11,520	1,200	238	146
DEAE-cellulose	26	510	3,530	31	430
Hydroxyapatite	15	177	8,200	26	1000
Heated	25	75	22,300	29	2700
Sephadex G-200	7	33	35,600	21	4340

[a] One unit of kinase activity is defined as that amount of enzyme which will catalyze the incorporation of 1 pmole of ^{32}P into calf thymus histone per minute at 37°. Specific activity is expressed in units per milligram of protein.

pH 6.8, containing 1 mM dithiothreitol. The total volume of the gradient is 500 ml, and 10-ml fractions are collected at a flow rate of about 30 ml per hour. The active fractions are pooled, concentrated by precipitation with 50% saturated ammonium sulfate, dissolved in buffer C, and dialyzed overnight against 2 liters of the same buffer.

Heat Treatment. The enzyme solution is made 4 mM in $MgCl_2$ and 0.2 mM in ATP and heated with constant stirring at 53° for 7 minutes. Under these conditions, kinase I is stable toward heat.[19] The denatured proteins are removed by centrifugation and the supernatant concentrated by Diaflo ultrafiltration (UM-20E membrane, Amicon Corp.) to about 3 ml.

Sephadex G-200 Gel-filtration. The concentrated enzyme solution is then applied to a Sephadex G-200 column (2.5 × 90 cm, Pharmacia Fine Chemicals) which has been previously equilibrated with buffer C. The elution is conducted upward at a flow rate of about 13 ml per hour and fractions of 6.5 ml are collected. The active fractions are pooled, concentrated by Diaflo ultrafiltration, and stored in liquid nitrogen. A summary of the results of the purification of kinase I is shown in Table I. The overall procedure represent a purification of greater than 4000-fold.

Purification of Kinases IIa and IIb

QAE-Sephadex Chromatography. The fraction II obtained from DEAE-cellulose chromatography is applied to a 3.8 × 36 cm QAE-Sephadex (A-50, Pharmacia Fine Chemicals) column previously equilibrated with buffer E. The column is washed with two column volumes of buffer

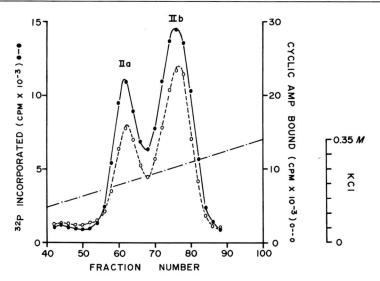

FIG. 2. Separation of kinases IIa and IIb by QAE-Sephadex chromatography. Experimental conditions are given in the text. The specific activities of [γ-^{32}P]ATP and cyclic [^{3}H]AMP are 26 cpm/pmole and 1.8×10^4 cpm/pmole, respectively.

E and the enzymes eluted with a linear gradient of 0–0.35 M KCl in buffer E. The elution is carried out using a total volume of 2 liters of the solutions at a flow rate of about 80 ml per hour; 20-ml fractions are collected. Figure 2 shows the resolution of fraction II into two kinase activity peaks, referred to as IIa (fractions 56–68) and IIb (fractions 69–84). Each of the pooled fractions are concentrated by precipitation with 50% ammonium sulfate, dissolved in buffer D, and dialyzed overnight against 2 liters of the same buffer. In order to obtain a good separation of IIa and IIb, it is desirable to adsorb the enzyme to the column within the upper 4–8 cm of the ion exchange bed.

Hydroxypatite Chromatography. Kinases IIa and IIb are further fractionated by chromatography on hydroxyapatite column. Each enzyme solution is applied to a 2.8×8 cm hydroxyapatite column equilibrated with buffer D. The column is washed with two column volumes of buffer D, and the enzyme is eluted with a linear gradient of 50 mM to 0.3 M potassium phosphate buffer, pH 6.8, containing 1 mM dithiothreitol. The elution is carried out at a flow rate of about 25 ml per hour using a total gradient of 500 ml, and 10-ml fractions are collected. Alternate fractions are assayed for both kinase and binding activities. The active fractions are pooled, concentrated by precipitation with 50% saturated ammonium sulfate, dissolved in buffer C, and dialyzed against the same

TABLE II
Purification of Cyclic AMP-Dependent Protein Kinase IIa

Fraction	Volume (ml)	Total protein (mg)	Specific activity (units/mg)	Yield (%)	Purification (x-fold)
Crude lysate	8250	701,250	8.2	100	1
Ammonium sulfate	360	11,520	1,200	238	146
DEAE-cellulose	32	1,024	2,800	50	342
QAE-Sephadex	10	150	2,900	7.6	354
Hydroxyapatite	3.2	28	9,100	4.5	1100
Sephadex G-200	5.3	5.3	16,200	1.6	1980

buffer. Although chromatography on the hydroxyapatite column shows substantal purification of both kinases IIa and IIb, it does not resolve IIa from IIb.

Sephadex G-200 Gel Filtration. The IIa and IIb enzyme solutions are each applied to a 2.5 × 90 cm Sephadex G-200 column equilibrated with buffer C. Application and elution of the enzyme are conducted upward at a flow rate of about 13 ml per hour; fractions of 6–7 ml are collected. Alternate fractions are assayed for both kinase and binding activities. The active fractions are pooled, concentrated by Diaflo ultrafiltration and stored in liquid nitrogen. Filtration on Sephadex G-200 gives considerable purification for IIa but not for IIb. However, owing to a difference in their molecular weights, this procedure further separates IIa from IIb where complete separation of the two kinases is not obtained on the QAE-Sephadex column. The purification of kinases IIa and IIb are summarized in Tables II and III, respectively. They are both purified to about 2000-fold.

TABLE III
Purification of Cyclic AMP-Dependent Protein Kinase IIb

Fraction	Volume (ml)	Total protein (mg)	Specific activity (units/mg)	Yield (%)	Purification (x-fold)
Crude lysate	8250	701,250	8.2	100	1
Ammonium sulfate	360	11,520	1,200	238	146
DEAE-cellulose	32	1,024	2,800	50	342
QAE-Sephadex	10	176	5,700	17	695
Hydroxyapatite	12	54	12,000	11	1500
Sephadex G-200	10	22.3	12,420	5	1510

Comments

The procedure described above gives reproducible resolution of cAMP-dependent protein kinases from rabbit red blood cells into three active fractions, kinases I, IIa, and IIb. No detectable cAMP phosphodiesterase activity is present in any of these fractions. Analysis by polyacrylamide gel electrophoresis reveals that all three fractions are functionally homogeneous, each fraction giving rise to a single cAMP-dependent protein kinase activity peak.[21] On sucrose density gradient centrifugation using *Escherichia coli* alkaline phosphatase (6.3 S) and ovalbumin (3.6 S) as markers, the sedimentation rates of I, IIa, and IIb are estimated to be 7.4, 5.2, and 7.2, respectively.

All three kinases are stimulated by low concentrations of cAMP with K_m values between 10 nM and 0.1 μM. However, the degree of stimulation varies, with kinase I being affected the most (about 10–40-fold) and IIa and IIb about equally stimulated (2–5-fold). The degree of stimulation is also dependent on the protein substrate used. The three kinases share some common properties, with K_m values for calf thymus histone and ATP at about 0.25 mg/ml and 10 μM, respectively. Lysine-rich histone, arginine-rich histone, protamine, casein, and albumin also serve as phosphoryl acceptors for these enzymes but with varying degree of effectiveness. No appreciable phosphotransferase activity is observed when GTP replaces ATP. The pH optimum is at 8.5 and all require Mg^{2+} for activity. Other cyclic nucleotide such as cUMP, cCMP, and cGMP at high concentration ($K_m \sim 10\ \mu M$) may replace cAMP to give maximal stimulation. On the other hand, cTMP has no effect on these enzyme reactions.

The kinases differ somewhat in their heat stability. Kinases IIa and IIb are relatively more stable than kinase I.[21] However, the stability of kinase I toward heat (53° for 7 minutes) may be greatly enhanced in the presence of both Mg^{2+} and ATP.[19] No similar protective effect is observed with IIa and IIb. The enzymes may be stored in liquid nitrogen for several months with no detectable loss of either phosphotransferase or cAMP binding activity.

[21] M. Tao and P. Hackett, *J. Biol. Chem.* **248**, 5324 (1973).

[46] Cytoplasmic Hepatic Protein Kinases

By LEE-JING CHEN *and* DONAL A. WALSH[1]

Studies of the protein kinase activity of liver have been approached from several points of view and in consequence have led to the description of several potentially different protein kinase activities. However, comparative and detailed data on the various protein kinases is still sparse, and so it is not possible to clearly identify each of the activities as separate molecular entities. In particular, the exact correlation between the nuclear and cytoplasmic activities is uncertain. Properties of some of the former are described by Kish and Kleinsmith.[2] This presentation is restricted to the enzymes that are found principally in the cytoplasm.

I. Cytosol Cyclic AMP-Regulated Protein Kinase

Mechanism of Cyclic AMP Action

Cyclic AMP (cAMP) activates cAMP-dependent protein kinase according to the mechanism delineated for the enzyme from many tissues, Eq. (1), and described elsewhere in this volume.[3]

$$RC \text{ (inactive)} + cAMP \rightleftharpoons R \cdot cAMP + C \text{ (active)} \qquad (1)$$

In consequence, this discussion of cAMP-dependent protein kinase includes not only the properties of the holoenzyme (RC) but also of the active catalytic fraction (C) and of the cAMP binding unit (R).

Assay Method

The procedure is essentially that of Reimann *et al.*[4] for rabbit muscle protein kinase with a slight modification.

Reagents

Assay buffer: 50 mM 2-(N-morpholino)ethanosulfonate, 50 mM magnesium chloride, 50 mM sodium chloride, 10 mM aminophylline, pH 5.9

---

[1] Established Investigator of American Heart Association.
[2] V. Kish and L. Kleinsmith, *J. Biol. Chem.* **249,** 750 (1974); this series Vol. 40, [16].
[3] J. A. Beavo, P. J. Bechtel, and E. G. Krebs, this volume [43].
[4] E. M. Reimann, D. A. Walsh, and E. G. Krebs, *J. Biol. Chem.* **246,** 1986 (1971).

Histone f₂b, prepared by the methods of Johns,[5] 1 mg/ml
[γ-³²P]Adenosine triphosphate, 10 mM, specific activity 4–9 × 10⁹ cpm/mmole, pH 7.0
cAMP, 10 μM (when added)

Procedure. A cocktail is prepared fresh before each assay containing equal amounts of the above reagents. An aliquot of 100 μl is pipetted to a test tube (12 × 75 mm). The reaction is started by adding 20 μl of enzyme preparation and incubated for 25 minutes at 30° in a water bath. The reaction is terminated by removing 50 μl of the mixture with an Eppendorf pipette onto a filter paper (2.0 × 2.0 cm of Whatman No. 31ET). The filter paper is dropped immediately into cold 10% TCA solution with a ratio of one paper per 10 ml of TCA solution. The paper is washed in cold 10% TCA for 30 minutes, cold 5% TCA for 30 minutes, and twice in 5% TCA for 30 minutes at room temperature. The paper is washed with 95% ethanol (enough to cover the papers) 5 minutes and rinsed with ether, dried, and transferred to a toluene-based scintillation fluid for counting.

Purification

A. Purification of Holoenzyme[6]

Procedure 1

Step 1. Preparation of Tissue Extract. All purification procedures are carried out at 4° unless otherwise stated. Minced rat liver (4 g) is homogenized with 20 ml of 0.25 M sucrose containing 5 mM Tris·Cl (pH 7.5), 1 mM EDTA, and 10 mM β-mercaptoethanol, utilizing a Potter-Elvehjem homogenizer with a Teflon pestle. The homogenate is centrifuged at 105,000 g for 35 minutes and the resultant supernatant solution is passed through two layers of cheesecloth to remove lipid.

Step 2. Fractionation on DEAE-Sephadex A-25. The filtrate (12 ml) is applied to a 1.8 × 22 cm column of DEAE-Sephadex A-25, which has been previously equilibrated with 5 mM Tris·Cl buffer (pH 7.5), containing 1 mM EDTA, and 10 mM β-mercaptoethanol (TME buffer). The column is eluted with a linear gradient of NaCl in the same buffer. The reservoir contains 50 ml of 0.5 M NaCl in TME buffer, and the mixing flask contains 50 ml of the same buffer without NaCl. Fractions of 2 ml are collected at a rate of 1 ml per minute. Three enzymatically active fractions are resolved by this procedure. The activity of fraction I, eluted between 0 and 0.06 M NaCl, is not stimulated by cAMP, and is free

[5] E. W. Johns, *Biochem. J.* **9**, 55 (1964).
[6] L. J. Chen and D. A. Walsh, *Biochemistry* **10**, 3614 (1972).

catalytic subunit. Fractions II and III are eluted between 0.08 and 0.25 M, and between 0.27 and 0.36 M NaCl, respectively. The activities of each of these two fractions are dependent on cAMP. Fraction II consisted of more than a single enzymatic component (see below).

Step 3. *Isoelectrofocusing Electrophoresis.* Technique of isoelectrofocusing electrophoresis is performed according to the method originally described by Svensson[7] using a 110-ml column (LKB Instruments, Inc.) maintained at 0° by a circulating water bath. A 2% carrier ampholyte with a pH range of 5 to 8 is used. The pH gradient is established during electrophoresis following the sequential addition of ampholyte solution in a 0 to 47% (w/v) sucrose gradient. Fractions II and III separated by chromatography on the DEAE-Sephadex A-25 column are each dialyzed against 10 volumes of TME buffer with two changes for 3 hours. The dialyzed solution (II or III) is applied in the center of the column. Electrophoresis is initiated at 200 V, increased to 800 V over a period of 24 hours, and continued at this voltage for at least another 16 hours for equilibration. Upon completion of the electrophoresis, fractions of 2 ml are collected and assayed for protein kinase activity and pH determination. Fractions II and III are each eluted as single peaks of enzymatic activity at pH of 5.2. They may be stored at 4° for a few days following dialysis against 0.25 M sucrose in TME buffer.

Procedure 2

The method described here is an alternative method originally described by Kumon *et al.*[8] and Yamamura *et al.*[9]

Step 1. *Tissue Extract.* The rat liver (12 g) is homogenized with 5 volumes of 0.25 M sucrose containing 6 mM β-mercaptoethanol and 3.3 mM CaCl$_2$ utilizing a Potter-Elvehjem homogenizer. Diisopropyl fluorophosphate, suspended in 5 volumes of isopropanol, is added to the homogenate to give a final concentration of 10 mM, and the mixture is stirred for 20 minutes at 0°. The solution is then centrifuged at 20,000 g for 20 minutes. The supernatant (50 ml) is brought to 70% saturation with 23.6 g of ammonium sulfate. The precipitate is collected by centrifugation and dissolved in 20 ml of 20 mM Tris·Cl buffer, pH 7.5, containing 6 mM β-mercaptoethanol. The solution is dialyzed against 2 liters of the same buffer for 15 hours with two changes.

Step 2. *DEAE-Sephadex A-50 Chromatography.* The dialyzate is ap-

---

[7] H. Svensson, *Acta Chem. Scand.* **16**, 456 (1962).

[8] A. Kumon, H. Yamamura, and Y. Nishizuka, *Biochem. Biophys. Res. Commun.* **41**, 1290 (1970).

[9] H. Yamamura, Y. Inoue, R. Shimomura, and Y. Nishizuka, *Biochem. Biophys. Res. Commun.* **46**, 589 (1972).

plied to a DEAE-Sephadex A-50 column (2 × 35 cm) that is equilibrated with 20 mM Tris·Cl, pH 7.5, containing 50 mM NaCl and 6 mM β-mercaptoethanol. After the column is washed with 250 ml of the same buffer, enzyme elution is carried out with a linear concentration gradient of NaCl. The mixing flask and reservoir each contains 200 ml of 50 mM and 0.6 M NaCl, respectively, in 20 mM Tris·Cl, pH 7.5, containing 6 mM β-mercaptoethanol. Fractions of 5 ml are collected. Three fractions with protein kinase activities are resolved by this procedure. Fraction A, eluted between fractions 21 and 36, fraction B eluted between fractions 40 and 54, the activity of each was stimulated by cAMP. Fraction C, eluted between tubes 56 and 70, is not stimulated by cAMP and is more active with casein or phosvitin as substrates than with histone. The last enzyme is possibly the same species as phosvitin kinase which is described in the following section.

Step 3. Hydroxyapatite Chromatography. Each of the first two fractions (A and B) is further purified on a column of hydroxyapatite (2 × 5 cm). The column is equilibrated with 30 mM potassium phosphate, pH 7.5, containing 6 mM β-mercaptoethanol. After the column is washed with 100 ml of the same buffer, elution is carried out with a linear gradient of potassium phosphate. The mixing chamber contains 200 ml of 50 mM potassium phosphate, pH 7.5, in 6 mM β-mercaptoethanol, and the reservoir contains 0.25 M potassium phosphate, pH 7.5, and in 6 mM β-mercaptoethanol. Fractions of 4 ml are collected. Fraction A is resolved into two enzymatically active fractions. The activity in the first peak (tubes 20 through 45) is cAMP dependent, whereas the second peak (tubes 46 through 50) is cAMP independent. The chromatogram of fraction B exhibits a broad peak (tubes 51–75) which does not correspond to either peak of fraction I. The identity of these fractions with those prepared by procedure 1 has not been determined.

B. Preparation of Catalytic Subunit

Modification of procedure 1 for the preparation of holoenzyme can yield 3 catalytic subunits that are distinct on the basis of isoelectric point. Fraction I obtained from chromatography on DEAE-Sephadex A-25 has an isoelectric point of 8.2. Isoelectrofocusing electrophoresis of fraction II, after preincubation with cAMP, yields two catalytic subunits (II_A and II_B) of pI 7.6 and 8.5, respectively. These various forms of catalytic subunits (I, II_A and II_B) can be identified as such on the basis of interaction with regulatory subunit or the heat stable inhibitor protein.[10]

[10] J. A. Traugh, C. D. Ashby, and D. A. Walsh, this volume [42].

C. Preparation of Regulatory Subunit

Several methods are available for preparation of protein kinase regulatory subunit.[6,8,11] The procedure described here is that of Kumon et al.[8,11] Rat liver (16 g) is homogenized in 90 ml of 0.25 M sucrose, pH 7.5, containing 3.3 mM $CaCl_2$ with a Potter-Elvehjem homogenizer with a Teflon pestle. The homogenate is filtered through four layers of cheesecloth to remove lipid, and centrifuged at 105,000 g for 45 minutes. Fifteen grams of finely powdered $(NH_4)_2SO_4$ is added slowly to 48 ml of the above supernatant. The precipitate is collected by centrifugation after stirring for 30 minutes, and then dissolved in 7.5 ml of 10 mM Tris·Cl buffer, pH 7.5, containing 10% glycerol and 6 mM β-mercaptoethanol (TMG buffer). After dialysis against 5 liters of TMG buffer overnight, 8.8 ml of dialyzate is added to 24 ml calcium phosphate gel suspension (27.3 mg/ml in TMG buffer). The gel is washed three times with 24 ml TMG buffer, and the regulatory subunit is eluted from the gel with 56 ml of 0.2 M potassium phosphate buffer, pH 8.1, containing 10% glycerol and 6 mM β-mercaptoethanol. The eluate is dialyzed against 3 liters of TMG buffer with 2 changes for 14-16 hours and applied to a DEAE-cellulose column (2 × 18 cm), which has been previously equilibrated with TMG buffer. The column is initially washed with 100 ml of TMG buffer containing 50 mM NaCl and subsequently eluted with a linear NaCl concentration gradient (300 ml, 50 mM to 0.5 M). The flow rate is maintained at 18 ml per hour and fractions of 2.5 ml are collected. Fractions which contain protein kinase activity and regulatory subunit, tubes 100 through 150, are pooled and dialyzed against 5 liters of TMG buffer overnight with two changes. The solution is chromatographed on a hydroxyapatite column (1 × 4 cm) equilibrated with the same buffer. After the column is washed with 250 ml of 10 mM potassium phosphate, pH 7.5, containing 6 mM β-mercaptoethanol and 10% glycerol, the regulatory protein is eluted as a sharp peak with 250 ml of 30 mM potassium phosphate, pH 7.5, containing 6 mM β-mercaptoethanol and 10% glycerol. The regulatory protein thus obtained is free of protein kinase.

Properties

The various hepatic cAMP-dependent protein kinases have to date only been minimally characterized. To the limited extent investigated, hepatic protein kinase shows a rather broad substrate specificity catalyzing the phosphorylation of histone, rabbit skeletal muscle glycogen phos-

[11] A. Kumon, K. Nishiyama, H. Yamamura, and Y. Nishizuka, *J. Biol. Chem.* **247**, 3726 (1972).

phorylase b,[9] and ribosomal protein,[12] and converting muscle glycogen synthetase I to the D form.[9] Salmon sperm protamine and bovine casein are approximately 12% and 3% as active as calf thymus histone, respectively. Egg yolk phosvitin, bovine albumin, and human γ-globulin are not substrates. The rate of phosphorylation of different histone fractions is in the following order: $f_2b > f_2a$, $f_3 > f_1$ with a slightly different activity ratio between different catalytic subunits.[6] The different forms of catalytic subunits exhibit similar kinetic properties. The K_m for ATP is 20 μM and the optimum pH is 7.0. Each species exhibits an $s_{20,w}$ of 4.0. The forms of holoenzyme are identical in size ($s_{20,w} = 6.8$), charge ($pI = 5.2$), and apparent K_a for cyclic AMP (40 nM).

II. Cytoplasmic Phosvitin Kinase

Assay Method

Reagents

Assay buffer: 0.1 M 2-(N-morpholino)ethanosulfonate, 0.1 M NaF, 0.1 M theophylline, pH 7.5

Phosvitin, 10 mg/ml (Nutritional Biochemicals, Cleveland)

[γ-^{32}P]Adenosine triphosphate, 6 mM, specific activity 4–8 × 10^9 cpm/mmole; magnesium acetate, 18 mM, adjusted to pH 7.0

Procedure. Each tube contains 10 μl of MES buffer, 10 μl of phosvitin solution, and 20 μl of ATP/Mg^{2+}. The reaction is initiated by adding 20 μl of enzyme. The mixture is incubated at 30° for 10 min and the reaction is terminated by pipeting 50 μl of the reaction mixture to a filter paper. The remainder of the procedure is as described in the above section.

Purification Procedure

The procedure described here is that of Baggio *et al.*[13] Rat liver (80 g) is homogenized with 400 ml of 0.25 M sucrose with a Potter-Elvehjem homogenizer. The homogenate is centrifuged for 60 minutes at 105,000 g. The supernatant (about 500 ml) is dialyzed against 5 liters of 50 mM Tris·Cl buffer, pH 7.5, for 4 hours with two changes. The dialyzate is applied to a phosphocellulose column (4.5 × 15 cm) previously equilibrated with the same buffer. The column is washed with 150 ml of the same buffer containing 0.25 M NaCl at a flow rate of 1.5 ml per minute.

[12] C. Eil and I. G. Wool, *Biochem. Biophys. Res. Commun.* **43**, 1001 (1971).

[13] B. Baggio, L. Pinna, V. Moret, and N. Siliprandi, *Biochim. Biophys. Acta* **212**, 515 (1970).

The bulk of protein is removed by this procedure, but the enzyme is retained on the column and is subsequently eluted by 300 ml of 0.75 M NaCl. The enzyme is purified more than 1000-fold with a nearly 100% recovery by this simple procedure.

Properties

The properties of the enzyme have not yet been well characterized; the phosphorylation of several proteins has been described. The relative activity with different substrates is casein (1.0) > histone f_2b (0.27) > histone f_3 (0.12) > histone f_2a (0.10) > histone f_1 (0.08). The enzyme is not stimulated by cAMP nor is it inhibited by regulatory subunit of cAMP-dependent protein kinase nor by heat-stable protein inhibitor. Thus, according to the criteria presented elsewhere in this volume,[10] this enzyme is classified as a type III protein kinase and is distinguished from the various forms of catalytic subunit of cAMP-dependent protein kinase.

Acknowledgments

This work was supported by Grant AM 13613 from the U.S. Public Health Service.

[47] Purification and Characterization of Cyclic GMP-Dependent Protein Kinases

By J. F. Kuo and Paul Greengard

After the discovery of cyclic AMP (cAMP)-dependent protein kinase in mammalian skeletal muscle,[1] this class of enzyme was subsequently found in liver,[2] and in numerous other vertebrate[3] and invertebrate[3-5] tissues. In addition, the occurrence of cyclic GMP (cGMP)-dependent protein kinases, which are activated by low concentrations of cGMP rather than by cAMP, was established in various lobster,[4] insect,[5] and mammalian tissues.[6] In general, cAMP-dependent enzymes appear to be

[1] D. A. Walsh, J. P. Perkins, and E. G. Krebs, *J. Biol. Chem.* **243,** 3763 (1968).
[2] T. A. Langan, *Science* **162,** 579 (1968).
[3] J. F. Kuo and P. Greengard, *Proc. Nat. Acad. Sci. U.S.* **64,** 1349 (1969).
[4] J. F. Kuo and P. Greengard, *J. Biol. Chem.* **245,** 2493 (1970).
[5] J. F. Kuo, G. R. Wyatt, and P. Greengard, *J. Biol. Chem.* **246,** 7159 (1971).
[6] P. Greengard and J. F. Kuo, *in* "Role of Cyclic AMP in Cell Function" (P. Greengard and E. Costa, eds.), p. 287. Raven, New York, 1970.

present in a much higher level than cGMP-dependent enzymes in mammalian tissues, in which only trace amounts of the latter class of enzyme activity are detectable. In contrast, many arthropod tissues represent exceptionally rich sources for cGMP-dependent protein kinases. The distribution pattern of the two classes of protein kinases in various species and tissues suggests a prominent role for the cAMP system in mammals and for the cGMP system in arthropoda. It also suggests distinctive roles for the two classes of cyclic nucleotides and protein kinases in regulating cellular function.

The natural substrates of cGMP-dependent protein kinase are not yet known in any tissue. We have used the commercial preparations of histones, whose phosphorylation by ATP is effectively catalyzed by both cGMP-dependent and cAMP-dependent protein kinases from all tissues studied, as an artificial phosphate acceptor for assaying protein kinase activity.[3-5]

Standard Assay Procedure

Cyclic nucleotide-dependent protein kinase activity is assayed as described elsewhere.[5-7] The standard incubation mixture contains, in a final volume of 0.2 ml, sodium acetate buffer, pH 6.0, 10 μmoles; magnesium acetate, 2 μmoles; histone mixture, 40 μg; [γ-^{32}P]ATP, 1 nmole, containing about $1-2 \times 10^6$ cpm; with or without added cyclic nucleotide. The incubation is carried out at 30° for 5 minutes. The amount of cyclic nucleotide added to the incubation mixture for the routine assay of protein kinase is 100 pmoles. At this concentration (0.5 μM), cGMP maximally or near-maximally stimulates the cGMP-dependent class of protein kinase while having little or no effect on most cAMP-dependent enzymes. cAMP at the same concentration, on the other hand, greatly stimulates cAMP-dependent enzyme activity with little or no effect on cGMP-dependent enzyme activity. One unit of protein kinase activity is defined as that amount of enzyme that transfers 1 pmole of ^{32}P from [γ-^{32}P]ATP to the recovered substrate protein under the standard assay conditions.

Purification

Method 1

Lobster tail muscle, which contains substantial amounts of both cGMP-dependent and cAMP-dependent protein kinase activity, is used as an example for enzyme purification.

[7] J. F. Kuo and P. Greengard, this volume [12].

Step 1. Preparation of Crude Extract. About 200–300 g of tail muscle from three live lobsters (Maine) obtained from a local fish market are cut into small pieces, and then homogenized with 3–4 volumes of neutral 4 mM EDTA solution for 2 minutes in a Waring blender. The homogenate is centrifuged at 27,000 g for 30 minutes in the cold. All procedures used for the purification of the enzyme are performed at 4°. All buffers used in the succeeding steps of the purification contain 2 mM EDTA.

Step 2. Acid Precipitation. The supernatant solution obtained after centrifugation is adjusted to pH 4.8 by dropwise addition of ice-cold 1 N acetic acid, with stirring. After the enzyme solution is allowed to stand for 10 minutes, the precipitate is removed by centrifugation. The pH of the clear supernatant solution is then adjusted to 6.8 with 1 M potassium phosphate buffer, pH 7.2.

Step 3. Ammonium Sulfate Precipitation. Solid ammonium sulfate (33 g/100 ml) is added to the resultant supernatant solution from the previous step. After 20 minutes of stirring, the precipitate is collected by centrifugation and dissolved in 6% of the crude extract volume of 5 mM potassium phosphate buffer, pH 7.0. The resulting solution is dialyzed overnight against 20 volumes of the same buffer, with two changes of the buffer. After dialysis, the solution is centrifuged for 30 minutes at 27,000 g, and the precipitate is discarded.

Step 4. DEAE-Cellulose Chromatography. The dialyzed enzyme solution obtained from the ammonium sulfate step (containing both cGMP-dependent and cAMP-dependent protein kinases) is then applied to a column (2.4 × 14 cm) of DEAE-cellulose previously washed with 300 mM potassium phosphate buffer, pH 7.0, and equilibrated with 5 mM potassium phosphate buffer, pH 7.0. Protein is eluted from the column by stepwise application of 100 ml each of 5, 50, 100, and 200 mM potassium phosphate buffer, pH 7.0. cGMP-dependent protein kinase activity is associated with the protein peak eluted by 5 mM phosphate buffer, and cAMP-dependent enzyme activity is found in the protein peak eluted by 50 mM phosphate buffer. No significant cyclic nucleotide-dependent enzyme activity is found in protein peaks eluted by 100 and 200 mM phosphate buffer. The elution patterns of the enzymes from the column are illustrated in Fig. 1. The active fractions eluted with 5 and 50 mM phosphate are individually pooled, and dialyzed extensively against 5 mM potassium phosphate buffer, pH 7.0. They are designated peak 1 and peak 2, respectively. The contamination by cAMP-dependent activity in peak 1 and by cGMP-dependent activity in peak 2 is low. Table I summarizes the purification of these lobster enzymes. The preparations of the two classes of protein kinases at this stage of purity are satisfac-

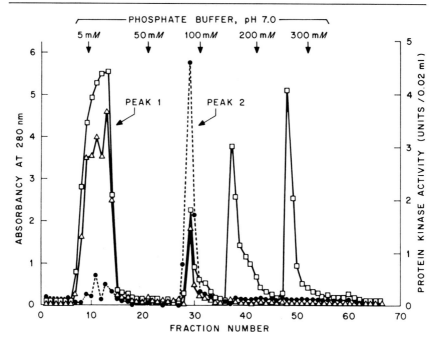

Fig. 1. Separation of cyclic GMP-dependent protein kinase activity from cyclic AMP-dependent protein kinase activity by chromatography on DEAE-cellulose. The optical density (□——□) and protein kinase activity assayed in the presence of 0.5 μM cyclic GMP (△——△) or 0.5 μM cyclic AMP (●····●) are shown as a function of fraction number. Fraction size was 9 ml. Protein kinase activity has been corrected for activity which occurred in the absence of added cyclic nucleotide. Taken from J. F. Kuo and P. Greengard, *J. Biol. Chem.* **245**, 2493 (1970).

tory for many studies of their properties. Peak 1 is also satisfactory as an analytic agent for assaying cGMP levels in tissues.[7,8]

Step 5. Calcium Phosphate Gel Treatment. The cGMP-dependent enzyme activity in peak 1 from the previous step can be readily purified about 2-fold further by the use of calcium phosphate gel. The amount of calcium phosphate gel suspension added is about 0.4–0.6 mg dry weight per 1.0 optical density unit (at 280 nm) of the enzyme solution. After gentle stirring of the mixture for 5 minutes in ice, it is centrifuged at 27,000 g for 10 minutes, and the precipitate is discarded. More than 95% of the cGMP-dependent enzyme activity present in the starting solution is recovered in the supernatant. The sensitivity to cGMP and the extent of activation by the cyclic nucleotide are comparable to the preparation from the DEAE-cellulose step.

[8] J. F. Kuo, T. P. Lee, P. L. Reyes, K. G. Walton, T. E. Donnelly, Jr., and P. Greengard, *J. Biol. Chem.* **247**, 16 (1972).

TABLE I
PREPARATION OF CYCLIC GMP (cGMP)-DEPENDENT AND CYCLIC AMP (cAMP)-DEPENDENT PROTEIN KINASES FROM LOBSTER MUSCLE[a,b]

Fraction	Volume (ml)	Protein (mg)	Protein kinase specific activity (units/mg protein) in presence of							
			No addition	cAMP	cIMP	cGMP	cUMP	cCMP	Cyclic dTMP
Crude extract	880	11,616	0.6	0.6	0.7	0.6	0.7	0.6	0.6
pH 4.9 supernatant	880	5,456	0.7	0.8	1.0	0.9	0.9	0.7	0.6
$(NH_4)_2SO_4$ precipitate	86	705	7.0	31.8	22.3	28.7	8.4	8.6	7.0
DEAE eluate, peak 1	60	330	10.4	14.0	13.1	38.8	10.6	10.1	10.1
DEAE eluate, peak 2	30	87	15.3	79.3	49.8	27.6	22.9	21.4	14.5

[a] Taken from J. F. Kuo and P. Greengard, *J. Biol. Chem.* **245**, 2493 (1970).
[b] Muscle (239 g) collected from the tails of three live lobsters was used as starting material. Cyclic nucleotides were present, where indicated, at a concentration of 0.5 μM.

Method 2

Steps 1, 2, and 3 are the same as those used in method 1. The starting materials, usually consisting of 0.5–10 g of fresh or frozen arthropod (e.g., lobster or insect) tissues, are homogenized with a small glass homogenizer, or an Omni-Mixer with or without the attachment for small samples.

Step 4. Calcium Phosphate Gel Treatment. This step replaces the DEAE-cellulose chromatography step in method 1 for the separation of cAMP-dependent enzyme activity (if also present in the same tissue) from cGMP-dependent enzyme activity. To the dialyzed enzyme solution from the ammonium sulfate step is added calcium phosphate gel. The amount of gel (30 mg, dry weight, per milliliter of gel suspension) added represents about 5 mg of dry gel per 1.0 optical density unit (at 280 nm) of the enzyme solution. After gentle stirring of the mixture for 5 minutes in ice, the gel is recovered by centrifugation. The supernatant fluid contains little or no cyclic nucleotide-dependent protein kinase activity and is discarded. cGMP-dependent enzyme activity is eluted first from the gel by suspending it in an appropriate volume of 50 mM potassium phosphate buffer, pH 7.0, and stirring gently for 5 minutes in ice. The gel, collected by centrifugation, is eluted once again with the same buffer. The two eluates, obtained with 50 mM phosphate buffer, are pooled and designated gel eluate 1. For the recovery of cAMP-dependent activity, an appropriate volume of 200 mM potassium phosphate buffer, pH 7.0, is then added to the gel, and the elution procedure is carried out twice as described above. The two eluates obtained with 200 mM phosphate are pooled and designated gel eluate 2. The volumes of 50 and 200 mM phosphate buffer used for eluting the enzymes are such that the absorbancies of the eluates at 280 nm are between 0.2 and 1.5. This procedure gives effective and simple separation and purification of the two classes of protein kinases. The entire procedure of calcium phosphate gel treatment is routinely performed even for those tissue preparations whose ammonium sulfate fractions (and earlier fractions) indicate the occurrence of only cAMP-dependent or cGMP-dependent enzyme activity. Sample protocols of the purification of enzymes from three representative tissue preparations—one exhibiting exclusively cGMP-dependent protein kinase activity (cecropia silkmoth pupal fat body), one exhibiting only cAMP-dependent activity (lobster gill tissue) and one with both types of enzyme activity (cecropia silkmoth larval body wall)—are shown in Table II.

The calcium phosphate gel step has certain advantages over the DEAE-cellulose step; it can be used, with a minimum loss of enzyme activity, for small-scale enzyme preparations from tissues which are not

TABLE II
Purification of Cyclic GMP-Dependent and Cyclic AMP-Dependent Protein Kinases from Three Representative Tissues of Arthropoda[a,b]

	Specific activity (units/mg protein)		
Source and fraction of protein kinases	No addition	cGMP	cAMP
Cecropia silkmoth, larval body wall			
Crude extract	9	32	36
pH 4.9 supernatant	16	67	60
$(NH_4)_2SO_4$ fraction	55	234	300
Gel eluate 1	138	660	167
Gel eluate 2	128	166	720
Cecropia silkmoth, pupal fat body			
Crude extract	5	15	7
pH 4.9 supernatant	10	31	12
$(NH_4)_2SO_4$ fraction	35	90	47
Gel eluate 1	182	702	242
Gel eluate 2	10	11	12
Lobster gill			
Crude extract	4	5	19
pH 4.9 supernatant	10	10	33
$(NH_4)_2SO_4$ fraction	9	28	109
Gel eluate 1	14	21	14
Gel eluate 2	49	54	425

[a] Taken from J. F. Kuo, G. R. Wyatt, and P. Greengard, *J. Biol. Chem.* **246**, 7159 (1971).

[b] The concentration of cyclic nucleotides used was 0.5 μM for the fractions from cecropia pupal fat body and lobster gill and 50 nM for cecropia larval body wall (due to the exceptionally low K_a values for cyclic nucleotides of the enzymes from this source; see Table III).

available in large quantities. Furthermore, the gel step can be substituted effectively for the DEAE-cellulose step in the large-scale separation of cGMP-dependent and cAMP-dependent protein kinases from tissues such as lobster tail muscle.

Characterization

Cyclic Nucleotide Specificity. The relative ability of varying concentrations of cGMP and cAMP to stimulate the cGMP-dependent protein kinase from lobster tail muscle is shown in Fig. 2. The apparent K_a for cGMP (i.e. the concentration causing a half-maximal increase in enzyme activity) was 75 nM, and that for cAMP was 3.6 μM, i.e., about 50 times greater. cAMP, at sufficiently high concentrations, could stimulate protein kinase activity maximally, i.e., to the same extent as did cGMP. The

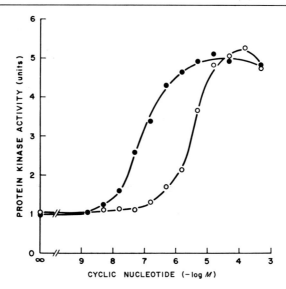

FIG. 2. Protein kinase activity of DEAE eluate, peak 1, in the presence of varying concentrations of cyclic GMP (●——●, approx. K_a 75 nM) and cyclic AMP (○——○, approx. K_a 3.6 μM). The amount of enzyme used was 120 μg. Taken from J. F. Kuo and P. Greengard, *J. Biol. Chem.* **245**, 2493 (1970).

relative ability of varying concentrations of cAMP and cGMP to stimulate cAMP-dependent protein kinase from lobster tail muscle is shown in Fig. 3. With this enzyme fraction, the apparent K_a for cAMP was 18 nM, and that for cGMP was 1.2 μM, i.e., about 65 times greater; cGMP, at sufficiently high concentrations, was able to activate the enzyme maximally. The relative ability of cAMP and cGMP to activate the cAMP-dependent protein kinase from lobster tail muscle is similar to that observed with cAMP-dependent protein kinases isolated from a variety of bovine tissues[9]; cAMP was about 50–200 times more effective than cGMP in activating this class of enzyme.

Table III presents the K_a values, for cGMP and cAMP, of cGMP-dependent (eluate 1) and cAMP-dependent (eluate 2) protein kinases isolated from various arthropod tissues by means of calcium phosphate gel treatment. Some representative mammalian cAMP-dependent enzymes are also included for comparison. The apparent affinities of the enzymes of different tissue origin for their respective activating cyclic nucleotides varied considerably. Among the arthropod preparations examined, the cGMP-dependent protein kinases ranged in their apparent

[9] J. F. Kuo, B. K. Krueger, J. R. Sanes, and P. Greengard, *Biochem. Biophys. Acta* **212**, 79 (1970).

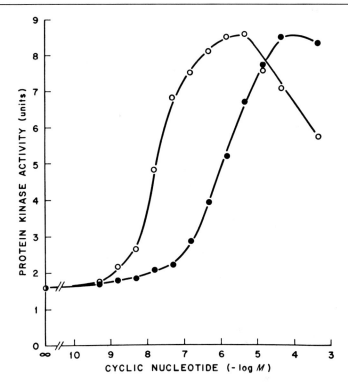

FIG. 3. Protein kinase activity of DEAE eluate, peak 2, in the presence of varying concentrations of cyclic AMP (○——○, app. K_a 18 nM) and cyclic GMP (●——●, app. K_a 1.2 μM). The amount of enzyme used was 40 μg. Taken from J. F. Kuo and P. Greengard, J. Biol. Chem. **245**, 2493 (1970).

K_a values for cGMP from 25 nM to 0.48 μM, or about 20-fold, and cAMP-dependent protein kinases ranged in their apparent K_a values for cAMP from 4.2 nM to 33 nM, or about 8-fold. In every case, maximal activation could be obtained by the "heterologous" cyclic nucleotide when present at sufficiently high concentration, but the ratio of K_a for heterologous cyclic nucleotide to that for "homologous" cyclic nucleotide varied greatly, from 5 for cockroach fat body cGMP-dependent enzyme to about 2000 for lobster green gland cAMP-dependent enzyme.

Some analogs of cGMP, cAMP, and cIMP were compared for their effectiveness in stimulating cGMP-dependent and cAMP-dependent protein kinases. The results[10] are presented in Table IV. 8-Bromo-cGMP was found to be more effective than cGMP in activating cGMP-dependent protein kinase from lobster tail muscle, whereas the 8-benzylamino

[10] J. F. Kuo, E. Miyamoto, and P. Greengard, unpublished data.

TABLE III
COMPARISON OF K_a VALUES OF CYCLIC GMP-DEPENDENT AND CYCLIC AMP-DEPENDENT PROTEIN KINASES WITH RESPECT TO CYCLIC GMP AND CYCLIC AMP[a]

Species and tissue source	K_a (μM)			
	cGMP-dependent kinase		cAMP-dependent kinase	
	cGMP	cAMP	cGMP	cAMP
Cecropia silkmoth				
Larval fat body	0.063	5.1		
Larval body wall	0.048	16.0	0.16	0.0042
Pupal fat body	0.051	10.3		
Pupal wing	0.12	19.3		
Pupal midgut	0.33	7.9		
Adult thorax	0.19	4.9	0.52	0.019
Polyphemus silkmoth				
Pupal fat body	0.025	4.0	0.11	0.0081
Cockroach				
Adult fat body	0.48	2.5		
Adult muscle			0.25	0.016
Drosophila				
Larvae, whole				0.019
Adults, whole			0.63	0.033
Lobster				
Tail muscle	0.082	3.2	1.42	0.022
Green gland			24.9	0.013
Gill			2.00	0.032
Cow				
Brain			7.9	0.11
Heart			5.0	0.040
Rat				
Isolated adipocytes			1.0	0.039

[a] Taken from J. F. Kuo, G. R. Wyatt, and P. Greengard, *J. Biol. Chem.* **246**, 7159 (1971).

and N^2-2′-O-dibutyryl derivatives of cGMP were 40 and 6000 times less effective, respectively, than the parent compound in stimulating the cGMP-dependent enzyme. Several of the 8-substituted analogs of cAMP were found to be slightly more effective than cAMP in activating the cAMP-dependent enzyme, and considerably more effective than cAMP in activating the cGMP-dependent enzyme. The 8-benzylamino derivatives of cAMP and cIMP were less effective than their parent compounds in activating the cAMP-dependent enzyme, but more effective in activating the cGMP-dependent enzyme.

Substrate Protein Specificity. Experiments were carried out to determine whether arthropod cGMP-dependent, arthropod cAMP-dependent, and mammalian cAMP-dependent enzymes showed characteristic patterns of substrate specificity. For this purpose histone type 3 (Sigma), histone

TABLE IV
SUMMARY OF APPARENT K_a VALUES OF PROTEIN KINASES FOR CYCLIC NUCLEOTIDE ANALOGS[a]

	Apparent K_a (M) for		
Compound	cGMP-dependent protein kinase, lobster muscle	cAMP-dependent protein kinase	
		Bovine heart	Bovine brain
8-Bromo-cGMP	2×10^{-8}	3×10^{-6}	ND[b]
cGMP	5×10^{-8}	2×10^{-6}	2×10^{-5}
8-Benzylamino-cAMP	2×10^{-7}	1×10^{-7}	4×10^{-7}
8-Bromo-cAMP	3×10^{-7}	1×10^{-8}	ND
8-Thio-cAMP	3×10^{-7}	2×10^{-8}	5×10^{-8}
8-Methylthio-cAMP	5×10^{-7}	2×10^{-8}	6×10^{-8}
8-(β-Hydroxyethylthio)cAMP	7×10^{-7}	3×10^{-8}	ND
Cyclic TuMP[c]	8×10^{-7}	3×10^{-8}	2×10^{-7}
8-Methylamino-cAMP	1×10^{-6}	8×10^{-8}	2×10^{-7}
8-Oxo-cAMP	2×10^{-6}	2×10^{-8}	1×10^{-8}
8-Amino-cAMP	2×10^{-6}	2×10^{-8}	4×10^{-8}
cAMP	2×10^{-6}	3×10^{-8}	1×10^{-7}
8-Bromo-cIMP	2×10^{-6}	2×10^{-7}	ND
6-Benzoyl-cAMP	2×10^{-6}	1×10^{-6}	ND
8-Benzylamino-cGMP	2×10^{-6}	2×10^{-5}	ND
BAPR[d]	3×10^{-6}	4×10^{-8}	ND
AICAR[e]	3×10^{-6}	1×10^{-6}	ND
8-Benzylamino-cIMP	3×10^{-6}	1×10^{-6}	ND
8-(β-Hydroxyethylamino)cAMP	6×10^{-6}	1×10^{-7}	ND
cIMP	8×10^{-6}	2×10^{-7}	2×10^{-6}
N^2-2'-O-Dibutyryl-cGMP	3×10^{-4}	3×10^{-3}	ND
Adenosine 3',5'-cyclothiophosphate	Not active[f]	2×10^{-5}	ND

[a] Unpublished data of J. F. Kuo, E. Miyamoto, and P. Greengard.
[b] ND: not determined.
[c] Tubercidin 3',5'-monophosphate; taken from J. F. Kuo and P. Greengard, *Biochem. Biophys. Res. Commun.* **40**, 1032 (1970).
[d] 6-Benzylaminopurine riboside 3',5'-monophosphate.
[e] Aminoimidazolecarboxamide riboside 3',5'-monophosphate.
[f] Highest concentration tested, 1 mM.

mixture (Schwarz/Mann), casein, and protamine were compared, at arbitrary concentrations, for their relative abilities to serve as substrates for various protein kinases (Table V). It is clear that all the arthropod cAMP-dependent protein kinases used protamine more readily than histone mixture as substrate. In contrast, the arthropod cGMP-dependent protein kinases and the mammalian cAMP-dependent protein kinases preferred histone mixture to protamine as substrate. Casein was generally the poorest of the substrates with all three types of enzyme. It can also be seen that the lobster muscle cGMP-dependent kinase, like that from cecropia larval body wall, phosphorylated histone mixture more readily than did the cAMP-dependent enzyme from the same source. Results using protein kinases from a variety of other tissue sources, for which data are not shown, supported these generalizations.

TABLE V
Protein Substrate Specificity of Cyclic GMP-Dependent and Cyclic AMP-Dependent Protein Kinases[a,b]

		Relative protein kinase activity in presence of			
Source of protein kinase	Cyclic nucleotide specificity	Histone type 3 (200 $\mu g/ml$) (%)	Histone mixture (50 $\mu g/ml$) (%)	Casein (3000 $\mu g/ml$) (%)	Protamine (200 $\mu g/ml$) (%)
Cecropia silkmoth					
Larval body wall	cGMP	100	164	4	79
	cAMP	100	74	22	195
Pupal fat body	cGMP	100	134	5	46
Pupal thorax	cGMP	100	78	12	5
	cAMP	100	51	35	284
Lobster					
Tail muscle	cGMP	100	88	6	36
	cAMP	100	49	66	151
Cow					
Brain	cAMP	100	154	26	16
Heart	cAMP	100	150	35	19
Rat					
Isolated adipocytes	cAMP	100	201	47	80

[a] Taken from J. F. Kuo, G. R. Wyatt, and P. Greengard, *J. Biol. Chem.* **246,** 7159 (1971).
[b] Relative protein kinase activity is expressed as the percentage of that obtained with histone type 3, in the presence of 0.5 μM of either cGMP or cAMP, as indicated. The enzyme activities were corrected for the values observed in the absence of added protein substrate.

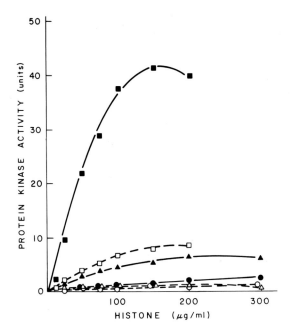

Fig. 4. Ability of various purified histone fractions to serve as substrate for cyclic AMP (cAMP)-dependent protein kinase from lobster tail muscle. ○---○, lysine-rich histone (histone Ib), −cAMP; ●——●, lysine-rich histone, +5 μM cAMP; △---△, slightly lysine-rich histone (histone IIb), −cAMP; ▲——▲, slightly lysine-rich histone, +5 μM cAMP; □---□, arginine-rich histone (histone IV), −cAMP; ■——■, arginine-rich histone, +5 μM cAMP. Activities have been corrected for values obtained in the absence of added histone, in the absence and in the presence of added cyclic nucleotide, respectively. Taken from J. F. Kuo and P. Greengard, *Biochim. Biophys. Acta* **212**, 434 (1970).

A comparison of the ability of purified preparations of histone Ib, histone IIb, and histone IV (all kindly provided by Professor James Bonner of the California Institute of Technology) to serve as phosphate acceptor in the reaction catalyzed by lobster tail muscle cAMP-dependent and cGMP-dependent protein kinases is presented in Figs. 4 and 5, respectively.[11] As can be seen in Fig. 4, the relative ability of the three histones to serve as substrate for the cAMP-dependent enzyme decreased in the order IV > IIb > Ib, at all concentrations of histone tested, while the apparent K_m values decreased in the order Ib > IIb > IV. The stimulation by the cyclic nucleotide of enzyme activity appeared to be associated with an increase in V_{max} with little or no change in the concentration of histone required for half-maximal velocity (the apparent K_m). Results

[11] J. F. Kuo and P. Greengard, *Biochim. Biophys. Acta* **212**, 434 (1970).

similar to those obtained with the lobster muscle cAMP-dependent protein kinase were also observed with cAMP-dependent protein kinase purified from bovine brain, from bovine heart and from cecropia silkmoth larval body wall.

In the case of the lobster muscle cGMP-dependent protein kinase (Fig. 5), the V_{max} values for histone Ib and histone IIb were approximately equal to each other and about half of the V_{max} value observed for histone IV. In addition, the apparent K_m of the cGMP-dependent protein kinase for histone IIb decreased from a value greater than 150 μg/ml in the absence of cGMP to a value of about 8.5 μg/ml in the presence of 5 μM cGMP. Moreover, this apparent K_m value of 8.5 μg/ml for histone IIb in the presence of cGMP was considerably less than that for either histone Ib or histone IV, which was not observed with any of the cAMP-dependent protein kinases. Results similar to those obtained with the lobster muscle cGMP-dependent enzyme were also observed with a cGMP-dependent protein kinase purified from cecropia silkmoth larval body wall.

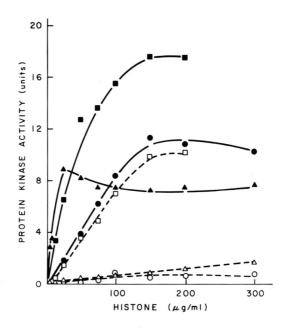

FIG. 5. Ability of various purified histone fractions to serve as substrate for cyclic GMP-dependent protein kinase from lobster tail muscle. Other details were as in Fig. 4, except that 5 μM cyclic GMP was substituted for 5 μM cyclic AMP. Taken from J. F. Kuo and P. Greengard, *Biochim. Biophys. Acta* **212**, 434 (1970).

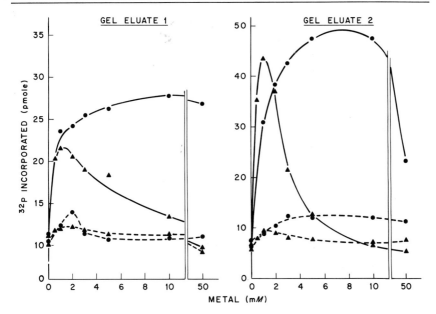

FIG. 6. Activity of cyclic GMP-dependent protein kinase (gel eluate 1) and cyclic AMP-dependent protein kinase (gel eluate 2) from larval body wall of cecropia silkmoths in the presence of varying concentrations of metal ions. The concentration of cyclic GMP and cyclic AMP used for eluate 1 and eluate 2, respectively, was 0.5 μM. Mg^{2+}, in the absence (●---●) and in the presence (●——●) of cyclic nucleotide; Co^{2+}, in the absence (▲---▲) and in the presence (▲——▲) of cyclic nucleotide. Taken from J. F. Kuo, G. R. Wyatt, and P. Greengard, *J. Biol. Chem.* **246**, 7159 (1971).

Metal Ion Requirement. The effects of Mg^{2+} and Co^{2+} on the activities of the protein kinases from silkmoth larval body wall[5] are illustrated in Fig. 6. For both cGMP-dependent and cAMP-dependent protein kinases, the stimulatory effect of Mg^{2+} in the presence of the cyclic nucleotide reached a maximum at about 5 mM; with the cAMP-dependent enzyme only, inhibition was observed at 50 mM Mg^{2+}. For both types of enzyme, Co^{2+} activated maximally at about 1 mM. Activity fell off at higher concentrations, more sharply with the cAMP-dependent than with the cGMP-dependent enzyme, the extent of activation by the cyclic nucleotides decreasing as the Co^{2+} concentration was raised.

In tests of the effects of metal ions on the activities of others of the protein kinase preparations, it was found that Mg^{2+} was generally slightly more effective at 10 mM than at 2 mM. Co^{2+} at 2 mM was a highly effective activator for almost every enzyme tested. Co^{2+} at 10 mM was strongly inhibitory to all of the cAMP-dependent kinases of insect origin, but to none of the cAMP-dependent enzymes of mammalian origin. Of the cGMP-dependent enzymes of insect origin, some were strongly inhibited by high Co^{2+} and others were not. Mn^{2+} was a very inefficient activator for most of the enzymes, but with a few preparations it approached Mg^{2+} in effectiveness. Ca^{2+} supported neither the basal activities of the protein kinases nor their stimulation by the cyclic nucleotides. Ca^{2+} strongly antagonized the stimulatory effect of Mg^{2+} for all protein kinases studied in the presence or in the absence of added cyclic nucleotides.

Subunit Structure. As in the case of cAMP-dependent protein kinases from various mammalian tissues[12-17] and lobster tail muscle,[18] cGMP-dependent protein kinase from lobster tail muscle was found to consist of analogous regulatory (cGMP-binding) and catalytic subunits.[18] The cGMP-independent catalytic subunit derived from lobster tail muscle cGMP-dependent protein kinase was conveniently prepared by incubating the holoenzyme with cGMP followed by chromatography on an Enzite CM-cellulose-protamine column.[18] The catalytic subunit of the lobster muscle cGMP-dependent enzyme, isolated by column chromatography, sedimented in a sucrose density gradient at a position corresponding to 3.6 S (MW 40,000), and the holoenzyme from which it was derived sedimented at a position of 7.7 S (MW 140,000) (Fig. 7A). The isolated catalytic subunit and the holoenzyme of lobster muscle cAMP-dependent protein kinase, on the other hand, sedimented at positions corresponding to 4.5 S (MW 60,000) and 5.7 S (MW 90,000), respectively (Fig. 7B).

It is interesting that the catalytic subunits derived from the lobster muscle cGMP-dependent enzyme and from bovine brain cAMP-dependent enzyme[18] had the same sedimentation coefficient (3.6 S) which differed from that of the catalytic subunit (4.5 S) from the lobster muscle cAMP-dependent enzyme. These findings correlate with our earlier observations[5] that, with respect to certain kinetic properties, cGMP-dependent enzymes from arthropods resemble mammalian cAMP-dependent enzymes, and are dissimilar to arthropod cAMP-dependent enzymes.

[12] M. A. Brostrom, E. M. Reimann, D. A. Walsh, and E. G. Krebs, *Advan. Enzyme Regul.* **8**, 191 (1970).
[13] G. N. Gill and L. D. Garren, *Biochem. Biophys. Res. Commun.* **39**, 335 (1970).
[14] M. Tao, M. L. Salas, and F. Lipmann, *Proc. Nat. Acad. Sci. U.S.* **67**, 408 (1970).
[15] A. Kumon, H. Yamamura, and Y. Nishizuka, *Biochem. Biophys. Res. Commun.* **41**, 1290 (1970).
[16] J. Erlichman, A. H. Hirsch, and O. M. Rosen, *Proc. Nat. Acad. Sci. U.S.* **68**, 731 (1971).
[17] E. Miyamoto, G. L. Petzold, J. S. Harris, and P. Greengard, *Biochem. Biophys. Res. Commun.* **44**, 305 (1971).
[18] E. Miyamoto, G. L. Petzold, J. F. Kuo, and P. Greengard, *J. Biol. Chem.* **248**, 179 (1973).

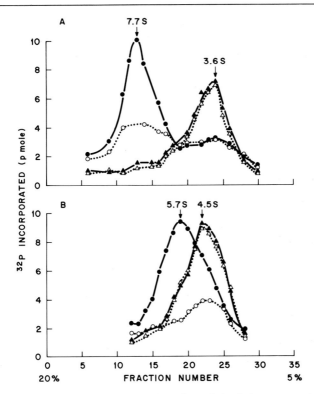

FIG. 7. Sucrose density gradient centrifugation of the holoenzymes and catalytic subunits of lobster muscle cyclic GMP (cGMP)-dependent and cyclic AMP (cAMP)-dependent protein kinases. The catalytic subunits were prepared by column chromatography on Enzite CM-cellulose-protamine. Holoenzyme (2.0 mg) and catalytic subunit (50 μg) of cGMP-dependent protein kinase (A), and holoenzyme (2.9 mg) and catalytic subunit (57 μg) of cAMP-dependent protein kinase (B), each in 0.22 ml of solution containing 50 mM sodium acetate buffer, pH 6.0, 0.3 mM EGTA and 2.5 mM 2-mercaptoethanol, were separately layered onto 4.8 ml of a 5–20% sucrose density gradient, containing the same concentration of acetate buffer, EGTA, and 2-mercaptoethanol. Protein kinase activity in fractions was assayed in the absence (····) or in the presence (———) of 5 μM cGMP (A) or 5 μM cAMP (B), respectively. The data for holoenzymes are represented by circles and those for the catalytic subunits by triangles. Taken from E. Miyamoto, G. L. Petzold, J. F. Kuo, and P. Greengard, *J. Biol. Chem.* **248**, 179 (1973).

When lobster muscle cGMP-dependent protein kinase was preincubated and then centrifuged in the presence of histone mixture (1 mg/ml), a partial dissociation of the enzyme as observed[18]; the peak of catalytic activity, which had been at 7.7 S and had been cGMP-dependent, decreased in size, and a new peak of catalytic activity, which was cGMP-independent, appeared in a position of 3.6 S. When the cGMP-dependent enzyme was preincubated and then centrifuged in the presence of 50 μM cGMP, the enzyme was again only partially dissociated, two peaks of catalytic activity again appearing at positions corresponding to 7.7 S and 3.6 S. cGMP was more effective than cAMP in dissociating and in activating the cGMP-dependent enzyme. When the cGMP-dependent enzyme was preincubated with both histone mixture (1 mg/ml) and cGMP (50 μM), an almost complete dissociation of the enzyme into the 3.6 S component was observed.

Recombination of the regulatory and catalytic subunits derived from the cAMP-dependent class of protein kinases has been reported for a number of mammalian tissues.[12-18] It has also been found[18] that a regulatory subunit derived from a bovine brain cAMP-dependent protein kinase was able to inhibit the enzymatic activity of a catalytic subunit derived from the lobster muscle cGMP-dependent protein kinase, accompanied by a concomitant restoration of cyclic nucleotide dependence of the resultant holoenzyme. The activity of the free catalytic subunit, measured in the presence or absence of added cAMP, and that observed with the reconstituted holoenzyme in the presence of added cyclic nucleotide, were comparable. cAMP was found to be more effective than cGMP in activating the reconstituted "hybrid" holoenzyme consisting of the regulatory subunit from a cAMP-dependent enzyme and the catalytic subunit from a cGMP-dependent enzyme; these results indicate that the cyclic nucleotide specificity of the regulatory subunit was not affected by its combination with the "heterologous" catalytic subunit. It would be interesting to study the effect of the regulatory subunit from a cGMP-dependent protein kinase on the activity of the catalytic subunit derived from cAMP-dependent and cGMP-protein kinases. Unfortunately, the regulatory subunit isolated from the lobster muscle cGMP-dependent enzyme has proved to be too unstable to date to permit such an experiment to be performed reliably.

Effects of Protein Kinase Modulator. Walsh *et al.*[19] have purified and studied a heat-stable factor from skeletal muscle which inhibited cAMP-dependent protein kinase activity from the same tissue. They used casein or phosphorylase *b* kinase as the substrate in their studies. We have investigated[20,21] the effect of crude and purified preparations of this heat-stable factor, obtained from various sources including *Escherichia coli*

[19] D. A. Walsh, C. D. Ashby, C. Gonzalez, D. Calkins, E. H. Fischer, and E. G. Krebs, *J. Biol. Chem.* **246**, 1977 (1971); see also this volume [48].

[20] J. F. Kuo and P. Greengard, *Fed. Proc., Fed. Amer. Soc. Exp. Biol.* **30**, 1089 (1971).

[21] T. E. Donnelly, Jr., J. F. Kuo, P. L. Reyes, Y. P. Liu, and P. Greengard, *J. Biol. Chem.*, **248**, 190 (1973).

and lobster tail muscle, on the activity of cAMP-dependent and cGMP-dependent protein kinases. We have found that the factor could either stimulate or inhibit the activity of both classes of protein kinases, depending upon the kind of substrates used, as well as upon several other assay conditions. Because of the dual effects, we proposed that this factor be called a protein kinase "modulator."

The effect of the modulator from lobster tail muscle on the activity of cyclic GMP-dependent protein kinase from the same tissue, using arginine-rich histone as substrate, in the presence of varying concentrations of cGMP, is shown in Fig. 8. When cGMP was absent, or was present in low concentrations, the modulator had negligible effect on the activity of the enzyme. However, when the cGMP concentration was 0.5 μM or higher, the modulator caused a pronounced stimulation of the enzyme activity. For example, at 5 μM cGMP, the enzyme activity was about twice as high in the presence of the modulator as in its absence, and the maximal stimulation by cGMP of the enzyme activity increased from 4-fold in the absence of the modulator to 8.5-fold in its presence. Results similar to those obtained with the cGMP-dependent enzyme from lobster

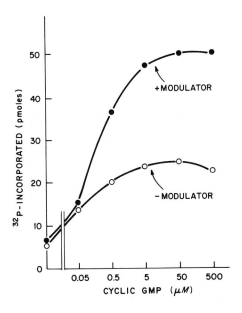

FIG. 8. The effect of protein kinase modulator on the activity of cyclic GMP-dependent protein kinase from lobster tail muscle in the presence of varying concentrations of cyclic GMP. The amounts of the enzyme and modulator used were 85 μg and 55 μg, respectively. Taken from T. E. Donnelly, Jr., J. F. Kuo, P. L. Reyes, Y. P. Liu, and P. Greengard, *J. Biol. Chem.* **248**, 190 (1973).

tail muscle were also obtained with the same class of protein kinase prepared from cecropia silkmoth pupal fat body.

In the case of cAMP-dependent protein kinase from bovine heart, the modulator depressed the enzyme activity in the absence and in the presence of cAMP when arginine-rich histone was used as substrate. Results similar to those obtained with the bovine heart enzyme were observed with cAMP-dependent protein kinases prepared from lobster tail muscle and several bovine and rat tissues.

The effects of the modulator on the cGMP-stimulated activity of cGMP-dependent protein kinase, and on the cAMP-stimulated activity

TABLE VI
THE EFFECTS OF PROTEIN KINASE MODULATOR FROM LOBSTER TAIL MUSCLE ON THE PHOSPHORYLATION OF VARIOUS SUBSTRATES BY LOBSTER TAIL MUSCLE CYCLIC GMP (cGMP)-DEPENDENT PROTEIN KINASE AND BOVINE HEART CYCLIC AMP (cAMP)-DEPENDENT PROTEIN KINASE[a,b]

Substrate and amount	cGMP-dependent enzyme activity (units)			
	− Modulator		+ Modulator	
	−cGMP	+cGMP	−cGMP	+cGMP
Arginine-rich histone, 40 μg	4.5	23.2	4.2	51.9
Lysine-rich histone, 40 μg	1.1	21.5	0.9	15.6
Histone mixture, 40 μg	3.8	19.2	1.6	16.5
Protamine, 40 μg	5.2	4.9	6.2	8.6
Casein, 600 μg	1.4	2.0	1.0	2.2

Substrate and amount	cAMP-dependent enzyme activity (units)			
	− Modulator		+ Modulator	
	−cAMP	+cAMP	−cAMP	+cAMP
Arginine-rich histone, 40 μg	20.9	241.4	16.4	129.0
Lysine-rich histone, 40 μg	32.8	111.4	12.7	95.6
Histone mixture, 40 μg	35.0	307.5	2.2	33.0
Protamine, 40 μg	8.6	16.6	18.1	57.3
Casein, 600 μg	1.8	40.4	3.4	36.3

[a] Modified from T. E. Donnelly, Jr., J. F. Kuo, P. L. Reyes, Y. P. Liu, and P. Greengard, *J. Biol. Chem.* **248**, 190 (1973).

[b] The amounts of modulator, and cGMP-dependent and cAMP-dependent enzyme were 85 μg, 85 μg, and 10 μg, respectively. The concentration of cGMP or cAMP, when present, was 5 μM. All values have been corrected for activity in the absence of added substrate.

TABLE VII
EFFECTS OF MODULATOR ON THE PHOSPHORYLATION OF VARIOUS SUBSTRATE PROTEINS BY CATALYTIC SUBUNIT ISOLATED FROM LOBSTER TAIL MUSCLE CYCLIC GMP-DEPENDENT PROTEIN KINASE (G-PK) OR FROM BOVINE BRAIN CYCLIC AMP-DEPENDENT PROTEIN KINASE (A-PK)[a,b]

Substrate	Enzyme activity (units)			
	G-PK catalytic subunit		A-PK catalytic subunit	
	− Modulator	+ Modulator	− Modulator	+ Modulator
Protamine, 40 μg	1.03	1.41	0.91	2.80
Arginine-rich histone, 20 μg	2.23	4.35	6.81	4.53
Histone mixture, 40 μg	2.25	0.92	6.98	1.90

[a] Taken from T. E. Donnelly, Jr., J. F. Kuo, E. Miyamoto, and P. Greengard, *J. Biol. Chem.* **248,** 199 (1973).

[b] The amounts of G-PK catalytic subunit, A-PK catalytic subunit, and protein kinase modulator (lobster tail muscle) used were 40, 15, and 60 μg, respectively. No cyclic nucleotides were added. All values were corrected for activity in the absence of added substrate.

of cAMP-dependent protein kinase, were studied in the presence of various protein substrates. The modulator stimulated the activity of the cGMP-dependent enzyme when arginine-rich histone or protamine was used as substrate, but inhibited the enzyme activity when lysine-rich histone or histone mixture was used as substrate (Table VI). In the case of the cAMP-dependent enzyme, the modulator inhibited enzyme activity in the presence of any histone substrate, but stimulated the phosphorylation of protamine. The effects of the modulator on the activity of either class of protein kinase were not pronounced when casein was used as substrate.

Studies of the effect of the modulator on the kinetic properties of subunits obtained from lobster muscle cGMP-dependent protein kinase and from bovine brain cAMP-dependent protein kinase indicate that the modulator effects observed with the holoenzymes are attributable to interaction of the modulator with the respective catalytic subunits.[22] The effects of the modulator on substrate specificity, observed with the lobster muscle cGMP-dependent holoenzyme and the bovine brain cAMP-dependent holoenzyme in the presence of added cyclic nucleotide (Table VI), were also observed using the isolated catalytic subunits of the enzymes in the absence of added cyclic nucleotide (Table VII).

[22] T. E. Donnelly, Jr., J. F. Kuo, E. Miyamoto, and P. Greengard, *J. Biol. Chem.* **248,** 199 (1973).

Concluding Remarks

Many arthropod tissues are rich in cGMP-dependent protein kinases and should provide favorable material for study of the properties and function of this class of enzymes. In contrast, it has not yet been possible to detect cGMP-dependent protein kinase activity in most vertebrate tissues examined.

A correlation of the relative tissue levels of cAMP and cGMP with the apparent relative tissue levels of cAMP-dependent and cGMP-dependent protein kinases would appear to be lacking.[5,6,8] For instance, cecropia silkmoth larval fat body, a tissue in which only cyclic GMP-dependent protein kinase activity has been detected, has a ratio of cGMP to cAMP of about 0.4; in contrast, rat cerebellum and lung, tissues in which only cAMP-dependent protein kinase has been detected, have a cGMP to cAMP ratio of about 0.7, highest of any of the mammalian or arthropod tissues examined. This lack of correlation is not entirely surprising, when one considers the great difficulty of estimating levels of enzyme activity *in vivo* from studies of broken cell preparations.

cGMP-dependent protein kinases from arthropods have many properties in common with the cAMP-dependent class of enzymes, especially those of mammalian origin. Distinctive differences, however, have been noted for these two classes of protein kinase. The wide variation in the apparent amount of cGMP-dependent enzyme activity amongst various arthropod tissues suggests that this class of enzymes is probably concerned with regulation of specific functions of individual tissues.

Acknowledgment

This work was supported by Grants HL-13305, NS-08440, and MH-17387 from the United States Public Health Service. One of us (J. F. K.) is the recipient of a Research Career Development Award (GM-50165) from the United States Public Health Service. This paper is also Publication Number 1159 of the Division of Basic Health Sciences, Emory University.

[48] Purification and Characterization of an Inhibitor Protein of Cyclic AMP-Dependent Protein Kinases

By C. Dennis Ashby and Donal A. Walsh

In studies to establish an assay system for cyclic AMP based on the activation of phosphorylase kinase, Posner *et al.*[1] first noted the presence

[1] J. B. Posner, R. Stern, and E. G. Krebs, *J. Biol. Chem.* **240**, 982 (1965).

in boiled tissue extracts of an inhibitor of this system. Appleman et al.[2] demonstrated that this trypsin-labile inhibitor blocked another cyclic AMP (cAMP)-dependent process, the conversion of glycogen synthetase I to glycogen synthetase D. The inhibitor protein has been shown *in vitro* to interact directly with the cAMP-dependent protein kinase which catalyzes both of these reactions.[3] The inhibitor specifically blocks the activity of all cAMP-dependent protein kinases examined to date, and as a result it has a general applicability in the study of phosphorylation reactions catalyzed by this class of enzymes. As a tool the inhibitor can also serve to characterize the various classes of protein kinases[4] and to elucidate cAMP-dependent processes which are potentially mediated by cAMP-dependent protein kinases.

Assay

The initial assay for the inhibitor was based on the inhibition of the activation of phosphorylase kinase as catalyzed by a cAMP-dependent protein kinase, which was a contaminant in the phosphorylase kinase preparation (method 1).[5] A less cumbersome and more reproducible method of assay is described here (method 2).[6]

Principle. The inhibitor can be quantitatively assayed based on its ability to inhibit the phosphorylation of casein as catalyzed by purified cyclic AMP-dependent protein kinase.

Reagents

Assay buffer: 0.25 M sodium glycerol phosphate; 0.1 M sodium fluoride; 1.5 mM ethylene glycol bis(β-aminoethyl ether) N,N'-tetraacetic acid (EGTA); and 10 mM aminophylline; pH 5.9

Casein is suspended in water at a concentration of 60 mg/ml, boiled for 10 minutes while the pH is maintained at 9.5 with 0.1 N sodium hydroxide, cooled, and adjusted to pH 5.9. The protein concentration of the supernatant solution obtained following centrifugation is adjusted to 30 mg/ml as based on biuret determination with bovine serum albumin as standard.

Adenosine 3',5'-monophosphate (cAMP), 20 μM

[2] M. M. Appleman, L. Birnbaumer, and H. N. Torres, *Arch. Biochem. Biophys.* **116**, 39 (1966).
[3] T. R. Soderling, J. P. Hickenbottom, E. M. Reimann, F. L. Hunkeler, D. A. Walsh, and E. G. Krebs, *J. Biol. Chem.* **245**, 6317 (1970).
[4] J. A. Traugh, C. D. Ashby, and D. A. Walsh, this volume [42].
[5] D. A. Walsh, C. D. Ashby, C. Gonzalez, D. Calkins, E. H. Fischer, and E. G. Krebs, *J. Biol. Chem.* **246**, 1977 (1971).
[6] C. D. Ashby and D. A. Walsh, *J. Biol. Chem.* **247**, 6637 (1972).

Inhibitor protein dilution buffer, which consists of sodium glycerol phosphate, 5 mM, and (ethylenedinitrilo)tetraacetic acid (EDTA), 1 mM; pH 7.0

A mixture of magnesium acetate, 18 mM; and [γ-^{32}P]ATP, 6 mM, specific activity 4–8 $\times$ 10^9 cpm/mmole. The pH of the solution is adjusted to 6.8.

cAMP-dependent protein kinase from rabbit muscle, purified through the first DEAE-cellulose chromatography step of Walsh et al.[7] or as described elsewhere in this volume [43]

Bovine serum albumin (BSA), 6.25 mg/ml

Trichloroacetic acid (TCA), 5% and 10%

Procedure. The assay buffer (0.02 ml), casein (0.02 ml), and cAMP (0.01 ml) solutions are added to tubes maintained in an ice bath. The inhibitor solution to be assayed is diluted appropriately in dilution buffer so as to contain 1–5 inhibitor units per 0.04 ml, the volume which is added to the reaction mixture. A standard inhibitor curve is run with each assay. Purified protein kinase (0.02 ml) is added at a concentration that will catalyze the incorporation of 7 $\pm$ 1 pmoles of ^{32}P per minute into casein per 0.13 ml reaction mixture in the absence of inhibitor. The reaction is initiated by the addition of [γ-^{32}P]ATP, magnesium acetate mixture (0.02 ml) to the prewarmed tubes. After a 35-minute incubation at 30°, the reaction is terminated by the rapid, sequential addition of BSA (0.2 ml), 10% TCA (0.5 ml), and 5% TCA (1.5 ml) all maintained at 0°. The precipitated protein suspension is kept at 0° for 10 minutes and then is collected by centrifugation in a clinical centrifuge. The protein precipitate obtained is suspended and recentrifuged in 2.5 ml of cold 5% TCA three times. The final protein pellet is dissolved in 1 ml of 90% formic acid and then counted in 15 ml of a dioxane-naphthalene based scintillant.

Unit of Activity. One unit of inhibitor is defined as that amount which will inhibit the protein kinase activity by 1 pmole of ^{32}P incorporated into casein per minute under the standard assay conditions.

General Comments. The inhibition curve presented in Fig. 1 indicates that the decrease in ^{32}P incorporation from 0 to 2.5 units of inhibitor is linear and results in an inhibition of 0–36%. From this point to 5.0 units the decrease in incorporation is not linear with an overall inhibition of about 63% at 5.0 units. The bars indicate the degree of precision obtained from 50 separate experiments.

It was found that the sensitivity of the assay for the inhibitor protein was increased by using the low level of cAMP-dependent protein kinase

---

[7] D. A. Walsh, J. P. Perkins, and E. G. Krebs, *J. Biol. Chem.* **243**, 3763 (1968).

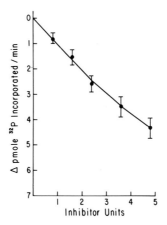

FIG. 1. Standard inhibitor curve. The ordinate indicates the decrease in the rate of ^{32}P incorporation into casein as compared to the protein kinase activity assayed in the absence of inhibitor. The bars indicate ±SD from 50 experiments. From C. D. Ashby and D. A. Walsh, *J. Biol. Chem.* **243**, 3763 (1968).

mentioned above. It is essential to maintain the protein kinase activity within narrow limits if accurate inhibitor quantitizations are to be obtained. This follows from the demonstration that the degree of inhibition is dependent not only on the amount of inhibitor present but also on the activity of the protein kinase.[6]

In order to assay crude extracts for the inhibitor protein, it is necessary to eliminate the protein kinase activity present. This is achieved most readily by boiling the extract for 5 minutes to destroy the protein kinase. The extract is then cooled, homogenized, and centrifuged at 27,000 g for 10 minutes. Assay of the resultant supernatant for inhibitor should be accompanied by a verification that the inhibitory material is labile to trypsin. (See this volume [40] for discussion of a nontrypsin labile inhibitor.)

Purification Procedure

Step 1. Tissue Extraction. The inhibitor protein is purified from rabbit skeletal muscle which contains 6000 units of inhibitor per gram wet weight of tissue. Four New Zealand white female rabbits, weighing 7–10 pounds, are sacrificed by intravenous Nembutal administration (60 mg per kilogram of body weight) and are exsanguinated. The muscle (about 2.4 kg) from the hind legs and back is iced, ground in a chilled meat grinder and homogenized in 2.5 volumes (ml/g) of cold 4 mM EDTA, pH 7.0 in a Waring blender at top speed for 1 minute. After centrifugation at 7000 g for 30 minutes, the supernatant fluid is decanted through glass wool to remove lipid material and this solution is used as the source of the inhibitor.

Step 2. Heat Treatment. The clear extract (5500 ml) is heated to 95° in a stainless steel bucket using two Fisher burners. An overhead stirrer fitted with a paddle is used to ensure adequate mixing and to prevent sticking. Temperature is reached in approximately 20 minutes. The suspension is then cooled to 10° in an ice bath and filtered first through a double layer of cheesecloth and then through filter paper (Eaton Dickeman No. 615). The denatured protein in the cheesecloth should be thoroughly squeezed to maximize the yield of inhibitor protein.

Step 3. Trichloroacetic Acid Precipitation. This and all subsequent procedures are performed at 4°. Trichloroacetic acid (100%, w/v) is added dropwise to the heat filtrate with stirring to give a final TCA concentration of 15%, w/v. After stirring on ice for 15 minutes, the precipitate, which contains the inhibitor, is collected by centrifugation at 10,000 g for 20 minutes. This precipitate is suspended in 25 ml of 50 mM glycerol phosphate, 2 mM EDTA, pH 6.8, and homogenized in a Potter-Elvehjem homogenizer with the addition of 6 N NH_4OH to adjust the pH to 6.8. Prolonged exposure of the suspension to alkaline pH should be avoided. The resultant material is extensively dialyzed against three changes (12 liters each) of 5 mM potassium phosphate, 1 mM EDTA, pH 7.0. The dialyzate is clarified by centrifugation at 34,000 g for 20 minutes and the supernatant solution containing the inhibitor can be stored at −15°. Inhibitor protein purified through this step is suitable for use in the Gilman binding assay for cAMP.[8]

Step 4. DEAE-Cellulose Chromatography. At this point in the purification several preparations purified through the TCA precipitation step may be pooled. The solution is adjusted to pH 5.0 with 1.0 N acetic acid at 4° and stirred for 15 minutes. After centrifugation at 34,000 g for 10 minutes, the supernatant is applied to a column of DEAE-cellulose (1.5 × 24 cm) equilibrated[9] with 5 mM sodium acetate, pH 5.0, containing 1 mM EDTA. The column is developed with a linear gradient (1000 ml) between 5 mM and 0.3 M sodium acetate buffer, pH 5.0, containing 1 mM EDTA. The inhibitor protein is eluted at a sodium acetate concentration of 0.14 M, well separated from the bulk of ultraviolet-absorbing material. Fractions containing the inhibitor are pooled, neutralized, and dialyzed against 0.5 mM glycerol phosphate, pH 7.0, containing 20 μM

[8] A. G. Gilman, *Proc. Nat. Acad. Sci. U.S.* **67**, 305 (1972); this volume [7].

[9] The DEAE-cellulose (Sigma) is washed with a solution of 0.5 N sodium hydroxide and 0.5 N sodium chloride prior to equilibration in the acetate buffer.

SUMMARY OF INHIBITOR PURIFICATION DATA

Fraction	Yield (%)	Purification[a] (x-fold)
1. Rabbit muscle extract		1
2. Filtrate from heat treatment	100	50
3. 15% Trichloroacetic acid precipitate	113	76
4. Eluate from DEAE-cellulose column	42	780
5. Eluate from Sephadex	34	760

[a] Based on per milligram of protein as assayed by biuret.

EDTA. This purification step should be completed as rapidly as possible in order to avoid prolonged exposure of the inhibitor protein to the low pH.

Step 5. Sephadex G-75 Chromatography. After the above dialysis, the inhibitor solution is concentrated by lyophilizing to dryness. The white powder is dissolved in approximately 2 ml of water and centrifuged at 27,000 g for 15 minutes. The supernatant solution is applied to a column (1.5 × 26 cm) of Sephadex G-75 equilibrated in 5 mM Tris-chloride, pH 7.5, containing 1 mM EDTA. The inhibitor is eluted between 20 and 35 ml of eluate. Although this purification step does not increase the specific activity of the inhibitor preparation, as shown in the table, it does eliminate contaminating ultraviolet-absorbing material.[5] The final preparation can be stored at −20° with little loss of activity.

A summary of the purification data of the inhibitor purified from rabbit skeletal muscle is presented in the table. A survey of other rabbit tissues for inhibitory activity that had the same characteristics of heat stability, chromatography on DEAE-cellulose, and trypsin lability as the material present in skeletal muscle has been made.[6] This study indicated that the purification procedure described here is applicable to tissues other than muscle. Most tissues examined, however, had less inhibitor present on a wet weight basis than did skeletal muscle. Brain tissue, on the other hand, had 50% more inhibitor than muscle and therefore may serve as a richer source of the inhibitor protein.

Properties

Purity. The purified inhibitor protein preparation exhibits two major bands[5] when examined by the disc gel electrophoresis system of Ornstein and Davis. The minor protein band was demonstrated to possess the inhibitor activity. The inhibitor preparation possesses no ATPase[6] or phosphoprotein phosphatase activity.[5]

Spectral Properties. The extinction coefficient at 280 nm of the inhibitor preparation is lower than that typical of most proteins. For this reason absorption at 235 nm is recommended to monitor column elutions.

Stability. Inhibitor solutions obtained from the standard preparation can be stored at $-20°$ for longer than one year with essentially no loss of activity.

Physiochemical Properties. The molecular weight of the inhibitor is 26,000 as determined by a gel filtration method using a column of Sephadex G-75.[5] Sucrose gradient sedimentation of the inhibitor gave an $s_{20,w} = 1.5$.[5] The isoelectric point is 4.2 as determined by isoelectrofocusing.

Interaction of the Inhibitor with cAMP-Dependent Protein Kinase

The subunit composition and the mechanism of cyclic nucleotide activation of cAMP-dependent protein kinases is now well established (see articles in this volume). This class of enzymes is composed of regulatory (R) and catalytic (C) subunits. As shown in Eq. (1) cAMP activates holoenzyme, designated by RC, by combining with regulatory subunit to form a noncovalent complex of regulatory subunit·cyclic AMP and free catalytic subunit.[10] This latter species is considered to be the active form of the protein kinase.

$$RC + cAMP \rightleftharpoons C + R \cdot cAMP \tag{1}$$

The inhibitor protein (designated I) does not interact with protein kinase prior to its activation but is effective against cAMP-dependent protein kinase only after dissociation of holoenzyme into free catalytic subunit.[6,11] Inhibitor combines directly with catalytic subunit as described by Eq. (2)[10]:

$$C + I \rightleftharpoons CI \tag{2}$$

Thus, a combination of Eqs. (1) and (2) account for the mechanism of action of inhibitor with cAMP-dependent protein kinase. Inhibitor protein is quantitatively as effective in inhibiting isolated catalytic subunit derived from holoenzyme as it is in blocking holoenzyme activity assayed in the presence of cAMP.[11] Qualitatively, regulatory subunit and inhibitor depress isolated catalytic subunit activity in a similar manner. Both effectors titrate catalytic subunit activity to the same plateau level.[6] A common binding site for inhibitor and regulatory subunit on the catalytic subunit species would be consistent with the data obtained to date.

The inhibitor is essentially equally effective against cAMP-dependent

---

[10] The stoichiometry of the reaction is unclear.
[11] C. D. Ashby and D. A. Walsh, *J. Biol. Chem.* **248**, 1255 (1973).

protein kinases present in crude extracts from thymus, heart, brain, kidney, liver, adipose tissue, and muscle. Greater than 90% of this cAMP-dependent activity is titratable by the inhibitor.[11] The multiple forms of cAMP-dependent protein kinases which have been noted in rat liver and rabbit muscle (see [43] and [46] this volume) are also susceptible to inhibition. Although the inhibitor is effective against all holoenzyme-derived catalytic subunits examined, it is not effective against other cAMP-independent protein kinases.[4] The specificity of this effector is such that it can be used in differentiating catalytic subunit activity from protein kinase activity that is unrelated to cAMP regulation.[4]

Interaction of the Inhibitor with Protein Kinase Substrates and Activators

Cyclic Nucleotides. The inhibitor does not destroy, bind, or compete with cAMP.[5] At concentrations of cAMP, cIMP, or cGMP that are sufficient to fully activate the protein kinase, the inhibitor is equally effective against protein kinase activity.[11] At submaximal concentrations of the cyclic nucleotides, however, this is not the case. Under conditions of low cAMP concentration, for example, the inhibitor is less effective against protein kinase activity.[11] The explanation for this is apparent upon examination of the mode of action of the inhibitor. When inhibitor combines with catalytic subunit produced by the cyclic nucleotide-promoted dissociation of holoenzyme, the equilibrium of Eq. (1) is shifted to the right. This results in the dissociation of available holoenzyme and the production of catalytic subunit which compensates for a portion of the inhibition. This same explanation accounts for the increase in the binding constant of cAMP to cAMP-dependent protein kinase without affecting total binding capacity.[5] The shift in equilibrium would favor the production of the cAMP·regulatory subunit complex. This phenomenon is used to increase the sensitivity of the Gilman assay for cAMP.[8]

Protein Substrates. When casein, histones f_1, f_2a, or f_3 are used as substrates for the protein kinase, there is no difference in the degree of inhibition observed.[6] In contrast, when protamine is used as substrate, the inhibitor is less effective in blocking protein kinase activity.[6] This is not due to an altered interaction of the inhibitor with the protein kinase per se, but is due rather to the binding of substrate to the inhibitor and thus decreasing the effective concentration of the latter. The acidic inhibitor also binds to polyarginine, which is not a substrate for protein kinase, to produce the same decreased effectiveness against protein kinase activity assayed with histone f_3.[6] It is thus necessary to proceed with caution in using inhibitor to examine protein kinase activity assayed on highly basic protein substrates. Casein, on the other hand, does not bind inhibitor protein, as evidence by the noncompetitive interaction shown by kinetic analysis.[5]

Other Reaction Constituents. The interaction of inhibitor and protein kinase is noncompetitive with respect to ATP.[5]

Acknowledgments

This research was supported by grant AM 13613 from United States Public Health Service.

[49] Techniques for the Study of Protein Kinase Activation in Intact Cells

By THOMAS R. SODERLING, JACKIE D. CORBIN, and CHARLES R. PARK

Since the isolation of a cAMP[1]-dependent protein kinase in 1968,[2] information has accumulated regarding the molecular properties and regulation of the purified enzyme.[3] Little direct information is available, however, documenting control in intact cells. Treatment of rat diaphragm with epinephrine or insulin has been reported to alter the cAMP dependency of the protein kinase.[4,5] Knowing the mechanism of *in vitro* activation of protein kinase by cAMP as depicted in Eq. (1), we have investigated the hormonal regulation of this enzyme in adipose tissue.[6,7] Binding of cAMP to the regulatory subunit of the inactive protein kinase (RC) causes dissociation of the enzyme into a regulatory subunit·cAMP complex (R·cAMP) and an active catalytic subunit (C). The stoichiometry and equilibrium constant of Eq. (1) have not been determined.

$$\text{RC (inactive)} + \text{cAMP} \rightleftharpoons \text{R} \cdot \text{cAMP} + \text{C (active)} \tag{1}$$

General Considerations

The methodology for the estimation of protein kinase activation in intact cells has been developed using adipose tissue. Its applicability to

[1] Abbreviations used: adenosine 3′,5′-monophosphate, cAMP; adrenocorticotropic hormone, ACTH.
[2] D. A. Walsh, J. P. Perkins, and E. G. Krebs, *J. Biol. Chem.* **243**, 3763 (1968).
[3] E. G. Krebs, *Curr. Top. Cell. Regul.* **5**, 99 (1972).
[4] L. C. Shen, C. Villar-Palasi, and J. Larner, *Physiol. Chem. Phys.* **2**, 536 (1971).
[5] O. Walaas, E. Walaas, and O. Grønnerød, *Isr. J. Med. Sci.* **8**, 353 (1972).
[6] J. D. Corbin, T. R. Soderling, and C. R. Park, *J. Biol. Chem.* **248**, 1813 (1973).
[7] T. R. Soderling, J. D. Corbin, and C. R. Park, *J. Biol. Chem.* **248**, 1822 (1973).

other tissues will be discussed. In adipose tissue the effects of lipolytic and antilipolytic agents on cAMP metabolism have been studied extensively,[8] and protein kinase is involved in the regulation of lipolysis.[9,10] To study the activation of protein kinase in intact cells, the equilibrium position of Eq. (1) can be determined in a tissue homogenate by assaying for the various components. Such an approach entails several potential sources of error. In this regard, factors involved in the homogenization such as the effect of dilution on Eq. (1) and the stability of cAMP and protein kinase in the homogenate have been examined. Errors due to the presence of several cell types and problems in the assays for the components of Eq. (1) have also been considered.[6,7] Because of these uncertainties, the measurements of protein kinase activation by our procedure may be semiquantitative.

Materials

Type II-A histone from calf thymus, ATP, and cAMP were obtained $^{32}P_i$ from International Chemical and Nuclear. Epinephrine was from from Sigma. [γ-^{32}P]ATP was prepared as described elsewhere[11] using Parke, Davis; insulin, 10 times recrystallized, was donated by Novo Laboratorium (Copenhagen); 1-methyl-3-isobutylxanthine was a gift of G. D. Searle and Co.; glucagon was donated by Eli Lilly; and ACTH was given by Wilson Laboratory. Sephadex gels were obtained from Pharmacia.

Methodology for Adipose Tissue

Protein Kinase Assay. The various methods that can be used to assay the cAMP-dependent protein kinase are discussed elsewhere in this volume.[12] The present study employs a 75-μl reaction mixture at 30° and pH 6.7 containing 6.6 mg/ml histone (type II-A), 11.1 mM potassium phosphate, 4.2 mM magnesium acetate, 0.22 mM [γ-^{32}P]ATP, 33.3 mM sodium fluoride, 1.33 μM cAMP (when added), and protein kinase. Reactions are initiated by addition of enzyme and terminated by pipetting

---

[8] G. A. Robison, R. W. Butcher, and E. W. Sutherland, "Cyclic AMP." Academic Press, New York, 1971.

[9] J. D. Corbin, E. M. Reimann, D. A. Walsh, and E. G. Krebs, *J. Biol. Chem.* **245**, 4849 (1970).

[10] J. K. Huttunen, D. Steinberg, and S. E. Mayer, *Biochem. Biophys. Res. Commun.* **41**, 1350 (1970).

[11] T. R. Soderling, J. P. Hickenbottom, E. M. Reimann, F. L. Hunkeler, D. A. Walsh, and E. G. Krebs, *J. Biol. Chem.* **245**, 6317 (1970).

[12] J. D. Corbin and E. M. Reimann, this volume [41].

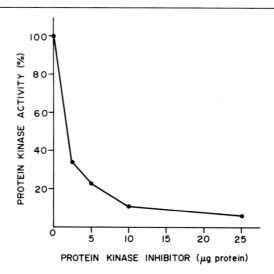

Fig. 1. Inhibition of adipose tissue protein kinase by the protein kinase inhibitor. The infranatant fraction from a fat pad homogenate was prepared as described in the text. This fraction was diluted 1:6 with 10 mM potassium phosphate, 10 mM EDTA, pH 6.5, and the protein kinase activity was assayed (plus cAMP) in the presence of the indicated amount of protein kinase inhibitor. The protein kinase inhibitor was partially purified from skeletal muscle up to the DEAE column step as described in D. A. Walsh, C. D. Ashby, C. Gonzalez, D. Calkins, E. H. Fischer, and E. G. Krebs, *J. Biol. Chem.* **246**, 1977 (1971).

50-μl aliquots of the assay mixtures onto numbered Whatman 3 MM filter paper disks (1 cm × 1 cm) which are immediately immersed in a beaker of cold 10% trichloroacetic acid. The filter papers are processed as described elsewhere.[12] One unit of protein kinase activity is that amount of enzyme which catalyzes incorporation of 1 pmole $^{32}P_i$ into histone per minute.

Specificity of Assay. It is important that the protein kinase assay utilizing histone as a substrate be specific for the cAMP-dependent protein kinase and not measure any other protein kinase(s) (e.g., phosphorylase kinase, phosvitin kinase, etc.) that may be present in the tissue homogenate. The specificity of the protein kinase assay can be determined in at least two ways. The first method utilizes the heat-stable inhibitor of cAMP-dependent protein kinase[13,14] which specifically binds to and inhibits the C subunit.[15] As shown in Fig. 1, the inhibitor is able

---

[13] D. A. Walsh, C. D. Ashby, C. Gonzalez, D. Calkins, E. H. Fischer, and E. G. Krebs, *J. Biol. Chem.* **246**, 1977 (1971).
[14] C. D. Ashby and D. A. Walsh, this volume [48].
[15] C. D. Ashby and D. A. Walsh, *J. Biol. Chem.* **247**, 6637 (1972).

to block up to 90–95% of the protein kinase activity of a crude adipose tissue extract. The second method to determine the specificity of the assay is based on physical resolution of the two forms of protein kinase on a Sephadex G-100 column (Fig. 2). The first peak of activity is dependent on cAMP and thus represents RC; the second peak, which is independent of cAMP for activity, is the C subunit. If an adipose tissue extract is chromatographed in the absence of exogenous cAMP and in a buffer of low ionic strength, essentially all the kinase activity elutes as RC (Fig. 3B). Chromatography in the presence of cAMP shifts 95% of the kinase activity to the C subunit (Fig. 3A).

Protein Kinase Activity Ratio. Experimental results may be expressed as the protein kinase activity ratio, that is, the fraction of total protein kinase present in the C form. This activity ratio can be determined by two methods. The first, designated —cAMP/+cAMP, is the ratio of protein kinase activities of the infranatant fraction (see below) assayed in the absence and presence of saturating exogenous cAMP. The second method, designated C/(RC + C), is based on the protein kinase activity profiles obtained by Sephadex G-100 column chromatography (Fig. 2). The protein kinase activity profile is plotted on graph paper, cut out, and weighed to obtain the value of RC + C. The second peak is then cut from this profile and weighed to obtain the value of C. It is necessary to include 0.5 M NaCl in the buffer equilibrating the column to prevent association of the kinase subunits.[6]

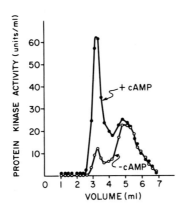

FIG. 2. Sephadex G-100 chromatography of protein kinase in an adipose tissue extract. Fifty microliters of an infranatant fraction to which 0.3 μM cAMP had been added was chromatographed on a Sephadex G-100 column (0.9 × 10 cm). Protein kinase assays were performed in the absence (open circles) and presence (filled circles) of cAMP. These data are from J. D. Corbin, T. R. Soderling, and C. R. Park, *J. Biol. Chem.* **248,** 1813 (1973).

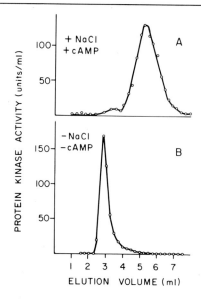

FIG. 3. Analysis of the two forms of protein kinase by Sephadex G-100 chromatography. cAMP was added to a concentration of 10 μM to an infranatant fraction prepared from a fat pad homogenate, and 50-μl aliquots were chromatographed on 0.9 cm diameter columns composed of a 14 cm lower layer of Sephadex G-100 and a 2 cm upper layer Sephadex G-25. The columns were equilibrated with 10 mM potassium phosphate, 10 mM EDTA, pH 6.5 plus the following: (A) 0.5 M NaCl and 50 μM cAMP; (B) no additions. Protein kinase activity was assayed in the presence of cAMP. The data are taken from J. D. Corbin, T. R. Soderling, and C. R. Park, *J. Biol. Chem.* **248**, 1813 (1973).

Tissue Incubation and Homogenization. Normal, fed Sprague-Dawley rats (140–180 g) are anesthetized with Nembutal, and the epididymal fat pads are excised, segmented into three pieces, and randomized into tared flasks containing 5 ml of Krebs-Ringer phosphate buffer (pH 7.4) modified to contain 1.3 mM CaCl$_2$, 1 mg/ml bovine serum albumin, and no added glucose. After being weighed, the fat pads are first incubated for 15 minutes with shaking at 37° and are then transferred to fresh buffer and incubated for another 10 minutes unless indicated otherwise. Hormones are added at the beginning of the 10-minute incubation except that insulin, when present, was included in both incubations. Following the second incubation, the fat pads are homogenized in cold buffer (1 ml/g tissue) with three passes of a motor-driven Teflon-pestle glass tube homogenizer. The homogenization buffer contains 10 mM potassium phosphate, 10 mM EDTA, 0.5 mM 1-methyl-3-isobutylxanthine, and 0.5 M NaCl at pH 6.5. The infranatant fraction is obtained by centrifugation at 12,000 g for 5 minutes at 5°.

TABLE I
Effect of Epinephrine and Caffeine on Protein Kinase and cAMP[a]

Epinephrine (μM)	Caffeine (mM)	Protein kinase activity ratio		cAMP ($M \times 10^7$)
		$\dfrac{-\text{cAMP}}{+\text{cAMP}}$	$\dfrac{C}{RC + C}$	
0.0	0.0	0.25	0.15	1.1
0.1	0.0	0.38	0.34	1.4
0.3	0.0	0.44	0.37	1.3
1.1	0.0	0.59	0.45	1.9
11.0	0.0	0.73	0.57	2.6
11.0	1.0	1.03	0.83	8.0

[a] Fat pads were treated with the designated concentrations of epinephrine and caffeine for 10 minutes, homogenized, and centrifuged. Protein kinase activity ratios and cAMP concentrations were determined in the infranatant fraction.

cAMP Assay. A 200-μl aliquot of the fat pad infranatant fraction is deproteinized by addition of 20 μl of 100% trichloroacetic acid and is then centrifuged at 12,000 g for 10 minutes. The trichloroacetic acid in the supernatant is extracted into ether as described by Wastila *et al.*[16] The concentration of cAMP in the supernatant is determined by the protein kinase binding assay described by Gilman.[17] The protein kinase utilized in the binding assay is purified from rabbit skeletal muscle by the use of a protamine-Sepharose column as described elsewhere.[18]

Treatment of the ether extracted supernatants with partially purified heart cyclic nucleotide phosphodiesterase at 30° for 2 hours and then 100° for 5 minutes reduces the cAMP concentration of the extracts to zero.

Hormonal Regulation of Adipose Tissue Protein Kinase

The above methodology has been utilized to determine the effects of certain hormones on the components of Eq. (1). Incubation of adipose tissue for 10 minutes with varying concentrations of epinephrine produces dose-dependent increases in both the protein kinase activity ratios and the cAMP concentrations (Table I). It is of interest that a maximum

[16] W. B. Wastila, J. T. Stull, S. E. Mayer, and D. A. Walsh, *J. Biol. Chem.* **246**, 1966 (1971).
[17] A. G. Gilman, *Proc. Nat. Acad. Sci. U.S.* **67**, 305 (1970).
[18] J. D. Corbin, C. O. Brostrom, R. L. Alexander, and E. G. Krebs, *J. Biol. Chem.* **247**, 3736 (1972).

rate of lipolysis is attained with only small increments in cAMP and the protein kinase activity ratio. It should also be pointed out that the two independent methods of determining the protein kinase activity ratio give similar values. Activity ratios determined by the chromatographic method are generally slightly lower than those determined kinetically. This may be due to the small amount of reassociation of the kinase subunits which occurs on the G-100 column,[6] or to some dissociation of the kinase subunits which might occur during the kinetic determination. The lipolytic hormones glucagon and ACTH also activate the protein kinase.[7]

The ability of the antilipolytic hormone insulin to lower hormone-elevated cAMP levels in adipose tissue has been well documented.[8] From Eq. (1) it should be expected that insulin will decrease the protein kinase activity ratio. As shown in Fig. 4 insulin rapidly decreases both the protein kinase activity ratio and the cAMP level by a value of 70% when the control values are subtracted. The quantitative correlation between the changes in cAMP concentration and activity ratio is very close. Further evidence that the effects of epinephrine and insulin on protein kinase are mediated by cAMP is the observation that the effects of these

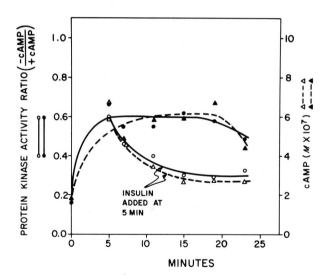

Fig. 4. Time course of epinephrine and insulin effects on protein kinase and cAMP. Flasks containing randomized fat pad segments (see text) were incubated with epinephrine (11 μM) for the times indicated. At 5 minutes insulin (final concentration of 6.9 nM) was added to some flasks. The solid symbols represent incubation with epinephrine only; the open symbols, epinephrine plus insulin. The data are taken from T. R. Soderling, J. D. Corbin, and C. R. Park, *J. Biol. Chem.* **248**, 1822 (1973).

TABLE II
EFFECT OF SEPHADEX CHROMATOGRAPHY ON THE
PROTEIN KINASE ACTIVITY RATIO[a]

Hormone treatment	Protein kinase activity ratio $\frac{-cAMP}{+cAMP}$	
	Before Sephadex G-25 chromatography	After Sephadex G-25 chromatography
None	0.25	0.22
Epinephrine (11 μM)	0.94	0.23
Epinephrine (11 μM) plus insulin (6.9 nM)	0.44	0.25

[a] Fat pads were treated for 10 minutes with the indicated concentrations of hormones, homogenized, and the infranatant fractions were prepared. The protein kinase activity ratios were determined before and after chromatography on a Sephadex G-25 column (0.9 × 15 cm) equilibrated with 10 mM potassium phosphate, 10 mM EDTA, pH 6.5. These data are taken from T. R. Soderling, J. D. Corbin, and C. R. Park, *J. Biol. Chem.* **248,** 1822 (1973).

hormones on protein kinase are abolished upon removal of the cAMP by Sephadex G-25 chromatography (Table II).

Other Tissues

The above methodology has also been successfully employed in an investigation of the effects of gonadotrophic hormones on testicular protein kinase.[19] Although extensive studies on other tissues have not been completed to date, preliminary experiments indicate that some of these methods may have a general applicability. For example, the data of Table III show that 90% or more of the "histone" kinase activity in several tissue extracts is due to the cAMP-dependent protein kinase on the basis of its inhibition by the protein kinase inhibitor. An exception appears to be the unfractionated homogenate of HeLa cells where only 60% of the kinase activity was inhibited. Furthermore, it has been shown that the RC and C forms of protein kinase in crude extracts of brain, heart, and liver can be resolved by Sephadex G-100 column chromatography.[6]

[19] A. R. Means, E. MacDougal, T. R. Soderling, and J. D. Corbin, *J. Biol. Chem.* **249,** 1231 (1974).

TABLE III
INHIBITION OF PROTEIN KINASE ACTIVITY IN TISSUE HOMOGENATES[a]

Tissue homogenate	Protein kinase inhibitor (μg protein)	Protein kinase activity (%)
Adipose tissue	0	100
	10	5
	25	1
Liver	0	100
	10	15
	25	10
Skeletal muscle	0	100
	10	9
	25	1
Heart	0	100
	10	9
	25	4
Cerebrum	0	100
	10	15
	25	14
Testis	0	100
	10	17
	25	13
HeLa cells[b] (unfractionated)	0	100
	10	41
	25	40
HeLa cells (27,000 g supernatant)	0	100
	10	24
	25	19

[a] Tissues were homogenized 1:3 (w/v) in 10 mM potassium phosphate, 10 mM EDTA, pH 6.5 and centrifuged at 27,000 g for 20 minutes. The supernatants were diluted 1:11 with the homogenization buffer and assayed for protein kinase activity (plus cAMP) in the presence of the indicated amount of skeletal muscle protein kinase inhibitor.

[b] Protein kinase activity was assayed in a 1:11 dilution of a HeLa cell homogenate without any centrifugation.

Initial studies on skeletal muscle extracts, however, indicate differences from adipose tissue extracts in the dissociation and recombination properties of the kinase subunits that may require modification of the above procedures.

Note Added in Proof. In studies on hormonal activation of protein kinase in rat heart, it has been determined that homogenization should be performed using a low salt buffer [S. L. Keely, J. D. Corbin, and C. R. Park, *Fed. Proc., Fed. Amer. Soc. Exp. Bio.* **32**, 643 (1973)] Furthermore, the protein kinase assay for this tissue should be done in the absence of NaF.

[50] The Purification and Analysis of Mechanism of Action of a Cyclic AMP-Receptor Protein from *Escherichia coli*

By IRA PASTAN, MARIA GALLO, and WAYNE B. ANDERSON

Both cyclic AMP (cAMP) and a specific cAMP receptor protein are needed by *Escherichia coli* to make normal amounts of a variety of inducible enzymes as well as other dispensable proteins.[1,2] cAMP levels are controlled by the carbon source the organism uses for growth. The levels are particularly low when the organism grows on glucose.[1,3] The inability of glucose-grown cells to make adequate amounts of inducible enzymes has been called the "glucose effect" or "catabolite repression." It is now established that catabolite repression is caused by low intracellular levels of cAMP.

Two classes of mutants have been isolated which are unable to make inducible enzymes as well as some other dispensable proteins. One class of mutants has lower or absent levels of cAMP due to a defective adenylate cyclase.[4] The inability of these cells to synthesize inducible enzymes is corrected by adding exogenous cAMP. The second class of mutants produces a defective cAMP binding protein.[5,6] This protein has been called the cAMP receptor (CRP), or the catabolite gene activator protein (CAP or CGA). The activity of CRP can be measured in crude cell extracts by its capacity to bind cAMP or by its ability to stimulate β-galactosidase synthesis in cell-free extracts prepared from a *crp*-strain.[6] The cAMP binding assay is preferred because of its simplicity.[7]

---

[1] I. Pastan and R. Perlman, *Science* **169**, 339 (1970).
[2] R. Perlman and I. Pastan, *Curr. Top. Cell. Regul.* **3**, 117–134 (1971).
[3] R. S. Makman and E. W. Sutherland, *J. Biol. Chem.* **240**, 1309 (1965).
[4] R. Perlman and I. Pastan, *Biochem. Biophys. Res. Commun.* **37**, 151 (1969).
[5] M. Emmer, B. de Crombugghe, I. Pastan, and R. Perlman, *Proc. Nat. Acad. Sci. U.S.* **66**, 480 (1970).
[6] G. Zubay, D. Schwartz, and J. Beckwith, *Proc. Nat. Acad. Sci. U.S.* **66**, 104 (1970).
[7] W. B. Anderson, A. B. Schneider, M. Emmer, R. Perlman, and I. Pastan, *J. Biol. Chem.* **246**, 5929 (1971).

cAMP Binding Assay Method

The binding activity of CRP is measured by incubating the protein sample with cyclic [^{3}H]AMP, precipitating the cyclic [^{3}H]AMP-CRP complex with ammonium sulfate and determining the isotope content of the precipitated complex.

Reagents

cAMP, 10 μM and 0.1 M
Cyclic [^{3}H]AMP, 20 Ci/mmole, 1 $\times$ 10^7 cpm/ml
5'-AMP, 0.1 M in 10 mM potassium phosphate, pH 7.7
Casein, 10 mg/ml, 10 mM potassium phosphate buffer, pH 7.7
Saturated solution of (NH$_4$)$_2$SO$_4$
NCS-Nuclear Chicago Solvent Liquifluor
Toluene

Procedure. The assay is performed at 1°. To a 10 $\times$ 75 mm disposable glass test tube add 10 μl 10 μM cAMP, 10 μl cyclic [^{3}H]AMP, 10 μl 5'AMP, CRP sample, H$_2$O to a final volume of 100 μl. To control tubes add 10 μl of 0.1 M cAMP. cAMP is usually obtained in the acid form, but a 0.1 M solution can be readily made by neutralizing the nucleotide solution with NaOH. Casein, 20 μl, is added to the assay tubes as a carrier protein for measurement of the activity of more highly purified preparations (after phosphocellulose). After a 5-minute incubation, add 0.4 ml of cold saturated (NH$_4$)$_2$SO$_4$ and mix. The precipitates are collected by centrifugation at 10,000 rpm (12,000 g) for 10 minutes in an SS-34 rotor in a Sorvall RC 2-B centrifuge. Carefully aspirate off the supernatant with a disposable pipette and then wipe the walls of each tube thoroughly with cotton-tipped applicator sticks. Dissolve the pellet in 0.5 ml of NCS, transfer to a counting vial containing 10 ml of Liquifluor-toluene scintillation cocktail (42 ml concentrated Liquifluor in 1000 ml of toluene). Wash pipette and tube by washing back and forth 4 times with scintillation mixture. Sufficient CRP is added to bind 10,000 to 20,000 cpm of cyclic [^{3}H]AMP. One unit of cAMP binding activity is defined as that amount of protein required to bind 1 pmole of cAMP under the conditions of the assay.

Bacteria

For the purification of CRP it is convenient to start with a partial diploid strain carrying two copies of the *crp* gene, because these cells contain twice as much CRP as haploid cells. *E. coli* strain KLF 41/JC1553 is plated on McConkey maltose and grown overnight at 37°. Maltose-positive colonies are diploid for *crp*, since the episome is *mal⁺crp⁺* and the chromosome is *mal⁻crp⁺*. The maltose-positive colonies are grown in a yeast-glucose medium containing 10 g of yeast extract, 5.6 g KH_2PO_4, 28.9 g K_2HPO_4, 10 mg of thiamine-HCl, and 10 g of glucose per liter to late-log phase. From each liter, 3–5 g of cells can be harvested. The cells are harvested in a continuous flow centrifuge and stored frozen at −70° until used. Cells 1 year old have been successfully used.

Purification of CRP

Reagents

Cell suspension buffer: 0.01 M Tris-acetate, pH 8.2; 0.01 M magnesium acetate; 0.06 M potassium acetate; 0.5 mM dithiothreitol; 0.1 mM sodium EDTA

Buffer A: 10 mM potassium phosphate pH 7.7; 0.5 mM dithiothreitol; 0.1 mM sodium EDTA

Buffer B: 10 mM potassium phosphate pH 7.0; 0.5 mM dithiothreitol; 0.1 mM sodium EDTA

Deoxyribonuclease, pancreatic

Ribonuclease, T_1

DEAE cellulose, Whatman DE-52

Phosphocellulose, Whatman P1

Sephadex G-100

Step 1. Preparation of Cell Extract. Five hundred grams of unwashed, frozen cells are suspended in 1200 ml of suspension buffer at low speed in a Waring Blendor. This suspension is passed through a French pressure cell at 10,000 psi twice to disrupt the cells and then centrifuged at 10,000 rpm (20,000 g) for 2 hours in a Sorvall RC2-B centrifuge using a GS-3 rotor to remove cell debris.

Step 2. Nuclease Treatment. To each milliliter of the supernatant, add 5 μg of pancreatic deoxyribonuclease and 5 μg of T_1 ribonuclease. Incubate at 4° for 16 hours. If CRP is to be used for transcription studies, do not use pancreatic ribonuclease because it cannot be separated from CRP in this purification.

Step 3. Dialysis. The extract is then dialyzed at 4° for at least 72 hours against 20-liter volumes of buffer A. The dialysis tubing is first heated at 80° for 20 minutes in 10 mM Na EDTA and stored at 4° in 1 mM Na EDTA.

Step 4. DEAE-Cellulose Treatment. DE-52 cellulose is prepared as described below and equilibrated with buffer A. The protein concentration of the dialyzed extract is determined by the method of Lowry *et al.* The DEAE is mixed with the extract in a large beaker at a ratio of 40 mg of protein per milliliter of wet packed resin. Pour the slurry into a large sintered-glass funnel and allow to drain. The effluent is collected and retained. The resin is then washed with two or three 200-ml aliquots of buffer A to recover at least 90% of the CRP binding activity. A poor recovery of CRP at this stage is usually due to inadequate ribonuclease treatment. This treatment should remove 60–70% of the total protein. Pool the effluent and all washes and adjust to pH 7.0 with 0.5 N acetic acid.

Step 5. Chromatography on Phosphocellulose. The procedure used to precycle and equilibrate phosphocellulose is described below. Prepare a 4.5 × 30 cm phosphocellulose column and equilibrate with buffer B. In general, use 1 ml of resin for each 40 mg of protein. After applying the sample wash the column with buffer B containing 0.3 M KCl until the absorbance at 280 nm of the eluate is reduced to 0.1–0.2. Elute the CRP with a linear gradient starting with 1000 ml of buffer B containing 0.3 M KCl in the mixing chamber and 1000 ml of buffer B containing 1.0 M KCl in the reservoir. Collect 10-ml fractions. CRP usually emerges at a conductivity of 30 ohm^{-1} (0.6 M KCl). The active fractions are pooled and concentrated by precipitation with ammonium sulfate. Solid ammonium sulfate (390 mg for each milliliter of pooled solution) is added slowly with stirring at 4°. The precipitate is collected by centrifugation at 20,000 g for 30 minutes and resuspended in 4 ml of buffer B containing 0.1 M KCl.

Step 6. Sephadex G-100. Four milliliters of the ammonium sulfate suspension are applied to a 2.5 × 95 cm Sephadex G-100 column equilibrated with buffer B containing 0.1 M KCl. The column is eluted with the same buffer at a flow rate of 0.5 ml per minute and 4-ml fractions are collected. The initial major protein peak contains cAMP binding activity. At this stage CRP is over 50% pure and occasionally over 95% pure. The last step is done if impurities remain.

Step 7. Second Phosphocellulose Column. The active Sephadex fractions are pooled and applied to a 0.9 × 10 cm phosphocellulose column equilibrated with buffer B containing 0.1 M KCl. Elute the CRP with a linear gradient starting with 75 ml buffer B containing 0.1 M KCl in the mixing chamber and 75 ml buffer B containing 1 M KCl in the reservoir. Collect 0.5-ml fractions. The first protein peak is CRP. The active fractions are pooled, and stored frozen in small aliquots.

A typical purification is summarized in the table.

PURIFICATION OF CYCLIC AMP RECEPTOR

Fraction	Total volume (ml)	Total protein (mg)	Total activity (units)	Activity (units/mg)	Yield (%)
French press	1600	64,000	250,000	3.9	100
Supernatant	1430	45,500	217,400	4.8	87
Dialyzate	1570	42,000	234,400	5.6	94
DEAE effluent	1960	18,000	215,000	12	86
Phosphocellulose	280	280	126,000	450	50
Sephadex G-100	268	67	106,400	1588	43
Phosphocellulose II	65	12	46,000	3833	18

Preparation of Ion-Exchange Resins

DEAE Cellulose. As purchased Whatman DE52 cellulose does not require initial HCl and NaOH washes. The DEAE is degassed to remove CO_2 by suspending the resin in 2 volumes of 0.1 M KH_2PO_4 and adjusting to pH 4.5 with H_3PO_4. This suspension is transferred to a vacuum flask and a good vacuum applied. The slurry is stirred constantly under vacuum until bubbles no longer form. A stoppered vacuum flask connected to a vacuum supply and placed on a magnetic stirrer is all that is required for this step. Adjust the degassed suspension to pH 7.7 with 0.5 N KOH, allow it to stand for 10–15 minutes and then filter in a large coarse sintered-glass funnel. Resuspend the DEAE in several volumes of buffer B and allow the resin to settle; pour off the supernatant which contains unwanted fine particles. Fine particles may be removed in this manner at intervals while equilibrating the DEAE with buffer A. Continue to resuspend and wash the resin in equilibration buffer A and to collect the DEAE by filtration until the effluent is of the same pH (7.7) and conductivity as buffer A. The equilibrated DEAE cellulose is stored in buffer B at 4°. For prolonged storage 0.03% toluene is added. The toluene must be thoroughly washed out with buffer B prior to use.

Phosphocellulose. Stir the weighted cellulose into 10 volumes of 0.5 NaOH. Filter in a large coarse sintered-glass funnel overlaid with filter paper, or use a large Büchner funnel. Wash with deionized water to pH 8. Stir phosphocellulose into 10 volumes of 0.5 M HCl and leave for 45 minutes. Repeat HCl treatment and wash with water until filtrate is near neutral. Wash with 0.1 M potassium phosphate buffer, pH 7, until filtrate is pH 7.0. Equilibrate by washing with copious amounts of buffer B. Store at 4°.

Transcription Assay Method

For 1.0 ml of transcription mix, add 0.06 ml of Tris · HCl, 1 M pH 7.9; 0.01 ml of $MgCl_2$, 1 M; 0.100 ml of KCl, 3 M; 0.03 ml of NaEDTA, 10 mM; 0.03 ml DTT, 10 mM; 0.025 ml ATP, 0.02 M; 0.025 ml GTP, 0.02 M; 0.025 ml UTP, 0.02 M; 0.0125 ml CTP, 0.02M; 0.02 ml [^{3}H]CTP (3×10^7 cpm); 0.03 ml bovine serum albumin, 10 mg/ml; 0.2 ml of λp*gal* DNA, 270 μg/ml; water to a final volume of 1 ml. To a 13×100 mm test tube in ice add: 0.03 ml of transcription mix, 0.005 ml 2 mM cAMP, an aliquot of CRP, and water to a final volume of 0.1 ml. CRP is added to a final concentration of 10–20 μg/ml. Control tubes are without cAMP. The tubes are preincubated for 3 minutes at 37°. Then add 2 μl of RNA polymerase (1.5 mg/ml) to start the reaction. The incubation is at 37° for 10 minutes. The reaction is terminated by adding 1 ml of ice-cold stopping solution containing 1 mM ATP and 0.1 mg per milliliter of heat-denatured salmon sperm DNA followed immediately by 1 ml of cold trichloroacetic acid. The resulting precipitate is collected on glass-fiber filters (Whatman) under vacuum and washed with three 5-ml aliquots of 1.5% perchloric acid. The filters are dried under a heating lamp and placed in counting vials. Add 0.5 ml of NCS followed by 10 ml of Liquifluor-toluene scintillation cocktail (42 ml concentrated Liquifluor in 1000 ml of toluene). Heat the vials of 60° for 10 minutes to lower fluorescence caused by the NCS. The vials are then cooled in ice and their content of radioactivity determined.

RNA polymerase is prepared by the method of Chamberlin.[8] With low concentrations of RNA polymerase only one round of transcription occurs and the total amount of RNA made is proportional to the concentration of CRP. When higher concentrations of polymerase are used then the assay is a rate assay. Some commercial preparations of *E. coli* RNA polymerase may be used, but frequently these preparations are deficient in sigma factor and inactive.

The conditions of the transcription assay have been devised to obtain maximal stimulation by cAMP and CRP. The data of Fig. 1 show the effect of altering the pH and the concentrations of KCl and $MgCl_2$. In addition Mg^{2+} can be replaced by Mn^{2+}.

Preparation of λp*gal*25 DNA

λcI857p*gal*25S*am*7 carries a normal operator, promoter, epimerase and transferase gene, but has a deletion of part of the kinase gene.[9] A lysogen

[8] D. Berg, K. Barrett, and M. Chamberlin, this series, Vol. 21, p. 506.
[9] P. Nissley, W. B. Anderson, M. Gallo, and I. Pastan, *J. Biol. Chem.* **247**, 4264 (1972).

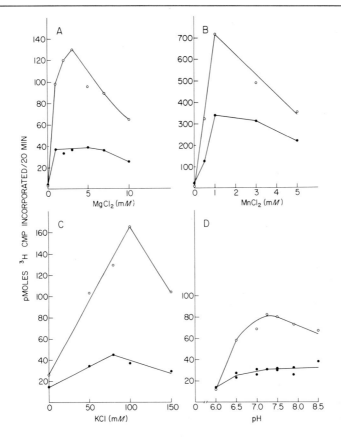

FIG. 1. Transcription was determined as described. (A) Effect of Mg^{2+} concentration on [^{3}H]RNA synthesis; ○——○, +CRP; ●——●, no CRP. (B) Effect of Mn^{2+} concentration on RNA synthesis; ○——○, +CRP; ●——●, no CRP. (C) Effect of KCl concentrations; ○——○, CRP; ●——●, no CRP. (D) Effect of pH on RNA synthesis; ○——○, +CRP; ●——●, no CRP. CRP concentration when present is 8.7 μg/ml.

with a normal kinase (λpgal8) may also be used. The S7 mutation renders the cell unable to lyse.

Growth of Bacterial Culture

Inoculate 1200 ml of tryptone broth (10 g of tryptone, 5 g of NaCl, 0.24 g of $MgSO_4$, and 2 mg of thiamine per liter) with 50 ml of an overnight culture of the bacterium [HfrH, galE⁻ PL-2(λcI857pgal25Sam7)] grown in the same medium at 34°. Incubate at 34° with aeration to an OD at 650 nm (Gilford) of 0.25. Raise the temperature of the culture to 41° and incubate for 15 minutes (check the temperature of the culture medium with a thermometer to be sure it is 41°). After 15 minutes lower the temperature to 37° and incubate for 4 hours, continuing to aerate the culture. Lysis should not occur. Harvest the cells by centrifugation at 5000 rpm in a refrigerated centrifuge for 10 minutes. Carefully suspended the cell pellet in 15 ml TMG buffer (10 mM Tris·HCl, pH 7.5; 10 mM MgSO$_4$, 100 mg of gelatin per milliliter). Add 0.5 ml of chloroform to lyse the cells, and let stand at room temperature for 20 minutes with frequent mixing. Add 150 µg of pancreatic deoxyribonuclease, mix, and let stand for 15 minutes with frequent stirring. Centrifuge at 10,000 rpm for 10 minutes to remove cell debris. Retain supernatant containing bacteriophage. To maintain integrity of the bacteriophage maintain magnesium concentration at 10 mM to 1 mM.

Purify phage by banding in a cesium chloride gradient. To each 5.8 ml of supernatant add 4.5 g of cesium chloride. Mix thoroughly to dissolve all the cesium chloride. The density of the suspension should be between 1.3764 and 1.3811 as determined by a refractometer. Fill transparent Oak Ridge polycarbonate tubes completely with cesium chloride-phage mixture or add paraffin oil to fill the tubes. Centrifuge overnight at 36,000 rpm in a Beckman 40 rotor in an ultracentrifuge. To recover banded phage, pierce a hole in the bottom of the polycarbonate tube and collect in drops. Dialyze banded phage for 16 hours at 4° against 0.1 M sodium phosphate, pH 7.1 containing 0.1 M NaCl.

To extract the phage DNA adjust the dialyzed suspension to 13–14 OD units/ml at 260 nm with the dialysis buffer. Divide into 12 ml portions or less and work at about 23°C. Add 25% sodium dodecyl sulfate to a final concentration of 0.5%. Add a volume of freshly distilled phenol equal to the volume of the phage suspension. (Freshly distill about 3 volumes of phenol for each volume of phage suspension, collect the phenol in water; equilibrate phenol first with 0.2 M Tris pH 7.9, then with 10 mM Tris pH 7.9.) Mix gently for 2 minutes. Centrifuge at 2000 rpm for 10 minutes in a refrigerated International centrifuge. Remove lower layer (phenol phase) with a syringe to which is attached polyethylene tubing or a long needle. Add an equal amount of phenol, mix gently for 2 minutes, and centrifuge as before. Discard phenol layer. Boil some dialysis tubing in 5% sodium carbonate for 20 minutes. Rinse thoroughly in deionized water. Add DNA to dialysis bag with a large-bore pipette or funnel. Dialyze against sodium EDTA, 20 mM pH 8.0 about 48 hours (3–4 changes). Dialyze against Tris·HCl, 10 mM pH 8.2, for 4–16 hours to remove all phenol. Measure OD of dialyzed DNA at 280, 270, 260 nm. If phenol is still present, redialyze.

Mechanism of Action of CRP

cAMP combines with CRP and stimulates the transcription of genes containing "cAMP"-sensitive promoters. Such promoters occur in the *lac* and *gal* operons which have been subjected to detailed study.[10,11]

The steps in cAMP action are indicated below:

$$\text{cAMP} + \text{CRP} \rightleftharpoons \text{cAMP–CRP} \quad (1)$$

$$\text{cAMP–CRP} + \text{DNA} \rightleftharpoons \text{cAMP–CRP–DNA} \quad (2)$$

$$\text{cAMP–CRP–DNA} + \text{RNP} \rightleftharpoons \text{cAMP–CRP–DNA–RNP} \quad (3)$$

$$\text{cAMP–CRP–DNA–RNA} \xrightarrow[\text{Mg}^{2+}]{\text{XTPs}} \text{RNA} \quad (4)$$

The cAMP-CRP complex binds to DNA and apparently alters the DNA at or close to a promoter site so that RNA polymerase can bind. This complex of cAMP, CRP, DNA, and RNA polymerase is termed a preinitiation complex; transcription begins when nucleoside triphosphates and Mg^{2+} are added. For the cAMP and CRP to stimulate transcription RNA polymerase must contain sigma factor. *Xenopus laevis* RNA polymerase I or II will not substitute for *E. coli* RNA polymerase. In animal cells all the actions of cAMP that have been investigated at a biochemical level are mediated *via* a protein kinase. However, the mechanism by which cAMP and CRP stimulate transcription does not involve protein phosphorylation. Indeed, ATP can be replaced by adenylyl imidodiphosphate, an ATP analog containing a terminal P–N–P linkage, and cAMP and CRP will still promote *gal* transcription. AMP-PNP does not serve as a phosphate donor for protein kinase but does serve as a substrate for RNA polymerase.

cAMP binds to CRP with a K_d of 10^{-5} mole/liter when measured by equilibrium dailysis.[6,7] A variety of other cyclic nucleotides will bind to CRP, but only tubercidin cyclic 3′,5′-phosphate will act like cAMP and stimulate transcription.[12] cGMP and 8-bromo-cAMP are inhibitors of cAMP action.

The cAMP–CRP complex binds tightly to DNA.[9,13] cGMP, 8-bromo-cAMP, and all other analogs that fail to stimulate transcription also do not stimulate binding of CRP to DNA. Surprisingly, the binding of the cAMP–CRP complex to DNA has not been shown to be specific for the type of DNA employed. For example, the cAMP–CRP complex binds

---

[10] B. de Crombrugghe, B. Chen, W. B. Anderson, M. Gottesman, R. Perlman, and I. Pastan, *J. Biol. Chem.* **246**, 7343 (1971).

[11] S. P. Nissley, W. B. Anderson, M. Gottesman, R. Perlman, and I. Pastan, *J. Biol. Chem.* **246**, 4671 (1971).

[12] W. B. Anderson, R. Perlman, and I. Pastan, *J. Biol. Chem.* **247**, 2717 (1972).

[13] A. D. Riggs, G. Reiness, and G. Zubay, *Proc. Nat. Acad. Sci. U.S.* **68**, 1222 (1971).

tightly to calf thymus DNA and poly(dAT), although it does not promote transcription from these templates.

Properties of CRP

As mentioned above, CRP binds to DNA in the presence of cAMP. CRP is a very basic protein with an isoelectric point of 9.2. The high isoelectric point appears to be due to a large content of asparagine and glutamine. CRP has a molecular weight of 45,000 and is composed of two identical subunits.[7] To date only one cAMP binding site per dimer has been demonstrated, but this point needs further investigation. Each dimer contains four free sulfhydryl groups, and these account for the four half-cysteines observed by amino acid analysis.

[51] The Detection and Characterization of Cyclic AMP-Receptor Proteins in Animal Cells

By GORDON N. GILL *and* GORDON M. WALTON

Receptor protein which has a strong affinity and high specificity for adenosine 3′,5′-monophosphate (cAMP) exists in eukaryotic cells.[1] The presently identified function of the cAMP receptor is the regulation of cAMP-dependent protein phosphokinase (EC 2.7.1.37) activity.[2-6] cAMP-dependent protein kinase exists as a molecular complex consisting of regulatory receptor and catalytic kinase subunits. Binding of cAMP to the inhibitory receptor results in dissociation of the complex with consequent activation of the free catalytic kinase subunit. This reaction may be expressed as:

$$R:K + cAMP \rightleftharpoons cAMP:R + K \qquad (1)$$

where R:K is the inactive receptor:kinase complex, cAMP:R is the cAMP:receptor complex, and K is the active kinase subunit. In many cells free receptor may also exist:

$$R + cAMP \rightleftharpoons cAMP:R \qquad (2)$$

[1] G. N. Gill and L. D. Garren, *Proc. Nat. Acad. Sci. U.S.* **63**, 512 (1969).
[2] D. A. Walsh, J. P. Perkins, and E. G. Krebs, *J. Biol. Chem.* **243**, 3763 (1968).
[3] G. N. Gill and L. D. Garren, *Biochem. Biophys. Res. Commun.* **39**, 335 (1970).
[4] M. Tao, M. L. Salas, and F. Lipmann, *Proc. Nat. Acad. Sci. U.S.* **67**, 408 (1970).
[5] G. N. Gill and L. D. Garren, *Proc. Nat. Acad. Sci. U.S.* **68**, 786 (1971).
[6] E. M. Reimann, C. O. Brostrom, J. D. Corbin, C. A. King, and E. G. Krebs, *Biochem. Biophys. Res. Commun.* **42**, 187 (1971).

The binding activity of the cAMP receptor protein can be measured by the retention of the cAMP:receptor complex on a cellulose ester membrane filter.[7] Receptor protein interacts with the membrane filter and is retained; cAMP complexed to receptor is similarly retained, and free cAMP passes through the filter. The pore size of the filter (0.45 μm) exceeds greatly the molecular size of the receptor protein so that retention on the basis of size does not occur.

Reagents

Reaction buffer: 0.1 M potassium phosphate pH 6.5; 16 mM theophylline; 20 mM MgCl$_2$
Wash buffer: 25 mM Tris·HCl pH 7.5 at 25°; 10 mM MgCl$_2$ (kept on ice)
[^{3}H]cAMP, 5 μM
Millipore filters, 0.45 μm pore size, 24 mm diameter, presoaked in cold wash buffer
Scintillation fluid[8]: 60 g of naphthalene, 200 mg of p-bis[2-(5-phenyloxazolyl)]benzene, 4 g of 2,5-diphenyloxazole; 100 ml of methyl alcohol, 20 ml of ethylene glycol, and dioxane to 1 liter

Procedure. To a small test tube (6 $\times$ 50 mm) add 0.1 ml of reaction buffer, 0.02 ml of 5 μM [^{3}H]cAMP, receptor protein, and water to give a final reaction volume of 0.2 ml. Duplicate reaction mixtures are initiated by adding, and vigorously mixing, up to 200 μg of a crude receptor protein preparation. Mixtures are incubated on ice for 60 minutes, then quantitatively transferred to a filter reservoir containing 5 ml of cold wash buffer. For multiple determinations a multifilter box is convenient (Hoefer Scientific Instruments Inc.). Free cAMP is removed by filtering the solution under vacuum followed by a thorough washing with cold wash buffer. The amount and specific activity of [^{3}H]cAMP used in the assay will determine the extent of washing necessary to remove free cAMP and can be quickly determined under assay conditions in the absence of receptor protein. A final rinse of the filter, with the reservoir removed, with 1–2 ml of wash buffer from a plastic wash bottle aids in reducing background levels of radioactivity. The filters are dissolved in 10 ml of Bray's scintillation solution, and radioactivity is determined. A convenient binding unit is 1 pmole of cAMP bound per milligram of protein.

The receptor assay can be adapted easily to measure the kinetics of the cAMP:receptor interaction under a variety of reaction conditions. The rate of association is determined in standard reaction mixtures equil-

[7] G. M. Walton and L. D. Garren, *Biochemistry* **9**, 4223 (1970).
[8] G. A. Bray, *Anal. Biochem.* **1**, 279 (1960).

ibrated at the experimental temperature. At time = 0 the receptor protein preparation is added to the incubation mixture with stirring; the association reaction is stopped at the desired times by diluting into 5 ml of cold wash solution in the filter reservoir; the bound complex is rapidly isolated by filtering. Because of the rapid association rates observed, convenient concentrations of cAMP and receptor sites are in the range of 10 to 0.1 nM.

The rate of dissociation is determined by allowing the standard receptor assay to reach equilibrium with [^{3}H]cAMP saturating the receptor. At time = 0, a 1000-fold or more excess of unlabeled cAMP is added; after further incubation for varying time periods, the dissociation reaction is stopped by filtering the incubation mixture as in the standard assay.

Comments. This procedure offers a simple and rapid method for detection and quantification of the cAMP receptor protein at any stage of protein purification. For quantitative purposes, excess levels of cAMP must be maintained relative to receptor protein to ensure maximal saturation of available sites. To ensure saturation, less than 25% of the total cAMP added should be bound at equilibrium. Because the activity measured is dependent on the adsorption of the complex to the filter, the presence of other proteins which also adsorb to the filter will influence the amount of protein which can be used in the assay. The reaction is linear only to about 200 μg of total protein per filter but can be extended to 400 μg by using two filters. The buffer and pH (4.8–8.5) can be varied if desired as well as the incubation temperature (0–37°). Incubation time must be determined for each condition to ensure complete complex formation. Magnesium ion is not required for binding, but a slight stimulation (15%) is observed with a concentration of 10 mM MgCl$_2$. The assay can be utilized not only for quantitation of receptor protein, but also as a competition-displacement assay for quantitation of cAMP.[7,9]

Characteristics of the cAMP: Receptor Protein Interaction

Even in crude protein fractions, only a single class of receptor sites with a strong affinity and high specificity for cAMP exist with an equilibrium dissociation constant K_D = 13 nM (Fig. 1).

As written [Eqs. (1) and (2)] the cAMP:receptor protein interaction is predicted to follow second-order reaction kinetics. Studies of both crude and purified receptor protein from adrenal cortical tissue demonstrated second order kinetic with a $k_a = 7.4 \times 10^5 \, M^{-1} \, \text{sec}^{-1}$ and $k_d = 10^{-3} \, \text{sec}^{-1}$ at

---

[9] A. G. Gilman and F. Murad, this volume [40].

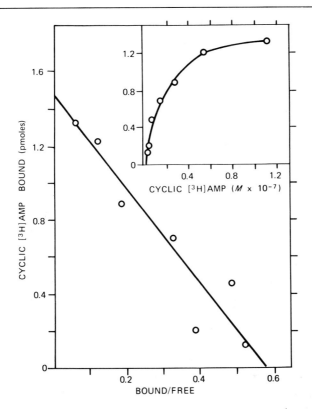

FIG. 1. Inset: The interaction of cAMP with receptor protein as a function of the total concentration of nucleotide added. The amount of nucleotide bound is also plotted as a function of the ratio of the concentration of bound to free cAMP. Single-site binding kinetics follow the equation: Amount bound = number of sites − K_m (bound/free). From G. M. Walton and L. D. Garren, *Biochemistry* **9**, 4223 (1970), copyright by the American Chemical Society, reprinted by permission.

25°. The reaction is highly temperature dependent; association is 14-fold faster at 37° than at 0°, and dissociation is 100-fold faster at 37° than at 0°. The thermodynamic parameters of the cAMP receptor interaction have been calculated from the change of the rate constants and equilibrium constant with temperature. The Arrhenius activation energy is 15 kcal M^{-1} for association and 22 kcal M^{-1} for dissociation. The enthalpy change for binding was determined as 9 kcal M^{-1}. The free energy change, was calculated to be −11 kcal M^{-1} at 25°; and the entropy change, 68 cal M^{-1} °K^{-1}. The increase in entropy is the driving force for the binding reaction. Decreasing pH increases the rate of association but does not alter the rate of dissociation. Changes in ionic strength over a wide range

SURVEY OF cAMP RECEPTOR ACTIVITY IN VARIOUS TISSUES[a]

Tissue	Units receptor activity
Adrenal cortex	7.9
Adrenal medulla	3.8
Heart	4.7
Muscle	3.0
Spleen	2.6
Brain	2.3
Liver	2.1
Kidney	1.7

[a] Weighed rat tissues were diced and homogenized in glass with a Teflon pestle in 4 volumes of 50 mM Tris · HCl (pH7.4). The homogenates were centrifuged at 12,000 g for 30 minutes, and the supernatants were dialyzed against the same buffer. All procedures were performed at 5°. Standard assays contained 90–180 μg of protein.

(0.01–0.4 M) does not affect the reaction. Second order reaction plots are not linear when holoenzyme is studied under conditions of equimolar cAMP and receptor sites.[10] Under pseudo-first-order conditions where cAMP exceeds receptor sites, the rate constants approximate those found with the isolated receptor. The presence of the kinase subunit thus affects the interaction of cAMP with receptor; the subunit interaction is significant at low cAMP concentrations.

The affinity of the receptor protein for cAMP is highly specific; other 3′,5′-nucleotides act as competitive inhibitors of cAMP binding but only at concentration excesses of 100-fold or greater. Other 5′-nucleotides require concentrations of 10^5 excess of cAMP to noncompetitively reduce the binding of cAMP. The bound cAMP can be quantitatively recovered unaltered after denaturation of the cAMP receptor.

Detection of cAMP Receptor in Tissues

A survey of cAMP receptor activity performed on cytosol fractions from various tissues of the rat are shown in the table. Since the amount of endogenously bound cAMP was not determined for the various tissues assayed, the receptor activity must represent a minimal value for each tissue.

Studies of subcellular localization of cAMP receptor in adrenal cortical tissue indicate that the cytosol contains the majority of the total receptor activity (60%). Microsomes which contain about 25% of the

[10] B. M. Sanborn, R. C. Bhalla, and S. G. Korenman, *J. Biol. Chem.* **248**, 3593 (1973).

total activity bind the nucleotide with the highest specific activity. Within the microsome cAMP receptor activity is highest in the smooth membrane fraction (6.4-fold greater than that present on free ribosomes).

Characteristics of the cAMP Receptor Protein

The cAMP receptor protein is heat labile, losing 50% activity at 42° in 15 minutes. Binding activity is inhibited by 1 mM p-chloromercuribenzoate. This inhibition is reversed by 1 mM 2-mercaptoethanol indicating a requirement for intact sulfhydryl groups. Receptor protein preparations were most stable in buffers containing 10% glycerol and were stable for several months when stored at −70°.

Several purification procedures for isolation of the cAMP receptor from various tissues have been published.[1,3,5,11] Generally the cAMP receptor has been isolated as a receptor:protein kinase enzyme complex. Purification from adrenal cortical tissue has yielded a free receptor protein preparation as well as a receptor–kinase complex. Though the kinase and receptor subunits may differ in molecular size in various tissues, in all instances the model of activation of cAMP-dependent protein kinase illustrated in Eq. (1) appears valid.

Using bovine adrenal cortical tissue as starting material the free cAMP receptor and cAMP receptor:protein kinase complex have been isolated. The cAMP receptor sediments at 4.6 S in rate zonal centrifugation and migrates in polyacrylamide gel electrophoresis with a molecular weight of 92,000. The receptor:kinase complex sediments at 7 S and by polyacrylamide gel electrophoresis and analytical ultracentrifugation has a molecular weight of 145,000 and 152,000, respectively. The free kinase subunit has an estimated molecular weight of 60,000.

Incubation of the complex with saturating levels of cAMP prior to isolation on sucrose gradients or polyacrylamide gels results in separation of the receptor and kinase subunits. The free kinase subunit is fully active and no longer binds cAMP, nor is it stimulated by the nucleotide. Addition of receptor to kinase subunits again results in a cAMP-dependent protein kinase.

Application

The filter binding assay can be used to quantitate the level of cAMP receptor protein present in a variety of tissues under a variety of growth

[11] J. A. Beavo, P. J. Bechtel, and E. G. Krebs, this volume [43]; C. S. Rubin, J. Erlichman, and O. M. Rosen, this volume [44]; M. Tao, this volume [45]; L. J. Chen and D. A. Walsh, this volume [46].

and hormonal conditions. This method provides a convenient assay for detection in tissue extracts and for purification. It can be modified conveniently to study the kinetics of the interaction and the factors influencing this. Comparison of receptor activity with cAMP-dependent protein kinase activity is useful in evaluating the latter reaction. This method of receptor assay has also been modified to determine cAMP levels by competitive protein binding techniques.

Section VI

Synthetic Derivatives of Cyclic Nucleotides and Their Precursors

[52] The Preparation and Use of Cyclic AMP Sepharose

By MEIR WILCHEK and ZVI SELINGER

The nucleotide cyclic 3′,5′-adenosine monophosphate (cAMP) has been shown to exert many of its effects via activation of the enzyme protein kinase. The cAMP dependent reaction involves binding of the nucleotide to a regulatory subunit, which dissociates from the enzyme and thus activates the catalytic subunit.[1]

$$R-C + cAMP \rightarrow [R-cAMP] + C$$
Inactive enzyme   Active enzyme

Preparation of the catalytic subunit of protein kinase free of the regulatory unit which binds cAMP is achieved using cAMP covalently linked to Sepharose.

Principle. The cAMP was bound to Sepharose as the N^6-ε-aminocaproyl derivative using the cyanogen bromide method.[2]

The introduction of the hydrocarbon side chain with a free amino group facilitated the binding of cAMP to Sepharose and caused the nucleotide to extend from the gel matrix. Filtration of protein kinases from different sources through a cAMP-Sepharose column yielded preparations that were fully active in the absence of cAMP.[3] No free cAMP could be detected in the column eluates. Covalent binding of cAMP to soluble and insoluble carriers via the 2-OH of the ribose moiety[4] and via the C-8 of the purine ring[5] were recently described.

[1] M. A. Brostrom, E. M. Reimann, D. A. Walsh, and E. G. Krebs, *in* "Advances in Enzyme Regulation" (G. Weber, ed.), p. 191. Pergamon, Oxford, 1970.
[2] J. Porath, *Nature (London)* **218**, 834 (1968).
[3] M. Wilchek, Y. Salomon, M. Lowe, and Z. Selinger, *Biochem. Biophys. Res. Commun.* **45**, 1177 (1971).
[4] A. L. Steiner, D. M. Kipnis, R. Utiger, and C. W. Parker, *Proc. Nat. Acad. Sci. U.S.* **64**, 367 (1969).
[5] G. I. Tesser, H. V. Fisch, and R. Schwyzer, *FEBS Lett.* **23**, 56 (1972).

cAMP-Sepharose Column

Preparation

Carbobenzoxy-ε-Aminocaproic Acid Anhydride. CBZ-ε-aminocaproic acid (5.3 g) and dicyclohexylcarbodiimide (2.9 g) were dissolved in 40 ml of absolute ethyl acetate. After 1 hour of incubation at room temperature, the dicyclohexylurea was removed by filtration, and the anhydride was precipitated by the addition of petroleum ether. The yield was quantitative, mp 75–77°.

N^6-Carbobenzoxy-ε-Aminocaproyl-cAMP. cAMP (250 mg) was dissolved in a minimal volume of cold aqueous 0.4 M triethylamine. The solution was evaporated to dryness under reduced pressure, taken up in an anhydrous pyridine, reevaporated, and dried for 5 hours in a vacuum desiccator. The dried material was dissolved in 9 ml of hot anhydrous pyridine, to which 4 g of CBZ-ε-aminocaproic anhydride were added. This mixture was boiled for 3–4 minutes and was left at room temperature for 8 days. At the end of this incubation period the mixture was evaporated to dryness, the residue was dissolved in ethyl acetate and was extracted with cold water containing sodium bicarbonate in an amount of two equivalent of the cAMP. The aqueous solution was cooled in ice and the pH adjusted to 13. After 5 minutes the alkaline solution was acidified to pH 2 with 2 M HCl and extracted with ether. The aqueous solution was brought to pH 6.2 with 2 N NaOH and evaporated to dryness. The dried residue was extracted several times with boiling absolute ethanol until the last ethanol extract showed negligible absorption at 272 nm. The combined ethanol extracts were concentrated by evaporation, and the N^6-CBZ-ε-aminocaproyl-cAMP was precipitated with dry ether. The yield was 180 mg. The material showed λ_{max} at 272 nm and gave one spot on thin-layer chromatography in a solvent system of ethanol (5 volumes), 0.5 M ammonium acetate (2 volumes), R_f 0.77.

N^6-ε-Aminocaproyl-cAMP. Palladium on charcoal (10% w/w) was added to a methanolic solution of N^6-CBZ-ε-aminocaproyl-cAMP (0.5 g/30 ml). Reduction was carried out in a Parr hydrogenation apparatus for 1.5 hour at room temperature. After hydrogenation, the palladium charcoal was removed by filtration and the solution was evaporated to dryness. The residue was redissolved in a small volume of methanol and was precipitated by addition of ether. The yield was quantitative. The product showed one spot of ultraviolet absorption with an R_f of 0.38 by thin-layer chromatography using the same solvent as above. The λ_{max} of this compound was 272 nm and EM^{-1} cm^{-1} 14.500.

Sepharose-ε-Aminocaproyl-cAMP. Cyanogen bromide activation of Sepharose is based on the method of Porath.[2]

Sepharose is mixed with an equal volume of water, and cyanogen bromide (125 mg per milliliter of settled Sepharose) is added. The pH is adjusted to and maintained at 10.5–11 by titration with 4 N NaOH.

After 8–10 minutes, the activated Sepharose is poured on a Büchner funnel and washed with cold water followed by cold 0.1 M NaHCO$_3$. The washed Sepharose (no detectable smell of cyanide) is suspended in cold 0.1 M NaHCO$_3$, pH 8.5, and the ε-aminocaproyl-cAMP is added. The mixture is stirred gently or shaken so as not to break the beads, at 4° for 16 hour. The cAMP-Sepharose is washed extensively with water and phosphate buffer pH 7.0, and stored at 4°.

Procedure

Separation of the Protein Kinase Catalytic Subunit Free of cAMP Binding Protein. Muscle protein kinase (22 mg) purified to the first DEAE-cellulose column[6] was applied to the cAMP-Sepharose column (1 ml volume) preequilibrated with 20 mM phosphate buffer, pH 7.4, and 1 mM EDTA. On elution of the column with the above buffer the breakthrough protein fraction contained 70% of the original protein kinase activity, which was fully activated in the absence of cAMP and did not bind cAMP. No free cAMP was present in this peak, and no regulatory subunit could be detected.

[6] D. A. Walsh, J. P. Perkins, and E. G. Krebs, *J. Biol. Chem.* **243**, 3763 (1968).

[53] The Preparation and Use of Diazomalonyl Derivatives of Cyclic AMP in Isolating and Identifying Cyclic AMP Receptor Sites

By BARRY S. COOPERMAN and DAVID J. BRUNSWICK

In order to provide a general tool for the isolation and identification of receptor sites for cAMP, we have synthesized three diazomalonyl derivatives of cAMP that are of potential use as photoaffinity reagents [$O^{2'}$-(ethyl 2-diazomalonyl)cAMP (I); $O^{2'},N^6$-di(ethyl 2-diazomalonyl)-cAMP (II); N^6-(ethyl 2-diazomalonyl)cAMP (III)].[1,2] These reagents offer three principal advantages for receptor site studies[3]: (1) their

[1] D. J. Brunswick and B. S. Cooperman, *Proc. Nat. Acad. Sci. U.S.* **68**, 1801 (1971).
[2] D. J. Brunswick and B. S. Cooperman, *Biochemistry* **12**, 4074 (1973).
[3] J. R. Knowles, *Accounts Chem. Res.* **5**, 155 (1972).

stability in the absence of light permits them to equilibrate across cell walls and membranes prior to photoactivation; (2) this same stability allows their noncovalent binding properties to be measured prior to covalent bond formation; (3) on photolysis they generate highly reactive carbenes, capable of insertion into any covalent bond, even those of aliphatic side chains. This eliminates the need for a nucleophile to be present in order for labeling to be achieved and increases the likelihood that covalent bond formation will occur at or very close to the point of photogeneration, i.e., at the noncovalent binding site.

In this article we present: (i) detailed procedures for the synthesis of compounds (I), (II), and (III) including methods for the preparation of samples of high specific radioactivity; (ii) a discussion of the consequences of the Dimroth rearrangement for the use of (II) or (III); (iii) procedures for photolysis, measurement of incorporation, and iterative methods for quantitative labeling; (iv) brief discussions of applications of these compounds to the study of rabbit muscle phosphofructokinase and to the endogenous cAMP-dependent protein kinase in erythrocyte ghosts.

Syntheses

Principle. Ethyl 2-diazomalonyl derivatives of cAMP are prepared using ethyl 2-diazomalonyl chloride as the acylating agent in pyridine as solvent. The yield of the $O^{2'}$ derivative is maximized using a short reaction time and a small excess of ethyl 2-diazomalonyl chloride over cAMP. The $O^{2'},N^6$-di derivative is prepared using long reaction times and a large excess of ethyl 2-diazomalonyl chloride over cAMP. The N^6 derivative is prepared by selective base hydrolysis of $O^{2'},N^6$-di(ethyl 2-diazomalonyl)cAMP.

Reagents and Methods

cAMP (Sigma)
Ethyl 2-diazomalonyl chloride (prepared as described previously[1])
Pyridine (reagent grade, dried over 4 Å molecular sieves) (Fisher)
Triethylamine, reagent grade, redistilled before use
Ascending paper chromatography was performed on either Whatman No. 40 or 3MM paper, with an overnight development, using ethanol:0.5 M ammonium acetate (pH 7.0) (5:2, v/v) as solvent. Ascending thin layer chromatography was carried out on Macherey–Nagel PEI–cellulose sheets with 1% LiCl as solvent.

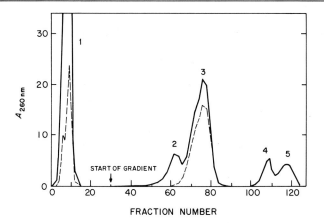

Fig. 1. Chromatographic purification of compound (I). Flow rate, 1.5 ± 0.5 ml/minute; fraction volume, 16 ml; ——, before photolysis - - -, after photolysis. Other conditions are described in text.

Synthesis of $O^{2'}$-(Ethyl 2-Diazomalonyl) Adenosine 3',5-Cyclic Monophosphate (I)

cAMP (0.1 g, 0.3 mmole) and [^{3}H]cAMP (25 μCi)[4] were dissolved in 3.4 ml of 0.38 M aqueous triethylamine, and the solution was evaporated to dryness. The residue was dissolved in dry pyridine and pyridine was removed by evaporation. The dried residue was dissolved with gentle heating and shaking in 2.0 ml of dry pyridine. This solution was brought to 0°, and to it was added, dropwise and with stirring, ethyl 2-diazomalonyl chloride (0.15 ml, 1.14 mmoles). Addition took 5 minutes, after which the reaction mixture was allowed to stand at room temperature for 10 minutes and then cooled to 0°. Absolute ethanol (2 ml) was added to quench the reaction, and after standing at room temperature for 30 minutes the solution was evaporated to dryness. The residue was dissolved in two 5-ml portions of absolute ethanol and evaporated to dryness each time. The residue was dissolved in 30 ml of 40 mM aqueous triethylammonium bicarbonate buffer (pH 7.5), giving a final pH of 5.6, and this solution was chromatographed on a DEAE-cellulose (HCO$_3^-$ form) column (3.4 × 15.0 cm), eluting first with 500 ml of H$_2$O followed by a linear gradient of triethylammonium bicarbonate at pH 7.2 (0.00–0.07 M, 1.61 total). The elution pattern is shown in Fig. 1. Peak 1 contained

[4] Inclusion of small amounts of [^{3}H]cAMP in the "cold" synthetic reaction mixture considerably simplified the problem of identifying cAMP derivatives, which was otherwise made very difficult by the presence of several UV absorbing species found on addition of ethyl 2-diazomalonyl chloride to pyridine.

a small amount of unidentified radioactive material, pyridine, diethyl diazomalonate, and other nonradioactive materials. Peak 2 contained ethyl-2-diazomalonic acid. Peak 3 contained (I) and cAMP. Peak 4 contained (II). Peak 5 contained an unidentified radioactive material. The proportion of cAMP and (I) in each fraction in peak 3 could be determined from the ratio of A^{260} before and after photolysis. For (I) this ratio is 1.33, while for cAMP it is 1.00. In this way fractions 69–78 were shown to contain essentially pure (I). These fractions were combined and rotary-evaporated under reduced pressure. The residue was contaminated by small amounts of cAMP and ethyl-2-diazomalonic acid. Both of these impurities are present in part because of partial hydrolysis of (I) on removal of triethylammonium bicarbonate.[2] Pure samples of (I) were obtained by preparative paper chromatography. In some cases a second paper chromatography was necessary to remove the last traces of ethyl-2-diazomalonic acid. The final yield of the triethylammonium salt of (I) was 55 mg (39%). In other preparations the yield varied from 20 to 50%.

Synthesis of $N^6,O^{2'}$-Di(Ethyl 2-Diazomalonyl) Adenosine 3',5-Cyclic Monophosphate (II)

A solution of the dried triethylammonium salt of cAMP (0.5 g, 1.5 mmoles, ^{3}H-labeled, 25 μCi)[4] in 26 ml of dry pyridine, prepared as described above, was brought to 0°, and to it was added, dropwise and with stirring, 2.25 ml (17.1 mmoles) of ethyl 2-diazomalonyl chloride. Addition took 10 minutes, after which the reaction mixture was allowed to stand overnight at 4°. To quench the reaction, 11 ml of water was added, and the resulting solution was allowed to stand at room temperature for 1 hour before being taken to dryness. The residue was dissolved in two 10-ml portions of absolute ethanol and evaporated to dryness each time. The residue was dissolved in 75 ml of 0.1 M potassium phosphate buffer (pH 7.5) and the solution was brought to pH 5–6 with 1 M NaOH (approximately 10 ml), and extracted with five 50-ml portions of $CHCl_3$. The chloroform extracts were evaporated to dryness, the residue was dissolved in 25 ml of 40 mM triethylammonium bicarbonate (final pH 5.2), and the solution was chromatographed on a DEAE-cellulose (HCO_3^- form) column (4.3 × 23 cm), eluting first with water and then with a linear gradient of triethylammonium bicarbonate (pH 7.2) (0.00 to 0.15 M, 41 total). The elution pattern is shown in Fig. 2. Peak 1 contains the same materials as in peak 1 of Fig. 1. Peak 2 contains (I) and unidentified radioactive material. Peak 3 contains (II). Peak 4 contains unidentified radioactive material. Fractions 66–75 were com-

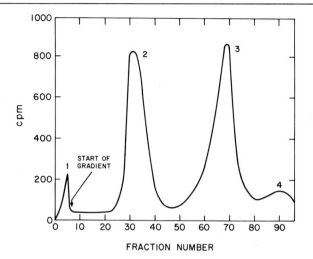

Fig. 2. Chromatographic purification of compound (II). Flow rate 2 ml/minute; fraction volume, 50 ml. Other conditions as described in text.

bined and rotary evaporated under reduced pressure,[5] giving the triethylammonium salt of (II) in essentially pure form (185 mg, 20%). Small amounts of a nonradioactive impurity were removed by preparative paper chromatography.[6] In two other preparations the yield was 35%.

Synthesis of N^6-(Ethyl 2-Diazomalonyl) Adenosine 3′,5-Cyclic Monophosphate (III)

To a stirred, ice-cold solution of the triethylammonium salt of (II) (50 mg, 70 μmoles) in 0.5 ml of water was added 0.5 ml of ice-cold 2 M NaOH. After 4 minutes, 1.2 ml of 1 M acetic acid was added, and the solution was subjected to preparative paper chromatography,[6] giving pure (III) in almost quantitative yield.

Synthesis of [^{3}H]cAMP Derivatives

[^{3}H]$O^{2′}$-(Ethyl 2-Diazomalonyl)Adenosine 3′,5′-Cyclic Monophosphate (I). [^{3}H]cAMP (0.5 mCi, 20 nmoles) was allowed to react with

[5] Residual triethylammonium bicarbonate was removed by two evaporations to dryness with 20-ml portions of water. During evaporation of triethylammonium bicarbonate, the pH rose above pH 9, leading to hydrolysis of (I) back to cAMP, and of (II) to (III). If CO_2 was bubbled through the solution prior to each rotary evaporation, this effect could be minimized.

[6] On paper and thin-layer (PEI) chromatography, compounds (II) and (III) could be recognized by their characteristic blue fluorescence on UV irradiation of the dried chromatograms.

ethyl 2-diazomalonyl chloride (15 μl, 0.11 mmole) in 0.2 ml of dry pyridine as described above in the synthesis of (I). ^{3}H-labeled compound (I) was purified on a small DEAE-cellulose column followed by paper chromatography and obtained in 30–40% yield.

[^{3}H]N^6-(Ethyl 2-Diazomalonyl)Adenosine 3',5'-Cyclic Monophosphate (III). [^{3}H]cAMP (0.1 mCi, 4 nmoles) was allowed to react with ethyl 2-diazomalonyl chloride (15 μl, 0.11 mmole) in 0.2 ml of dry pyridine as described above in the synthesis of (II). The reaction was quenched with water, taken to dryness, ethanol was added and evaporated, and the residue was dissolved in 0.4 ml of ice-cold 1 M NaOH. After 4 minutes, 0.5 ml of 1 M acetic acid was added, and the reaction mixture was paper chromatographed. The radioactive band cochromatographing with authentic (III) was eluted and further purified by thin-layer chromatography, giving ^{3}H-labeled compound (III) in 20% yield.

The Dimroth Rearrangement

The UV spectra of (II) and (III) are pH dependent in the neutral pH range, with apparent spectrophotometric pK_a's of 6.4 for (II) and 6.0 for (III). This pH dependence is due to a Dimroth rearrangement characteristic of α-diazomides.[2]

R = purine 3',5'-cyclic ribonucleotide

The rate at which equilibration is achieved between these two species is strongly dependent on pH. A plot of k_{obs} for (III) [(II is similar] as a function of final pH at 19° is given in Fig. 3. These data are most simply explained by the reversible reactions (1) and (2), where DH represents the diazo form and T$^-$ is the enolate ion of the 1,2,3-triazole form. k_{obs} is given by Eq. (3),

$$\text{DH} + \text{OH}^- \underset{k_{-1}}{\overset{k_1}{\rightleftharpoons}} \text{T}^- \tag{1}$$

$$\text{DH} \underset{k_{-2}}{\overset{k_2}{\rightleftharpoons}} \text{T}^- + \text{H}^+ \tag{2}$$

$$k_{obs} = k_1(\text{OH}^-) + k_{-1} + k_2 + k_{-2}(\text{H}^+) \tag{3}$$

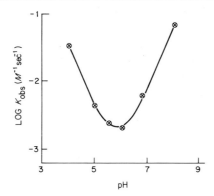

Fig. 3. Rates of α-diazoamide–triazole equilibration for compound (III) as a function of pH at 19°. For rates measured at pH > 6, III was preincubated at pH 4.0 and an aliquot was added to a large excess of buffer. For rates measured at pH < 6, (III) was preincubated at pH 8.0; an aliquot was added to a large excess of buffer. The equilibration process was monitored at 280 nm.

where the rate constants have the following approximate values: k_1, $1.21 \times 10^5 \ M^{-1} \ \text{sec}^{-1}$; k_2, $3.3 \times 10^2 \ M^{-1} \ \text{sec}^{-1}$; $k_{-1} \approx k_{-2} = 0.5 \times 10^{-3} \ \text{sec}^{-1}$.

Failure of the triazole forms of compounds (II) and (III) to efficiently photolyze to yield a carbene poses a problem for practical use of (II) and (III) as cAMP photoaffinity labels, at least at pH values greater than 6.5–7.0, where the triazole form predominates. In some cases this problem can be overcome by adding (II) or (III) in the diazo form to a solution of pH > 7, and photolyzing before isomerization to the triazole becomes important. At 19°, $t_{1/2}$ for equilibration of (III) in 90 seconds at pH 7 and 10 seconds at pH 8. However, at 5°, $t_{1/2}$ is 140 seconds at pH 8, thus increasing the useful pH range. This compares with a $t_{1/2}$ for photolysis of 0.02 mM (III) (diazo form) of about 20 seconds, using the Rayonet reactor as the light source.

Procedures

Irradiation

Equipment

Light sources: We have used a Rayonet photochemical reactor, which is suitable for irradiation at either 253.7 nm, 300 nm, or 350 nm. Most of our work has been at 253.7 nm, and at this wavelength a UVS 11 (Mineralight) lamp provides an inexpensive alternate source of adequate brightness.

Vessels. Irradiations at 253.7 nm are done in quartz vessels. Ordinary quartz test tubes or cuvettes are suitable.

Procedure and Comments. Vessels are irradiated for measured time periods. The extent of photolysis is determined by monitoring the disappearance of the diazo band in the ultraviolet (in the absence of a biological sample) or by monitoring incorporation of radioactive material into a biological sample. Our work has been done mostly at 253.7 nm because of the inconveniently long time needed for photolysis in the Rayonet at 350 nm. For example, photolysis of 0.1 mM (I) is half over in 7 seconds at 253.7 nm versus 45 minutes at 350 nm (the UV spectrum of α-diazoketones has, characteristically, one intense band with λ_{max} 255 nm, ϵ_{max} 7000, and one weak band, λ_{max} 350 nm, ϵ_{max} 25). In recent experiments using light of wavelength >300 nm from high intensity (1000 W) mercury arc lamps, photolysis of diethyl diazomalonate has been effected with a half-life of approximately 1 minute. Use of the mercury arc lamp will be of great importance for problems in which irradiation at 253.7 nm is undesirable because of resultant damage to the biological sample.[7]

Measurement of Incorporation. Following irradiation covalent incorporation of radioactive ethyl 2-diazomalonyl derivatives of cAMP into protein has been measured in two ways. The first method has been to precipitate protein by addition of an equal volume of 20% trichloroacetic acid followed by two washings with 10% trichloroacetic acid. The precipitate is redissolved in 0.1 M NaOH, protein concentration is determined from A_{280},[8] and an aliquot is counted for radioactivity. The second method has been to use a Sephadex G-25 filtration to separate protein from all but covalently bound cAMP derivative, and then determine protein concentrations and radioactivity as above. Similar procedures could be used for other biological target materials.

Iterative Methods for Quantitative Labeling

Principle. A common feature of studies with photoaffinity reagents has been that quantitative labeling of the target site has not been achieved. This result can be understood in terms of the simplified scheme presented below, where E is enzyme, D is diazo compound, A is the product of reaction of photolyzed D with solvent (presumably the hydroxy

[7] B. S. Cooperman and D. J. Brunswick, *Biochemistry* 12, 4079 (1973).
[8] In our studies with rabbit muscle phosphofructokinase,[1,7] the contribution to A_{280} from covalently bound cAMP derivative was negligible. In cases where this is not so, a ratio method (e.g., A_{260}/A_{280}) could be used to measure protein concentration.

compound), EA and ED are noncovalent complexes, and EA' is the product of reaction of photolyzed D with enzyme, i.e., labeled enzyme.

$$E + D \underset{}{\overset{K_D}{\rightleftharpoons}} ED \qquad (4)$$
$$E + A \underset{}{\overset{K_A}{\rightleftharpoons}} EA \qquad (5)$$
$$h\nu + D \to A \qquad (6)$$
$$h\nu + ED \to EA \qquad (7)$$
$$h\nu + ED \to EA' \qquad (8)$$

Before photolysis, E, D, and ED are present in solution. On irradiation, D is converted into A and ED partitions into EA and EA'. With the light sources we have used, photolysis has a half-life of seconds, which should be slow compared to the time needed for reequilibration among E, EA, and ED as D is converted to A. Thus, as photolysis proceeds, EA replaces ED and since EA' can only arise from ED, labeling is incomplete. From this analysis, it is clear that quantitative labeling can in principle be achieved by an iterative method which removes A from solution and introduces fresh D. Two such methods are discussed below.

Iterative Labeling—Column Photolysis

Equipment

Narrow diameter quartz column: In our work columns of dimensions 5×26 mm were used, which allowed incident UV light to pass right through the column.

Procedure and Comments. DEAE-A50 Sephadex (Pharmacia), equilibrated in slightly alkaline buffer (we used 0.1 M Tris phosphate (pH 8.0) in our work with rabbit muscle phosphofructokinase), is mixed with target protein and packed into a quartz column. cAMP derivative is passed through the column, dissolved in buffer of sufficiently high ionic strength to prevent retention of derivative by the column, but of sufficiently low ionic strength to retain protein. The column is irradiated continuously over its entire length. The flow rate is adjusted so that not more than half of the derivative is irradiated while on the column, to allow incorporation to take place into protein bound to the bottom of the column. Protein may be removed from the column by increasing the ionic strength or decreasing the pH of the eluting buffers, and the amount of covalent incorporation can be determined on an aliquot of eluate. When several identical columns are made up, it is possible to measure incorporation as a function of volume of cAMP derivative-containing buffer pased over the column, and thus to establish the maximum possible incorporation.

Iterative Labeling—Dialysis Photolysis

Equipment

Flow dialysis cell with quartz window cut into upper half (Bel-Art Products)
Peristaltic pump

Procedure and Comments. Protein dissolved in buffer containing cAMP derivative is placed into the half of the flow dialysis cell which has the quartz window. The other, flow-through, half of the cell is connected to a reservoir of cAMP derivative in buffer, which is pumped through the cell by means of a peristaltic pump. The cell is irradiated continuously, but, since it is made of a plastic opaque to UV light, photolysis takes place only in the protein-containing half. As photolysis proceeds, photolyzed derivative not covalently incorporated into protein is dialyzed out and unphotolyzed derivative is dialyzed in. Eventually all the derivative is photolyzed and the reservoir is emptied and refilled with fresh buffer containing unphotolyzed derivative. The extent of covalent incorporation is measured in aliquots taken from the protein-containing half of the cell and this process is continued until full labeling is achieved.

Results

The table summarizes results obtained with the two iterative techniques described above, with rabbit muscle phosphofructokinase as a test

ITERATIVE LABELING OF PHOSPHOFRUCTOKINASE WITH COMPOUND (I)[a]

Column photolysis[b]		Dialysis photolysis[d]	
Volume of (I)-containing buffer passed through column (ml)	% cAMP sites labeled	Number of cycles[e]	% cAMP sites labeled
2.5	25.5 ± 2.0	1	42 ± 3
6	34.0 ± 2.8	2	59 ± 4
15	52.8 ± 4.3	3	76 ± 5
30	63.8 ± 5.2	5	67 ± 5
75	66.2 ± 5.3	7	74 ± 5
—[c]	16.9 ± 1.4	—[f]	22 ± 2

[a] Data taken from B. S. Cooperman and D. J. Brunswick, *Biochemistry* **12**, 4079 (1973). Reprinted with permission. Copyright by the American Chemical Society.
[b] Concentration of (I) in buffer; 48 μM.
[c] Labeling obtained on photolysis on column without replacement of (I).
[d] Concentration of (I) in buffer, 500 μM.
[e] A new cycle begins when photolyzed (I) in the reservoir is replaced with fresh (I).
[f] Labeling obtained on simple photolysis in a quartz cuvette.

protein. The apparent failure to achieve 100% labeling is almost certainly an artifact of calculation, reflecting the presence of contaminating protein in the preparation used. Of the two techniques, the column method has been found easier to do and more reproducible, but it is dependent on conditions being found where protein sticks to the column and cAMP derivative does not. For work with larger biological targets, such as whole cells, iterative labeling could be simply accomplished by centrifuging after photolysis and resuspending the cells in buffer containing unphotolyzed cAMP derivative.

Applications

Soluble Enzyme—Rabbit Muscle Phosphofructokinase

Phosphofructokinase is subject to inhibition by high ATP concentrations. Addition of cAMP to ATP-inhibited enzymes leads to activation.[9-11] The noncovalent binding of cAMP derivatives (I), (II), and (III) to PFK was examined by their ability either to substitute for cAMP in reactivating enzyme or to prevent activation by cAMP. By these criteria only (I) bound noncovalently with an affinity approximately five times less than that of cAMP. (I) was also the only derivative to become covalently incorporated into PFK on photolysis.

Simple incorporation experiments[1] with phosphofructokinase showed that incorporation (a) is specific for the cAMP site, (b) depends on carbene formation, (c) is irreversible, and (d) reaches a saturating value corresponding to considerably less than 100% labeling. Above we have discussed methods of increasing the extent of labeling. When phosphofructokinase, labeled to the extent of approximately 90%, was prepared via the column dialysis method described above, it was found to differ from native PFK in two respects. First, although still subject to ATP inhibition with virtually unchanged ATP affinity, the extent of the inhibition was much reduced; in native enzyme the activity at high ATP concentration falls to about 3% of the maximal activity, while in modified enzyme the corresponding figure is 20%. Second, as expected, ATP inhibited modified enzyme is subject to only a minor activation by cAMP, attributable perhaps to incomplete labeling of the cAMP site. Separate control experiments showed these changes to be due to chemical modification as opposed to irradiation effects.[7] That ATP can substantially inhibit

[9] K. Uyeda and E. Racker, *J. Biol. Chem.* **240**, 4682 (1965).
[10] R. G. Kemp and E. G. Krebs, *Biochemistry* **6**, 423 (1967).
[11] R. G. Kemp, *J. Biol. Chem.* **246**, 245 (1971).

enzyme whose cAMP site is blocked supports earlier suggestions[12] of separate cAMP and ATP binding sites. That ATP inhibition is less complete in the modified enzyme can be rationalized in two simple ways: (1), that complete inhibition of native enzyme requires a second molecule of ATP to bind to the blocked site in modified enzyme; (2) that ATP binding to a single inhibitory site results in a transition from catalytically active to inactive enzyme states, and that the transition is less favorable in modified enzymes.

Intact Systems—Human Erythrocyte Ghosts

Human erythrocyte ghosts contain a cAMP dependent protein kinase which catalyzes phosphorylation of an endogenous protein.[13–15] In the absence of irradiation, (III) is found to mimic the noncovalent and activating properties of cAMP, with an approximately 3-fold weaker affinity. (I) and (II) are found to bind much more poorly, if at all. On photolysis ^{3}H-labeled compound (III) is incorporated into the ghosts. Covalent incorporation is found to be dependent on photolysis and is largely blocked in the presence of cAMP. Electrophoresis of total ghost protein on an SDS polyacrylamide gel shows a single ^{3}H peak, whose position on the gel coincides exactly with that of the endogenous protein acceptor labeled with ^{32}P (from $[\gamma^{32}\text{-P}]$ATP).[16] This coincidence strongly suggests that the ^{3}H-labeled protein is the regulatory subunit of the membrane-bound cAMP dependent protein kinase, by analogy with work on the soluble cAMP-dependent protein kinase from heart muscle, which has been shown to catalyze phosphorylation of its own cAMP-binding (regulatory) subunit.[17] This work illustrates the potential of derivatives (I), (II), and (III) for specifically labeling cAMP receptor proteins in quite complex systems.

---

[12] M. M. Mathias and R. G. Kemp, *Biochemistry* **11**, 578 (1972).
[13] C. E. Guthrow, Jr., J. E. Allen, and H. Rasmussen, *J. Biol. Chem.* **247**, 8145 (1972).
[14] C. S. Rubin and O. M. Rosen, *Biochem. Biophys. Res. Commun.* **50**, 421 (1973).
[15] A. D. Roses and S. H. Appel, *J. Biol. Chem.* **248**, 1408 (1973).
[16] C. E. Guthrow, Jr., H. Rasmussen, D. J. Brunswick, and B. S. Cooperman, *Proc. Nat. Acad. Sci. U.S.* **70**, 3344 (1973).
[17] O. M. Rosen, C. S. Rubin, and J. Erlichman, manuscript in preparation.

[54] The Preparation of Acylated Derivatives of Cyclic Nucleotides[1]

By T. Posternak and G. Weimann

cAMP is in general a relatively inert compound when injected into animals or applied to intact cells and tissues. It is usually thought that nucleotides and other anionic phosphorylated compounds penetrate cell membranes poorly. Moreover, exogenous cAMP may be destroyed by the specific phosphodiesterase which converts it to AMP. In the hope of obtaining compounds which might penetrate cell membranes more readily, a number of derivatives of cAMP were prepared. The basic idea was in fact to introduce into the molecule liposoluble fatty acid residues of variable length that could facilitate passage across cellular membranes. Compounds carrying an acyl group either in position N^6 (III) or in position 2'-O (I) or simultaneously in both positions (II) were therefore pre-

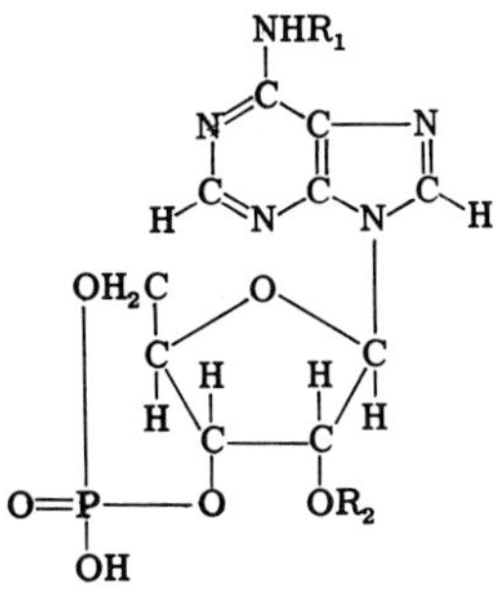

(I) $R_1 = H$, $R_2 = $ acyl
(II) $R_1 = R_2 = $ acyl
(III) $R_1 = $ acyl, $R_2 = H$

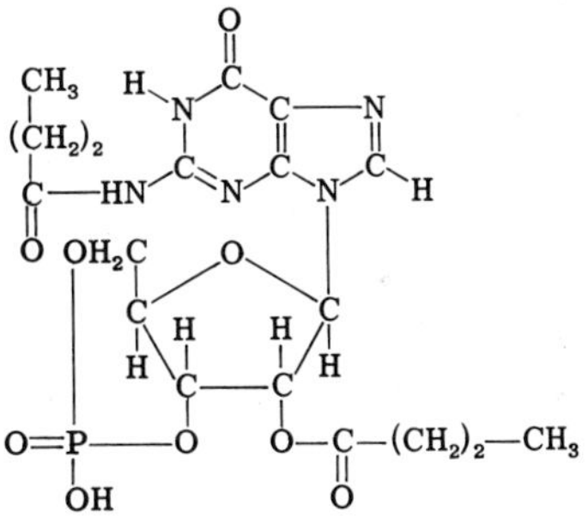

(IV)

pared. Practically all the biological experiments were carried out with the butyryl derivatives, which proved more active than other corresponding acyl derivatives, for instance, for the stimulation of dog liver phosphorylase.[2,3] In numerous tests involving intact cells, the effects of the dibutyryl derivative were found in general to be qualitatively the same as those of cAMP itself, but in addition, in many cases, the greater

---

[1] Abbreviations used: cAMP, adenosine 3',5'-phosphate; cGMP, guanosine 3',5'-phosphate; cUMP, uridine 3',5'-phosphate; cIMP, inosine 3',5'-phosphate; cyclic iso-AMP, iso-adenosine 3',5'-phosphate.

[2] T. Posternak, E. W. Sutherland, and W. F. Henion, *Biochim. Biophys. Acta* **65**, 558 (1962).

[3] W. F. Henion, E. W. Sutherland, and T. Posternak, *Biochim. Biophys. Acta* **148**, 106 (1967).

effectiveness of the dibutyryl derivative could be demonstrated. Although there is no direct evidence that the compounds penetrate cell membranes more readily than does cAMP, this may explain their greater potency. In addition, their greater resistance to phosphodiesterase plays certainly an important role. Comparing the relative potencies in stimulating the activation of dog liver phosphorylase, it was found that the N^6-monobutyryl[4] and the N^6-2'-O-dibutyryl[4] derivatives of cAMP were approximatively 45–50 times more active on liver slices than cAMP itself; the N^6 derivative in this case was more active than the 2'-O derivative. On the other hand, the N^6-monoacyl compounds, and especially the 2'O-monoacyl compounds, were less active when added to liver or heart extracts.[2,3] The intact liver and heart cells probably contain enzymes capable of removing the acyl groups from the 2'-O position, apparently these enzymes are lost or inactivated during the preparation of the extracts.

cGMP is the only other cyclonucleotide known to occur in nature. It is present in most tissues in concentrations at least 10-fold lower than those of cAMP. Although cGMP effectively mimics a number of effects of cAMP in several intact cell systems, its biological role has not yet been established. In addition, it has been found that cIMP and cUMP at high concentrations can mimic certain effects of cAMP, such as stimulation of lipolysis and activation of phosphorylase.[5] N^2-2'-O-Dibutyryl cGMP (IV), 2'-O-monobutyryl cUMP (V) and 2'-O-monobutyryl cIMP (VI) have been prepared; these acylated derivatives have not yet

(V)   (VI)

been studied in biological systems involving intact cells, but they are potentially interesting.

Until now this review was limited to acylated cyclic nucleotides de-

---

[4] J.-G. Falbriard, T. Posternak, and E. W. Sutherland, *Biochim. Biophys. Acta* **148**, 99 (1967).

[5] M. DuPlooy, G. Michal, G. Weimann, and M. Nelböck, *Biochim. Biophys. Acta* **230**, 30 (1971).

rived from nucleotides which occur in nature. It should be pointed out that substitution of the purine skeleton in N^6 by alkyl residues[6–8] and in C-8 or in C-2 by certain groups (OH, SH, SCH_3, NH_2, NHR)[6,9] may enhance the effects in several biological tests. For instance, 8-mercapto cAMP proved more active than cAMP itself in stimulating the release of growth hormone by hypophysis slices *in vitro*. The conversion of this compound to its N^6-$2'$-O-dibutyryl derivative resulted in a still higher effectiveness.[6,9] In addition, cyclic iso-AMP which contains the sugar moiety in N-3 instead of N-9 proved more effective than cAMP in certain tests, such as the dispersion of melanin granules within the melanophores of *Anolis* or the release of growth hormone and thyroid-stimulating hormone from hypophysis slices *in vitro*.[6,7,10] The monobutyryl derivative of cyclic iso-AMP showed in these tests about the same effectiveness as cyclic iso-AMP itself.

Principle of Synthetic Methods

The general preparative method is as follows[2,4]: A suitable salt of cAMP is treated in anhydrous pyridine by an acid anhydride. A considerable difference was found in the speeds of acylation: the reaction on the $2'$-O-oxygen took place much more rapidly than that on the N^6-nitrogen. Brief treatment by the acylating agent gave therefore a $2'$-O-monoacyl cAMP (I); more prolonged treatment produced a N^6-$2'$-O-diacyl derivative (II). These diacyl derivatives can also be obtained by treatment of a salt of cAMP in pyridine solution with an acid chloride; this alternative method possesses in general no advantages over the one previously described.

The two acyl residues have very different susceptibility to the action of alkalis, and for this reason the $2'$-O-acyl group may be selectively eliminated by a brief treatment at room temperature (about 5 minutes in 0.1 M NaOH) with the formation of a N^6-monoacyl cAMP (III).

The N^2-$2'$-O-dibutyryl derivative (IV) of cGMP, the $2'$-O-monobutyryl derivatives of cUMP (V) and of cIMP (VI) were obtained again by treatment of the cyclonucleotides by butyric anhydride in anhydrous

[6] T. Posternak and G. Cehovic, *Ann. N.Y. Acad. Sci.* **185,** 42 (1971).
[7] T. Posternak, I. Marcus, A. Gabbai, and G. Cehovic, *C. R. Acad. Sci. Ser. D* **269,** 2409 (1969).
[8] G. Cehovic, I. Marcus, A. Gabbai, and T. Posternak, *C. R. Acad. Sci. Ser. D* **271,** 1399 (1970).
[9] T. Posternak, I. Marcus, and G. Cehovic, *C. R. Acad. Sci. Ser. D* **272,** 622 (1971).
[10] G. Cehovic, I. Marcus, S. Vengadabady, and T. Posternak, *C. R. Soc. Phys. Hist. Nat. Geneve* **3,** 135 (1968).

pyridine. By butyrylation of cyclic iso-AMP, one obtains a derivative with an unmodified λ_{max}: this means that a $2'$-O-monobutyryl cyclic iso-AMP was obtained and that the N^6 position resists butyrylation.

Purification of all the acylated cyclic nucleotides can be effected on a small scale by paper chromatography; on a larger scale, column chromatography must be used. The derivatives are isolated as barium salts by precipitation from a concentrated aqueous solution either with alcohol or preferably with alcohol–ether. For biological tests, the barium salts are converted to sodium salts by treatment with sodium sulfate. The sodium salts can also be obtained by passage through a column of Dowex-50 (Na^+ form), or by precipitation from a concentrated aqueous solution of a salt of a suitable organic base by addition of sodium perchlorate dissolved in anhydrous acetone.

Preparative chromatography was performed in the system ethanol–0.5 M ammonium acetate (5:2, v/v) on Whatman 3 MM paper which had been thoroughly washed with dilute HCl, then with twice-distilled water. The bands were located by their fluorescence in ultraviolet light using lamp NN 27 X (Quarzlampen Ges., Hanau). These bands were cut out and extracted three times by a total weight of solvent corresponding to eight times the weight of paper. The usual solvent employed was water. For the extraction of the dioctanoate, dihexanoate, and monostearate of cAMP, 80% ethanol was used. The extracts thus obtained were employed sometimes directly, after phosphorus estimation, for biochemical assays.

Analytical chromatography of the acylated derivatives was usually performed in system A (ethanol–0.5 M ammonium acetate 5:2, v/v) on Whatman No. 1 paper or in system B (isopropanol–1 M ammonium acetate 5:2, v/v) on Schleicher & Schuell No. 2043 b paper (descending technique). After drying, spots were located by their fluorescence in ultraviolet light.

The R_f values of the acyl derivatives of cAMP increase regularly with length and number of acyl residues. All the $2'$-O-acyl derivatives have practically the same λ_{max} as cAMP itself (258–260 nm) and ϵ_{max} values from 13,000 to 15,000. When there is substitution on N^6, the λ_{max} is usually displaced toward 270–273 nm. An N^6-benzoyl group produces a λ_{max} of 280 nm (see table).

Acylated Derivatives of cAMP[2,4]

Preparation of $2'$-O-Acyl Derivatives (I)

The general method of acylation used was as follows. Equimolar quantities of cAMP (250 mg) and $4'$-morpholino-N,N'-dicyclohexylcar-

FORMULAS, R_f, λ_{max}, AND ϵ_{max} OF CYCLIC NUCLEOTIDES AND OF THEIR ACYLATED DERIVATIVES

Compounds	Formulas	R_f	λ_{max} (nm)	ϵ_{max}
cAMP	$C_{10}H_{12}N_5O_6P \cdot H_2O$	0.34	258^c	14,100
2'-O-Monoacetyl	$C_{12}H_{13}N_5O_7PBa/2 \cdot 2\ H_2O^a$	0.47^b	260^c	14,005
N^6-Monoacetyl	$C_{12}H_{13}N_5O_7PBa/2 \cdot 2\ H_2O^a$	0.48^b	273^c	14,100
N^6-2'-O-Diacetyl	$C_{14}H_{15}N_5O_8PBa/2 \cdot 2\ H_2O^a$	0.61^b	273^c	14,030
2'-O-Monobutyryl	$C_{14}H_{17}N_5O_7PBa/2 \cdot 2\ H_2O^a$	0.61^b	258^c	13,400
N^6-Monobutyryl	$C_{14}H_{17}N_5O_7PBa/2 \cdot 2\ H_2O^a$	0.63^b	272^c	15,100
N^6-2'-O-Dibutyryl	$C_{18}H_{23}N_5O_8PBa/2 \cdot 3\ H_2O^a$	0.80^b	270^c	14,840
N^6-Monohexanoyl	$C_{16}H_{21}N_5O_7PBa/2 \cdot 2\ H_2O^a$	0.63^b	274^c	16,900
N^6-2'-O-Dihexanoyl	—	0.80^b	274^c	—
2'-O-Monooctanoyl	$C_{18}H_{26}N_5O_7P \cdot 2\ H_2O^a$	0.75^b	259^c	14,600
N^6-Monooctanoyl	$C_{18}H_{25}N_5O_7PBa/2 \cdot 2\ H_2O^a$	0.78^b	273^c	18,460
N^6-2'-O-Dioctanoyl	—	0.88^b	272^c	14,800
N^6-Monolauryl	$C_{22}H_{33}N_5O_7PBa/2 \cdot 2\ H_2O^a$	0.82^b	273^c	17,200
N^6-Monostearyl	—	0.84^b	273^c	24,000
N^6-Monobenzoyl	$C_{17}H_{15}N_5O_7PBa/2 \cdot 2\ H_2O^a$	0.60^b	280^c	19,450
cGMP	$C_{10}H_{11}N_5O_7PCa/2 \cdot 5\ H_2O$	0.43^b	254^c	12,950
N^2-2'-O-Dibutyryl	$C_{18}H_{23}N_5O_9PNa$	0.91^d	260^e	16,700
cIMP	$C_{10}H_{11}N_4O_7P \cdot H_2O$	—	249^e	12,200
2'-O-Monobutyryl	$C_{14}H_{17}N_4O_8PNa$	0.75^d	249^e	12,200
cUMP	$C_9H_{11}N_2O_3P \cdot C_6H_{15}N$	0.64^b	261^c	9,940
2'-O-Monobutyryl	$C_{13}H_{16}N_2O_9PNa$	0.83^d	257^e	10,250
Cyclic iso-AMP	$C_{10}H_{11}N_5O_6PAg \cdot 3\ H_2O^a$	0.43^b	278^c	12,400
2'-O-Monobutyryl	—	0.80^b	278^c	14,100
8-Mercapto-cAMP	$C_{10}H_{12}N_5O_6PS^a$	0.25^b	297^f	24,000
N^6-2'-O-dibutyryl	$C_{18}H_{23}N_5O_8PSBa/2^a$	0.85^b	246^c	19,800
			323^c	25,500

[a] After drying at 110° *in vacuo* over P_2O_5.
[b] Solvent system: ethanol–0.5 M ammonium acetate (5:2, v/v) Whatman 1.
[c] pH 6.0.
[d] Solvent system: isopropanol–1 M ammonium acetate (5:2, v/v), Schleicher & Schuell 2043b.
[e] pH 7.0.
[f] pH 11.0.

boxamidine[11] (220 mg) were dissolved in hot anhydrous pyridine (7.5 ml). After cooling, 3.75 g of acid anhydride were added, and the resulting mixture was left at room temperature with moisture excluded. From time to time, 0.02-ml samples were taken, to which 0.03 ml of H_2O was added; after 4 hours 0.01 ml of this mixture was taken for paper chromatog-

[11] M. Smith, G. I. Drummond, and H. G. Khorana, *J. Amer. Chem. Soc.* **83**, 698 (1961).

raphy. This procedure has shown the nearly quantitative formation of the 2′-O-acyl derivatives according to the following times, respectively: acetyl derivative, 2 hours; butyryl derivative, 8 hours; octanoyl derivative, 12 hours. The mixture was cooled in ice; 3.75 ml water was added slowly and left for 5 hours at 4°. During this last process any excess anhydride reagent and any mixed anhydrides formed between the carboxylic acid and the organic phosphate derivative were destroyed by hydrolysis. The mixture was then evaporated (30–40°), first with a water pump, then for 24 hours under high vacuum. Acetic acid, butyric acid, and most of octanoic acid are thus removed.

In order to isolate the 2′-O-monoacetate and the 2′-O-monobutyrate, a solution of 150 mg of $BaI_2 \cdot 2H_2O$ in 1 ml of H_2O was added. When solution was complete, 10 ml of absolute alcohol and 40 ml of freshly distilled anhydrous ether were added. After some hours at 4°, the precipitate was recovered by centrifugation and washed with an alcohol–ether (1:4, v/v) mixture. It was purified by dissolving in 1.5–2 parts of water, followed by the addition of alcohol and ether. Found after drying *in vacuo* over H_2SO_4: 270 mg of 2′-O-acetate and 296 mg 2′-O-butyrate (as their barium salts).

In order to isolate the 2′-O-octanoate, the residue was taken up in 10 ml H_2O, 0.07 ml 2 M HCl and 30 ml ether. After stirring, the precipitate formed (270 mg) was collected by centrifugation and washed with ether. It was redissolved in 0.9 ml 1 M $NaHCO_3$, any insoluble matter eliminated by centrifugation, and acidified to pH 2 with 3 M HCl. This leads to precipitation of the product in the form of the free acid.

Preparation of N^6-2′-O-Diacyl Derivatives (II)

cAMP, 250 mg, was dissolved in the minimum quantity of cold aqueous 0.4 M triethylamine. The solution was evaporated to dryness *in vacuo*, taken up in anhydrous pyridine, reevaporated to dryness, and the whole operation repeated once more. Finally, it was dried for 5 hours under high vacuum. The residue was taken up by 7.5 ml of hot anhydrous pyridine; after cooling, 3.75 g of acid anhydride were added and the mixture was boiled for 3–4 minutes, during which operation the product dissolved almost completely. The solution was immediately cooled and left at room temperature, protected from light and moisture. The reaction was followed by paper chromatography of samples taken as previously described. After 5–8 days, the formation of the diacyl derivative was practically quantitative. Attempts to accelerate the reaction by using higher temperatures led to complications; after boiling for 3 hours in the presence of pyridine and butyric anhydride, the major product formed differed by its R_f from the expected dibutyrate. The reaction mixture was cooled in ice and 3.5 ml water added in small quantities with stirring. After 5 hours at 4°, it was evaporated (30–40°), first using the vacuum of a water pump and then for 24 hours under high vacuum. In the case of the N^6-2'-O-dioctanoate, the octanoic acid was removed by heating for 20 minutes at 110° under a vacuum of 0.3–0.4 Torr. Judging by the chromatography, the product was not decomposed to a marked extent.

The procedure for isolating the N^6-2'-O-diacetate and N^6-2'-O-dibutyrate as barium salts was exactly as previously described for the isolation of the barium salts of the 2'-O-monoacyl derivatives. In the case of the dibutyryl derivative, whose precipitation was sometimes incomplete, the united supernatants from the first precipitation were evaporated to dryness and taken up by a solution of 75 mg $BaI_2 \cdot 2H_2O$ in 0.5 ml of H_2O. By adding 6 ml of alcohol and 25 ml of dry ether, a second precipitation of crude barium salt was obtained. This was added to the first precipitate and purified by dissolving in water and reprecipitating with alcohol and ether. The yield, after drying *in vacuo* over H_2SO_4, was 0.315 g N^6-2'-O-diacetate and 0.210 g of N^6-2'-O-dibutyrate.

To obtain the N^6-2'-O-dioctanoate, the residue after evaporation under high vacuum was extracted several times with boiling light petroleum (bp 30–45°) and dried under high vacuum. After dissolving in the minimum quantity of hot water, 150 mg solid $BaI_2 \cdot 2H_2O$ were added, and the barium salt precipitated as previously described, with alcohol and ether. After standing at 4°, the product was collected by centrifugation, washed with a mixture of ether and light petroleum (1:1, v/v), and left at 4° under light petroleum. A partial hydrolysis may take place during these operations. The dry barium salt was therefore dissolved in anhydrous pyridine and submitted to a second acylation. After hydrolysis of the residual anhydride and elimination of pyridine and octanoic acid as above, the product was precipitated when taken up in a sufficient quantity of light petroleum and was collected by centrifugation; it was then chromatographically homogeneous.

Preparation of N^6-Monoacyl Derivatives (III)

cAMP, 250 mg, as triethylammonium salt was converted to N^6-2'-O-diacyl derivatives, as previously described, by treatment with an acid anhydride in pyridine. After hydrolysis of the excess anhydride, evaporation of the pyridine and fatty acid with a water pump, and then high vacuum, the residue was taken up in 20 ml of alcohol. It was cooled in ice and neutralized to phenolphthalein with $2\ M$ NaOH. Then excess $2\ M$ NaOH was added to give a final concentration of $0.1\ M$. The solution was kept for 4–5 minutes out of the ice-bath, returned to the ice-bath and the pH was rapidly adjusted to 2 with $2\ M$ HCl. This solution was extracted 4 times with freshly distilled ether. The aqueous phase, adjusted to pH 6.1–6.5 by 2 M NaOH, was evaporated to dryness at a low temperature. The well dried residue was thoroughly extracted with boiling absolute alcohol. NaCl was removed by filtration on a sintered-glass filter. The alcoholic extracts containing, according to the nature of the fatty acid employed, 50 to 90% of the original quantity of P, were evaporated to dryness. The residue was dissolved in 1 ml of water (warming if necessary), and the theoretical quantity of solid barium acetate was added. The barium salt precipitated on cooling; in some cases the precipitation was completed by addition of alcohol.

The acid chloride rather than the anhydride had to be used for the preparation of the N^6-benzoyl derivative. For this reason this preparation is described here with some details.

Two hundred milligrams cAMP were dissolved in 8 ml of 0.38 M triethylamine. After evaporating to dryness, the residue is left for 3 hours under high vacuum, then taken up in 8 ml of anhydrous pyridine. The solution is cooled in ice and 1.36 ml benzoyl chloride added. After 75 minutes at room temperature in the dark, 20 ml of water were introduced slowly, the solution being kept cold in ice. After 3 hours at 0°, it was extracted with chloroform (100 ml). The chloroform extract was dried over Na_2SO_4, evaporated to dryness, and the residue taken up with 16 ml of a mixture of pyridine–H_2O (2:1, v/v). Sixteen milliliters of 2 M NaOH were added, and the mixture was left for 5 minutes at room temperature and then treated rapidly with 50 ml of H_2O and 30 g of Dowex-50 (H^+ form), which lowered the pH to 7.0. The solution was filtered and concentrated to 50 ml *in vacuo*. The pH was further lowered to 2.0 by addition of 20 g of Dowex, and the solution was extracted with ether. The pH was adjusted to 7.0 with pyridine and the solution was evaporated to dryness. The residue, containing 74% of the original quantity of P, was dissolved in 2 ml of H_2O and 0.7 ml of a 10% barium acetate solution were added. After reducing the volume to 1 ml, the barium salt (112 mg) was precipitated with 4 volumes of ethanol. A second, less pure, fraction may be obtained by addition of acetone to the supernatant.

Butyryl Derivatives of Other Cyclonucleotides

Preparation of N^2-2'-O-Dibutyryl-cGMP (IV)[12]

Ten grams of cGMP (free acid)[11,13] were dissolved under stirring in a mixture of 600 ml of distilled water and 18.9 ml of a 40% aqueous

---

[12] Unpublished procedures communicated by Boehringer Mannheim G.m.b.H. (1972).

[13] cGMP was obtained as free acid in crystalline state by Boehringer-Mannheim G.m.b.H.

solution of tetrabutylammonium hydroxide. The solution was lyophilized. The freezed-dried powder was suspended in 1.5 liters of dry pyridine and, after addition of 46 ml of butyric anhydride, the mixture was refluxed for 5 hours in an oil bath. Paper chromatography indicated a practically quantitative yield, the amount of monobutyryl cyclic GMP being less than 3%.

The solvent was removed *in vacuo*, the residue was taken up in 300 ml of ice water and reevaporated to dryness, and the whole operation was repeated once more. The residue was taken up in about 400 ml of distilled water and extracted 10 times with ether. The organic phase was discarded, and traces of ether in the water layer were removed in a rotary evaporator. The solution was treated with 100 mg of charcoal and 50 mg of bleaching earth (Merck A. G., Darmstadt).

The filtered solution was passed through a column (32 cm long, 2 cm wide, 100 ml bed volume) of Dowex-50 W X8 (Na^+ form). The column was washed with water until the extinction at 260 nm was lower than 1. The treatment with charcoal and bleaching earth was repeated once more. The filtered solution was brought to 250 ml and freeze-dried. Yield was 14 g (85% of theory).

Preparation of 2'-O-Monobutyryl-cUMP (V)[12]

cUMP,[11] 4.1 g, was dissolved in 50 ml of distilled water and passed through a column (bed volume 20 ml) of Dowex-50 W X8 (H^+ form). The column was washed thoroughly with distilled water. The eluate was mixed with 8.1 ml of a 40% aqueous solution of tetrabutylammonium hydroxide and concentrated *in vacuo* to an oil. For removal of residual water, the oil was taken up in 20 ml of anhydrous pyridine (dried over calcium hydride for 24 hours), and the solution was evaporated under high vacuum; these operations were repeated twice. The oily residue was taken up in 125 ml of dry pyridine, 12.8 ml of butyric anhydride were added, and the reaction mixture was left at room temperature for 48 hours. Paper chromatography showed then a quantitative conversion to the 2'-O-monobutyryl derivative.

The solvent was removed *in vacuo;* the residue was taken up in 30 ml of ice water and the solution was evaporated *in vacuo;* this operation was repeated once more. After addition of 100 ml of water, the solution was extracted 10 times with 30 ml of ether. The aqueous layer was treated with 100 mg of charcoal, filtered, and passed through a column of Dowex-50 W X8 (Na^+ form). The solution was concentrated *in vacuo* to about 50 ml and lyophilized after an additional charcoal treatment. Yield is 3.6 g (70% of theory).

Preparation of 2'-O-Monobutyryl-cIMP (VI)[12]

A mixture of 5 g of cIMP (free acid),[14] 24.2 ml of triethylamine, 30.6 ml of butyric anhydride, and 250 ml of anhydrous pyridine was shaken for 3 hours at room temperature. The nucleotide went into solution, and paper chromatography indicated complete conversion to the 2'-O-butyryl derivative (VI). The isolation procedure is analogous to that given for the preparation of the 2'-O-butyryl-cUMP. Yield was 4 g (65% of theory).

Preparation of N^6-2'-O-Dibutyryl-8-mercapto-cAMP (VII)[15]

A mixture of 350 mg of 8-mercapto-cAMP[9,15,16] suspended in a mixture of 12.0 ml of anhydrous pyridine and 6.0 ml of butyric anhydride was shaken at room temperature for 5 days. Heptane was added to the solution until no precipitation occurred. The compound was collected by filtration on a thin layer of silica gel (Merck A.G., Darmstadt) and washed with heptane. It was then dissolved in water. Na_2CO_3 in 10% excess of the theoretical amount (determined by UV absorption) was added, and the mixture was fractionated (fractions of 10 ml each) on a column of silica gel (80 cm long, 3 cm wide) by elution with the mixture acetone–ethanol–H_2O (8:1.5:0.5, v/v). The compound was detected in the fractions by UV spectroscopy; 290 mg of dibutyryl compound was obtained. Some of the fractions were eventually discarded because they contained an impurity that could be detected only by paper chromatography in system A (pink fluorescence). The yield was 50 mg of analytically pure sodium salt, $C_{18}H_{23}N_5O_8PSNa$.

The dibutyryl compound (VII) could also be obtained by a modification of the procedures described above. Starting from 75 mg of 8-mercapto-cAMP, the butyric acid was removed under high vacuum after hydrolysis of the acid anhydride. The residue was taken up in 5 ml of H_2O and fractionated on a column of DEAE-cellulose (HCO_3^- form; 40 cm long, 2.8 cm wide). The column was washed with 500 ml of distilled water. The compounds were eluted using a linear gradient (1.5 liter H_2O–1.5 liter 0.35 M triethylammonium hydrogenocarbonate). The fractions, detected by UV spectroscopy, which contained the dibutyryl compound were mixed and evaporated to dryness and maintained 15 hours under high vacuum. After three coevaporations with methanol, the resi-

---

[14] R. B. Meyer, D. A. Shuman, R. K. Robins, R. J. Bauer, M. K. Dimmit, and L. N. Simon, *Biochemistry* 11, 2704 (1972).

[15] I. Marcus, Thesis, University of Geneva (1971).

[16] K. Muneyama, R. J. Bauer, D. A. Shuman, R. K. Robins, and L. N. Simon, *Biochemistry* 10, 2390 (1971).

(VII) (VIII)

due was taken up in 2 ml of methanol, and the theoretical amount of BaI·2H$_2$O dissolved in 1 ml ethanol was added. The barium salt (29 mg) was collected by centrifugation and washed with absolute ethanol.

Preparation of 2'-O-Monobutyryl Cyclic Iso-AMP (VIII)[15]

Cyclic iso-AMP[10] (triethylammonium salt), 25 mg, was dissolved in a mixture of 0.75 ml of anhydrous pyridine and 0.375 ml of butyric anhydride. Paper chromatography indicated after 3 hours a complete conversion at room temperature to a new compound which was not modified after 7 days. The solution was then cooled in ice and 0.35 ml H$_2$O were added with stirring. After evaporation under vacuum, the residue was taken up in water and fractionated on a column of DEAE-cellulose (HCO$_3^-$ form; 30 cm long, 1.8 cm wide). The column was washed with 500 ml of distilled water and the compounds were eluted using a linear concentration gradient (1.5 liter of H$_2$O to 1.5 liters of triethylammonium hydrogenocarbonate 0.1 M). Fractions (10 ml each) were collected which showed the same λ_{max} at 278 nm as the starting material. They were mixed and evaporated to dryness under vacuum. The residue was coevaporated several times *in vacuo* with methanol. It was taken up in 0.5 ml of methanol; 0.5 ml of a 1 M solution of sodium perchlorate in anhydrous acetone was added. The precipitation of the sodium salt was completed by introduction of 5 ml of anhydrous acetone. The solid (13 mg) was collected by centrifugation and washed with anhydrous acetone.

[55] The Synthesis of [^{32}P]Adenosine 3′,5′-Cyclic Phosphate and Other Ribo- and Deoxyribonucleoside 3′,5′-Cyclic Phosphates

By R. H. Symons

The general procedure described here allows the two-step chemical synthesis from [^{32}P]orthophosphoric acid of the eight common ribo- and deoxyribonucleoside 3′,5′-cyclic monophosphates. The method is simple and reliable and provides any of the cyclic nucleotides with equal facility in yields of 30–60% relative to starting ^{32}P$_i$ and with a specific activity of greater than 1 mCi/μmole. The procedure given here has been modified from that originally published,[1] but the basic chemical steps are the same.

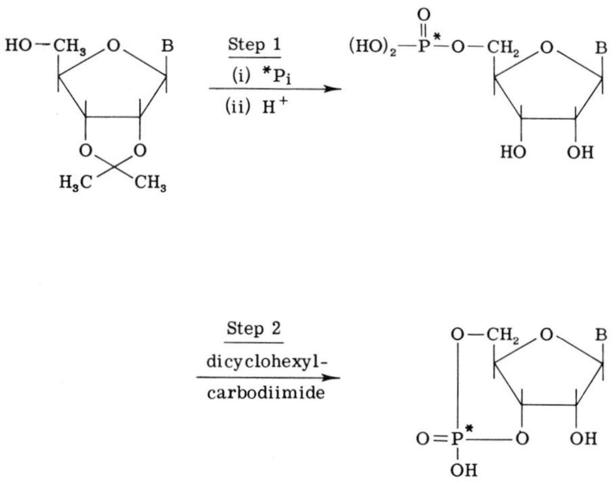

FIG. 1. Route for the synthesis of [^{32}P]ribonucleoside cyclic 3′,5′-phosphates.

Principle

The two chemical steps involved in the preparation of the labeled cyclic nucleotides are outlined in Figs. 1 and 2. The first step is the coupling of 1.0 μmole of ^{32}P$_i$ (1–10 mCi) to an excess (40–50 μmoles) of the appropriate 2′,3′-O-isopropylidene ribonucleoside or unprotected deoxynucleoside using trichloroacetonitrile as the condensing agent in the

[1] R. H. Symons, *Biochem. Biophys. Res. Commun.* **38**, 807 (1970).

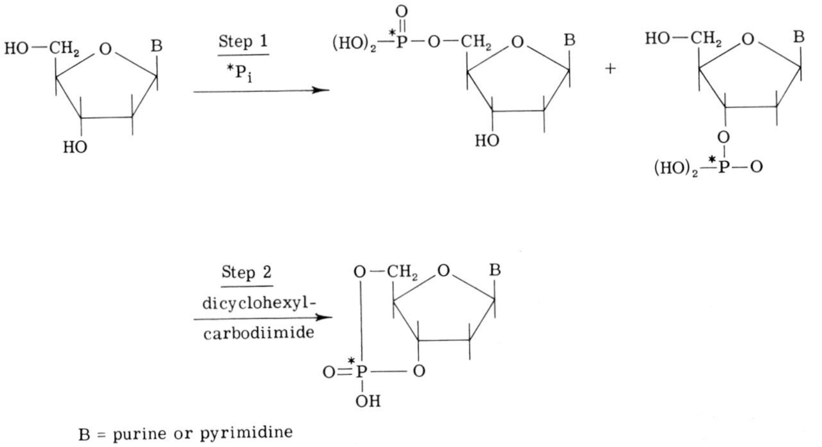

FIG. 2. Route for the synthesis of [^{32}P]deoxyribonucleoside cyclic 3',5'-phosphates.

presence of triethylamine and with dimethyl sulfoxide as solvent.[2-5] The large excess of nucleoside is necessary to ensure high yields, but rigorous drying of the components of the reaction mixture is unnecessary. The amino groups of adenine, cytosine, and guanine are left unprotected as they are not phosphorylated.

Protected Ribonucleosides (Fig. 1). The commercially available 2',3'-O-isopropylidene ribonucleosides are used to ensure phosphorylation only on the 5'-hydroxyls; the protecting group is readily removed under acid conditions prior to the second step. Yields of mononucleotides usually vary from 70 to 90% relative to ^{32}P$_i$ added.

Unprotected Deoxyribonucleosides (Fig. 2). When unprotected deoxynucleosides are used, both hydroxyl groups can be phosphorylated but, under the conditions used, only monophosphorylated products are obtained. Yields of total mononucleotide vary from 65 to 90% relative to starting ^{32}P$_i$; about 65% is 5'-dNMP and 35% 3'-dNMP.

The second chemical step (Figs. 1 and 2) is the cyclization reaction which is carried on directly from step 1 in the same reaction flask and without purification of the reaction mixture. It is achieved by refluxing the triethylammonium salt of the mononucleotide under dilute conditions in anhydrous pyridine in the presence of dicyclohexylcarbodiimide.[6] It

[2] R. H. Symons, *Biochem. Biophys. Res. Commun.* **24,** 872 (1966).
[3] R. H. Symons, *Biochim. Biophys. Acta* **155,** 609 (1968).
[4] R. H. Symons, *Biochim. Biophys. Acta* **190,** 548 (1969).
[5] R. H. Symons, this series, Vol. 29, p. 102.
[6] M. Smith, G. I. Drummond, and H. G. Khorana, *J. Amer. Chem. Soc.* **83,** 698 (1961).

is necessary to add 0.6 ml of dimethyl sulfoxide to the pyridine solution (8 ml) to solubilize the mononucleotides which are otherwise poorly soluble in anhydrous pyridine.[6] There appears to be insignificant dimethyl sulfoxide–dicyclohexylcarbodiimide oxidation of free nucleotide hydroxyls,[7] presumably because of the presence of excess pyridine and of nucleoside remaining from step 1.

After removal of unreacted nucleoside and dimethyl sulfoxide by adsorption of the reaction mixture on and elution from a small column of DEAE-cellulose, the 3',5'-cyclic mononucleotides are isolated by a single paper chromatography step in yields of 30–60% relative to $^{32}P_i$ and with a specific activity of 1–10 mCi/µmole and a purity greater than 96%.

General Experimental Details

Purification and Drying of Solvents and Reagents

It is considered advisable to distill all solvents and volatile reagents. Once distilled, they can be stored in the dark in screw-capped bottles for at least two years at room temperature. For convenience, small volumes of triethylamine and trichloroacetonitrile can be kept in screw-capped tubes.

Acetonitrile is dried by distillation from CaH_2 at normal pressure (bp 80°) with protection from moisture. It is distilled again from P_2O_5 and stored in a dark bottle over lumps of CaH_2. Dimethyl sulfoxide is dried for at least 1 day after the addition of CaH_2 and is then distilled under reduced pressure from CaH_2 and stored over CaH_2. Triethylamine is dried by distillation from CaH_2 at normal pressure (bp 89–90°) and stored over CaH_2 but not over molecular sieves. Trichloroacetonitrile (Aldrich or K and K Laboratories) is distilled in the absence of moisture at normal pressure and stored in a well-sealed bottle. *WARNING:* this reagent is extremely irritating to the nose and eyes.

Reagents

Carrier-free [^{32}P]orthophosphoric acid in dilute (0.01 N) HCl is obtained from the Australian Atomic Energy Commission, Lucas Heights, New South Wales, and is used directly. The polyphosphate content can be readily checked by ascending chromatography of a sample on a small piece of polyethyleneimine cellulose thin-layer (Macharey-Nagel or Brinkman Polygram CEL 300 PEI/UV$_{254}$) using freshly prepared 0.4

---

[7] K. E. Pfitzner and J. G. Moffatt, *J. Amer. Chem. Soc.* **87**, 5661 (1965).

M NH$_4$HCO$_3$ as solvent and AMP and ATP as markers. P$_i$ runs with AMP and polyphosphates from ATP to the origin; after drying, the thin-layer strip is cut up and counted in a scintillation counter.

It is considered important that samples be removed from the stock ^{32}P$_i$ vial only with glass pipettes; circumstantial evidence has indicated that the use of disposable plastic syringes with metal needles can lead to low or negligible yields during the chemical phosphorylation step.

2',3'-O-Isopropylidene ribonucleosides and 2'-deoxynucleosides are obtained from Sigma Chemical Co. and are stored dry at 4° in screw-capped bottles over silica gel.

Apparatus

Long-necked, 50 ml, round-bottom flasks are recomended. They can be obtained commercially or constructed with a B14 Quickfit socket (standard taper 14/35), a neck 5 cm long and a bulb 5 cm in diameter. All evaporations are carried out with a rotary evaporator (Buchi) which is connected to a water pump with no special precautions taken for the removal of moisture.

All glassware and pipettes required for anhydrous reaction mixtures are kept in an oven at 100°. Just before use, they are removed and used as soon as they have cooled.

For the descending paper chromatography step described, a chromatography tank or cabinet is required that will take large sheets of chromatography paper (about 50 × 50 cm). A large guillotine (50 cm × 50 cm) is used for cutting strips from the paper chromatograms.

Handling of Radioactive Material

Wherever possible, medium weight lead impregnated radiologists' gloves which extend as far up to the elbow as possible are used for radiation protection (not contamination protection). All work is carried out behind Perspex (Lucite) screens (1.0 cm thick × 50 cm × 70 cm) which can either have fixed supports or be held by a retort stand. Chromatograms are carried behind or, preferably, between these screens. Strips of paper are carried between smaller pieces of Perspex. Reaction flasks are always carried and handled in Perspex (Lucite) beakers (cylindrical, 10 cm deep, 5.5–6.0 cm internal diameter, walls 0.8–1.0 cm thick with a thick base of 1.5–2 cm).

Spotting of paper chromatograms is done behind Perspex screens through a narrow slit between two strips of Perspex held together at each end by a cross bar and running well over each side of the chromatogram.

The paper is supported between two glass rods held together at each end with rubber bands and is handled with these rods. Spotting is done with a Pasteur pipette, and the solvent is removed by a stream of warm air from one or two hair dryers. Nucleotide bands located under ultraviolet light are cut out with a large guillotine and one end of the strip cut to a point for elution; the marker nucleotide is not included. Lead-impregnated gloves are always worn and paper strips are handled with large forceps (30 cm long). Radioactive waste paper is stored in metal bins for at least 6 months before disposal. Nucleotide bands are eluted by descending chromatography in a simple enclosed elution rack with 0.1 mM EDTA, pH 8, as solvent. Nucleotides may break down rapidly when stored in the absence of EDTA after elution from paper. A strip 30 cm long can be eluted in 3–4 hours if it is shortened once or twice as the solvent moves down the paper.

All contaminated glassware is thoroughly rinsed in tap water and then immersed in alcoholic KOH (95% ethanol saturated with KOH) overnight. It is rinsed well in tap water, then distilled water and dried. Residual radioactivity should be negligible.

Paper Chromatography Solvents

Solvent A: Propan-1-ol–conc. NH_4OH–water; 55:20:25, v/v/v
Solvent B: Propan-2-ol–conc. NH_4OH–water; 7:1:2, v/v/v
Solvent C: Butan-1-ol–acetic acid–water; 4:1:1, v/v/v

Preparation of Cyclic Nucleotides

Step 1. [^{32}P]Adenosine 5′-Monophosphate—Chemical Synthesis Procedure

Reaction Mixture. Into a reaction flask is weighed 40–50 µmoles (12–15 mg) of 2′,3′-O-isopropylidene adenosine. About 1 ml of ethanol is added, then 1.0 µmole of H_3PO_4 (0.02 ml of 50 mM H_3PO_4) and 1–10 mCi of carrier-free $^{32}P_i$ in 0.01 N HCl. One drop of triethylamine is added to neutralize the HCl, and the flask contents are taken to dryness on the rotary evaporator. The flask contents are further dried by the addition and evaporation of two 1 ml lots of acetonitrile, although one drying is probably quite sufficient.

To the stoppered flask are added 0.4 ml of dimethyl sulfoxide and 3 µl (21 µmoles) of triethylamine; the flask contents are dissolved by shaking. Finally, 3 µl (30 µmoles) of trichloroacetonitrile are added and mixed in by shaking; the stoppered flask is incubated at 37° for 30–60

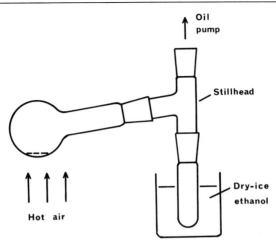

FIG. 3. Apparatus for removal of dimethyl sulfoxide and excess volatile reagents from the reaction mixture.

minutes (the reaction is complete after about 20 minutes, but it can be left for at least 2 hours with no effect on yields). The solution becomes pale yellow-orange.

The amounts of triethylamine and trichloroacetonitrile added are not critical and some variation is possible with little effect on yields. However, if there is insufficient triethylamine, the reaction mixture does not become colored.

Workup of Reaction Mixture. The removal of dimethyl sulfoxide and excess volatile reagents from the reaction mixture is carried out as shown in Fig. 3. The reaction flask is connected to a stillhead, and the volatile flask contents are distilled under vacuum from an oil pump into the tube immersed in dry ice–ethanol. The flask is heated with a hair dryer, and distillation is complete in a few minutes. Three milliliters of 2 N acetic acid are added, and the stoppered flask is heated in a boiling water bath for 1 hour to hydrolyze the isopropylidene group. The acetic acid is then taken off on the rotary evaporator, and complete removal is ensured by the addition and evaporation of two 5-ml lots of water. The flask contents of [^{32}P]5'-AMP, ^{32}P$_i$, and adenosine are then dried by the evaporation of two 5-ml portions of dry acetonitrile in preparation for the cyclization reaction of step 2.

Comments on Step 1 Procedure

The method described above for the synthesis of [^{32}P]5'-AMP and that described below for the other mononucleotides has been modified slightly[5] from that originally published.[1-4] The main change is in the preparation of the reaction mixture. Since reasonable amounts of triethylammonium chloride do not affect the phosphorylation reaction (see below), the HCl added with the $^{32}P_i$ is neutralized with excess triethylamine rather than being removed by repeated evaporation with water. This gives a significant saving in time and some simplification of the initial preparation. This approach also allows the neutralization at this stage of any protected or unprotected nucleoside added as its hydrochloride, e.g., deoxycytidine hydrochloride.

The effect on mononucleotide yield of adding various compounds to the reaction mixture has been investigated in trial runs using several microcuries of $^{32}P_i$. Up to 100 μmoles of HCl, and hence triethylammonium chloride, added with the $^{32}P_i$ had no effect on yields whereas 200 μmoles caused a 20% decrease. Water, 2 μl (110 μmoles), added to the final reaction mixture before the trichloroacetonitrile had no effect whereas 10 μl (555 μmoles) caused a 10–20% decrease. Up to 90% decrease in yields was obtained on the addition of 1.0 μmole of KCl, $MgCl_2$, or $FeCl_3$ with the $^{32}P_i$.

Step 2. Cyclization of [^{32}P]5-AMP to [^{32}P]3',5'-Cyclic AMP

To the dried residue of the triethylammonium salt of 5'-AMP and unreacted P_i from step 1 are added 0.6 ml of dry dimethyl sulfoxide, 8 ml of dry pyridine, and 40 μl (8 μmoles) of a stock solution of 0.2 M dicyclohexylcarbodiimide in dry pyridine. (This stock solution can be kept in a tightly sealed tube inside a jar over silica gel for several months in the dark at room temperature.) The mixture is then heated under reflux on a hot plate for 30 minutes with protection from moisture by a silica gel guard tube at the top of the water-cooled condenser. After removal of the pyridine on the rotary evaporator (the dimethyl sulfoxide does not evaporate), about 5 ml of water are added, and the solution is passed through a column of DEAE-cellulose (bicarbonate; 0.7 × 2 cm) previously washed with 0.1 N NaOH, 1.0 M NH_4HCO_3, and finally water. The column is washed with about 10 ml of water to complete the removal of solvent and unreacted nucleoside, then the cyclic AMP and unreacted $^{32}P_i$ are eluted with 5 ml of freshly prepared 0.15 M NH_4HCO_3 at a flow rate of about 0.5 ml per minute. The eluate is collected in a round-bottom reaction flask; the completeness of the elution can be checked with a hand radiation monitor. A few drops of triethylamine are added to convert the NH_4HCO_3 to the more volatile triethylammonium salt, and the flask contents are taken to dryness. The residue is dissolved in about 2 ml of warm water and then dried as a band 30 cm long on a sheet of Whatman 3 MM chromatography paper. Markers of cyclic 3',5'-AMP and 5'-AMP are added on one side of the chromatogram. After descending chromatography overnight in either solvent A or B (see the table), the chromatogram is dried, the band of cyclic AMP is located under ultraviolet light, cut out, eluted with 0.1 mM EDTA, pH 8, and

R_f VALUES AND ELECTROPHORETIC MOBILITIES

Compound	Paper chromatography[a]			Paper electrophoresis[e] in 25 mM NH$_4$HCO$_3$ (mobilities relative to cyclic 3',5'-AMP)
	Solvent A[b]	Solvent B[c]	Solvent C[d]	
5'-AMP	0.06	0.32	0.04	1.75
Cyclic 3',5'-AMP	0.38	0.60	0.07	1.00
Adenosine	0.50	0.66	0.26	−0.04
5'-CMP	0.05	0.32	0.05	1.98
Cyclic 3',5'-CMP	0.32	0.54	0.06	1.16
Cytidine	0.44	0.62	0.18	−0.08
5'-GMP	0.02	0.22	0.02	1.68
Cyclic 3',5'-GMP	0.16	0.38	0.03	0.96
Guanosine	0.22	0.48	0.15	0.00
5'-UMP	0.05	0.28	0.05	1.96
Cyclic 3',5'-UMP	0.22	0.52	0.07	1.34
Uridine	0.38	0.58	0.20	0.00
5'-dAMP	0.09	0.38	0.07	1.62
Cyclic 3',5'-dAMP	0.40	0.62	0.10	0.98
Deoxyadenosine	0.60	0.74	0.36	−0.04
5'-dCMP	0.08	0.36	0.06	1.90
Cyclic 3',5'-dCMP	0.36	0.60	0.08	1.22
Deoxycytidine	0.55	0.76	0.24	−0.08
5'-dGMP	0.03	0.26	0.02	1.68
Cyclic 3',5'-dGMP	0.19	0.44	0.04	1.00
Deoxyguanosine	0.34	0.54	0.20	0.00
5'-dTMP	0.10	0.38	0.08	1.90
Cyclic 3',5'-dTMP	0.40	0.62	0.12	1.24
Deoxythymidine	0.65	0.76	0.40	−0.04

[a] Descending chromatography on Whatman 3 MM chromatography paper.
[b] Propan-1-ol–conc. NH$_4$OH–water; 55:20:25 by volume.
[c] Propan-2-ol–conc. NH$_4$OH–water; 7:1:2 by volume.
[d] Butan-1-ol–acetic acid–water; 4:1:1 by volume.
[e] Voltage at 40 V/cm for 45–50 minutes on Whatman 3 MM chromatography paper. Mobilities relative to cyclic AMP were determined by measurement from the center of the neutral marker, uridine, to the front of each ultraviolet light absorbing area.

stored at $-15°$. The unreacted $^{32}P_i$ remains at the origin although some may run in the region of AMP.

Cyclic [^{32}P]3',5'-AMP is obtained in a yield of 40–50% relative to $^{32}P_i$ and with a specific activity of 1–10 mCi/μmole. It has chromatographic, electrophoretic, and spectral properties similar to those of unlabeled cyclic AMP, and it has a radiochemical purity of greater than 96% on paper electrophoresis in 25 mM NH_4HCO_3, pH 8.

Alternative Purification Procedures

Two alternative procedures can be used for the purification of the cyclization reaction mixture after reflux. In the first, the DEAE-cellulose column step is omitted and the solution is dried directly on a sheet of chromatography paper and run overnight in solvent C. The chromatogram is then dried and run overnight in solvent A *without* eluting the cyclic AMP. This two-step chromatography is necessary to separate the cyclic AMP completely from the large excess of unreacted adenosine. Although this procedure takes an extra day, it has the advantage of simplicity and the minimum of effort.

The second alternative is to purify the cyclic [^{32}P]AMP from reaction mixture after reflux by gradient elution from DEAE-cellulose (bicarbonate).[6,8,9]

Comments on Overall Procedure

The method described allows the preparation of cyclic [^{32}P]3',5'-AMP with a specific activity of 1–10 mCi/micromole. A total working time of about 4 hours spread over 2 days is required for each preparation.

In the original published method,[1] [^{32}P]5'-AMP was purified by paper chromatography prior to its cyclization to cyclic AMP. In the method described here, the two-step procedure is carried out in the one reaction flask without an intermediate workup and this has led to a considerable saving in time and in the amount of handling required of radioactive material.

The final product will be contaminated with small amounts of material eluted from the chromatography paper. If necessary, further purification can be obtained by adsorption of the nucleotide onto a small column of DEAE-cellulose (bicarbonate), washing with 5 mM NH_4HCO_3, then elution with freshly prepared 0.1 M NH_4HCO_3. The eluted nucleotide is

---

[8] M. Smith and H. G. Khorana, this series, Vol. 6, p. 645.
[9] R. K. Borden and M. Smith, *J. Org. Chem.* **31**, 3247 (1966).

then desalted on the rotary evaporator after the addition of a few drops of triethylamine. A cleaner product will also be obtained by the use in step 2 of chromatography paper that has been washed by chromatography in solvent A or B for 1–2 days before use.

Determination of Specific Activity

If an accurate measure of the specific activity and concentration of the cyclic [^{32}P]AMP is required, the following procedure can be used. The specific activity of the starting ^{32}P$_i$ is measured at the beginning of step 1, and this is taken as the specific activity of the final cyclic AMP. Further, the actual *concentration* of the product can be readily determined by counting an appropriately diluted sample and calculating from the known specific activity. These calculations assume that the starting ^{32}P$_i$ and the final product are 100% pure; appropriate corrections can be made if necessary.

Procedure. After the nucleoside and ^{32}P$_i$ have been added to the reaction flask in step 1 and taken to dryness, 3.0 ml of water are added to dissolve the contents. A 10 μl sample is removed and added to 5.0 ml of water to give the stock reference solution; 10 μl of this solution are dried on a GF/C glass fiber filter and counted. After correcting for dilution, the specific activity is calculated from the amount of cold H_3PO_4 added. This stock reference solution is kept for the determination of the specific activity at any time subsequent to the initial preparation.

A sample of the final cyclic AMP solution is appropriately diluted, and a measured sample is counted. From the actual dilution used and the specific activity of the starting ^{32}P$_i$ determined at the same time, the actual concentration of cyclic AMP can be calculated.

Other [^{32}P]3′,5′-Cyclic Ribonucleotides

The method described above for the preparation of cyclic AMP can be used for the preparation of cyclic 3′,5′-CMP, GMP, and UMP starting with the appropriate 2′,3′-O-isopropylidene nucleoside. Yields obtained vary from 40–60% relative to ^{32}P$_i$ for cyclic CMP and UMP but are lower (20–30%) for cyclic GMP because of the greater insolubility of guanine nucleotides during reflux in step 2.

Cyclic nucleotides which are not available commercially can be prepared in small quantities for use as markers by refluxing 1–2 μmoles of the triethylammonium salt of the appropriate nucleoside 5′-monophosphate as described in step 2. They can be prepared on a larger scale by the methods of Smith et al.[6] or Borden and Smith.[9]

[^{32}P]3′,5′-Cyclic Deoxyribonucleotides

These can also be readily prepared essentially as described for cyclic AMP, but starting with the unprotected deoxynucleosides. The overall procedure is simpler because an acid hydrolysis step is not required.

Procedure. The reaction mixture of step 1 is prepared as described for cyclic AMP using the appropriate deoxynucleoside. At the end of the incubation to produce the mixture of [^{32}P]5′- and 3′-mononucleotides, residual triethylamine and trichloroacetonitrile are removed from the reaction mixture by the addition and evaporation on the rotary evaporator of two 5 ml portions of dry acetonitrile. The cyclization reaction of step 2 is then carried out as described above.

Yields of cyclic [^{32}P]3′,5′-dAMP, dCMP, dGMP, and dTMP vary from 35 to 60% relative to starting $^{32}P_i$.

Note Added in Proof: Some improvements on the methods given here have been published.[10]

[10] R. H. Symons, *Biochim. Biophys. Acta* **320**, 535 (1973).

[56] Adenylylimidodiphosphate and Guanylylimidodiphosphate

By RALPH G. YOUNT

The replacement of the β-γ bridge oxygen of ATP[1] or GTP[2] with an —NH-group yields analogs which are resistant to enzymatic cleavage of the β-γ linkage[3] but which are effective substrates for the enzymes adenylate cyclase[4] or guanylate cyclase which cleave the α-β linkage. These latter enzymes, which are normally membrane bound, may then be assayed and characterized in the presence of the ubiquitous membrane phosphohydrolases. Thus, adenylylimidodiphosphate (AMP-PNP) has been used to locate adenylate cyclase cytochemically by lead precipitation of the imidodiphosphate released.[5] This procedure eliminates problems of lead phosphate precipitates which predominate when ATP is used

[1] R. G. Yount, D. Babcock, W. Ballantyne, and D. Ojala, *Biochemistry* **10**, 2484 (1971).
[2] F. Eckstein, M. Kettler, and A. Parmeggiani, *Biochem. Biophys. Res. Commun.*, **45**, 1151 (1971).
[3] R. G. Yount, D. Ojala, and D. Babcock, *Biochemistry* **10**, 2490 (1971).
[4] M. Rodbell, L. Birnbaumer, S. L. Pohl, and H. M. J. Krans, *J. Biol. Chem.* **246**, 1877 (1971).
[5] R. Wagner, P. Kreiner, R. J. Barrnett, and M. W. Bitensky, *Proc. Nat. Acad. Sci. U.S.* **69**, 3175 (1972).

as a substrate. In addition, AMP-PNP competes with ATP with equal facility for adenylate cyclase from liver plasma membranes[4] and hence can be used to determine K_m and V_m values of the unpurified enzyme. The analogous carbon analog adenylylmethylenediphosphonate is ineffective as a substrate for this enzyme.

Preparation of Tetrasodium Imidodiphosphate[6]

Inorganic tetrasodium imidodiphosphate (PNP_i)[7] is a key intermediate necessary to prepare either AMP-PNP or GMP-PNP.

Its preparation involves four separate steps:

$$(C_6H_5O)_2P(O)Cl + 2NH_3 \rightarrow (C_6H_5O)_2P(O)NH_2 + NH_4Cl \xrightarrow{PCl_5}$$
$$\text{(I)} \hspace{4cm} \text{(II)}$$

$$(C_6H_5O)_2P(O)N{=}PCl_3 \xrightarrow{H_2O} (C_6H_5O)_2P(O)NHPO_3H_2 \xrightarrow[C_6H_5OH]{NaOH}$$
$$\text{(III)} \hspace{3cm} \text{(IV)}$$

$$Na_2PO_3\text{—}NH\text{—}PO_3Na_2$$
$$\text{(V)}$$

Absolute ethanol (350 ml) in a tared beaker is cooled in an ice-ethanol bath to $-10°$ in a hood and gaseous NH_3 bubbled in through a sintered-glass diffuser until approximately 3 moles of NH_3 have dissolved (about 2 hours). Diphenyl chlorophosphate (I) (180 g, ~ 0.7 mole) is added to this solution (kept cooled to below 30°) dropwise using overhead stirring. The almost solid reaction mixture is mixed with 2.5 liters of water and filtered on a Büchner funnel; the product, diphenylphosphoamide (II), is washed with water until essentially free of chloride ion ($AgNO_3$ test). Briefly air dry and recrystallize (II) from 500 ml of hot absolute ethanol. The final crystals are collected and dried at 100° for 2 hours. Yield = 85–95%; mp = 148°.

Diphenylphosphoamide (II) (26.3 g, 0.105 mole) and freshly powdered phosphorus pentachloride (22 g, 0.105 mole) are mixed intimately in a 1 liter round-bottom flask protected with a drying tube. The reaction mixture slowly liquefies and is allowed to stand overnight at room temperature before heating at 50° for 30 minutes and then at 90° until the evolution of HCl ceases (wet pH paper at drying tube outlet). The liquid trichloride product (III) is transferred to a beaker, 53 ml of water are added, and the solution is heated on a hot plate to about 85° with stirring until HCl is vigorously evolved. Remove from heat,

---

[6] M. Nielsen, R. Ferguson, and W. Coakley, *J. Amer. Chem. Soc.* **83**, 99 (1961).

[7] Abbreviations used are PNP_i, imidodiphosphate; AMP-PNP, adenylylimidodiphosphate; GMP-PNP, guanylylimidodiphosphate; $ADP-NH_2$, adenylylphosphoamide; HMP, hexamethylphosphoric triamide.

cool 1 hour and grind the resulting diphenylimidodiphosphoric acid (IV) with water using a mortar and pestle. Filter and repeat this process until the filtrate is essentially free of chloride ion. Dry the crystals of (IV) in a vacuum desiccator at <1 mm for 1–2 hours over magnesium perchlorate. Recrystallize by dissolving (IV) in 3–10 ml of boiling dioxane, partially cooling, and adding 50 ml of diethyl ether with vigorous stirring. Cool for 1–2 hours at 4°, filter, and dry the crystals as above overnight. Yield = 14.5 g (40%); mp = 167–168°.

The diphenylimidodiphosphoric acid (IV) is saponified by a modification of method B of Nielsen et al.[6] Diphenylimidodiphosphoric acid (14.5 g, 0.045 mole) is mixed with 1 mole sodium hydroxide (40 g) and 0.45 mole phenol (39 g) in 108 ml of water in a 1-liter round-bottom flask, and heated as rapidly as possible in a preheated oil bath (170–180°) to a temperature of 140° and held there for 10 minutes. Immediately pour the hot solution into a beaker and allow the crystals of tetrasodium imidodiphosphate (V) to form. Filter and recrystallize from water (250 ml), adding 95% ethanol to the point of incipient precipitation. Cool at 2–4° overnight, filter, and recrystallize until the product gives a negative phenol spot test.[8] The product $Na_4P_2O_6NH \cdot 10H_2O$ (MW 445) (V) is air dried and stored, capped in a desiccator, at −10°. It is stable for many months under these conditions. The molecular weight can be confirmed by acid-labile phosphate determination (1 N HCl, 100°, 30 minutes) using standard P_i analytical reagents. Note that P—N—P bonds are *more* stable in strong acid than P—O—P bonds but less stable in weakly acidic solutions.

5'-Adenylylimidodiphosphate

Principle

The reaction between AMP and diphenylchorophosphate in the presence of aprotic solvents and hindered tertiary amine bases gives P^1-adenosine-5' P^2-diphenylpyrophosphate (VI) in high yield.[9]

$$Ad-O-\underset{O^-}{\overset{\overset{O}{\|}}{P}}-O^- + (C_6H_5O)_2P(O)Cl \rightarrow Ad-O-\underset{O^-}{\overset{\overset{O}{\|}}{P}}-O-\underset{OC_6H_5}{\overset{\overset{O}{\|}}{P}}-OC_6H_5$$
$$(VI)$$

Subsequent reaction of (VI) with a slight excess of the tributylammonium salt of imidodiphosphate in the presence of pyridine yields adenylyl-

---

[8] The major impurity is sodium phenoxide. Add a small amount of sodium nitrate to concentrated sulfuric acid. Phenol but not imidodiphosphoric acid gives a dark brown to purple color with this reagent.

[9] A. M. Michelson, *Biochim. Biophys. Acta* **91**, 1 (1964).

imidodiphosphate (VII).

$$\text{Ad—O—}\overset{\overset{O}{\|}}{\underset{\underset{O^-}{|}}{P}}\text{—O—}\overset{\overset{O}{\|}}{\underset{\underset{OC_6H_5}{|}}{P}}\text{—OC}_6\text{H}_5 + O_3P\text{—NH—}PO_3^{-4} \longrightarrow$$

$$\text{Ad—O—}\overset{\overset{O}{\|}}{\underset{\underset{O^-}{|}}{P}}\text{—O—}\overset{\overset{O}{\|}}{\underset{\underset{O^-}{|}}{P}}\text{—NH—}\overset{\overset{O}{\|}}{\underset{\underset{O^-}{|}}{P}}\text{—O}^- + {}^-\text{O—}\overset{\overset{O}{\|}}{\underset{\underset{OC_6H_5}{|}}{P}}\text{—OC}_6\text{H}_5$$
$$\text{(VII)}$$

This is a general method of synthesis and can be used to prepare the β-γ imido analogs of all the common nucleotides. It also can be used effectively to prepare the β-γ P—C—P nucleotide analogs by using methylene diphosphonate in place of imidodiphosphate.[1]

An alternate enzymatic method for synthesis of AMP-PNP exists. Rodbell and co-workers[4] used the mixed aminoacyl tRNA synthetases from *Escherichia coli* with ATP, 19 naturally occurring amino acids and an excess of imidodiphosphate to prepare AMP-PNP by the following reactions:

$$\text{ATP} + \text{a.a.} + \text{enzyme} \rightleftharpoons (\text{AMP} \sim \text{a.a.} \cdot \text{enzyme}) + \text{PP}_i$$
$$\text{PNP}_i + (\text{AMP} \sim \text{a.a.} \cdot \text{enzyme}) \rightarrow \text{AMP-PNP} + \text{enzyme} + \text{a.a.}$$

where a.a. = amino acids. The reaction goes essentially to completion because of contaminating pyrophosphatase activity and because AMP-PNP is a poor substrate for this class of enzymes. This method is especially useful for preparing the radioactive forms of AMP-PNP since ATP of high specific activity is readily available. It also has been scaled up 1000-fold to make micromolar amounts of AMP-PNP for kinetic studies.[10]

Procedure

Sodium imidodiphosphate (1.2 mmoles) was dissolved in 10 ml of cold water, and the resulting solution was added to a 2-cm diameter column contaminating 60 meq of freshly regenerated and washed Dowex 50 H$^+$ (X8, 50–100 mesh) at 2°. The free acid of PNP$_i$ was eluted with cold water, collecting all acidic fractions (pH paper) in a stirred solution of excess tributylamine[11] and water in an ice bath. This solution was evaporated to a gum on a rotatory evaporator using a dry ice–2-propanol cooled

---

[10] C. R. Bagshaw, J. Eccleston, D. R. Trentham, D. W. Yates, and R. S. Goody, *Cold Spring Harbor Symp. Quant. Biol.* **37**, 127 (1972).

[11] Tributylamine was redistilled under vacuum and stored at 4° in dark bottles. Pyridine was refluxed over sodium hydroxide pellets, distilled through a short column, and stored over molecular sieves. Chloroform and dimethylformamide were dried over molecular sieves (Linde, Type 4A, 1/16 pellets) for at least 10 days and used without further purification.

trap, a mechanical vacuum pump, and a bath temperature of 25–30°. The residue was dissolved in 10–20 ml of dry chloroform[11] and evaporated to dryness. The process was repeated twice, and the resulting dried tributylammonium salt was stored in a stoppered flask at −10° until used.

AMP · H_2O (1 mmole, free acid, Sigma grade) was dissolved in several milliliters of absolute methanol containing 2 mmoles of tributylamine (0.370 g) with stirring under reflux. The methanol was removed under reduced pressure as before, and the residue was dried by repeated addition and evaporation of three 5-ml portions of dry dimethylformamide.[11] The residue was again dissolved in dimethylformamide (5 ml) and 0.3 ml of diphenyl chlorophosphate (1.5 mmoles) and 0.47 ml of tributylamine (2.0 mmoles) were added and the solution was allowed to stand at room temperature for 2 hours protected from moisture. If a precipitate formed, it was dissolved by adding tributylamine dropwise. Solvents were removed under reduced pressure and dry ether (50 ml) was added to the residue with swirling. The flask was cooled in an ice bath for 20 minutes, and the ether was decanted. Remaining ether was removed under reduced pressure, and the residue was dissolved in 5 ml of dimethylformamide. This solution was added dropwise over a 30-minute period to a stirred solution of the tributylammonium salt of PNP_i dissolved initially in 5 ml of dimethylformamide and to which 15 ml of pyridine were added. [Addition of the solution of (VI) in reverse order or all at once led to greatly increased yields of the by-product containing 2 moles of AMP linked to 1 mole of PNP_i.] Solvents were removed under reduced pressure after the reaction mixture was allowed to stand at room temperature for 2 hours. The residue was extracted with ether (50 ml), and the ether was removed as before. The resulting gum was dissolved in 50 ml of cold water containing 0.4 ml of concentrated ammonium hydroxide. The final pH was near 9.0. If the solution was cloudy, it was extracted with ether, and the clear water layer was applied to a 2.5 × 75 cm DEAE-cellulose (HCO_3^-) column[12] at 2°. Purification followed the general procedure of Moffatt[13] using a linear gradient of 0–0.4 M triethylammonium bicarbonate[14] (8 liters total volume) as the elutant. A flow rate of 5 ml per

[12] Coarse DEAE-cellulose well washed to remove "fines" should be used. The microcrystalline cellulose derivatives (Reeve-Angel) give too slow flow rates with triethylammonium bicarbonate to be useful. DEAE-Sephadex A-25 may be used as it gives improved flow rates. However, the column must be repoured after each run because of the marked shrinkage of the DEAE-Sephadex beads with increasing ionic strength.

[13] J. G. Moffatt, Can. J. Chem. 42, 599 (1964).

[14] Triethylamine was redistilled at atmospheric pressure and 404 g (555 ml) added to 3.5 liters H_2O. CO_2 was bubbled through this mixture kept at 0° until the pH fell to 7.5. The final volume was adjusted to 4 liters to give a 1 M solution of trimethylammonium bicarbonate.

minute was maintained with a chromatographic pump, and AMP-PNP was eluted near fraction No. 240 if 25-ml fractions were collected. Acid-labile phosphate[15] to adenine ratios were determined (PNP$_i$ sometimes overlaps the front edge of the AMP-PNP peak), and those tubes with a ratio of 2.0 ± 0.1 were pooled and evaporated to dryness as described previously, keeping the bath temperature below 25°. Residual triethylammonium bicarbonate was removed by repeated evaporations with 25-ml portions of methanol. The residue was transferred to a 40-ml glass centrifuge tube with several rinses of small amounts of methanol. After the solution was concentrated to 4–5 ml with an air stream, a 1 M sodium iodide in acetone solution was added (6–8 Eq relative to the equivalents of adenine present) to precipitate the sodium salt of AMP-PNP. Additional cold acetone was added, and the precipitate was collected by centrifugation in the cold. The precipitate was washed three times with cold acetone, twice with cold ether, and then dried briefly with a gentle air stream. After drying over anhydrous magnesium perchlorate *in vacuo*, the free-flowing white powder was stored desiccated at −20°. Less than 3% decomposition in 8 months was observed with two different preparations as judged using the two solvent systems and the electrophoresis system given in the table. Overall isolated yields in eight preparations were 20–35% based on AMP. The major impurity, if any, was adenylylphosphoamide resulting from the hydrolysis of the terminal phosphate of AMP-PNP. Hydrolysis to yield ADP and inorganic phosphoramidate appears to be much less favored. Breakdown of AMP-PNP was reduced by minimizing the time it was kept in the triethylammonium bicarbonate solutions.

Analysis. Calculated for $C_{10}H_{13}N_6Na_4O_{12}P_3 \cdot 4H_2O$ (MW 666); P, 14.0; acid-labile ammonia,[16] 1.0; acid-labile phosphate, 2.0. Found: MW 660 (from OD$_{260}$ assuming a molar absorbancy index of 15,400); P, 14.1; acid-labile ammonia, 1.06; acid-labile phosphate, 1.97.

Guanylylimidodiphosphate

Principle

An alternate method to preparing the β-γ imido nucleotide analogs involves the activation of 5'-nucleotides with N,N'-carbonyldiimidazole and subsequent reaction with imidodiphosphate. This method has been

---

[15] Aliquots must be evaporated to dryness before acid hydrolysis (30 minutes, 1 N HCl, 100°) since triethylamine interferes with any phosphate analyses using acid molybdate reagents.

[16] W. Umbreit, R. Burris, and J. Staufer, "Manometric Techniques," 3rd ed., p. 238. Burgess, Minneapolis, Minnesota, 1962.

CHARACTERIZATION OF ADENYLYLIMIDODIPHOSPHATE BY PAPER CHROMATOGRAPHY AND PAPER ELECTROPHORESIS[a,b]

Compound	R_f A	R_f B	Mobility (cm) (I)
AMP-PNP	0.39	0.34	12.0
ATP	0.50	0.37	14.0
ADP	0.54	0.42	11.0
ADP-NH$_2$	0.69	0.55	9.5
PNP$_i$	0.20	0.29	19.0
P$_i$	0.30, 0.45 (two spots)	0.35	15.0

[a] Reprinted from R. G. Yount, D. Babcock, W. Ballantyne and D. Ojala, *Biochemistry* **10**, 2484 (1971). Copyright (1971) by the American Chemical Society. Reprinted by permission of the copyright owner.

[b] Ascending chromatography with Whatman No. 31 ET paper was used with solvent systems (A) 2-propanol–dimethylformamide–methyl ethyl ketone–water–concentrated ammonia (20:20:20:39:1, v/v) and (B) 1-propanol–concentrated ammonia–water (6:3:1, v/v). Whatman No. 31 ET paper was used for electrophoresis (I) on a water-cooled flat-plate apparatus using 50 mM sodium citrate (pH 6.8), 35 V/cm for 45 minutes. Adenine-containing compounds were detected by their ultraviolet quenching. PNP$_i$ was detected after acid hydrolysis on the paper.

used by Eckstein and co-workers to prepare guanylylimidodiphosphate (GMP-PNP)[2] and has been used successfully by us to prepare a number of PNP analogs. An important consideration is the use of the solvent, hexamethylphosphoric triamide (HMP), in place of dimethylformamide.[17] This step eliminates many side reactions and greatly increases the yields of the desired product.

Procedure

Guanosine monophosphate (Na$^+$ salt, 1.0 mmole) was converted to the tributylammonium salt by passage through a 2-cm column containing 40 meq of freshly regenerated and washed Dowex 50 H$^+$ (X8, 50–100 mesh) collecting all acidic fractions in a stirred solution of excess tributylamine[11] and water. The solution was evaporated to dryness on a rotatory evaporator, and the residue was dried by repeated addition and evaporation of three 5-ml portions of dry dimethylformamide.[11] The residue was dissolved in 2 ml of HMP (Pierce Chemical Co.) and 1.6 g (10

[17] D. Ott, V. Kerr, E. Hansbury, and F. N. Hanes, *Anal. Biochem.* **21**, 469 (1967).

mmoles) of 1,1'-carbonyldiimidazole (Pierce) in 5 ml of HMP were added. The mixture was stirred for 30 minutes (protected from moisture) and then stored for 5 hours in a desiccator. Anhydrous methanol (12 mmoles, 0.5 ml) in 1 ml HMP was added and allowed to react to 5 hours. Separately tetrasodium imidodiphosphate (2 mmoles) was converted to the dry tributylammonium salt as described previously in the preparation of AMP-PNP and dissolved in 4 ml of HMP. The solution containing activated GMP was added to the imidodiphosphate solution with vigorous mixing, and the stoppered reaction was allowed to stand for 12 hours. HMP was removed by extraction with chloroform (100 ml) in the presence of 25 ml of water, and the water phase containing the GMP-PNP was evaporated to dryness by rotary evaporation. The residue was dissolved in 20 ml of water, and the resulting solution was adjusted to pH 8 with 3 N KOH, and applied to 2.5 $\times$ 90 column of DEAE-Sephadex A-25 (HCO_3^-)[12] at 2°. The column was developed with an 8-liter (4 liters each reservoir) linear gradient of 0.05 to 0.5 M triethylammonium bicarbonate. GMP-PNP emerged at approximately 0.40 M buffer after small peaks of diguanosylylpyrophosphate and GMP. Triethylammonium bicarbonate was removed from the GMP-PNP by repeated evaporations to dryness with methanol (25 ml), and the sodium salt of GMP-PNP was isolated as described in the preparation of AMP-PNP. GMP-PNP was homogeneous by paper chromatography (Whatman 31 ET in 2-propanol–concentrated ammonia–water (55:10:35, v/v). R_f values: GMP-PNP, 0.30; GTP, 0.35; GMP, 0.42 and on paper electrophoresis using 50 mM sodium citrate, pH 6.8 (relative mobility; GTP, 1.0; GMP-PNP, 0.86; GDP, 0.80). The ultraviolet absorption spectrum was identical to that of GTP. Based on GMP, yields were from 45 to 60%. Acid labile phosphate to purine ratios were normally 2:1 for this method, but occasionally they were higher owing to the presence of imidodiphosphate and other condensed inorganic phosphates. If small amounts of contaminating inorganic phosphates are a problem, then the anion exchange procedure used to make AMP-PNP is recommended.

Acknowledgment

This research was supported by a grant (AM 05195) from the National Institute of Health.

[57] The Synthesis of 1,N^6-Ethenoadenosine 3′,5′-Monophosphate; A Fluorescent Analog of Cyclic AMP

By JOHN A. SECRIST III

Analogs of cyclic AMP (cAMP) are useful from the standpoint of potentially possessing biological properties significantly different from those of cAMP itself, as well as being able to supply valuable information concerning structure–activity relationships and mechanisms of action. Fluorescence techniques are an excellent method for obtaining information of this sort because of the sensitivity to small changes in environment as well as the low concentrations sufficient to obtain meaningful results. Desirable attributes of a fluorescent analog would be high quantum yield, long fluorescence lifetime, and an excitation wavelength at sufficiently low energy to allow excitation without interference from other ultraviolet-absorbing moieties in proteins and nucleic acids. 1,N^6-Ethenoadenosine 3′,5′-monophosphate (I) (abbreviated "cyclic εAMP")[1] has all these

(I)

qualities, and additionally is readily available from cAMP itself. The 1,N^6-etheno derivative was found to substitute for cAMP very well in the stimulation of purified muscle protein kinase, as assayed by the phosphorylation of histone, as well as by the phosphorylation of purified muscle glycogen synthase I and its conversion to synthase D.[2] In a comparison with cAMP, cyclic εAMP competes effectively with cyclic [³H]AMP for binding sites on protein kinase from bovine skeletal muscle.[2] These experiments indicate that in these systems, at least, cyclic εAMP is an excellent substitute for cAMP, and probably acts by the

---

[1] The abbreviation cyclic εAMP is derived from the name given in the text, where ε stands for the etheno bridge, and is also suggestive of the molar absorbance term and of fluorescence emission. According to the nomenclature based on the ring system, compound (I) is 3-β-D-ribofuranosylimidazo[2,1-*i*]purine 3′,5′-monophosphate.

[2] J. A. Secrist III, J. R. Barrio, N. J. Leonard, C. Villar-Palasi, and A. G. Gilman, *Science* **177**, 279 (1972).

same mechanism. Fluorescence techniques, ranging from simply examining the changes in quantum yield and wavelength of emission with respect to various parameters to the use of fluorescence polarization methods for obtaining information about the actual binding of the coenzyme are then available. An initial study utilizing fluorescence polarization techniques to examine the nature of the binding of 1,N^6-ethenoadenosine triphosphate (ϵATP) to pyruvate kinase has already demonstrated the potential in this area.[3] Comparable studies utilizing cyclic ϵAMP, together with further fluorescence methods, should prove extremely valuable.

Synthesis and Purification

The preparation of cyclic ϵAMP involves the treatment of cAMP with an aqueous solution of chloroacetaldehyde at room temperature or slightly above. The reaction can be followed easily, and the fluorescent product is readily isolated and purified.

An aqueous solution of chloroacetaldehyde was prepared as follows. One hundred grams of chloroacetaldehyde dimethylacetal (commercially available) in 500 ml of 50% H_2SO_4 (w/v) were refluxed gently for 45 minutes, and the mixture was evaporated under aspirator pressure (a rotary evaporator is suitable) while the receiving vessel was cooled with a dry ice–acetone bath. The upper layer of the distillate was adjusted to pH 4.5 with 1 N NaOH (the lower layer is a small amount of unreacted chloroacetaldehyde dimethylacetal) and distilled as before. The recovered distillate is 1.0 to 1.6 M chloroacetaldehyde in water with some methanol, with a pH near 4. The solution can be frozen and stored for long periods of time, simply thawing prior to use. The pH gradually lowers with time, however, and the pH adjustment-distillation procedure may be repeated if desired.

Reaction was carried out by dissolving cAMP in a sufficient quantity of the chloroacetaldehyde solution to afford at least a 20-fold excess of chloroacetaldehyde, and stirring at pH 4.0–4.5 at 37° for 24 hours. To ensure complete reaction, the disappearance of starting material and the appearance of the new bluish fluorescent spot can be monitored readily utilizing thin-layer chromatography with the system given in the next section, as well as by ultraviolet spectroscopy. Decolorization followed by evaporation to dryness, removing the volatile chloroacetaldehyde, affords cyclic ϵAMP. Purification is accomplished by reprecipitation from aqueous ethanol followed by washing with ethanol and drying. Isolated

---

[3] J. R. Barrio, J. A. Secrist III, Y. Chien, P. J. Taylor, J. L. Robinson, and N. J. Leonard, *FEBS Lett.* **29**, 215 (1973).

yields are 90–95% of analytically pure cyclic εAMP (mp 267–268° with decomp.). Both cAMP and cyclic εAMP are stable under the conditions of synthesis, and no hydrolysis was detected by a thin-layer chromatographic comparison (see next section) with 3'-εAMP and 5'-εAMP prepared in analogous fashion from 3'-AMP and 5'-AMP.

Characterization

Identification of cyclic εAMP can be carried out in a number of ways. The R_f values observed on Eastman Chromagram cellulose sheets without fluorescent indicator with the solvent system isobutyric acid:ammonium hydroxide:water (75:1:24, by volume), were as follows: cyclic εAMP, 0.27; 3'-AMP, 0.25; 5'-AMP, 0.19, compared with 0.40 for cAMP. This allows the product to be distinguished both from the starting material and the possible cyclic phosphate hydrolysis products. The bluish color of the "ε" derivatives on these chromatograms under ultraviolet light also allows them to be distinguished from the dark-colored adenine-containing precursors. As little as 0.5 μg can be readily detected under an ultraviolet lamp.

The ultraviolet absorption spectrum of cyclic εAMP in aqueous buffered solution at pH 7.0, as with all the "ε" derivatives, is very characteristic, showing maxima at 294 nm (ϵ 3100), 275 nm (ϵ 6000), 265 nm (ϵ 6000), and 258 nm (ϵ 5000). This provides an additional method for proof of total conversion to the fluorescent derivative. Not until all the starting material is consumed does the 275 nm band have an extinction coefficient equal to that of the 265 nm band, and thus examination of the ultraviolet absorption spectrum is an excellent method of monitoring the progress of the reaction.

Excitation of the lowest energy ultraviolet absorption band ($\simeq$300 nm in pH 7 buffer) causes fluorescence emission near 415 nm, with a quantum yield in the neighborhood of 0.59, allowing the possibility of detection at very low concentrations ($\sim$10 nM). The fluorescence lifetime is of the order of 20 nsec. These facts taken together provide the opportunity for using more detailed fluorescence techniques to look at the interaction of cyclic εAMP with proteins.

Author Index

Numbers in parentheses are reference numbers and indicate that an author's work is referred to, although his name is not cited in the text.

A

Albano, J. D. M., 49, 53
Alberici, A., 237
Alberty, R. A., 169
Alexander, R. L., 363
Allen, J. E., 398
Allen, J. M., 259
Ames, B. N., 269
Anderson, L. E., 44
Anderson, W. B., 367, 372, 375
Andrews, P., 168, 236, 269
Appel, S. H., 398
Appleman, M. M., 10, 16, 205, 208(4), 211, 212, 222, 223(9), 235, 236, 249, 257, 258, 259, 263, 351
Arai, K.-I., 85
Arnaiz, G. R. D. L., 237
Arnold, L. A., Jr., 188
Arnold, L. J., Jr., 185, 186(18)
Ashby, C. D., 49, 268, 269, 274, 282(7), 292, 293, 296, 326, 329(10), 346, 351, 353, 355(5, 6), 356, 357(4, 5, 6, 11), 358(5), 360
Aurbach, G. D., 101, 150, 151(5), 152(1, 2), 153, 192, 195(2), 283

B

Babcock, D., 120, 420, 423(1), 426
Bär, H.-P., 125
Baggio, B., 328
Bagshaw, C. R., 423
Ballantyne, W., 420, 423(1), 426
Ballentine, R., 190
Barnett, R. J., 260
Barrett, K., 372
Barrio, J. R., 428, 429
Barrnett, R. J., 420
Basch, J. J., 293
Bauer, R. J., 408

Beavo, J. A., 106, 195, 196, 205, 222, 237, 244, 257, 259(4), 296, 323, 381
Bechtel, P. J., 296, 323, 381
Beck, W. S., 200
Beckwith, J., 367, 375(6)
Berenson, J. A., 182
Berg, D., 372
Berman, J. D., 187
Bernard, M., 235
Bhalla, R. C., 380
Bibring, T., 197
Bingham, E. W., 293
Birnbaumer, L., 120, 174, 177(2), 351, 420, 421(4), 423(4)
Bitensky, M. W., 153, 154, 155, 420
Black, C. T., 39
Blonde, L., 96
Blumberg, P. M., 187
Boardman, N. K., 127
Bodley, J. W., 87
Böhme, E., 10, 16(3), 63, 72, 107, 112, 118, 121, 122, 123(2), 124(2), 192, 195, 197, 200
Bollum, F. J., 287
Borden, R. K., 418, 419
Bourne, G. H., 260
Bourne, H. R., 205
Braceland, B. M., 57
Bradham, L. S., 142
Braná, H., 249
Bratvold, G. E., 274, 282(6)
Bravard, L. J., 244
Bray, G. A., 44, 157, 317, 377
Breckenridge, B. M., 62, 237
Brewer, H. B., 274
Bright, H. J., 181
Broadus, A. E., 62
Brodie, B. B., 13, 39, 41(1), 123, 125, 132(8), 144, 156, 177, 205, 278
Brooker, G., 10, 16, 20, 22(6), 23, 24, 25(2), 126, 205, 208, 211, 212, 236, 257, 263

C

Broomfield, C. A., 187
Brostrom, C. O., 288, 289(13), 306, 316, 363, 376
Brostrom, M. A., 290, 344, 346(12), 385
Brown, B. L., 49, 53
Brunswick, D. J., 387, 390(2), 392(2), 394, 396, 397(1), 398
Buell, M. V., 218
Burger, M. M., 187
Burk, D. J., 212
Burris, R., 425
Butcher, R. W., 63, 106, 125, 135, 136(3), 155, 187, 218, 219(1), 221, 222(1), 223, 237, 263, 274, 275, 278, 281(1), 315, 359
Butler, L. G., 120

C

Cailla, H., 96, 99
Cailla, H. L., 105
Calkins, D., 49, 274, 282(7), 346, 351, 355(5), 356(5), 357(5), 358(5), 360
Carlson, S. F., 192
Castagna, M., 305
Caswell, M. C., 271
Catkins, D., 268
Catsimpoolas, N., 270
Cehovic, G., 401, 408(9), 409(10)
Cha, C.-J. M., 80, 81
Cha, S., 80, 81
Chamberlin, M., 372
Chan, P. S., 39, 63, 123, 129
Chang, M. L. W., 218
Chang, M. M., 4
Chang, Y. Y., 244
Changeux, S. P., 187
Chappell, J. B., 67, 107, 288
Chase, L. R., 150, 152(1), 153
Chassy, B. M., 245, 247
Chen, B., 375
Chen, L. J., 324, 327(6), 328(6), 381
Cheung, W. Y., 132, 213, 223, 225, 231, 232, 233(8, 9), 234(9), 235, 236(14), 237, 238, 239, 262, 263, 264, 268(5), 270, 271, 272, 273
Chi, Y.-M., 144, 147(8), 148
Chiang, M. H., 238
Chien, Y., 429

Chrisman, T. D., 120, 195, 196
Christian, W., 230, 268
Christiansen, R. O., 235
Chytil, F., 249
Clarke, J. T., 268
Coakley, W., 421
Cohn, E. J., 269, 270
Cohn, W. E., 126
Colowick, S. P., 128, 168, 169(16, 17)
Cook, W. H., 41, 123
Cooper, R. A., 205, 309
Cooperman, B. S., 387, 390(2), 392(2), 394, 396, 397(1), 398
Corbin, J. D., 289, 290, 306, 315, 316, 358, 359, 360(12), 361, 362, 363, 364, 365, 366, 376
Costa, E., 237
Covell, J. W., 5
Crestfield, A. M., 129, 130
Cuatrecasas, P., 181, 187

D

Davis, B. J., 260
Davis, J. W., 10, 14(2), 15(2), 103, 106, 109(1)
de Crombugghe, B., 367, 375
Delaage, M., 96, 99
Delaage, M. A., 105
DeLange, R. J., 205, 309
Denburg, J. L., 62
De Robertis, E., 237
Desalles, L., 125, 131(4), 144
Desbuquois, B., 101
Devries, J. R., 187
Dietz, S. B., 73, 87
Dimmit, M. K., 408
Dintzis, H. M., 181
Dixon, G. H., 287
Dixon, J. E., 180, 182, 188
Donnelly, T. E., Jr., 92, 93, 94, 95, 332, 346, 347, 348, 349, 350(8)
Drummond, G. I., 143, 145(2, 4), 146(2, 4), 147(2, 3, 4), 149(4), 197, 198(6), 199(6), 403, 406(11), 407(11), 411, 418(6), 419(6)
Duncan, L., 143, 145(2), 146(2), 147(2, 3)
Dunn, A. S., 257

AUTHOR INDEX

DuPlooy, M., 400
Dwyer, I. M., 271

E

Eccleston, J., 423
Eckstein, F., 420, 426
Edelman, G. M., 187
Edsall, J. T., 269, 270
Eil, C., 328
Ekins, R. P., 49, 53
Emmer, M., 367, 375(7)
Ensinck, J. W., 237
Entman, M. L., 144
Epstein, S. E., 144, 147, 148
Erlichman, J., 173, 309, 314, 316, 344, 346(16), 381, 398
Estensen, R., 73

F

Falbriard, J.-G., 96, 97(5), 400, 401(4), 402(4)
Fanestil, D. D., 187
Farr, A. L., 144, 161, 171, 190, 192, 245, 310, 317
Farrell, H. M., Jr., 293
Fedak, S. A., 150, 151(5), 153 283
Feinstein, H., 187
Ferguson, R., 421
Ferrendelli, J. A., 4
Fichman, M., 20
Fisch, H. V., 385
Fischer, E. H., 49, 268, 274, 282(7), 293, 346, 351, 355(5), 356(5), 357(5), 358(5), 360
Fiske, C. H., 218, 264
Florendo, N. T., 260
Forte, L. R., 135, 136(6)
Franks, D. J., 57
Freeman, J., 155
Friedman, D. L., 287
Fruchter, R. G., 129

G

Gabbai, 401
Gallo, M., 372, 375(9)

Garbers, D. L., 119, 198, 199
Gardner, T. L., 187
Garren, L. D., 49, 290, 316, 344, 346(13), 376, 377, 378(7), 379, 381(1, 3, 5)
Gay, M. H., 4
Gazdar, C., 158
Gerisch, G., 247, 248(8)
Germershausen, J., 187
Gill, G. N., 290, 316, 344, 346(13), 376, 381(1, 3, 5)
Gilman, A. G., 13, 26, 49, 51, 52(1), 54, 55, 56, 60, 61, 103, 136, 296, 309, 354, 357, 363, 378, 428
Glynn, I. M., 67, 107, 288
Goidl, E. A., 245
Goldberg, N. D., 62, 73, 75, 87, 103
Goldman, R., 180
Goldstein, L., 180
Gomori, G., 259
Gonzalez, C., 49, 268, 274, 282(7), 346, 351, 355(5), 356(5), 357(5), 358(5), 360
Goody, R. S., 423
Goren, E. N., 222, 235, 259, 261, 268
Gorman, R. E., 153, 154
Gottesman, M., 375
Gray, J. P., 119, 120, 123, 195, 196, 197, 198(6), 199(6, 7)
Gray, W. R., 270
Greene, R. C., 17
Greengard, P., 49, 57, 91, 92, 93, 94, 95, 168, 169(16, 17), 238, 239, 260, 290, 296, 315, 329, 330, 333, 335, 336, 337, 338, 339, 340, 341, 342, 343, 344(5), 345, 346, 347, 348, 349, 350(5, 6, 8)
Greenstein, J. P., 99
Greenwood, F. C., 99
Grønnerod, O., 358
Gunness, M., 271
Guthrow, C. E., Jr., 398

H

Haber, E., 187
Hachmann, J., 86
Hackett, P., 322
Hadden, E. M., 73
Hadden, J. W., 73
Haddox, M. K., 73, 75

Hammermeister, K. E., 274, 282(6)
Handler, J., 283
Hanes, F. N., 426
Hansbury, E., 426
Hardman, J. G., 10, 14(2), 15, 62, 63, 67, 72, 106, 109(1), 112, 119, 120, 192, 195, 196, 197, 198, 199, 200, 205, 222, 237, 244, 257, 259(4)
Harris, J. S., 296, 344, 346(17)
Hartle, D. K., 73
Hartree, E. F., 224
Harvey, E. B., 197
Hayaishi, O., 39, 125, 132(5), 160, 162, 168, 169(6, 16, 17)
Hechter, O., 125
Heersche, J., 150, 153(4), 283
Heilbronn, E., 187
Hemington, J. G., 257
Henion, W. F., 283, 399, 400(2, 3), 401(2), 402(2)
Hersh, L. S., 181
Hickenbottom, J. P., 287, 351, 359
Hirata, M., 39, 125, 132(5), 160
Hirs, C. H. W., 271
Hirsch, A. H., 235, 259, 261(1), 344, 346(16)
Ho, E. S., 288, 289(13)
Ho, R. H., 278
Holt, D. A., 142
Hopkins, J. T., 5
Horrisett, J. D., 187
Horvath, C. G., 20
Hrapchak, R. J., 222
Huang, Y.-C., 240, 241(1)
Huberman, A., 156, 158(6), 159(6)
Hunkeler, F. L., 287, 351, 359
Hunter, W. M., 99
Huttunen, J. K., 287, 359
Hyncik, G., 259
Hynie, S., 205

I

Ide, M., 160, 163, 217, 238, 256
Inamasu, M., 289
Inman, J. K., 181
Inoue, Y., 325, 328(9)
Ishikawa, E., 103

Ishikawa, S., 103
Iwai, H., 289

J

Jakobs, K. H., 121
Jakoby, W. B., 166
Janeček, J., 249
Jansons, V. K., 187
Jard, S., 235
Jergil, B., 287
Johns, E. W., 324
Johnson, R. A., 15, 62, 63, 119, 135, 136, 138, 139, 140, 141, 142, 143(15, 16)
Johnston, R. E., 237
Jung, R., 122

K

Kakiuchi, S., 233, 235, 268, 279
Kaplan, N. O., 128, 180, 182, 185, 186, 187, 188
Kapphahn, J. I., 83, 218
Karlsson, E., 187
Katchalski, E., 180
Katzen, H. M., 187
Kawakita, M., 85
Kaziro, Y., 85
Keely, S. L., 290, 366
Keilin, D., 224
Keirns, J. J., 153, 154, 155
Kemp, R. G., 205, 240, 241(1), 309, 397, 398
Kerr, V., 426
Kettler, M., 420, 426(2)
Khorana, H. G., 403, 406(11), 407(11), 411, 418, 419(6)
Kimura, H., 119, 123(6)
King, C. A., 306, 316, 376
Kipnis, D. M., 26, 96, 99(1, 2, 3), 101(1, 2), 385
Kish, V., 323
Klein, I., 147, 148, 174, 177(7), 178(7)
Kleinsmith, L., 323
Knowles, J. R., 387
Korenman, S. G., 380
Kozyreff, V., 174, 177(2)

AUTHOR INDEX

Krans, H. M. J., 120, 174, 177(2), 420, 421(4), 423(4)
Krebs, E. G., 5, 49, 66, 67, 205, 268, 274, 282(6, 7), 287, 288, 289(13), 290, 293, 296, 299, 301(2), 306, 309, 315, 316, 323, 329, 344, 346(12), 350, 351, 352, 355(5), 356(5), 357(5), 358(5), 358, 359, 360, 363, 376, 381, 385, 387, 397
Kreiner, P., 420
Krichevsky, M. I., 245, 247(6)
Krishna, G., 13, 39, 41, 123, 125, 132(8), 144, 156, 177, 205, 278
Krueger, B. K., 336
Kumon, A., 290(9), 291, 299, 306(3), 315, 316, 325, 327, 344, 346(15)
Kuo, J. F., 49, 57, 91, 92, 93, 94, 95, 239, 290, 315, 329, 330, 333, 335, 336, 337, 338, 339, 340, 341, 342, 343, 344, 345, 346, 347, 348, 349, 350(5, 6, 8)
Kurashina, Y., 160, 162, 168(6), 169(6)

L

Lang, M., 187
Langan, T. A., 287, 293, 315, 329
Laraia, P. J., 148
Larner, J., 62, 103, 287, 358
Lasser, M., 187
Lavin, B. E., 293
Lee, C.-L. T., 182
Lee, R. T., 62
Lee, T. P., 92, 93, 94, 95, 332, 350(8)
Lefkowitz, R. J., 187
Lehne, R., 65
Leonard, N. J., 428, 429
Levey, G. S., 135, 136(9, 10), 144, 147, 148, 149, 174, 175, 176, 177(5, 6, 7), 178, 179, 180(6)
Lewis, R. M., 5
Lin, L., 87
Lin, M. C., 63, 123, 129, 135, 136(7)
Lin, Y. M., 262, 270, 272(19), 273
Lineweaver, H., 212
Lipkin, D., 41, 123
Lipmann, F., 156, 290, 315, 316(10), 344, 346(14), 376
Lipsky, S. R., 20
Little, S. A., 195, 237
Liu, P. P.-C., 232

Liu, Y. P., 270, 272(19), 273, 346, 347, 348
Lopez, C., 73
Lorand, L., 125, 131(4), 144
Love, L. L., 245, 247(6)
Lowe, M., 385
Lowry, O. H., 3, 4, 7(1), 83, 144, 161, 171, 190, 192, 218, 245, 301, 310, 317
Ludens, J. H., 187
Lust, W. D., 6

M

Macchia, V., 174
McClure, W. O., 44
MacDougal, E., 365
McElroy, W. D., 62
Macmanus, J. P., 57
Maim, E., 187
Makman, R. S., 155, 367
Malchow, D., 247, 248(8)
Manganiello, V., 10, 57, 58, 59(12), 77, 88, 283
Mans, R. J., 66
Marcinka, K., 268
Marcus, F. R., 153, 154, 155
Marcus, I., 401, 408, 409(10, 15)
Marcus, R., 150, 153(4)
Markham, R., 41, 123
Maroko, P. R., 180, 188
Mars, R. J., 287
Martin, R. G., 269
Marx, S. J., 150, 151(5), 152(2), 153(2)
Massay, K. L., 232
Mathias, M. M., 398
Mayer, S. E., 3, 4, 5, 8, 66, 68, 72(1), 147, 148, 180, 188, 190, 287, 359, 363
Means, A. R., 365
Mechler, I., 122
Meich, R. P., 80
Meisler, J., 125, 131(4), 144
Mendoza, S., 283
Meng, H. C., 278
Menon, T., 125, 135, 136(2), 143(2), 144, 147(7), 149(7), 174
Mertens, R., 3
Meunier, J. C., 187
Meyer, R. B., 408
Michal, G., 400

Michelson, A. M., 422
Miki, N., 153, 154, 155
Milette, C. F., 187
Miller, D. L., 85, 86
Miller, W. H., 153, 154
Miyamoto, E., 49, 296, 337, 339, 344, 345, 346(17, 18), 349
Miyoshi, Y., 19, 37, 109
Modak, A. T., 6
Moe, O. A., 120
Moffatt, J. G., 412, 424
Mommaerts, W. F. H. M., 144
Monard, D., 249, 253
Monks, R., 118
Monn, E., 235
Moore, M. M., 140
Moore, S., 129, 130, 132, 269, 271
Moret, V., 328
Muneyama, K., 408
Munske, K., 192
Murad, F., 10, 13, 26, 57, 58, 59(12), 60, 61, 77, 88, 119, 123(6), 144, 147, 148, 274, 275(2, 3, 4, 5), 276, 277, 278, 280, 281(2), 282(2, 4), 283, 378

N

Nagele, B., 247, 248(8)
Nair, K. G., 232
Nelböck, M., 400
Neufeld, A. H., 154
Neufeld, E. F., 128
Neuhard, J., 32
Nielsen, L. D., 253
Nielsen, M., 421, 422
Nirenberg, M., 55
Nishiyama, K., 299, 306(3), 315, 327
Nishizuka, Y., 290(1), 291, 299, 306(3), 315, 316, 325, 327(8), 328, 344, 346(15)
Nissley, P., 372, 375
Norris, G. F., 5
Northup, S. J., 45, 195
Novelli, G. D., 66, 287

O

O'Dea, R. F., 75
O'Hara, D., 187

Ohga, Y., 293
Ojala, D., 120, 420, 423(1), 426
Okabayashi, T., 163, 217, 238, 256
Okamoto, H., 160, 162, 168(6), 169(6)
Oldham, K. G., 118
Olsen, R. W., 187
Orloff, J., 283
Ortiz, P. J., 160
Osborn, M., 168, 269
O'Toole, A. G., 62, 73, 87, 103
Ott, D., 426
Oye, I., 187, 188(17)

P

Painter, E., 236
Pannbacker, R. G., 244
Park, C. R., 289, 358, 359(6), 361, 362, 364, 365, 366
Parker, C. W., 26, 96, 99(1, 2, 3), 101(1, 2), 385
Parks, R. E., Jr., 80, 81
Parmeggiani, A., 420, 426(2)
Partridge, S. M., 127
Passonneau, J. V., 3, 4, 6, 7(1), 83
Pastan, I., 174, 367, 372, 375
Pataki, G., 27, 35(4)
Patterson, W. D., 195, 196
Pedersen, K. O., 270
Pennington, S. N., 126, 238
Perkins, J. P., 49, 66, 135, 136(4), 140, 142, 287, 288, 289(13), 290, 315, 329, 352, 358, 376, 387
Perlman, R., 367, 375
Perry, S. V., 106
Peterkofsky, A., 158
Petzold, G. L., 296, 344, 345, 346(17, 18)
Pfitzner, K. E., 412
Pilkis, S. J., 119, 142
Pinna, L., 328
Pöch, G., 205
Pohl, S. L., 120, 174, 177, 420, 421(4), 423(4)
Pool, P. E., 5
Porath, J., 385, 386
Posner, J. B., 274, 282, 350
Post, R. L., 91, 107
Posternak, T., 96, 97(5), 283, 399, 400, 401, 402(2, 4), 408(9), 409(10)

Preiss, B. A., 20
Price, T. D., 10, 20(1), 278

R

Rabinowitz, M., 125, 131(4), 144, 290
Racine-Weisbuch, M. S., 105
Racker, E., 397
Rafferty, M. A., 187
Rall, T. W., 125, 135, 136(2), 143(2), 144, 147(7, 8), 148, 149(7), 174, 223, 274, 275(2, 3), 276, 277, 278, 279, 280, 281(2), 282(2), 283(2)
Ramachandran, J., 125
Randall, R. J., 144, 161, 171, 190, 192, 245, 310, 317
Randerath, E., 27, 28(3), 32, 35(3), 38
Randerath, K., 27, 28(2, 3), 32, 35(2, 3), 38
Rasmussen, H., 222, 398
Reddy, W. J., 148
Reichard, D. W., 187
Reimann, E. M., 67, 287, 290, 296, 299, 301, 306, 315, 316, 323, 344, 346(12), 351, 359, 360(12), 376, 385
Reiness, G., 375
Reyes, P. L., 92, 93, 94, 95, 332, 346, 347, 348, 350(8)
Rickenberg, H. V., 249, 253
Riggs, A. D., 375
Riley, W. D., 205, 309
Ristau, O., 5, 6, 75
Roberts, N. R., 83, 218
Robins, R. K., 408
Robinson, G. A., 187
Robinson, J. L., 429
Robison, G. A., 5, 63, 125, 135, 136(3), 155, 315, 359
Rock, M. W., 83
Rodbell, M., 120, 125, 154, 174, 177(2), 420, 421(4), 423
Rodnight, R., 293
Rosebrough, N. J., 144, 161, 171, 190, 192, 245, 310, 317
Rosen, O. M., 170, 171, 173, 205, 222, 235, 237, 259, 261, 268, 309, 314, 316, 344, 346(16), 381, 398

Rosen, S. M., 170, 171, 173
Roses, A. D., 398
Ross, J., Jr., 180, 187, 188
Rubin, C. S., 309, 314, 316, 381, 398
Rudolph, S. A., 238
Russell, T. R., 222, 223(9), 257, 258, 259
Rutishauser, V., 187

S

Sachs, D. H., 236
Salas, M. L., 290, 315, 344, 346(14), 376
Salganicoff, L., 237, 238(28)
Salomon, Y., 385
Sandborn, B. M., 380
Sanes, J. R., 336
Santa, T. R., 260
Sasko, H., 62, 103
Schachman, H. K., 270
Schmidt, D. E., 5
Schmidt, J., 187
Schmidt, M. J., 5
Schneider, A. B., 367, 375(7)
Schneider, F. W., 259
Schoffa, G., 5, 6, 75
Schonhofer, P. S., 205
Schramm, M., 187
Schultz, G., 10, 14(2), 15, 16(3), 63, 67, 72, 106, 107, 109(1), 112, 118, 121, 123(2), 124(2), 136, 143, 192, 197, 200
Schultz, K., 10, 14(2), 15(2), 106, 109(1), 121
Schulz, D. W., 83
Schwartz, D., 367, 375(6)
Schwarz, H., 247, 248(8)
Schwyzer, R., 385
Secrist, J. A., III, 428, 429
Sedgwick, W. G., 4
Selinger, Z., 385
Sen, A. K., 91, 107
Seraydarian, K., 144
Severson, D. L., 143, 145(4), 146(4), 147(3, 4), 149(4)
Shen, L. C., 358
Sheppard, H., 121
Shimomura, R., 325, 328(9)
Shimoni, H., 106, 111(4)
Shuman, D. A., 408
Siliprandi, N., 328

Simon, L. N., 408
Sims, M., 142
Skelton, C. L., 148
Skidmore, I. F., 205
Skoog, W. A., 200
Smith, M., 403, 406(11), 407(11), 411, 418, 419
Smoake, J. A., 271, 272(21), 273
Soderling, T. R., 287, 289, 290, 351, 358, 359, 361, 362, 364, 365
Soderman, D. D., 187
Song, S. Y., 271, 272
Sqherzi, A. M., 49
Staufer, J., 425
Stavinoha, W. B., 6
Stein, W. H., 129, 130, 132, 269, 271
Steinberg, D., 287, 359
Steiner, A. L., 13, 26, 96, 99(1, 2, 3), 101(1, 2), 385
Stern, R., 350
Stokes, J. L., 271
Stolzenbach, F. E., 182
Strada, S., 65
Strauch, B. S., 274, 275(5)
Strominger, J. L., 187
Stull, J. T., 4, 8, 66, 68, 72(1), 190, 287, 363
Sturtevant, J. M., 238
Subba Row, Y., 218, 264
Suddath, J. L., 120, 199
Sugino, Y., 19, 37, 106, 109, 111(4)
Sulakhe, P. V., 143, 145(4), 146(4), 147(4), 149(4)
Sutherland, C. A., 15
Sutherland, E. W., 10, 14(2), 15(2), 16, 62, 63, 96, 97(5), 103, 106, 109(1), 120, 125, 135, 136(2, 3, 5), 138, 139, 140, 141, 142, 143(2, 15, 16), 144, 147(7, 8), 148, 149, 155, 174, 187, 188(17), 192, 195, 196, 197, 205, 218, 219(1), 221, 222, 223, 237, 244, 257, 259(4), 263, 274, 275, 278, 281(1), 283, 315, 359, 367, 399, 400, 401(2, 4), 402(2, 4)
Suzuki, C., 162, 168(6), 169(6)
Suzuki-Hori, C., 160
Svedberg, C. L., 270
Svensson, H., 325
Swislocki, N. I., 135, 136(8)
Symons, R. H., 410, 411, 415(5), 416(1, 2, 3, 4), 418(1), 420

T

Takahashi, K., 132
Takai, K., 160, 162, 168(6), 169(6)
Takeda, M., 293
Takeyama, S., 289
Tampion, W., 49
Tao, M., 156, 158(6), 159(6), 290, 296, 315, 316, 317, 319(19), 322, 344, 346(14), 376, 381
Taylor, P. J., 429
Tell, G. P., 187
Teo, T. S., 222, 268
Terasaki, W. L., 257, 258
Teshima, Y., 233, 235
Tesser, G. I., 385
Thomas, L. J., Jr., 10, 205, 236
Thompson, B., 8
Thompson, W. J., 16, 195, 205, 211, 222, 223(9), 235, 249, 257, 259, 263
Torres, H. N., 351
Tovey, K. C., 118
Traugh, J. A., 297, 326, 329(10), 351, 357(4)
Traut, R. R., 297
Trentham, D. R., 423
Tsuboi, K. K., 10, 20(1), 278

U

Udenfriend, S., 162
Ueki, A., 162, 168(6), 169(6)
Umbreit, W., 425
Utiger, R., 96, 99(1), 101(1), 385
Uyeda, K., 397
Uzunov, P., 235

V

Varner, J. E., 232
Vaughan, M., 10, 57, 58, 59(12), 77, 88, 147, 148, 274, 275(2, 5), 276, 277, 278, 280, 281(2), 282(2), 283
Veech, R. L., 6
Vengadabady, S., 401, 409(10)
Venter, B. R., 186
Venter, J. C., 180, 185, 186, 187, 188
Villar-Palasi, C., 358, 428

W

Wagner, R., 420
Waitzman, M. B., 260
Walaas, E., 358
Walaas, O., 358
Walsh, D. A., 49, 66, 67, 72(1), 190, 268, 269, 274, 282, 287, 288, 289, 290, 292, 293, 296, 299, 301(2), 305, 315, 323, 324, 326, 327(6), 328(6), 329, 344, 346, 351, 352, 353, 355(5, 6), 356, 357(4, 5, 6, 11), 358, 359, 360, 363, 376, 381, 385, 387
Walton, G. M., 49, 377, 378(7), 379
Walton, H. F., 32
Walton, K. G., 92, 93, 94, 95, 332, 350(8)
Wang, J. H., 222, 223, 268
Wang, T. H., 268
Warburg, O., 230, 268
Wastila, W. B., 8, 66, 68, 72(1), 190, 287, 363
Weber, K., 168, 269
Weetall, H. H., 180, 181
Wehmann, R. E., 96
Weibel, M. K., 181
Weichselbaum, T. E., 301
Weiland, G. A., 187, 188(16)
Weimann, G., 400
Weinryb, I., 98, 101(7), 105
Weintraub, S. T., 6
Weiss, B., 13, 39, 41(1), 65, 123, 125, 132(8), 144, 156, 177, 235, 237, 278
Weissbach, H., 85, 86
Wheeler, H., 3
White, A. A., 42, 45, 46(1), 88, 125, 131(6), 192, 195

White, J. G., 73
Widlund, L., 187
Wilchek, M., 385
Williams, B. J., 39
Williams, R. H., 195, 237
Winitz, M. A., 99
Wollenberger, A., 5, 6, 75
Woodard, C. J., 150, 152(2), 153(2)
Woods, W. D., 260
Wool, I. G., 328
Wyatt, G. R., 91, 329, 330(5), 335, 338, 340, 343, 344(5), 350(5)

Y

Yamamoto, M., 232
Yamamura, H., 290(1), 291, 293, 299, 306(3), 315, 316, 325, 327, 328(9), 344, 346(15)
Yamamura, H. I., 187
Yamazaki, R., 233, 235, 268
Yates, D. W., 423
Yong, M. S., 185
Yoshimoto, A., 163
Youdale, T., 57
Yount, R. G., 120, 420, 423(1), 426

Z

Zenser, T. V., 42, 45, 46(1), 88, 125, 131(6), 195
Zieve, F. J., 87
Zieve, S. T., 87
Zubay, G., 367, 375

Subject Index

A

Acetic acid, thin-layer chromatography and, 29–30, 34, 36
Acetone, phosphodiesterase precipitation by, 252
Acetonitrile, drying of, 412
Acetylcholine
  adenylate cyclase and, 173
  cyclic nucleotide levels and, 94
Acrylamide gel electrophoresis
  phosphodiesterase, 228–229, 260
    activator, 265–267
Adenine, phosphodiesterase and, 244
Adenosine
  adenylate cyclase and, 169
  phosphodiesterase and, 244
Adenosine 3',5'-cyclic monophosphate, see Cyclic adenosine monophosphate
Adenosine diphosphate, cGMP assay and, 112
Adenosine 5'-monophosphate
  labeled
    chemical synthesis, 414–416
    cyclization, 416–418
    determination of specific activity, 419
  phosphodiesterase and, 256
Adenosine triphosphatase
  cyclase preparations and, 143, 151
  protein kinase assay and, 289
Adenosine triphosphate
  cAMP assay and, 56
  cyclase assay and, 131, 141, 146
  cyclic nucleotide immunoassay and, 102, 103
  formation from cAMP, 62
  labeled
    chromatography of, 67–68
    impurities, 45, 46
    preparation of, 107–108
  phosphodiesterase and, 154, 155, 233, 248
  phosphofructokinase and, 397–398

purification of, 79
removal of, 25
Adenyl cyclase, see Adenylate cyclase
Adenylate cyclase
  adenylyliminodiphosphate and, 420–421
  assay of, 42–46, 72
  automated chromatographic assay, 125–126, 134–135
    cyclase preparation, 128–129
    identification of product, 132
    materials, 127–128
    nucleotide separation, 126–127
    procedure, 129–130
    sensitivity and reproducibility, 131
  bovine brain, preparation of, 128–129
  Brevibacterium liquefaciens, 160
    forward reaction assay, 161
    properties, 167–169
    purification, 163–167
    reverse reaction assay, 162–163
  cAMP inhibitor formation and, 274–275
  erythrocyte
    assay, 170–171
    properties, 173
    purification, 171–172
  Escherichia coli, 155
    assay, 156–158
    properties, 159–160
    purification, 159
  general principles of assays
    determination of residual substrate, 124–125
    divalent cations, 120
    enzyme concentration, 117
    labeled nucleoside triphosphates, 118–119
    linearity of cyclic nucleotide production, 122
    nucleoside triphosphate concentration and volume, 115–117
    protection of cyclase activity, 121–122

SUBJECT INDEX

purification, proof of identity and determination of product, 123-124
reduction of product degradation, 120-121
of substrate degradation, 119-120
termination of cyclase reaction, 122-123
heart and skeletal muscle
  assay, 143-144
  preparation, 144-146
  properties, 146-149
  solubilization, 149
hormone-sensitive
  assay of, 190
  binding to isoproterenol on glass beads, 189
  preparation of, 188
kidney, preparation, 150-152
myocardial, solubilization and role of phospholipids, 174-180
neonatal bone, 152-153
particulate from brain, 135-136
  assay, 136
  characteristics, 140
  dispersion of, 138-140
  preparation, 136-138
rod outer segments, 154-155
solubilized from brain, characteristics, 140-143
Adenylate kinase, cAMP assay and, 62, 63, 65
Adenylyl cyclase, see Adenylate cyclase
Adenylyliminodiphosphate
  adenylate cyclase and, 169
  DNA transcription and, 375
  preparation, 420-421
  principle, 422-423
  procedure, 423-425
  tetrasodium iminodiphosphate and, 421-422
Adipose tissue, protein kinase, hormonal regulation, 363-366
Adrenal gland, cAMP receptor in, 380-381
β-Adrenergic receptor-adenylate cyclase complex
  glass bead-immobilized proterenol, 187-188
  adenylate cyclase preparation, 188
  binding, 189

cyclase assay, 190
epinephrine-sensitive membrane fraction, 188-189
erythrocyte preparation, 188
properties, 190-191
Adrenocorticotropic hormone, protein kinase and, 364
Albumin, cyclase assays and, 122
Alkaline phosphatase, phosphodiesterase assay and, 260
Alkylamine glass
  long-chain, synthesis of, 183-184
  synthesis of, 181-182
Alumina C_γ, protein kinase and, 303, 311
Alumina column chromatography
  cyclic nucleotide separation, 44-42, 45-46, 88
  column preparation, 43-44
  cyclase assays, 42-43
  interfering compounds, 44
  material and methods, 42
  procedure, 44
Amino acid(s), phosphodiesterase activator composition, 269-270, 271
Aminoacyl transfer ribonucleic acid synthetases, adenylyliminodiphosphate synthesis and, 423
N^6-ϵ-Aminocaproyl-cyclic adenosine monophosphate, preparation of, 386
α-Aminopropyltriethoxysilane, catecholamine glass beads and, 181
Ammonium acetate, cyclic nucleotide separation and, 30, 31
Ammonium formate, cyclic nucleotide separation and, 31
Ammonium sulfate
  cyclic nucleotide separation and, 32-34, 35
  guanylate cyclase and, 193
Antigen, preparation, cyclic nucleotide assay and, 98
Antisera, cyclic nucleotides, sensitivity and selectivity, 104
Arginine vasopressin, adenylate cyclase and, 151-152
Arylamine glass, production of, 182-183
Atropine, cyclic nucleotide levels and, 94
Azide
  adenylate cyclase assays and, 119
  guanylate cyclase assay and, 196

B

Bacteriophage, DNA, preparation of, 374
Barium carbonate, cAMP isolation and, 40, 41
Barium sulfate, cAMP isolation and, 40, 41
N^6-Benzoyl cyclic adenosine monophosphate preparation of, 406
Bethanechol, cyclic nucleotide levels and, 94
Bio-Gel P-300, protein kinase and, 312
Biopsies, sampling methods, 4–5
BioRad AG-1-X2 resin, cyclic nucleotide purification and, 25
BioRad AG-1-X8, cGMP separation and, 76–77, 91
Blood cells, guanylate cyclase in, 200
Bone, neonatal, adenylate cyclase of, 152–153
Bovine serum albumin, adenylate cyclase and, 141, 168, 177
Brain
  adenylate cyclase, preparation, 128–129
  cAMP receptor in, 380
  enzymes, rapid inactivation of, 5–6
  particulate adenylate cyclase, 135–136
    assay, 136
    characteristics, 140
    dispersion of, 138–140
    preparation, 136–138
  phosphodiesterase, 223–224
    assay, 132–134, 224–225
    properties, 232–239
    purification, 225–232
    regional and subcellular distribution, 237–238
  phosphodiesterase activator, 262–263
    assay, 263–264
    properties, 267–273
    purification, 264–267
  protein kinases of, 298, 365, 366
  solubilized adenylate cyclase, characteristics of, 140–143
Brevibacterium liquefaciens
  adenylate cyclase, 160
    forward reaction assay, 161
    properties, 167–169
    purification, 163–167
    reverse reaction assay, 162–163
Buffers, alumina column chromatography and, 46

C

Caffeine
  cAMP inhibitor and, 274, 275
  cyclase assays and, 121
  phosphodiesterase and, 244
  protein kinase and, 363
Calcitonin, adenylate cyclase and, 151–152, 153
Calcium ions, phosphodiesterase activator and, 272–273
Calcium phosphate gel
  cGMP-dependent protein kinase and, 332, 334–335
  phosphodiesterase purification and, 221, 227
  protein kinase regulatory subunit and, 327
cAMP, *see* Cyclic adenosine monophosphate
Carbobenzoxy-ε-aminocaproic acid anhydride, preparation of, 386
N^6-Carbobenzoxy-ε-aminocaproyl-cyclic adenosine monophosphate, preparation of, 386
Carbon dioxide, high pressure ion-exchange chromatography and, 21–22
N,N'-Carbonyldiimidazole, guanylyliminodiphosphate preparation and, 425, 427
Carboxymethyl cellulose, adenylate cyclase and, 172
Carboxymethyl Sephadex
  protein kinase and, 302–303, 305–306
  catalytic subunit, 306, 307
Cardiolipin, adenylate cyclase and, 177, 180
Casein
  cAMP-binding assay and, 368
  protein kinase and, 328, 340, 348, 351, 357, 358
  purification and dephosphorylation of, 67
Catecholamines
  adenylate cyclases and, 147–149, 170, 173, 176

SUBJECT INDEX

cAMP inhibitor and, 275
glass bead-immobilized, 180–181
  alkylamine glass synthesis, 181–182
    catecholamine glass production, 184
  arylamine glass production, 182–183
  succinyl glass and long-chain alkylamine glass synthesis, 183–184
  washing procedure, 185–186
Cations
  divalent
    adenylate cyclase and, 120, 141–142, 146, 160, 169, 173
    guanylate cyclase and, 195, 198, 199, 201
    phosphodiesterase and, 223, 233–234, 243, 248, 256
    protein kinases and, 314, 322, 343–344
Cecropia silkmoth, tissues, protein kinases of, 335
cGMP, see Cyclic guanosine monophosphate
Charcoal
  cyclic nucleotide preparation and, 10, 20
  purification of, 79
Chelating agents, adenylate cyclase and, 142
Chloroacetaldehyde, 1,N^6-ethenoadenosine 3′,5′-monophosphate preparation and, 429
p-Chloromercuribenzoate
  adenylate cyclase and, 169
  cAMP-receptor protein and, 381
  phosphodiesterase and, 244
p-Chloromercuriphenyl sulfonate, guanylate cyclase and, 195
Cholera toxin, cAMP and, 3
Coenzyme A, purification of, 81
Creatine phosphokinase
  cyclase assays and, 119, 154
  cGMP assay and, 73, 78, 79
Crystallization, adenylate cyclase, 166–167
Cyanogen bromide, Sepharose activation and, 386–387
Cyclic adenosine monophosphate
  2′-O-acyl derivatives, preparation and properties, 402–404

assay
  luciferin–luciferase system and, 62–65
  protein kinase activation and, 66–73
  receptor protein binding and, 49–57
  radioimmunoassay and, 96–105
  sample preparation, 3–9, 9–20, 54–55, 100
  simultaneous, with cGMP assay, 60–61
cGMP-dependent protein kinase and, 335–338
cholera toxin and, 3
N^6-2′-O-diacyl derivatives, preparation of, 404–405
diazomalonyl derivatives, 387–388
  applications, 397–398
  Dimroth rearrangement, 392–393
  procedures, 393–397
  syntheses, 388–393
excretion of, 163
identification of, 132
inhibitor, 273–274, 282–283
  assay, 278–282
  formation in cell-free incubations, 274–276
  purification, 276–278
N^6-monoacyl derivatives, preparation of, 405–406
^{32}P-labeled, 410–412
  general experimental details, 412–414
  preparation of cyclic nucleotides, 414–420
protein kinase classification and, 290–292, 295–296
Cyclic adenosine monophosphate binding protein, preparation of, 49–52
Cyclic adenosine monophosphate-dependent protein kinases, see Protein kinases
Cyclic adenosine monophosphate-receptor protein
  animal cells, 376–377
  application, 381–382
  assay, 377–378
  characteristics of interaction, 378–380
  detection in tissues, 380–381
  properties, 381
  Escherichia coli, 367
  assay, 368

bacteria, 368-369
  DNA preparation, 372-373
  growth of bacteria, 373-374
  ion-exchange resin preparation, 371
  mechanism of action, 375-376
  properties, 376
  purification, 369-371
  transcription assay, 372
Cyclic adenosine monophosphate Sepharose, preparation of, 385-387
Cyclic adenylic acid, see Cyclic adenosine monophosphate
Cyclic AMP, see Cyclic adenosine monophosphate
Cyclic 8-benzylaminoguanosine monophosphate, protein kinase and, 337-339
Cyclic 8-bromoguanosine monophosphate, protein kinase and, 337, 339
Cyclic deoxyribonucleotides, preparation of, 420
Cyclic guanosine monophosphate
  cAMP-dependent protein kinases and, 336-338
  assay
    elongation factor Tu and, 85
    enzymatic cycling, 73-74
    labeled GDP formation, 106
    preparation of extracts, 85-87
    receptor protein binding and, 57-60
    radioimmunoassay and, 13
  phosphodiesterase and, 235, 237, 242-244, 259
  protein kinases and, 329-330
    characterization, 335-349
    purification, 330-335
    standard assay, 330
  purification of, 76
    column chromatography, 76-77
    thin-layer chromatography, 77-78
  separation of, 13
Cyclic guanosine monophosphate binding protein, preparation of, 57
Cyclic guanosine monophosphate-dependent protein kinases, 329
Cyclic guanylic acid, see Cyclic guanosine monophosphate
Cyclic inosine monophosphate, reactivity in immunoassay, 103

Cyclic isoadenosine monophosphate, activity of, 401
Cyclic nucleotides
  acylated derivatives, 399-401
    butyryl, 406-409
    cAMP, 402-406
    principle of synthetic methods, 401-402
  analogs, protein kinases and, 339
  ion-exchange resin chromatography, 9-10
    Dowex-50, 10-13
    polyethylene imine cellulose, 16-20
    quaternary aminoethyl Sephadex, 13-16
  protein kinase inhibitor and, 357
  purification, 9, 20, 27, 38, 41, 208
  radioimmunoassay methods
    preparation of materials, 96-101
    procedure, 101-105
  thin-layer chromatography, 158
Cyclic nucleotide phosphodiesterase, see Phosphodiesterase
Cytosol, cAMP receptor in, 380

D

Deoxyadenosine triphosphate, adenylate cyclase and, 160, 169, 173
Deoxyguanosine monophosphate, guanylate cyclase and, 199
Deoxyribonuclease
  adenylate cyclase purification and, 159, 172
  cAMP-binding protein preparation and, 369
  phosphodiesterase purification and, 251
Deoxyribonucleic acid
  λpgal 25Sam 7, preparation of, 372-374
  transcription, assay of, 372
Deoxyribonucleosides, phosphorylation of, 411-412
Deoxyribonucleotides, thin-layer chromatography of, 30
Dibenzyline, adenyl cyclase and, 173
N^2-2'-O-Dibutyryl cyclic guanosine monophosphate
  preparation of, 406-407
  protein kinases and, 338, 339

SUBJECT INDEX

N^6-2'-O-Dibutyryl-8-mercapto cyclic adenosine monophosphate, preparation of, 408–409
Dichlorodifluoromethane, tissue freezing and, 4
Dichloroisoproterenol, adenylate cyclases and, 147
Dictyostelium discoideum
  phosphodiesterase, 244–245
    assay, 245
    properties, 247–248
    purification, 245–247
Diethylaminoethyl cellulose
  acyl cAMP derivatives and, 408, 409
  adenylate cyclase and, 159, 165, 167, 176
  adenylyliminodiphosphate and, 424–425
  cAMP binding protein and, 50, 51, 370
  cAMP inhibitor and, 277
  cGMP-dependent protein kinase and, 331–332
  cyclic nucleotides and, 416, 418–419
  diazomalonyl derivatives of cAMP and, 389–390, 392
  guanylate cyclase and, 193–194
  phosphodiesterase
    brain, 230–232, 234
    Escherichia coli, 254
    heart, 219–220
    liver, 258
    muscle, 241, 242
    slime mold, 246–247
  phosphodiesterase activator and, 265
  preparation of, 371
  protein kinase(s) and, 297, 299–302, 312, 313, 318, 324–325
    catalytic subunit, 314–315
    preparation, 67
    regulatory subunit, 327
  protein kinase inhibitor and, 354–355
Diethylaminoethyl Sephadex
  column photolysis and, 395
  guanylyliminodiphosphate and, 427
  labeled ATP preparation and, 67–68
  protein kinase and, 311, 324–326
$N^6,O^{2'}$-Di(ethyl-2-diazomalonyl)adenosine 3',5'-cyclic monophosphate, synthesis of, 390–391

Digitonin
  adenylate cyclase preparation and, 128–129
  phosphodiesterase preparation and, 132
6-Dimethylaminopurine, phosphodiesterase and, 244
Dimethyl sulfoxide, drying of, 412
Dimroth rearrangement, cAMP derivatives and, 392–393
Diphenhydramine, adenylate cyclases and, 148, 178
Diphenylphosphoamide, preparation, of, 421
Diphosphopyridine nucleotide, cGMP assay and, 74, 78, 81–83
α,α'-Dipyridyl, phosphodiesterase and, 256
5,5'-Dithiobis(2-nitrobenzoate), phosphodiesterase and, 244
Dithionite, catecholamine glass beads and, 186
Dithiothreitol
  adenylate cyclase assays and, 122, 139, 140, 141, 173
  guanylate cyclase and, 202
  phosphodiesterase and, 246–248, 251, 255
Dowex-1
  cAMP chromatography and, 55
  cAMP inhibitor and, 277
  cGMP assay and, 59, 88
  nucleotide separation and, 126–128, 130
Dowex-2
  cAMP inhibitor and, 276
  phosphodiesterase assay and, 250
Dowex-50
  cAMP inhibitor and, 277
  cyclic nucleotide chromatography, 55, 144
    resin preparation, 10–11
    separation procedure, 11–13
  dibutyryl cGMP and, 407
  succinyl cAMP purification and, 97

E

Electrophoresis, cyclic nucleotides, 417
Elongation factor Tu, cGMP assay and, 85–90
Enthalpy, cAMP hydrolysis and, 238–239

Epinephrine
  adenylate cyclases and, 147–149, 170, 173
  cAMP inhibitor assay and, 279–281
  protein kinase and, 363–365
Erythrocytes
  adenylate cyclase, 188
    assay, 170–171
    properties, 173
    purification, 171–172
  ghosts, cAMP derivatives and, 398
Escherichia coli
  adenylate cyclase, 155
    assay, 156–158
    properties, 159–160
    purification, 159
  cAMP-receptor protein, 367
    assay, 368
    bacteria, 368–369
    DNA preparation, 372–373
    growth of bacteria, 373–374
    ion-exchange resin preparation, 371
    mechanism of action, 375–376
    properties, 376
    purification, 369–371
    transcription assay, 372
  elongation factor Tu, cGMP assay and, 85–90
  phosphodiesterase
    assay, 249–250
    culture, 250–251
    extract preparation, 251
    properties, 255–256
    purification, 251–255
$1,N^6$-Ethenoadenosine 3′,5′-monophosphate
  biological activities, 428–429
  characterization, 430
  synthesis and purification, 429–430
N^6-(Ethyl-2-diazomalonyl)adenosine 3′,5′-cyclic monophosphate
  preparation of, 391
  labeled, 392
$O^{2'}$(Ethyl-2-diazomalonyl)adenosine 3′,5′-cyclic monophosphate
  synthesis of, 389–390
  labeled, 391–392
Ethylenediaminetetraacetate
  cyclic nucleotide immunoassay and, 105
  phosphodiesterase and, 233, 243, 248, 256
Ethyleneglycol bis(β-aminoethylenether) N,N,N',N'-tetraacetic acid
  brain adenylate cyclase and, 142
  phosphodiesterase activator and, 273
N-Ethylmaleimide, guanylate cyclase and, 199

F

Fluorescence, $1,N^6$-ethenoadenosine 3′,5′-monophosphate, 430
Fluoride
  cAMP inhibitor and, 275
  cyclase assays and, 119, 140, 145–147, 149, 160, 170, 173, 190–191
  phosphodiesterase assay and, 133
  protein kinase assay and, 289
Fluorometer
  adenylate cyclase assay and, 162–163
  cAMP assay and, 62
  cGMP assay and, 82
Formic acid, thin-layer chromatography and, 29–30
Frog, erythrocytes, adenylate cyclase of, 170–173

G

β-Galactosidase, cAMP binding and, 367
Gallbladder, cAMP in, 3
Glass beads
  catecholamine immobilization, 180–181
    alkylamine glass synthesis, 181–182
    arylamine glass production, 182–183
    catecholamine glass production, 184
    succinyl glass and long-chain alkylamine glass synthesis, 183–184
    washing procedure, 185–186
Glucagon
  adenylate cyclase and, 147–149, 173, 176, 178, 180
  cAMP inhibitor and, 275
  cyclic nucleotide levels and, 94
  protein kinase and, 364
Glucose, catabolite repression and, 367

Glucose-6-phosphate dehydrogenase, adenylate cyclase assay and, 162
Glyceraldehyde-3-phosphate dehydrogenase, cGMP assay and, 106–107
Glycogen, cAMP inhibitor assay and, 279, 280
Glycogen synthetase, protein kinase and, 328, 351
Growth hormone, release, cAMP derivatives and, 401
Guanosine diphosphate
  labeled, cGMP assay and, 106–112
  preparation of, 86–87
  removal from EF Tu, 86
Guanosine 3',5'-monophosphate, see Cyclic guanosine monosphosphate
Guanosine monophosphate kinase, cGMP assay and, 73, 78, 80, 85, 89, 106, 109–111
Guanosine triphosphate, labeled, impurities, 45, 46
Guanylate cyclase
  bovine lung
    assay, 192
    properties, 195
    purification, 192–195
  general principles of assay
    determination of residual substrate, 124–125
    divalent cations, 120
    enzyme concentration, 117
    labeled nucleoside triphosphates, 118–119
    linearity of cyclic nucleotide production, 122
    nucleoside triphosphate concentration and volume, 115–117
    protection of cyclase activity, 121–122
    purification, proof of identity and determination of product, 123–124
    reduction of product degradation, 120–121
    substrate degradation, 119–120
    termination of cyclase reaction, 122–123
  human platelets
    apparent activation with time, 201–202
    assay, 199–200
    distribution among blood cells, 200
    divalent cations and, 201
    homogenate preparation, 200
    subcellular distribution, 201
  rod outer segments, 154–155
  sea urchin sperm
    assay, 196–198
    properties, 198–199
Guanyl cyclase, see Guanylate cyclase
Guanylyl cyclase, see Guanylate cyclase
Guanylyliminodiphosphate
  preparation
    principle, 425–426
    procedure, 426–427

H

Heart
  cAMP-receptor in, 380
  biopsy of, 4–5
  phosphodiesterase
    assay, 218–219
    properties, 222–223
    purification, 219–221
  protein kinase, 298, 365, 366
    assay, 308–310
    properties, 313–315
    purification, 310–313
HeLa cells, protein kinase of, 365, 366
Hemocyanin, coupling to succinylated cyclic nucleotides, 98
Hexamethonium, cyclic nucleotide levels and, 94
Hexamethylphosphoric triamide, guanylyliminodiphosphate preparation and, 426–427
Hexokinase, adenylate kinase assay and, 162
Histamine
  adenylate cyclases and, 147–148, 176, 178, 180
  cyclic nucleotide levels and, 94
Histone
  phosphorylation, cGMP assay and, 90, 92, 93
  protein kinase assay and, 287, 288, 313, 316–317, 322, 324, 327–328, 340–342, 357, 359
    cGMP-dependent, 330, 340–342, 348

Hormones, tissue incubations and, 362
Human serum albumin, coupling to succinylated cyclic nucleotides, 98
Hydrogen fluoride, catecholamine glass beads and, 186
Hydroxyapatite
  adenylate cyclase and, 165
  protein kinase and, 312, 318–321, 326
  regulatory subunit, 327
p-Hydroxymercuribenzoate
  guanylate cyclase and, 199
  phosphodiesterase and, 235
6-Hydroxypurine, phosphodiesterase and, 244

I

Imidazole, phosphodiesterase and, 223, 244, 248
Immunization, schedule, cyclic nucleotide assay and, 98
Inorganic salt coprecipitation
  cAMP, 38–39
    reagents and procedure, 39
    results and discussion, 39–41
Insulin
  adenylate cyclase and, 173
  protein kinase and, 364–365
Ion-exchange chromatography
  cyclic nucleotides
    Dowex-50, 10–13
    polyethyleneimine cellulose, 16–20
    quaternary aminoethyl cellulose, 13–16
  high pressure
    operation of equipment, 20–22
    preliminary prepurification, 24–26
    procedures, 22–24
Irradiation, cAMP diazomalonyl derivatives, 393–394
Isoelectric focusing electrophoresis, protein kinases and, 325
Isopropylnorepinephrine, adenylate cyclases and, 147–149
Isoproterenol
  adenyl cyclase and, 173
  cyclic nucleotide levels and, 94
  derivativization to glass beads, 189

Iterative labeling
  column photolysis and, 395
  dialysis photolysis and, 396–397
  principle, 394–395

K

α-Ketobutyrate, adenylate cyclase and, 169
α-Ketovalerate, adenylate cyclase and, 169
Kidney
  adenylate cyclase of, 150–152
  cAMP-receptor in, 380
  protein kinases of, 298

L

Lactate dehydrogenase, cGMP assay and, 74, 78, 81, 82
Lead nitrate, phosphodiesterase assay and, 260, 261
Lecithin, adenylate cyclase and, 177, 180
Light
  phosphodiesterase and, 154–155
  rod outer segment isolation and, 153–154
Lithium bromide, cyclase preparation and, 145
Lithium chloride, cyclic nucleotide separation and, 30, 31, 35, 36
Liver
  cAMP-receptor in, 380
  cytoplasmic phosvitin kinase of, 328–329
  cytosol protein kinase
    assay, 323–324
    mechanism of cAMP action, 323
    properties, 327–328
    purification, 324–327
  phosphodiesterase, of multiple forms
    alternate preparations, 258–259
    assay, 257
    chromatography, 258
    properties, 259
    tissue extract preparation, 257
  protein kinases of, 298, 365, 366

SUBJECT INDEX

Lobster, cGMP-dependent protein kinase of, 330–335
Lubrol-PX, adenylate cyclases and, 137–140, 149, 174–175
Lubrol-WX, cyclase and, 140
Luciferin–luciferase
  cAMP assay and
    enzyme decontamination, 64–65
    instrumentation, 62
    procedure, 63–64
    reagents, 63
Lung
  guanylate cyclase
    assay, 192
    properties, 195
    purification, 192–195
Lysolecithin, adenylate cyclase and, 177
Lysozyme, adenylate cyclase extraction and, 163–164

M

Manganese ions, phosphodiesterase assay and, 216–217
Melanin, dispersion, cAMP derivatives and, 401
Membranes, epinephrine-sensitive, preparation of, 188–189
8-Mercapto cyclic adenosine monophosphate, activity of, 401
Mercaptoethanol, phosphodiesterase and, 234–235
β,γ-Methylene adenosine triphosphate, adenylate cyclase and, 169
1-Methyl-3-isobutylxanthine, cyclase assays and, 72, 121, 154
Methylxanthines, phosphodiesterase and, 223, 244
Microsomes, cAMP receptor in, 380–381
Microwave radiation, rapid tissue fixation and, 5–6
N^6-Monobutyryl cyclic adenosine monophosphate, cAMP inhibitor and, 275, 282, 283
$2'$-O-Monobutyryl cyclic inosine monophosphate, preparation of, 408
$2'$-O-Monobutyryl cyclic isoadenosine monophosphate, preparation of, 409

$2'$-O-Monobutyryl cyclic uridine monophosphate, preparation of, 407
Muscle
  adenylate cyclase
    assay, 143–144
    preparation, 144–146
    properties, 146–149
    solubilization, 149
  cAMP-receptor in, 380
  phosphodiesterase
    assay, 240–241
    properties, 243–244
    purification, 241–243
  phosphofructokinase, cAMP derivatives and, 396–398
  protein kinases, 298–301, 366
    catalytic subunits, 306–308
    holoenzyme, 301–306
    inhibitor, 353–355
Myocardium, *see also* Heart
  adenylate cyclase, solubilization and role of phospholipids, 174–180
Myosin, cGMP assay and, 106, 109

N

Nicotinamide adenine dinucleotide phosphate, adenylate cyclase assay and, 162
p-Nitrobenzoyl chloride, arylamine glass production and, 183
Nitrogen, liquid, tissue freezing and, 4
Norepinephrine, adenylate cyclases and, 147–149, 173, 179–180
Nucleoside(s)
  separation on polyethyleneimine cellulose, 17–18
  thin-layer chromatography, 28–30, 158
Nucleoside triphosphatases, cyclase assays and, 116, 117
Nucleoside triphosphate(s), *see also* Adenosine triphosphate; Guanosine triphosphate
  concentration and volume, nucleotide cyclase assays and, 115–117
  degradation, reduction of, 119–120
  labeled, impurities in, 118
  thin-layer chromatography, 35–36

5'-Nucleotidase
  cGMP assay and, 88
  phosphodiesterase assay and, 133, 205–208, 210, 218, 224, 225, 240, 249, 250, 263–264
Nucleotides
  guanylate cyclase and, 195, 199
  ion-exchange chromatography, 126–128
  thin-layer chromatography, 28–30, 158
    cyclic nucleotides, 30–35

O

Oxaloacetate
  adenylate cyclase and, 169
  guanylate cyclase and, 195

P

Papaverine, phosphodiesterase and, 72
Paper chromatography
  adenylyliminodiphosphate, 426
  cAMP acyl derivatives, 402
  cyclic nucleotides, 417, 418
  solvents, 414
Parathyroid hormone, adenylate cyclase and, 151–153
pH
  adenylate cyclases and, 147, 159, 168, 173
  guanylate cyclase and, 195, 199
  phosphodiesterases and, 214–215, 232, 244, 247–248, 256
  protein kinase and, 314, 322
1,10-Phenanthroline, phosphodiesterase and, 248, 256
Phosphate, radioactive, precautions in handling, 413–414
Phosphatidylethanolamine, adenylate cyclase and, 176–177, 180
Phosphatidylinositol, adenylate cyclase and, 149, 177, 179–180
Phosphatidylserine, adenylate cyclase and, 149, 176–178, 180
Phosphocellulose
  cAMP-binding protein and, 370
  phosvitin kinase and, 328–329
  preparation of, 371
  protein kinases and, 297

Phosphodiesterase
  activator, 262–263
    assay, 263–264
    preparation, 254–255, 264–267
    properties, 267–273
    requirement for, 255
  activity stain for on gels
    assay method, 259–260
    electrophoresis procedure, 260
    staining procedure, 260–261
  assay by continuous titrimetric technique
    apparatus and experimental conditions, 214
    conclusions, 217
    dynamic aspects, 216–217
    pH selection, 214–215
    principle, 213
    reagents, 213–214
    sensitivity, 215–216
  assay with radioactive substrates
    one-step procedure, 210–211
    principle, 205–206
    reagents and constituents, 208
    theory, 206–207
    two-step procedure, 209–210
  bovine brain, 223–224
    assay, 132–134, 224–225
    properties, 232–239
    purification, 225–232
    regional and subcellular distribution, 237–238
  bovine heart
    assay, 218–219
    properties, 222–223
    purification, 219–221
  cAMP assay and, 62, 63
  cAMP authenticity and, 26
  cAMP inhibitor and, 275–276, 282
  cGMP assay and, 73, 78, 79, 83, 85, 88–89, 106, 108, 110, 111
  cyclase assays and, 120–121, 124
  cyclase preparations and, 143
  Dictyostelium discoideum, 244–245
    assay, 245
    properties, 247–248
    purification, 245–247
  Escherichia coli
    assay, 249–250
    culture, 250–251

extract preparation, 251
properties, 255–256
purification, 251–255
kinetics, activator and, 272
liver, multiple forms from
alternate preparations, 258–259
assay, 257
chromatography, 258
properties, 259
tissue extract preparation, 257
purification of, 79
rod outer segments, 154–155
skeletal muscle
assay, 240–241
properties, 243–244
purification, 241–243
Phosphoenol pyruvate
cAMP assay and, 62
cyclase assays and, 119, 120, 143
guanylate cyclase and, 195
Phosphofructokinase, iterative labeling of, 396–398
3-Phosphoglycerate kinase, cGMP assay and, 106–107
Phosphorylase
cAMP derivatives and, 399–400
inactive, cAMP inhibitor assay and, 279–281
Phosphorylase kinase, protein kinase inhibitor and, 350–351
Phosvitin kinase
cytoplasmic
assay, 328
properties, 329
purification, 328–329
Photometer, cAMP assay and, 62
Picrylsulfonic acid, alkylamine glass beads and, 181–183
Plasma, cyclic nucleotide assay of, 105
Platelets
guanylate cyclase, 122
apparent activation with time, 201–202
assay, 199–200
distribution among blood cells, 200
divalent cations and, 201
homogenate preparation, 200
subcellular distribution, 201
Polyarginine, protein kinase inhibitor and, 357

Polyethyleneimine cellulose
cyclic nucleotide column chromatography
column preparation, 16–17
fractionation procedures, 17–20
thin-layer chromatography, 27
biological materials, 37–38
cyclic nucleotides, 30–35
elutions and, 38
materials and general methods, 28
nucleoside triphosphates, 35–36
nucleotide separation, 28–30
pyrimidine nucleotides, 35
related purine nucleotides, 36–37
Poly-L-lysine, coupling to succinylated cyclic nucleotides, 98
Polyphosphates, detection of, 412–413
Potassium chloride, cyclic nucleotide separation and, 30, 31
Propranolol, adenylate cyclases and, 147, 149, 173, 179–180, 191
Prostaglandins, adenylate cyclases and, 147–148, 173
Protamine, protein kinase assay and, 309, 313, 327–328, 340, 348, 349, 357
Protein(s), phosphodiesterase assay and, 217
Protein kinase(s)
activation in intact cells, 358–359
hormonal regulation, 363–365
materials, 359
methodology for adipose tissue, 359–363
other tissues, 365–367
bovine heart
assay, 308–310
properties, 313–315
purification, 310–313
cAMP-dependent
assay of, 287–290
preparation of, 66–67
cAMP inhibitor and, 281
catalytic subunit preparation of, 306–308, 314–315, 326, 387
cGMP assay
materials, 90–92
principle, 90
procedure, 92–95
cGMP-dependent, 329–330
characterization, 335–349

conclusions, 350
preparation, 91
purification, 330–335
standard assay, 330
classification, 290–291
criteria, 291–293
generality of, 293
examples, 297–299
procedure, 293–296
cyclase preparations and, 143
erythrocyte, cAMP derivatives and, 398
liver cytosol
assay, 323–324
mechanism of cAMP action, 323
properties, 327–328
purification, 324–327
modulator, cGMP-dependent kinase and, 346–349
protein inhibitor, 296, 350–351, 360–361
assay, 351–353
interaction with enzyme, 356–357
with substrates and activators, 357–358
properties, 355–356
purification, 353–355.
rabbit reticulocyte, 297, 315–316
assay, 316–317
comments, 322
crude preparation, 317–318
purification of kinase I, 318–319
of kinase II, 319–321
rabbit skeletal muscle, 299–301
catalytic subunit, 306–308
holoenzyme, 301–306
regulatory subunit
inhibition by, 296
preparation of, 327
Protein kinase activation
cAMP assay
method, 66–70
modifications, 72–73
sensitivity, reproducibility and validation, 70–72
Protein kinase inhibitor
cAMP assay and, 49, 52, 53
preparation of, 52
Purine bases
elution of, 17–18
thin-layer chromatography, 28–30, 158

Pyrimidines, cyclic nucleotides, separation of, 35
Pyrophosphate
adenylate cyclase and, 146–147, 160
alumina column chromatography and, 44
Pyruvate, adenylate cyclase and, 160, 161, 168, 169
Pyruvate kinase
cAMP assay and, 62, 63, 65
cGMP assay and, 73–74, 78, 82
cyclase assays and, 119, 143

Q

Quaternary aminoethyl-Sephadex
cyclic nucleotide separation on
resin preparation, 13–14
separation techniques, 14–16
enzyme decontamination and, 65
protein kinase and, 319–320

R

Radioimmunoassay, cyclic nucleotides, 90–105
Reticulocytes
protein kinases of, 297, 315–316, 322
assay, 316–317
crude preparation, 317–318
purification of kinase I, 318–319
of kinases IIa and IIb, 319–321
Ribonuclease
cAMP-binding protein preparation and, 369
phosphodiesterase activator and, 254
Ribonucleic acid polymerase, DNA transcription assay and, 372
Ribonucleosides, protected, phosphorylation of, 410–411
Rod outer segments, cyclic nucleotide metabolizing enzymes, 153–155

S

Sarcolemma, adenylate cyclase and, 146
Sea water, artificial, 197

SUBJECT INDEX

Sephadex G-10, iodinated cyclic nucleotide derivatives and, 100
Sephadex G-75, protein kinase inhibitor and, 355, 356
Sephadex G-100
  cAMP-binding protein and, 370
  phosphodiesterase and, 254, 256
  phosphodiesterase activator and, 255, 269
  protein kinase subunits and, 306–307, 361–362, 364, 365
Sephadex G-200
  adenylate cyclase and, 166, 168
  guanylate cyclase and, 194–195
  phosphodiesterase
    brain, 234–236
    muscle, 242, 243
  protein kinase and, 319, 321
Sepharose 4B, phosphodiesterase and, 227–230, 234
Sepharose 6B, protein kinase and, 304
Sepharose-ε-aminocaproyl-cyclic adenosine monophosphate, preparation of, 386–387
Silica gel, thin-layer plates, cGMP separation and, 77-78, 91
Silver nitrate-sodium iodide, cAMP isolation and, 40
Sodium acetate, cyclic nucleotide separation and, 32
Sodium carbonate-barium chloride, cAMP isolation and, 40
Sodium carbonate-cadmium chloride, cAMP isolation and, 40
Sodium carbonate-calcium chloride, cAMP isolation and, 40, 41
Sodium formate, cyclic nucleotide separation and, 32
Sodium sulfate-barium chloride, cAMP isolation and, 40
Sperm, guanylate cyclase in, 197
Sphingomyelin, adenylate cyclase and, 176–177
Spleen, cAMP-receptorin, 380
Streptomycin sulfate, phosphodiesterase and, 252
Strongylocentrotus purpuratus
  guanylate cyclase
    assay, 196–198
    properties, 198–199

Subunits
  cAMP-dependent protein kinase and, 299
  cGMP-dependent protein kinase and, 344–346
Succinate thiokinase, cGMP assay and, 73–74, 78, 81, 82
2'-O-Succinyl cyclic adenosine monophosphate, synthesis of, 97, 105
2'-O-Succinyl cyclic guanosine monophosphate, synthesis of, 97–98
2'-O-Succinyl cyclic nucleotide(s)
  synthesis of, 96–97
  tyrosine methyl ester derivatives, 99
  iodination of, 99–100
Succinyl glass, synthesis of, 183
Sucrose density gradient centrifugation, adenylate cyclase, 150–151

T

Temperature, high pressure ion-exchange chromatography and, 21
Testis, protein kinase in, 365, 366
Tetramethylammonium, cyclic nucleotide levels and, 94
Tetrasodium iminodiphosphate, preparation of, 421–422
Theobromine, phosphodiesterase and, 244
Theophylline
  cAMP isolation and, 41
  cyclase assays and, 121, 136, 143, 171, 177
  guanylate cyclase and, 196
  phosphodiesterase and, 244, 256, 272
Thin-layer chromatography
  cAMP inhibitor and, 277
  cellulose, cAMP separation and, 156–158
  cGMP separation and, 77–78, 91
  polyethyleneimine cellulose, 27
    biological material, 37–38
    cyclic nucleotide separations, 30–35
    elutions from, 38
    materials and general methods, 28
    nucleoside triphosphates and, 35–36
    nucleotide separation, 28–30
    pyrimidine nucleotide separation, 35
    related purine nucleotides, 36–27

Thymus, protein kinases of, 298
Thyroxine, adenylate cyclases and, 148, 149, 173, 176
Tissues
　cyclic nucleotide levels in, 93, 94, 104
　extraction techniques, 7–9, 14, 24–25, 68, 75
　preparation for cAMP assay, 54–55, 100
　purification of extracts, cGMP assay and, 87–88, 100
　rapid fixation of, 3–6, 75
　storage of, 7
Transfer factor G, labeled GDP preparation and, 87
Trichloroacetonitrile, drying of, 412
Triethylamine, drying of, 412
Triiodothyronine, adenylate cyclases and, 148, 173, 176
Triton X-100
　adenylate cyclase and, 140, 149, 155, 167, 168
　guanylate cyclase and, 198, 199, 201
Trypsin
　phosphodiesterase and, 235, 236, 272
　phosphodiesterase activator and, 268
　protein kinase inhibitor and, 351, 353
Tubercidin cyclic 3′,5′-phosphate, DNA transcription and, 375
Tyrosine methyl ester, succinyl cyclic nucleotide derivatives, 99

U

Ultraviolet absorption, $1,N^6$-ethenoadenosine 3′,5′-monophosphate, 430
Ultraviolet source, high pressure ion-exchange chromatography and, 22
Uric acid
　elution of, 17–18
　thin-layer chromatography, 28–30
Urine, preparation for assay, 76

V

Vasopressin, adenylate cyclase and, 151–152, 173

Z

Zinc carbonate
　cAMP isolation and, 40, 41
　cGMP isolation and, 197
Zinc ions, cyclase assays and, 123
Zinc sulfate-barium chloride, cAMP isolation and, 40
Zinc sulfate-barium hydroxide, cAMP isolation and, 39–41
Zinc sulfate-calcium chloride, cAMP isolation and, 40
Zinc sulfate-sodium carbonate, cAMP isolation and, 40, 41